**Analytic Methods in Systems and Software Testing**

# Analytic Methods in Systems and Software Testing

*Edited by*

*Ron S. Kenett*
*KPA, Israel and Samuel Neaman Institute, Technion, Israel*

*Fabrizio Ruggeri*
*CNR-IMATI, Italy*

*Frederick W. Faltin*
*The Faltin Group and Virginia Tech, USA*

The right of Professor Ron S. Kenett, Dr Fabrizio Ruggeri and Frederick W. Faltin to be identified as the authors of the editorial material in this work has been asserted in accordance with law.

*Registered Office(s)*
John Wiley & Sons, Inc., 111 River Street, Hoboken, NJ 07030, USA
John Wiley & Sons Ltd, The Atrium, Southern Gate, Chichester, West Sussex, PO19 8SQ, UK

*Editorial Office*
9600 Garsington Road, Oxford, OX4 2DQ, UK

For details of our global editorial offices, customer services, and more information about Wiley products visit us at www.wiley.com.

Wiley also publishes its books in a variety of electronic formats and by print-on-demand. Some content that appears in standard print versions of this book may not be available in other formats.

*Library of Congress Cataloging-in-Publication Data*
Names: Kenett, Ron, editor. | Ruggeri, Fabrizio, editor. | Faltin, Frederick W., editor.
Title: Analytic methods in systems and software testing / edited by, Ron S. Kenett, KPA,
    Raanana, Israel and Neaman Institute, Technion, Haifa, Israel, Fabrizio Ruggeri, CNR-IMATI, IT,
    Frederick W. Faltin, The Faltin Group, USA.
Description: 1 edition. | Hoboken, NJ, USA : Wiley, [2018] | Includes bibliographical references and index. |
Identifiers: LCCN 2018008695 (print) | LCCN 2018017885 (ebook) | ISBN 9781119487364 (pdf) |
    ISBN 9781119487401 (epub) | ISBN 9781119271505 (cloth)
Subjects: LCSH: Computer software–Testing.
Classification: LCC QA76.76.T48 (ebook) | LCC QA76.76.T48 A52 2018 (print) | DDC 005.1/4–dc23
LC record available at https://lccn.loc.gov/2018008695

Cover design by Wiley
Cover image: © SergeyNivens/iStockphoto

Set in 10/12pt Warnock by SPi Global, Pondicherry, India

Printed in Singapore by C.O.S. Printers Pte Ltd

10   9   8   7   6   5   4   3   2   1

*To Jonathan, Alma, Tomer, Yadin, Aviv, Gili, Matan and Eden*

–Ron S. Kenett

*To Anna, Giacomo and Lorenzo*

–Fabrizio Ruggeri

*To Donna, Erin, Travis and Maddie*

–Frederick W. Faltin

# Contents

# List of Contributors

**Dani Almog**
Ben Gurion University of the Negev; and
Sami Shamoon College of Engineering
Beer Sheva, Israel

**Sarit Assaraf**
Israel Aerospace Industries Missile and
Space Group
Israel

**Matthew Avery**
Institute for Defense Analyses (IDA)
4850 Mark Center Drive
Alexandria, VA 22311
USA

**Xiaoying Bai**
Department of Computer Science and
Technology
Tsinghua University
Beijing, 100084
China

**Amanda M. Bonnie**
HPC Environments Group
Los Alamos National Laboratory
PO Box 1663 MS T084
Los Alamos, NM 87545
USA

**Jan Bosch**
Chalmers University of Technology
Lindholmspiren 5
41756 Göteborg
Sweden

**Hadas Chasidim**
Sami Shamoon College of
Engineering Beer Sheva
Israel

**Justace Clutter**
Institute for Defense Analyses (IDA)
4850 Mark Center Drive
Alexandria, VA 22311
USA

**Frank Coolen**
Department of Mathematical Sciences
Durham University
Durham, DH1 3LE
United Kingdom

**Laura A. Davey**
HPC Environments Group
Los Alamos National Laboratory
PO Box 1663 MS T080
Los Alamos, NM 87545,
USA

**Xiaoxu Diao**
Ohio State University
201 W. 19th Avenue
Scott Laboratory W382
Columbus, OH 43210
USA

**Tomas Docekal**
VSB-Technical University
of Ostrava, FEECS

Department of Cybernetics and
Biomedical Engineering
17. Listopadu 15
708 33 Ostrava-Poruba
Czech Republic

**Michael Felderer**
Department of Computer Science
University of Innsbruck
6020 Innsbruck
Austria

**Norman Fenton**
Queen Mary University of London
London; and
Agena Ltd, Cambridge
United Kingdom

**Laura J. Freeman**
Institute for Defense Analyses (IDA)
4850 Mark Center Drive
Alexandria, VA 22311
USA

**Michael Goldstein**
Department of Mathematical Sciences
Durham University
Durham, DH1 3LE
United Kingdom

**Jürgen Großmann**
Fraunhofer Institute for Open
Communication Systems FOKUS
Kaiserin-Augusta-Allee 31
10589 Berlin
Germany

**Avi Harel**
Ergolight
Haifa
Israel

**Kejia Hou**
Department of Computer Science and
Technology
Tsinghua University
Beijing, 100084
China

**Jun Huang**
Department of Computer Science
and Technology
Tsinghua University
Beijing, 100084
China

**Joshua R. Johnson**
Black Oak Analytics
3600 Cantrell Road, Suite 205
Little Rock, Arkansas, 72202
USA

**Thomas Johnson**
Institute for Defense Analyses (IDA)
4850 Mark Center Drive
Alexandria, VA 22311
USA

**Ron S. Kenett**
KPA, Raanana; and
Samuel Neaman Institute
Technion, Haifa
Israel

**Jiri Koziorek**
VSB-Technical University
of Ostrava, FEECS
Department of Cybernetics
and Biomedical Engineering
17. Listopadu 15
708 33 Ostrava-Poruba
Czech Republic

**Boyuan Li**
Ohio State University
201 W. 19th Avenue
Scott Laboratory W382
Columbus, OH 43210
USA

**V. Bram Lillard**
Institute for Defense Analyses (IDA)
4850 Mark Center Drive
Alexandria, VA 22311
USA

*Sarah E. Michalak*
Statistical Sciences Group
Los Alamos National Laboratory
PO Box 1663 MS F600
Los Alamos, NM 87545
USA

*Andrew J. Montoya*
HPC Environments Group
Los Alamos National Laboratory
PO Box 1663 MS T084
Los Alamos, NM 87545
USA

*Joseph Morgan*
SAS Institute Incorporated
Cary NC 27513
USA

*Thomas E. Moxley III*
HPC Systems Group
Los Alamos National Laboratory
PO Box 1663 MS T084
Los Alamos, NM 87545
USA

*Martin Neil*
Queen Mary University of London
United Kingdom; and
Agena Ltd, Cambridge
United Kingdom

*Seán Ó Ríordáin*
School of Computer Science and
Statistics
Trinity College
Dublin 2
Ireland

*Stepan Ozana*
VSB-Technical University of Ostrava,
FEECS
Department of Cybernetics and
Biomedical Engineering
17. Listopadu 15
708 33 Ostrava-Poruba
Czech Republic

*Edsel A. Peña*
Department of Statistics
University of South Carolina
Columbia, SC 29208
USA

*Daniel Pullen*
Black Oak Analytics
3600 Cantrell Road, Suite 205
Little Rock, Arkansas, 72202
USA

*Beidi Qiang*
Department of Mathematics and
Statistics
Vadalabene Center 1028
Southern Illinois University Edwardsville
Edwardsville, IL 62026-1653
USA

*Elena V. Ravve*
Ort Braude College of Engineering
Carmiel
Israel

*Brian J. Reich*
Department of Statistics
North Carolina State University
Campus Box 8203
5234 SAS Hall
Raleigh, NC, 27695
USA

*Steven E. Rigdon*
Saint Louis University
Salus Center
3545 Lafayette Ave., Room 481
St. Louis, MO 63104
USA

*Manuel Rodriguez*
Ohio State University
201 W. 19th Avenue
Scott Laboratory
Columbus, OH 43210
USA

**Fabrizio Ruggeri**
CNR-IMATI
Via Bassini 15
20131 Milano
Italy

**William N. Rust**
Statistical Sciences Group
Los Alamos National Laboratory
PO Box 1663 MS F600
Los Alamos, NM 87545
USA

**Ina Schieferdecker**
Fraunhofer Institute for Open
Communication Systems FOKUS
Kaiserin-Augusta-Allec 31
10589 Berlin
Germany

**Uri Shafrir**
KMDI, iSchool
University of Toronto
Toronto
Canada

**Gilli Shama**
Amdocs
8 HaPnina Street
Ra'anana 4321545
Israel

**Carol Smidts**
Ohio State University
201 W. 19th Avenue
Scott Laboratory E429
Columbus, OH 43210
USA

**Refik Soyer**
Department of Decision Sciences
The George Washington University
Funger Hall
2201 G Street NW
Washington, DC 20052
USA

**Vilem Srovnal**
VSB-Technical University of Ostrava,
FEECS
Department of Cybernetics and
Biomedical Engineering
17. Listopadu 15
708 33 Ostrava-Poruba
Czech Republic

**Daniel Ståhl**
Ericsson AB
Datalinjen 3
58330 Linköping
Sweden

**Curtis B. Storlie**
Department of Biomedical Statistics
and Informatics
Mayo Clinic College of Medicine
Rochester, MN, 55905
USA

**John R. Talburt**
Information Science Department
University of Arkansas Little Rock
2801 South University Avenue
Little Rock, Arkansas, 72204
USA

**Lawrence O. Ticknor**
Statistical Sciences Group
Los Alamos National Laboratory
PO Box 1663 MS F600
Los Alamos, NM 87545
USA

**Zeev Volkovich**
Ort Braude College of Engineering
Carmiel
Israel

**Pei Wang**
Biomedical Informatics
University of Arkansas for Medical
Sciences

4301 W Markham Street
Little Rock, Arkansas, 72205
USA

**Simon P. Wilson**
School of Computer Science and
Statistics
Trinity College
Dublin 2
Ireland

**David Wooff**
Department of Mathematical Sciences
Durham University

Durham, DH1 3LE
United Kingdom

**Mingli Yu**
Department of Computer Science and
Technology
Tsinghua University
Beijing, 100084
China

**Shelemyahu Zacks**
Binghamton University
Binghamton, NY
USA

# Preface

The objective of this edited volume is to compile leading edge methods and examples of analytical approaches to systems and software testing from leading authorities in applied statistics, computer science, and software engineering. The book provides a collection of methods for practitioners and researchers interested in a general perspective on this topic that affects our daily lives, and will become even more critical in the future. Our original objective was to focus on analytic methods but, as the many co-authors show, a contextual landscape of modern engineering is required to appreciate and present the related statistical and probabilistic models used in this domain. We have therefore expanded the original scope and offered, in one comprehensive collection, a state of the art view of the topic.

Inevitably, testing and validation of advanced systems and software is comprised in part of general theory and methodology, and in part of application-specific techniques. The former are transportable to new domains. While the latter generally are not, we trust the reader will share our conviction that case study examples of successful applications provide a heuristic foundation for adaptation and extension to new problem contexts. This is yet another example of where statistical methods need to be integrated with expert knowledge to provide added value.

The structure of the book consists of four parts:

Part I: Testing Concepts and Methods (Chapters 1–6)
Part II: Statistical Models (Chapters 7–12)
Part III: Testing Infrastructures (Chapters 13–17)
Part IV: Testing Applications (Chapters 18–21)

It constitutes an advanced reference directed toward industrial and academic readers whose work in systems and software development approaches or surpasses existing frontiers of testing and validation procedures. Readers will typically hold degrees in statistics, applied mathematics, computer science, or software engineering.

The 21 chapters vary in length and scope. Some are more descriptive, some present advanced mathematical formulations, some are more oriented towards system and software engineering. To inform the reader about the nature of the chapters, we provide an annotated list below with a brief description of each.

The book provides background, examples, and methods suitable for courses on system and software testing with both an engineering and an analytic focus. Practitioners will find the examples instructive and will be able to derive benchmarks and suggestions to build upon. Consultants will be able to derive a context for their work with clients and colleagues. Our additional goal with this book is to stimulate research at the theoretical and practical level. The testing of systems and software is an area requiring further developments. We hope that this work will contribute to such efforts.

The authors of the various chapters clearly deserve most of the credit. They were generous in sharing their experience and taking the time to write the chapters. We wish to thank them for their collaboration and patience in the various stages of writing this project. We also acknowledge the professional help of the Wiley team who provided support and guidance throughout the long journey that lead to this book.

To help the reader we provide next an annotated list of chapters giving a peak preview as to their content. The chapters are grouped in three parts but there is no specific sequence to them so that one can meander from topic to topic without following the numbered order.

## Annotated List of Chapters

| | | |
|---|---|---|
| 1 | Recent Advances in Classifying Risk-Based Testing Approaches | A taxonomy of risk-based testing aligned with risk considerations in all phases of a test process. It provides a framework to understand, categorize, assess, and compare risk-based testing approaches. |
| 2 | Improving Software Testing with Causal Modeling | An introduction to causal modeling using Bayesian networks and how it can be used to improve software testing strategies and predict software reliability. |
| 3 | Optimal Software Testing across Version Releases | Looks at models for bug detection in software with a regular schedule of version releases. Model inference is developed from a Bayesian perspective and applied to data on bug reports of Mozilla Firefox. |
| 4 | Incremental Verification and Coverage Analysis of Strongly Distributed Systems | A model of both software and hardware hierarchical systems as logical structures using a finite state machine. Properties to be tested are expressed with extensions of first-order logic. |
| 5 | Combinatorial Testing: An Approach to Systems and Software Testing Based on Covering Arrays | Combinatorial testing is an approach for sampling from the input test space. It exploits evidence that system failures are typically due to the interaction of just a few inputs. |

| 6 | Conceptual Aspects in Development and Teaching of System and Software Test Engineering | A novel approach for improving the learning of system and software testing in academic and lifelong learning programs. It builds on Meaning Equivalence Reusable Learning Objects (MERLO) used in formative assessment of conceptual understanding. |
|---|---|---|
| 7 | Non-homogeneous Poisson Process Models for Software Reliability | The non-homogeneous Poisson process is used as a model for the reliability of repairable hardware systems. The chapter discusses its applicability to software reliability. |
| 8 | Bayesian Graphical Models for High-Complexity Testing: Aspects of Implementation | The Bayesian graphical models approach to software testing is presented. It is a method for the logical structuring of software testing where the focus is on high reliability final stage integration testing. |
| 9 | Models of Software Reliability | Software reliability time domain and data domain models are discussed. Topics such as stopping time for fault detection, the Chernoff–Ray procedure, and capture–recapture methods are covered. |
| 10 | Improved Estimation of System Reliability with Application in Software Development | Shrinkage ideas are implemented to obtain improvements in the estimation of software reliability functions. This helps decide if software can be deployed in the field or further debugging is required. |
| 11 | Decision Models for Software Testing | Game- and decision-theoretic frameworks are used to develop optimal software testing strategies. Specifically, optimal choice of release time for software under different scenarios is provided. |
| 12 | Modeling and Simulations in Control Software Design | Control systems are studied through modeling and simulations. These designs have a major impact on the quality of control applications and the reliability, sustainability, and further extension of the system. |

| 13 | A Temperature Monitoring Infrastructure and Process for Improving Data Center Energy Efficiency with Results for a High Performance Computing Data Center | An approach to energy efficiency improvement is presented. It includes a temperature monitoring infrastructure and a process that enables changes to DC cooling systems, infrastructure, and equipment. |
|----|----|----|
| 14 | Agile Testing with User Data in Cloud and Edge Computing Environments | A framework for mitigating the risks of user errors is presented. It incorporates usability considerations in the design, testing, deployment, and operation of dynamic collaborative systems. |
| 15 | Automated Software Testing | This chapter presents model-based automatic software testing challenges and solutions. Building such models involves modeling language selection and operational profile development. |
| 16 | Dynamic Test Case Selection in Continuous Integration: Test Result Analysis Using the Eiffel Framework | Dynamic test case selection determines which tests to execute at a given time, rather than from predefined static lists. The Eiffel framework for continuous integration and delivery is presented. |
| 17 | An Automated Regression Testing Framework for a Hadoop-Based Entity Resolution System | A framework built to support a large-scale entity resolution application is presented. It addresses distributed processing environments such as Hadoop Map/Reduce and Spark with parallel processing. |
| 18 | Testing Defense Systems | Defense systems consistently push the limits of scientific understanding. The chapter highlights, with examples, the core statistical methodologies that have proven useful in testing defense systems. |
| 19 | A Search-Based Approach to Geographical Data Generation for Testing Location-Based Services | A framework to support automatic location-based services testing is presented from two aspects, geographical test data generation and query results. How to design and validate tests is discussed. |

The editors of *Analytic Methods in Systems and Software Testing*:
Ron S. Kenett, KPA Ltd. and Samuel Neaman Institute, Technion, Israel
Fabrizio Ruggeri, CNR-IMATI, Italy
Frederick W. Faltin, The Faltin Group, and Virginia Tech, USA

**Part I**

**Testing Concepts and Methods**

# 1

# Recent Advances in Classifying Risk-Based Testing Approaches

*Michael Felderer, Jürgen Großmann, and Ina Schieferdecker*

## Synopsis

In order to optimize the usage of testing efforts and to assess risks of software-based systems, risk-based testing uses risk (re-)assessments to steer all phases in a test process. Several risk-based testing approaches have been proposed in academia and/or applied in industry, so that the determination of principal concepts and methods in risk-based testing is needed to enable a comparison of the weaknesses and strengths of different risk-based testing approaches. In this chapter we provide an (updated) taxonomy of risk-based testing aligned with risk considerations in all phases of a test process. It consists of three top-level classes: contextual setup, risk assessment, and risk-based test strategy. This taxonomy provides a framework to understand, categorize, assess, and compare risk-based testing approaches to support their selection and tailoring for specific purposes. Furthermore, we position four recent risk-based testing approaches into the taxonomy in order to demonstrate its application and alignment with available risk-based testing approaches.

## 1.1  Introduction

Testing of safety-critical, security-critical or mission-critical software faces the problem of determining those tests that assure the essential properties of the software and have the ability to unveil those software failures that harm the critical functions of the software. However, for "normal," less critical software a comparable problem exists: Usually testing has to be done under severe pressure due to limited resources and tight time constraints with the consequence that testing efforts have to be focused and be driven by business risks.

Both decision problems can be adequately addressed by risk-based testing, which considers the risks of the software product as the guiding factor to steer all the phases of a test process, i.e., test planning, design, implementation, execution, and evaluation (Gerrard and Thompson, 2002; Felderer and Ramler, 2014a; Felderer and Schieferdecker, 2014). Risk-based testing is a pragmatic approach widely used in companies of all sizes (Felderer and Ramler, 2014b, 2016) which uses the straightforward idea of focusing test activities on those scenarios that trigger the most critical situations of a software system (Wendland et al., 2012).

*Analytic Methods in Systems and Software Testing*, First Edition.
Edited by Ron S. Kenett, Fabrizio Ruggeri, and Frederick W. Faltin.
© 2018 John Wiley & Sons Ltd. Published 2018 by John Wiley & Sons Ltd.

The recent international standard ISO/IEC/IEEE 29119 Software Testing (ISO, 2013) on testing techniques, processes, and documentation even explicitly specifies risk considerations to be an integral part of the test planning process. Because of the growing number of available risk-based testing approaches and its increasing dissemination in industrial test processes (Felderer et al., 2014), methodological support to categorize, assess, compare, and select risk-based testing approaches is required.

In this paper, we present an (updated) taxonomy of risk-based testing that provides a framework for understanding, categorizing, assessing, and comparing risk-based testing approaches and that supports the selection and tailoring of risk-based testing approaches for specific purposes. To demonstrate the application of the taxonomy and its alignment with available risk-based testing approaches, we position four recent risk-based testing approaches, the RASEN approach (Großmann et al., 2015), the SmartTesting approach (Ramler and Felderer, 2015), risk-based test case prioritization based on the notion of risk exposure (Yoon and Choi, 2011), and risk-based testing of open source software (Yahav et al., 2014a), in the taxonomy.

A *taxonomy* defines a hierarchy of classes (also referred to as categories, dimensions, criteria, or characteristics) to categorize things and concepts. It describes a tree structure whose leaves define concrete values to characterize instances in the taxonomy. The proposed taxonomy is aligned with the consideration of risks in all phases of the test process and consists of the top-level classes *context* (with subclasses risk driver, quality property, and risk item), *risk assessment* (with subclasses factor, estimation technique, scale, and degree of automation), and *risk-based test strategy* (with subclasses risk-based test planning, risk-based test design and implementation, and risk-based test execution and evaluation). The taxonomy presented in this chapter extends and refines our previous taxonomy of risk-based testing (Felderer and Schieferdecker, 2014).

The remainder of this chapter is structured as follows. Section 1.2 presents background on software testing and risk management. Section 1.3 introduces the taxonomy of risk-based testing. Section 1.4 presents the four selected recent risk-based testing approaches and discusses them in the context of the taxonomy. Finally, Section 1.5 summarizes this chapter.

## 1.2 Background on Software Testing and Risk Management

### 1.2.1 Software Testing

*Software testing* (ISTQB, 2012) is the process consisting of all lifecycle activities, both static and dynamic, concerned with planning, preparation, and evaluation of software products and related work products to determine that they satisfy specified requirements, to demonstrate that they are fit for purpose, and to detect defects. According to this definition it comprises static activities like reviews but also dynamic activities like classic black- or white-box testing. The tested software-based system is called the *system under test* (SUT). As highlighted before, *risk-based testing* (RBT) is a testing approach which considers the risks of the software product as the guiding factor to support decisions in all phases of the test process (Gerrard and Thompson, 2002; Felderer and Ramler, 2014a; Felderer and Schieferdecker, 2014). A *risk* is a factor that could result in future negative consequences and is usually expressed by its likelihood and

**Figure 1.1** Core test process steps.

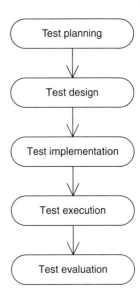

Test planning

Test design

Test implementation

Test execution

Test evaluation

impact (ISTQB, 2012). In software testing, the *likelihood* is typically determined by the probability that a failure assigned to a risk occurs, and the *impact* is determined by the cost or severity of a failure if it occurs in operation. The resulting *risk value* or *risk exposure* is assigned to a *risk item*. In the context of testing, a risk item is anything of value (i.e., an asset) under test, for instance, a requirement, a component, or a fault.

Risk-based testing is a testing-based approach to risk management that can only deliver its full potential if a test process is in place and if risk assessment is integrated appropriately into it. A *test process* consists of the core activities test planning, test design, test implementation, test execution, and test evaluation (ISTQB, 2012) – see Figure 1.1. In the following, we explain the particular activities and associated concepts in more detail.

According to ISO (2013) and ISTQB (2012), *test planning* is the activity of establishing or updating a test plan. A test plan is a document describing the scope, approach, resources, and schedule of intended test activities. It identifies, amongst others, objectives, the features to be tested, the test design techniques, and exit criteria to be used and the rationale of their choice. *Test objectives* are the reason or purpose for designing and executing a test. The reason is either to check the functional behavior of the system or its non-functional properties. *Functional testing* is concerned with assessing the functional behavior of an SUT, whereas *non-functional testing* aims at assessing non-functional requirements such as security, safety, reliability, or performance. The scope of the features to be tested can be components, integration, or system. At the scope of *component testing* (also referred to as unit testing), the smallest testable component, e.g., a class, is tested in isolation. *Integration testing* combines components with each other and tests those as a subsystem, that is, not yet a complete system. In *system testing*, the complete system, including all subsystems, is tested. *Regression testing* is the selective retesting of a system or its components to verify that modifications have not caused unintended effects and that the system or the components still comply with the specified requirements (Radatz et al., 1990). *Exit criteria* are conditions for permitting a process to be officially completed. They are used to report against and to plan when to stop testing. Coverage criteria aligned with the tested feature types and the applied test design techniques are typical exit criteria. Once the test plan has been established, test control begins. It is an ongoing activity in which the actual progress is compared against the plan, which often results in concrete measures.

During the *test design* phase the general testing objectives defined in the test plan are transformed into tangible test conditions and abstract test cases. *Test implementation* comprises tasks to make the abstract test cases executable. This includes tasks like preparing test harnesses and test data, providing logging support, or writing test scripts, which are necessary to enable the automated execution of test cases. In the *test execution* phase, the test cases are then executed and all relevant details of the execution are logged and monitored. Finally, in the *test evaluation* phase the exit criteria are evaluated and the logged test results are summarized in a test report.

### 1.2.2 Risk Management

*Risk management* comprises the core activities *risk identification, risk analysis, risk treatment*, and *risk monitoring* (Standards Australia/New Zealand, 2004; ISO, 2009). In the risk identification phase, risk items are identified. In the risk analysis phase, the likelihood and impact of risk items and, hence, the risk exposure is estimated. Based on the risk exposure values, the risk items may be prioritized and assigned to risk levels defining a risk classification. In the risk treatment phase the actions for obtaining a satisfactory situation are determined and implemented. In the risk monitoring phase the risks are tracked over time and their status is reported. In addition, the effect of the implemented actions is determined. The activities risk identification and risk analysis are often collectively referred to as *risk assessment*, while the activities risk treatment and risk monitoring are referred to as *risk control*.

## 1.3 Taxonomy of Risk-Based Testing

The taxonomy of risk-based testing is shown in Figure 1.2. It contains the top-level classes *contextual setup, risk assessment*, and *risk-based test process*, and is aligned with the consideration of risks in all phases of the test process. In this section, we explain these classes, their subclasses, and concrete values for each class of the risk-based testing taxonomy in depth.

### 1.3.1 Context

The *context* characterizes the overall context of the risk assessment and testing processes. It includes the subclasses *risk driver, quality property*, and *risk item* to characterize the drivers that determine the major assets, the overall quality objectives that need to be fulfilled, and the items that are subject to evaluation by risk assessment and testing.

#### 1.3.1.1 Risk Driver
A *risk driver* is the first differentiating element of risk-based testing approaches. It characterizes the area of origin for the major assets and thus determines the overall quality requirements and the direction and general setup of the risk-based testing process. *Business*-related assets are required for a successful business practice and thus often directly

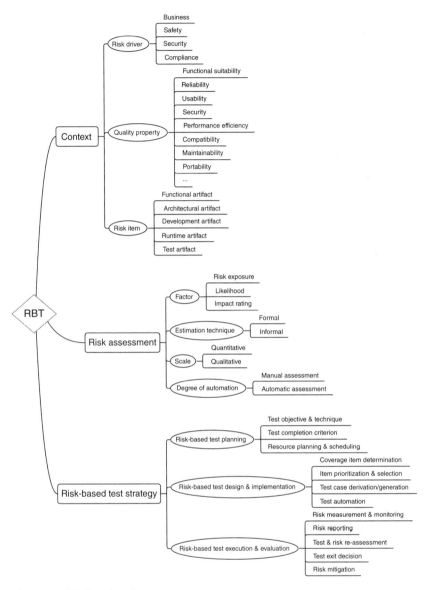

**Figure 1.2** Risk-based testing taxonomy.

relate to software quality properties like functionality, availability, security, and reliability. *Safety* relates to the inviolability of human health and life and thus requires software to be failsafe, robust, and resilient. *Security* addresses the resilience of information technology systems against threats that jeopardize confidentiality, integrity, and availability of digital information and related services. Finally, *compliance* relates to assets that are directly derived from rules and regulations, whether applicable laws, standards, or other forms of governing settlements. Protection of these assets often, but not exclusively, relates to quality properties like security, reliability, and compatibility.

### 1.3.1.2 Quality Property

A *quality property* is a distinct quality attribute (ISO, 2011) which contributes to the protection of assets, and thus is subject to risk assessment and testing. As stated in ISO (2000), risks result from hazards. Hazards related to software-based systems stem from software vulnerabilities and from defects in software functionalities that are critical to business cases, safety-related aspects, security of systems, or applicable rules and regulations.

One needs to test that a software-based system is

- functionally suitable, i.e., able to deliver services as requested;
- reliable, i.e., able to deliver services as specified over a period of time;
- usable, i.e., satisfies the user expectation;
- performant and efficient, i.e., able to react appropriately with respect to stated resources and time;
- secure, i.e., able to remain protected against accidental or deliberate attacks;
- resilient, i.e., able to recover in a timely manner from unexpected events;
- safe, i.e., able to operate without harmful states.

The quality properties considered determine which testing is appropriate and has to be chosen. We consider *functionality*, *security*, and *reliability* to be the dominant quality properties that are addressed for software. Together they form the reliability, availability, safety, security, and resilience of a software-based system and hence constitute the options for the risk drivers in the RBT taxonomy.

As reported by different computer emergency response teams such as GovCERT-UK, software defects continue to be a major, if not the main, source of incidents caused by software-based systems. The quality properties determine the test types and test techniques that are applied in a test process to find software defects or systematically provide belief in the absence of such defects. Functional testing is likewise a major test type in RBT to analyze reliability and safety aspects – see, e.g., Amland (2000). In addition, security testing including penetration testing, fuzz testing, and/or randomized testing is key in RBT (Zech, 2011; ETSI, 2015b) to analyze security and resilience aspects. Furthermore, performance and scalability testing focusing on normal load, maximal load, and overload scenarios analyze availability and resilience aspects – see, e.g., Amland (2000).

### 1.3.1.3 Risk Item

The *risk item* characterizes and determines the elements under evaluation. These risk items are the elements to which risk exposures and tests are assigned (Felderer and Ramler, 2013). Risk items can be of type *test case* (Yoon and Choi, 2011), i.e., directly test cases themselves as in regression testing scenarios; *runtime artifact*, like deployed services; *functional artifact*, like requirements or features; *architectural artifact*, like a component; or *development artifact*, like source code file. The risk item type is determined by the test level. For instance, functional or architectural artifacts are often used for system testing, and generic risks for security testing. In addition, we use the term *artifact* to openly refer to other risk items used in requirements capturing, design, development, testing, deployment, and/or operation and maintenance, which all might relate to the identified risks.

## 1.3.2    Risk Assessment

The second differentiating element of RBT approaches is the way risks are determined. According to ISTQB (2012), *risk assessment* is the process of *identifying* and subsequently *analyzing* the identified risk to determine its level of risk, typically by assigning likelihood and impact ratings. Risk assessment itself has multiple aspects, so that one needs to differentiate further the *factors* influencing risks, the risk *estimation technique* used to estimate and/or evaluate the risk, the *scale* type that is used to characterize the risk exposure, and the *degree of automation* for risk assessment.

### 1.3.2.1    Factor

The risk factors quantify identified risks (Bai et al., 2012). *Risk exposure* is the quantified potential for loss. It is calculated by the likelihood of risk occurrence multiplied by the potential loss, also called the impact. The risk exposure typically considers aspects like liability issues, property loss or damage, and product demand shifts. Risk-based testing approaches might also consider the specific aspect of *likelihood* of occurrence, e.g., for test prioritization or selection, or the specific aspect of *impact rating* to determine test efforts needed to analyze the countermeasures in the software.

### 1.3.2.2    Estimation Technique

The estimation technique determines how the risk exposure is actually estimated and can be *list based* or a *formal model* (Jørgensen et al., 2009). The essential difference between formal-model- and list-based estimation is the quantification step; that is, the final step that transforms the input into the risk estimate. Formal risk estimation models are based on a complex, multivalued quantification step such as a formula or a test model. On the other hand, list-based estimation methods are based on a simple quantification step – for example, what the expert believes is riskiest. List-based estimation processes range from pure "gut feelings" to structured, historical data including failure history and checklist-based estimation processes.

### 1.3.2.3    Scale

Any risk estimation uses a scale to determine the risk "level." This risk scale can be *quantitative* or *qualitative*. Quantitative risk values are numeric and allow computations; qualitative risk values can only be sorted and compared. A qualitative scale often used for risk levels is low, medium, and high (Wendland et al., 2012).

### 1.3.2.4    Degree of Automation

Risk assessment can be supported by automated methods and tools. For example, risk-oriented metrics can be measured *manually* or *automatically*. Manual measurement is often supported by strict guidelines, and automatic measurement is often performed via static analysis tools. Other examples for automated risk assessment include the derivation of risk exposures from formal risk models – see, for instance, Fredriksen et al. (2002).

## 1.3.3    Risk-Based Testing Strategy

Based on the risks being determined and characterized, RBT follows the fundamental test process (ISTQB, 2012) or variations thereof. The notion of risk can be used to optimize already existing testing activities by introducing risk-based strategies for prioritization, automation, selection, resource planning, etc. Depending on the approach, nearly all activities and phases in a test process may be affected by taking a risk-based perspective. This taxonomy aims to highlight and characterize the RBT specifics by relating them to the major phases of a normal test process. For the sake of brevity, we have focused on the phases *risk-based test planning, risk-based test design and implementation*, and *risk-based test execution and evaluation*, which are outlined in the following subsections.

### 1.3.3.1 Risk-Based Test Planning

The main outcome of test planning is a test strategy and a plan that depicts the staffing, the required resources, and a schedule for the individual testing activities. *Test planning* establishes or updates the scope, approach, resources, and schedule of intended test activities. Amongst other things, *test objectives, test techniques*, and *test completion criteria* that impact risk-based testing (Redmill, 2005) are determined.

***Test Objective and Technique*** *Test objectives and techniques* are relevant parts of a test strategy. They determine what to test and how to test a test item. The reason for designing or executing a test, i.e., a *test objective*, can be related to the risk item to be tested, to the threat scenarios of a risk item, or to the countermeasures established to secure that risk item; see also Section 1.3.3.2. The selection of adequate *test techniques* can be done on the basis of the *quality properties* as well as from information related to defects, vulnerabilities, and threat scenarios coming from risk assessment.

***Test Completion Criterion*** Typical exit criteria for testing that are used to report against and to plan when to stop testing include all tests running successfully, all issues having been retested and signed off, or all acceptance criteria having been met. Specific RBT-related exit criteria (Amland, 2000) add criteria on the residual risk in the product- and coverage-related criteria: all risk items, their threat scenarios, and/or countermeasures being covered. Risk-based metrics are used to quantify different aspects in testing such as the minimum level of testing, extra testing needed because of a high number of faults found, or the quality of the tests and the test process. They are used to manage the RBT process and optimize it with respect to time, effort, and quality (Amland, 2000).

***Resource Planning and Scheduling*** Risk-based testing requires focusing the testing activities and efforts based on the risk assessment of the particular product, or of the project in which it is developed. In simple words: if there is high risk, then there will be serious testing. If there is no risk, then there will be a minimum of testing. For example, products with high complexity, new technologies, many changes, many defects found earlier, developed by personnel with less experiences or lower qualification, or developed along new or renewed development processes may have a higher probability of failing and need to be tested more thoroughly. Within this context, information from risk assessment can be used to roughly identify high-risk areas or features of the SUT and thus determine and optimize the respective test effort, the required personnel and their qualifications, and the scheduling and prioritization of the activities in a test process.

### 1.3.3.2 Risk-Based Test Design and Implementation

*Test design* is the process of transforming test objectives into test cases. This transformation is guided by the coverage criteria, which are used to quantitatively characterize the test cases and often used for exit criteria. Furthermore, the technique of transformation depends on the test types needed to realize a test objective. These test types directly relate to the *quality property* defined in Section 1.3.1. *Test implementation* comprises tasks like preparing test harnesses and test data, providing logging support, or writing automated test scripts to enable the automated execution of test cases (ISTQB, 2012). Risk aspects are especially essential for providing *logging support* and for *test automation*.

***Coverage Item Determination*** Risk-based testing uses coverage criteria specific to the risk artifacts, and test types specific to the risk drivers on functionality, security, and safety. The classical code-oriented and model-based coverage criteria like path coverage, condition-oriented coverage criteria like modified condition decision coverage, and requirements-oriented coverage criteria like requirements or use case coverage are extended with coverage criteria to cover selected or all assets, threat scenarios, and countermeasures (Stallbaum et al., 2008). While *asset coverage* rather belongs to requirements-oriented coverage (Wendland et al., 2012), *threat scenario and vulnerability coverage* and *countermeasure coverage* can be addressed by code-oriented, model-based, and/or condition-oriented coverage criteria (Hosseingholizadeh, 2010).

***Test or Feature Prioritization Selection*** In order to optimize the costs of testing and/or the quality and fault detection capability of testing, techniques for prioritizing, selecting, and minimizing tests as well as combinations thereof have been developed and are widely used (Yoo and Harman, 2012). Within the ranges of intolerable risk and "as low as reasonably practicable" (ALARP)[1] risks, these techniques are used to identify tests for the risk-related test objectives determined before. For example, design-based approaches for test selection (Briand et al., 2009) and coverage-based approaches (Amland, 2000) for test prioritization are well suited to RBT. Depending on the approach, prioritization and selection can take place during different phases of the test process. Risk-based feature or requirement prioritization and selection selects the requirements or features to be tested. This activity is usually started during test planning and continued during test design. Test case prioritization and selection requires existing test specifications or test cases. It is thus either carried out before test implementation, to determine the test case to be implemented, or in the preparation of test execution or regression testing, to determine the optimal test sets to be executed.

***Test Case Derivation/Generation*** Risk assessment often comprises information about threat scenarios, faults and vulnerabilities that can be used to derive the test data, the test actions, probably the expected results, and other testing artifacts. Especially when addressing publicly known threat scenarios, these scenarios can be used to directly refer to predefined and reusable test specification fragments, i.e., so-called test pattern. These test patterns already contain test actions and test data that are directly applicable to either test specification, test implementation, or test execution (Botella et al., 2014).

---

1 The ALARP principle is typically used for safety-critical systems, but also for mission-critical systems. It says that the residual risk shall be as low as reasonably practical.

*Test Automation*   Test automation is the use of special software (separate from the software under test) to control the execution of tests and the comparison of actual outcomes with predicted outcomes (Huizinga and Kolawa, 2007). Experiences from test automation (Graham and Fewster, 2012) show possible benefits like improved regression testing or a positive return on investment, but also caveats like high initial investments or difficulties in test maintenance. Risks may therefore be beneficial in guiding decisions as to where and to what degree testing should be automated.

### 1.3.3.3   Risk-Based Test Execution and Evaluation

*Test execution* is the process of running test cases. In this phase, risk-based testing is supported by *monitoring* and *risk metric measurement*. *Test evaluation* comprises decisions on the basis of exit criteria and logged test results compiled in a test report. In this respect, risks are *mitigated* and may require a *reassessment*. Furthermore, risks may guide *test exit decisions* and *reporting*.

*Monitoring and Risk Metric Measurement*   Monitoring is run concurrently with an SUT and supervises, records, or analyzes the behavior of the running system (Radatz et al., 1990; ISTQB, 2012). Differing from software testing, which actively stimulates the system under test, monitoring only passively observes a running system. For risk-based testing purposes, monitoring enables additional complex analysis, e.g., of the internal state of a system for security testing, as well as tracking the project's progress toward resolving its risks and taking corrective action where appropriate. *Risk metric measurement* determines risk metrics defined in the test planning phase. A measured risk metric could be the number of observed critical failures for risk items where failure has a high impact (Felderer and Beer, 2013).

*Risk Reporting*   Test reports are documents summarizing testing activities and results (ISTQB, 2012) that communicate risks and alternatives requiring a decision. They typically report progress of testing activities against a baseline (such as the original test plan) or test results against exit criteria. In *risk reporting*, assessed risks that are monitored during the test process are explicitly reported in relation to other test artifacts. Risk reports can be descriptive, summarizing relationships of the data, or predictive, using data and analytical techniques to determine the probable future risk. Typical descriptive risk reporting techniques are risk burn down charts, which visualize the development of the overall risk per iteration, as well as traffic light reports, which provide a high level view on risks using the colors red for high risks, yellow for medium risks, and green for low risks. A typical predictive risk reporting technique is residual risk estimation, for instance based on software reliability growth models (Goel, 1985).

*Test and Risk Reassessment*   The *reassessment of risks* after test execution may be planned in the process or triggered by a comparison of test results against the assessed risks. This may reveal deviations between the assessed and the actual risk level and require a reassessment to adjust them. Test results can be explicitly integrated into a formal risk analysis model (Stallbaum and Metzger, 2007), or just trigger the reassessment in an informal way.

**Test Exit Decision**   The *test exit decision* determines if and when to stop testing (Felderer and Ramler, 2013), but may also trigger further risk mitigation measures. This decision may be taken on the basis of a test report matching test results and exit criteria, or ad hoc, for instance solely on the basis of the observed test results.

**Risk Mitigation**   *Risk mitigation* covers efforts taken to reduce either the likelihood or impact of a risk (Tran and Liu, 1997). In the context of risk-based testing, the assessed risks and their relationship to test results and exit criteria (which may be outlined in the test report) may trigger additional measures to reduce either the likelihood or impact of a risk occurring in the field. Such measures may be bug fixing, redesign of test cases, or re-execution of test cases.

## 1.4   Classification of Recent Risk-Based Testing Approaches

In this section, we present four recent risk-based testing approaches: the RASEN approach (Section 1.4.1), the SmartTesting approach (Section 1.4.2), risk-based test case prioritization based on the notion of risk exposure (Section 1.4.3), and risk-based testing of open source software (Section 1.4.4); we position each in the risk-based testing taxonomy presented in the previous section.

### 1.4.1   The RASEN Approach

#### 1.4.1.1   Description of the Approach

The RASEN project (www.rasen-project.eu) has developed a process for combining compliance assessment, security risk assessment, and security testing based on existing standards like ISO 31000 and ISO 29119. The approach is currently extended in the PREVENT project (www.prevent-project.org) to cover business-driven security risk and compliance management for critical banking infrastructure. Figure 1.3 shows an overview of the RASEN process.

The process covers three distinguishable workstreams that each consist of a combination of typical compliance assessment, security risk assessment activities, and/or security testing activities, emphasizing the interplay and synergies between these formerly independent assessment approaches.

1) The test-based security risk assessment workstream starts like a typical risk assessment workstream and uses testing results to guide and improve the risk assessment. Security testing is used to provide feedback on actually existing vulnerabilities that have not been covered during risk assessment, or allows risk values to be adjusted on the basis of tangible measurements like test results. Security testing should provide a concise feedback as to whether the properties of the target under assessment have really been met by the risk assessment.
2) The risk-based compliance assessment workstream targets the identification and treatment of compliance issues. It relies on security risk assessment results to identify compliance risk and thus systematize the identification of compliance issues. Moreover, legal risk assessment may be used to prioritize the treatment of security issues.

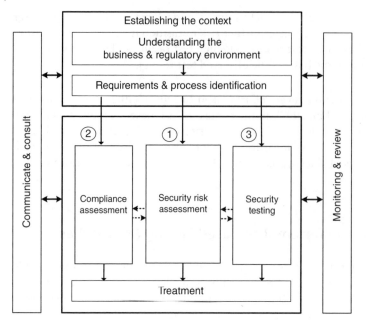

**Figure 1.3** Combining compliance assessment, security risk assessment, and security testing in RASEN.

3) The risk-based security testing workstream starts like a typical testing workstream and uses risk assessment results to guide and focus the testing. Such a workstream starts by identifying the areas of risk within the target's business processes, and building and prioritizing the testing program around these risks. In this setting risks help focus the testing resources on the areas that are most likely to cause concern, or support the selection of test techniques dedicated to already identified threat scenarios.

According ISO 31000, all workstreams start with a preparatory phase called *Establishing the Context* that includes preparatory activities like understanding the business and regulatory environment as well as the requirements and processes. During this first phase the high level security objectives are identified and documented, and the overall process planning is done. Moreover, the process shows additional support activities like *Communication and Consult* and *Monitoring and Review* that are meant to set up the management perspective, thus to continuously control, react, and improve all relevant information and the results of the process. From a process point of view, these activities are meant to provide the contextual and management-related framework. The individual activities covered in these phases might differ in detail depending on whether the risk assessment or testing activities are the guiding activities. The main phase, namely the *Security Assessment* phase, covers the definition of the integrated compliance assessment, risk assessment, and security testing workstreams.

**The Risk Assessment Workstream**   The overall risk assessment workstream is decomposed into the three main activities *Risk Identification*, *Risk Estimation*, and *Risk Evaluation*.

RASEN has extended the risk identification and risk estimation activities with security testing activities in order to improve the accuracy and efficiency of the overall workstream.

Risk identification is the process of finding, recognizing, and describing risks. This consists of identifying sources of risk (e.g., threats and vulnerabilities), areas of impacts (e.g., the assets), malicious events, their causes, and their potential impact on assets. In this context, security testing is used to obtain information that eases and supports the identification of threats and threat scenarios. Appropriate are testing and analysis techniques that yield information about the interfaces and entry points (i.e., the attack surface) like automated security testing, network discovery, web crawling, and fuzz testing.

Following risk identification, risk estimation is the process of expressing the likelihood, intensity, and magnitude of the identified risks. In many cases, the relevant information on potential threats is often imprecise and insufficient, so that estimation often relies on expert judgment only. This, amongst others, might result in a high degree of uncertainty related to the correctness of the estimates. Testing or test-based risk estimation may increase the amount of information on the target of evaluation. Testing might in particular provide feedback regarding the resilience of systems, i.e., it can support the estimation of the likelihood that an attack will be successful if initiated. Information from testing on the presence or absence of potential vulnerabilities has direct impact on the likelihood values of the associated threat scenarios. Similar to test-based risk identification, penetrating testing tools, model-based security testing tools, static and dynamic code analysis tools, and vulnerability scanners are useful for obtaining this kind of information.

***The Compliance Assessment Workstream***  The risk-based compliance assessment workstream consists of three major steps. The compliance risk identification step provides a systematic and template-based approach to identify and select compliance requirements that imply risk. These requirements are transformed into obligations and prohibitions that are the basis for further threat and risk modeling using the CORAS tool. The second step, compliance risk estimation, is dedicated to understanding and documenting the uncertainty that originates from compliance requirement interpretation. Uncertainty may arise from unclear compliance requirements or from uncertainty about the consequences in case of non-compliance. During compliance risk evaluation, compliance requirements are evaluated and prioritized based on their level of risk so that during treatment compliance resources may be allocated efficiently based on their level of risk. In summary, combining security risk assessment and compliance assessment helps to prioritize compliance measures based on risks, and helps to identify and deal with compliance requirements that directly imply risk.

***The Security Testing Workstream***  The risk-based security testing workstream is structured like a typical security testing process. It starts with a test planning phase, followed by a test design and implementation phase, and ends with test execution, analysis, and summary. The result of the risk assessment, i.e., the identified vulnerabilities, threat scenarios, and unwanted incidents, are used to guide the test planning and test identification, and may complement requirements engineering results with systematic information concerning the threats and vulnerabilities of a system.

Factors like probabilities and consequences can additionally be used to weight threat scenarios and thus help identify which threat scenarios are more relevant are thus the ones that need to be treated and tested more carefully. From a process point of view, the interaction between risk assessment and testing could be best described following the phases of a typical testing process.

1) Risk-based security test planning deals with the integration of security risk assessment in the test planning process.
2) Risk-based security test design and implementation deals with the integration of security risk assessment in the test design and implementation process.
3) Risk-based test execution, analysis, and summary deals with risk-based test execution and with the systematic analysis and summary of test results.

#### 1.4.1.2 Positioning in the Risk-Based Testing Taxonomy

*Context* The overall process (ETSI, 2015a; Großmann and Seehusen, 2015) is directly derived from ISO 31000 and slightly extended to highlight the integration with security testing and compliance assessment. The approach explicitly addresses *compliance*, but also *business* and in a limited way *safety*, as major *risk drivers*. It is defined independently of any application domain and independently of the level, target, or depth of the security assessment itself. It could be applied to any kind of technical assessment process with the potential to target the full number of *quality properties* that are defined in Section 1.3.1.2. Moreover, it addresses legal and compliance issues related to data protection and security regulations. Looking at risk-based security testing, the approach emphasizes executable *risk items*, i.e., *runtime artifacts*. Considering risk-based compliance assessment, the approach also addresses the other risk items mentioned in the taxonomy.

*Risk Assessment* The test-based risk assessment workstream uses test results as explicit input to various risk assessment activities. Risk assessment in RASEN has been carried out on the basis of the CORAS method and language. Thus, risk estimation is based on *formal* models that support the definition of *likelihood* values for events and *impact* values to describe the effect of incidents on assets. Both likelihood and impact values are used to calculate the overall *risk exposure* for unwanted incidents, i.e., the events that directly harm assets. CORAS is flexible with respect to the calculation scheme and to the scale for defining risk factors. It generally supports values with *qualitative scale* as well as with *quantitative scale.*

*Risk-Based Test Strategy* Security is not a functional property and thus requires dedicated information that addresses the (security) context of the system. While functional testing is more or less guided directly by the system specification (i.e., features, requirements, architecture), security testing often is not. The RASEN approach to *risk-based security test planning* especially addresses the risk-based selection of *test objectives and test techniques* as well as risk-based *resource planning and scheduling*. Security risk assessment serves this purpose and can be used to roughly identify high-risk areas or features of the SUT and thus determine and optimize the respective test effort. Moreover, a first assessment of the identified vulnerabilities and threat scenarios may help to select test strategies and techniques that are dedicated to deal with the most critical security risks.

Considering *security test design and implementation*, especially the selection and prioritization of the feature to test, the concrete test designs and the determination of *test coverage items* are critical. A recourse to security risks, potential threat scenarios, and potential vulnerabilities provide good guidance for improving *item prioritization and selection*. Security-risk-related information supports the selection of features and test conditions that require testing. It helps in identifying which coverage items should be covered to what depth, and how individual test cases and test procedures should look. The RASEN approach to risk-based security test design and implementation uses information on expected threats and potential vulnerabilities to systematically determine and identify coverage items (besides others, *asset coverage* and *threat scenario and vulnerabilities coverage*), test conditions (testable aspects of a system), and test purposes. Moreover, the security risk assessment provides quantitative estimations on the risks, i.e., the product of frequencies or probabilities and estimated consequences. This information is used to select and prioritize either the test conditions or the actual tests when they are assembled into test sets. Risks as well as their probabilities and consequence values are used to set priorities for the test selection, test case generation, and for the order of test execution expressed by risk-optimized test procedures. Risk-based test execution allows the prioritization of already existing test cases, test sets, or test procedures during regression testing. *Risk-based security test evaluation* aims to improve *risk reporting* and the *test exit decision* by introducing the notion of risk coverage and remaining risks on the basis of the intermediate test results as well as on the basis of the errors, vulnerabilities, or flaws that have been found during testing. In summary, we have identified the three activities that are supported through results from security risk assessment.

### 1.4.2 The SmartTesting Approach

#### 1.4.2.1 Description of the Approach

Figure 1.4 provides an overview of the overall process. It consists of different steps, which are either directly related to risk-based test strategy development (shown in bold font) or which are used to establish the preconditions (shown in normal font) for the process by linking test strategy development to the related processes (drawn with dashed lines) of defect management, requirements management, and quality management. The different steps are described in detail in the following subsections.

*Definition of Risk Items*   In the first step, the risk items are identified and defined. The risk items are the basic elements of a software product that can be associated with risks. Risk items are typically derived from the functional structure of the software system, but they can also represent non-functional aspects or system properties. In the context of testing it should be taken into account that the risk items need to be mapped to test objects (ISTQB, 2012), i.e., testable objects such as subsystems, features, components, modules, or functional as well as non-functional requirements.

*Probability Estimation*   In this step the probability values (for which an appropriate scale has to be defined) are estimated for each risk item. In the context of testing, the probability value expresses the likelihood of defectiveness of a risk item, i.e., the likelihood that a fault exists in a specific product component due to an error in a previous development phase that may lead to a failure. There are several ways to estimate or predict

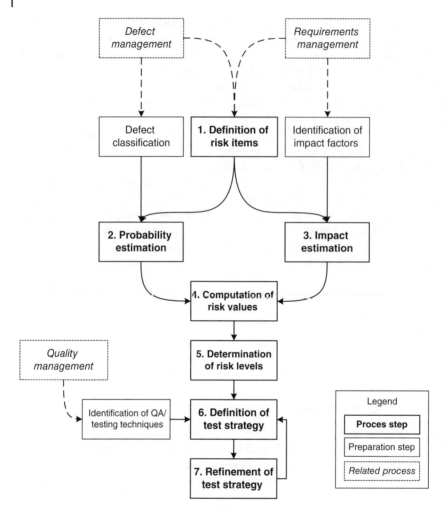

**Figure 1.4** SmartTesting process (Ramler and Felderer, 2015).

the likelihood of a component's defectiveness. Most of these approaches rely on historical defect data collected from previous releases or related projects. Therefore, defect prediction approaches are well suited to supporting probability estimation (Ramler and Felderer, 2016).

*Impact Estimation*   In this step the impact values are estimated for each risk item. The impact values express the consequences of risk items being defective, i.e., the negative effect that a defect in a specific component has on the user or customer and, ultimately, on the company's business success. The impact is often associated with the cost of failures. The impact is closely related to the expected value of the components for the user or customer. The value is usually determined in requirements engineering when eliciting and prioritizing the system's requirements. Thus, requirements management may be identified as the main source of data for impact estimation.

*Computation of Risk Values*   In this step risk values are computed from the estimated probability and impact values. Risk values can be computed according to the definition of risk as $R = P \times I$, where $P$ is the probability value and $I$ is the impact value. Aggregating the available information to a single risk value per risk item allows the prioritization of the risk items according to their associated risk values or ranks. Furthermore, the computed risk values can be used to group risk items, for example according to high, medium, and low risk. Nevertheless, for identifying risk levels it is recommended to consider probability and impact as two separate dimensions of risk.

*Determination of Risk Levels*   In this step the spectrum of risk values is partitioned into risk levels. Risk levels are a further level of aggregation. The purpose of distinguishing different risk levels is to define classes of risks such that all risk items associated with a particular class are considered equally risky. As a consequence, all risk items of the same class are subject to the same intensity of quality assurance and test measures.

*Definition of Test Strategy*   In this step the test strategy is defined on the basis of the different risk levels. For each risk level the test strategy describes how testing is organized and performed. Distinguishing different levels allows testing to be performed with differing levels of rigor in order to adequately address the expected risks. This can be achieved either by applying specific testing techniques (e.g., unit testing, use case testing, beta testing, reviews) or by applying these techniques with more or less intensity according to different coverage criteria (e.g., unit testing at the level of 100% branch coverage or use case testing for basic flows and/or alternative flows).

*Refinement of Test Strategy*   In the last step the test strategy is refined to match the characteristics of the individual components of the software system (i.e., risk items). The testing techniques and criteria that have been specified in the testing strategy for a particular risk level can be directly mapped to the components associated with that risk level. However, the test strategy is usually rather generic. It does not describe the technical and organizational details that are necessary for applying the specified techniques to a concrete software component. For each component, thus, a test approach has to be developed that clarifies how the test strategy should be implemented.

#### 1.4.2.2   Positioning in the Risk-Based Testing Taxonomy

*Context*   SmartTesting provides a lightweight process for the development and refinement of a risk-based test strategy. It does not explicitly address risk drivers, but – as in every risk-based testing process – it is implicitly assumed that a risk driver and a quality property to be improved are available. The risk drivers of the broad range of companies involved in the accompanying study (Ramler and Felderer, 2015) cover all types, i.e., business, safety, and compliance. Also, different quality properties of interest are covered, mainly as impact factors. For instance, the companies involved considered performance and security besides functionality as impact factors.

*Risk Assessment*   SmartTesting explicitly contains a step to define risk items, which can in principle be of any type from the taxonomy. For the companies involved, risk items were typically derived from the system's component structure. Via the process step computation of risk values, SmartTesting explicitly considers *risk exposure*, which is *qualitatively* estimated by a mapping of risk values to risk levels in the process step determination of risk levels. The risk value itself is measured based on a *formal model* in the process step computation of risk values, which combines values from probability and impact estimation. Probability estimation takes defect data into account, and impact estimation is based on impact factors, which are typically *assessed manually*.

**Risk-Based Test Strategy**   The process steps definition and refinement of the test strategy comprise *risk-based test planning* resulting in the assignment of concrete techniques, resource planning and scheduling, prioritization and selection strategies, metrics as well as exit criteria to the risk levels and further to particular risk items.

### 1.4.3   Risk-Based Test Case Prioritization Based on the Notion of Risk Exposure

#### 1.4.3.1   Description of the Approach

Choi et al. present different test case prioritization strategies based on the notion of *risk exposure*. In Yoon and Choi (2011), test case prioritization is described as an activity intended "to find the most important defects as early as possible against the lowest costs" (Redmill, 2005). Choi et al. claim that their risk-based approach to test case prioritization performs well against this background. They empirically evaluate their approach in a setting where various versions of a traffic conflict avoidance system (TCAS) are tested, and show how their approach performs well compared to the prioritization approach of others. In Hettiarachchi et al. (2016) the approach is extended using an improved prioritization algorithm and towards an automated risk estimation process using fuzzy expert systems. A fuzzy expert system is an expert system that uses fuzzy logic instead of Boolean logic to reason about data. Conducting risk estimation with this kind of expert system, Choi et al. aim to replace the human actor during risk estimation and thus avoid subjective estimation results. The second approach has been evaluated by prioritizing test cases for two software products, the electronic health record software *iTrust*, an open source product, and an industrial software application called *Capstone*.

#### 1.4.3.2   Positioning in the Risk-Based Testing Taxonomy

**Context**   Neither approach explicitly mentions one of the risk drivers from Section 1.3.1.1, nor do they provide exhaustive information on the quality properties addressed. However, in Yoon and Choi (2011) the authors evaluate their approach in the context of a safety critical application. Moreover, the authors emphasize that they refer to risks that are identified and measured during the product risk assessment phase. Such a phase is typically prescribed for safety critical systems. Both facts indicate that *safety* seems to be the major *risk driver*, and the safety-relevant attributes like *functionality*, *reliability*, and *performance* are the major quality properties that are addressed by testing. In contrast, the evaluation in Hettiarachchi et al. (2016) is carried out with *business* critical software, considering quality properties like *functionality* and *security*. Both approaches have in common that they do not address *compliance* as a risk driver.

**Risk Assessment**   The risk assessment process for both approaches aims to calculate *risk exposure*. The authors define risk exposure as a value with a *quantitative* scale that expresses the magnitude of a given risk. While in Yoon and Choi (2011) the authors explicitly state that they are intentionally not using their own testing-related equivalent for expressing risk exposure but directly refer to risk values coming from a pre-existing risk assessment, risk estimation in Hettiarachchi et al. (2016) is done automatically and tailored towards testing. The authors calculate risks on the basis of a number of indicators that are harvested from *development artifacts* like requirements. They use properties like requirements modification status and frequency as well as require-ments complexity and size to determine the risk likelihood and risk impact for each requirement. In addition, indicators on potential security threats are used to address and consider the notion of *security*. In contrast to Yoon and Choi (2011), Hettiarachchi et al. (2016) address the *automation* of the risk estimation process using an expert system that is able to aggregate the risk indicators and thus to automatically compute the overall risk exposure. While Yoon and Choi (2011) do not explicitly state whether the initial risk assessment relies on formal models or not, the approach in Hettiarachchi et al. (2016) is completely *formal*. However, since Yoon and Choi (2011) refer to safety critical systems, we can assume that the assessment is not just a list-based assessment.

**Risk-Based Test Strategy**   With respect to testing, both approaches aim for an efficient *test prioritization and selection* algorithm. Thus, they are mainly applicable in situations where test cases, or at least test case specifications, are already available. This first of all addresses regression testing, and also decision problems during test management, e.g., when test cases are already specified and the prioritization of test implementation efforts is required.

To obtain an efficient test prioritization strategy, both approaches aim to derive risk-related weights for individual test cases. In Yoon and Choi (2011), the authors propose two different strategies. The first strategy aims for simple *risk coverage*. Test cases that cover a given risk obtain a weight that directly relates to the risk exposure for that risk. If a test case covers multiple risks, the risk exposure values are summed. The second strategy additionally tries to consider the fault-revealing capabilities of the test cases. Thus, the risk-related weight for a test case is calculated by means of the risk exposure for a given risk correlated with the number of risk-related faults that are detectable by that test case, so that test cases with higher fault-revealing capabilities are rated higher. The fault-revealing capabilities of test cases are derived through mutation analysis; i.e., this strategy requires that the test cases already exist and that they are executable.

In Hettiarachchi et al. (2016), test cases are prioritized on the basis of their relationship to risk-rated requirements. Risk rating for requirements is determined by an automated risk rating conducted by the fuzzy expert system, and additional analysis of fault classes and their relation to the individual requirements. In short, a fault class is considered to have more impact if it relates to requirements with a higher risk exposure. In addition, a fault of a given fault class is considered to occur more often if that fault class relates to a larger number of requirements. Both values determine the overall risk rating for the individual requirements and thus provide the prioritization criteria for requirements. Finally, test cases are ordered by means of their relationship to the prioritized require-ments. During the evaluation of the approach, the authors obtained the relationship between test cases and requirements from existing traceability information.

While both approaches provide strong support for *risk-based item selection*, they do not support other activities during *risk-based test design and implementation*, nor do they establish dedicated activities in the area of *risk-based test execution and evaluation*.

### 1.4.4  Risk-Based Testing of Open Source Software

#### 1.4.4.1  Description of the Approach

Yahav et al. (2014a,b) provide an approach to risk-based testing of open source software (OSS) to select and schedule dynamic testing based on software risk analysis. Risk levels of open source components or projects are computed based on communication between developers and users in the open source software community. Communication channels usually include mail, chats, blogs, and repositories of bugs and fixes. The provided data-driven testing approach therefore builds on three repositories, i.e., a social repository, which stores the social network data from the mined OSS community; a bug repository, which links the community behavior and OSS quality; and a test repository, which traces tests (and/or test scripts) to OSS projects. As a preprocessing step, OSS community analytics is performed to construct a social network of communication between developers and users. In a concrete case study (Yahav et al., 2014b), the approach predicts the expected number of defects for a specific project with logistic regression based on the email communication and the time since the last bug.

#### 1.4.4.2  Positioning in the Risk-Based Testing Taxonomy

**Context**  The approach does not explicitly mention one of the risk drivers from Section 1.3.1.1, nor the quality properties. However, the authors state that the purpose of risk-based testing is mitigation of the significant failures experienced in product quality, timeliness, and delivery cost when adapting OSS components in commercial software packages (Yahav et al., 2014a). Therefore, risk drivers may be *business* to guarantee the success of a system, where the tested OSS component is integrated, or even safety, if the OSS component were to be integrated into a *safety* critical system. Due to the testing context, i.e., selection or prioritization of available tests of OSS components, the main quality property is supposed to be *functionality*. The risk item type is OSS components (developed in OSS projects), i.e., *architectural artifacts*.

**Risk Assessment**  The risk assessment approach quantifies the risk *likelihood*. For this purpose, a *formal model* is created to predict the number of bugs based on the communication in different communities and the time since the last bug. The scale is therefore *quantitative* as the approach tries to predict the actual number of bugs. The approach implements an *automatic assessment* as it uses monitors to automatically store data in repositories and then applies machine learning approaches, i.e., logistic regression, to predict the risk level.

**Risk-Based Test Strategy**  The approach explicitly supports risk-based test planning in terms of *test prioritization and selection* and *resource planning and scheduling*. The approach mainly addresses the allocation of available test scripts for dynamic testing and highlights that exhaustive testing is infeasible and that therefore selective testing techniques are needed to allocate test resources to the most critical components. Therefore, risk-based test design is not explicitly addressed. As *risk metrics*, the number of communication metrics as well as the time since the last defect are computed. The approach uses specific *logging support* to log and trace community and defect data. For *risk reporting* confusion matrices are used, which contrast the actual and predicted number of defects.

## 1.5 Summary

In this chapter, we presented a taxonomy of risk-based testing. It is aligned with the consideration of risks in all phases of the test process and consists of three top-level classes: contextual setup, risk assessment, and risk-based test strategy. The contextual setup is defined by risk drivers, quality properties, and risk items. Risk assessment comprises the subclasses factors, estimation technique, scale, and degree of automation. The risk-based test strategy then takes the assessed risks into account to guide test planning, test design and implementation, and test execution and evaluation. The taxonomy provides a framework to understand, categorize, assess, and compare risk-based testing approaches to support their selection and tailoring for specific purposes. To demonstrate its application and alignment with available risk-based testing approaches, we positioned within the taxonomy four recent risk-based testing approaches: the RASEN approach, the SmartTesting approach, risk-based test case prioritization based on the notion of risk exposure, and risk-based testing of open source software.

## References

Amland, S. (2000) Risk-based testing: Risk analysis fundamentals and metrics for software testing including a financial application case study. *Journal of Systems and Software*, 53(3), 287–295.

Bai, X., Kenett, R.S., and Yu, W. (2012) Risk assessment and adaptive group testing of semantic web services. *International Journal of Software Engineering and Knowledge Engineering*, 22(05), 595–620.

Botella, J., Legeard, B., Peureux, F., and Vernotte, A. (2014) *Risk-Based Vulnerability Testing Using Security Test Patterns*. Berlin: Springer, pp. 337–352.

Briand, L.C., Labiche, Y., and He, S. (2009) Automating regression test selection based on UML designs. *Information and Software Technology*, 51(1), 16–30.

ETSI (2015a) TR 101 583: *Methods for Testing and Specification (MTS); Risk-Based Security Assessment and Testing Methodologies* v1.1.1 (2015-11), Tech. Rep., European Telecommunications Standards Institute.

ETSI (2015b) TR 101 583: *Methods for Testing and Specification (MTS); Security Testing; Basic Terminology* v1.1.1 (2015-03), Tech. Rep., European Telecommunications Standards Institute.

Felderer, M. and Beer, A. (2013) Using defect taxonomies to improve the maturity of the system test process: Results from an industrial case study, in *Software Quality: Increasing Value in Software and Systems Development*, Berlin: Springer, pp. 125–146.

Felderer, M., Haisjackl, C., Pekar, V., and Breu, R. (2014) A risk assessment framework for software testing, in *International Symposium On Leveraging Applications of Formal Methods, Verification and Validation*, Berlin: Springer, pp. 292–308.

Felderer, M. and Ramler, R. (2013) Experiences and challenges of introducing risk-based testing in an industrial project, in *Software Quality: Increasing Value in Software and Systems Development*, Berlin: Springer, pp. 10–29.

Felderer, M. and Ramler, R. (2014a) Integrating risk-based testing in industrial test processes. *Software Quality Journal*, 22(3), 543–575.

Felderer, M. and Ramler, R. (2014b) A multiple case study on risk-based testing in industry. *International Journal on Software Tools for Technology Transfer*, 16(5), 609–625.

Felderer, M. and Ramler, R. (2016) Risk orientation in software testing processes of small and medium enterprises: An exploratory and comparative study. *Software Quality Journal*, 24(3), 519–548.

Felderer, M. and Schieferdecker, I. (2014) A taxonomy of risk-based testing. *International Journal on Software Tools for Technology Transfer*, 16(5), 559–568.

Fredriksen, R., Kristiansen, M., Gran, B.A., Stølen, K., Opperud, T.A., and Dimitrakos, T. (2002) The CORAS framework for a model-based risk management process, in *SAFECOMP, Lecture Notes in Computer Science*, vol. 2434 (eds S. Anderson, S. Bologna, and M. Felici), Berlin: Springer, pp. 94–105.

Gerrard, P. and Thompson, N. (2002) *Risk-Based E-Business Testing*. London: Artech House Publishers.

Goel, A.L. (1985) Software reliability models: Assumptions, limitations, and applicability. *IEEE Transactions on Software Engineering*, 11(12), 1411–1423.

Graham, D. and Fewster, M. (2012) *Experiences of Test Automation: Case Studies of Software Test Automation*, Reading, MA: Addison-Wesley Professional.

Großmann, J., Mahler, T., Seehusen, F., Solhaug, B., and Stølen, K. (2015) RASEN D5.3.3 Methodologies for Legal, Compositional, and Continuous Risk Assessment and Security Testing V3, Tech. Rep., RASEN, FP7-ICT-2011-8, Project Nr. 316853.

Großmann, J. and Seehusen, F. (2015) *Combining Security Risk Assessment and Security Testing Based on Standards*. New York: Springer International Publishing, pp. 18–33.

Hettiarachchi, C., Do, H., and Choi, B. (2016) Risk-based test case prioritization using a fuzzy expert system. *Information and Software Technology*, 69, 1–15.

Hosseingholizadeh, A. (2010) A source-based risk analysis approach for software test optimization, in *Computer Engineering and Technology (ICCET)*, 2nd International Conference on, vol. 2, IEEE, pp. 601–604.

Huizinga, D. and Kolawa, A. (2007) *Automated Defect Prevention: Best Practices in Software Management*. Chichester: Wiley.

ISO (2000) *ISO 14971: Medical Devices – Application of Risk Management to Medical Devices*, Geneva: ISO.

ISO (2009) ISO 31000 – Risk Management. http://www.iso.org/iso/home/standards/iso31000.htm [accessed 30 January, 2018].

ISO (2011) *ISO/IEC 25010:2011 Systems and Software Engineering – Systems and Software Quality Requirements and Evaluation (SQuaRE) – System and Software Quality Models*. Geneva: ISO.

ISO (2013) ISO/IEC/IEEE 29119: Software Testing. www.softwaretestingstandard.org [accessed 30 January, 2018].

ISTQB (2012) *Standard Glossary of Terms Used in Software Testing*. Version 2.2, Tech. Rep., ISTQB.

Jørgensen, M., Boehm, B., and Rifkin, S. (2009) Software development effort estimation: Formal models or expert judgment? *IEEE Software*, 26(2), 14–19.

Radatz, J., Geraci, A., and Katki, F. (1990) *IEEE Standard Glossary of Software Engineering Terminology*. IEEE Std, 610121990 121 990.

Ramler, R. and Felderer, M. (2015) A process for risk-based test strategy development and its industrial evaluation, in *International Conference on Product-Focused Software Process Improvement*, New York: Springer, pp. 355–371.

Ramler, R. and Felderer, M. (2016) Requirements for integrating defect prediction and risk-based testing, in *42nd EUROMICRO Conference on Software Engineering and Advanced Applications (SEAA 2016)*, IEEE, pp. 359–362.

Redmill, F. (2005) Theory and practice of risk-based testing. *Software Testing, Verification and Reliability*, 15(1), 3–20.

Stallbaum, H. and Metzger, A. (2007) Employing requirements metrics for automating early risk assessment. *Proc. MeReP07, Palma de Mallorca, Spain*, pp. 1–12.

Stallbaum, H., Metzger, A., and Pohl, K. (2008) An automated technique for risk-based test case generation and prioritization, in *Proc. 3rd International Workshop on Automation of Software Test*, ACM, pp. 67–70.

Standards Australia/New Zealand (2004) *Risk Management,* AS/NZS 4360:2004.

Tran, V. and Liu, D.B. (1997) A risk-mitigating model for the development of reliable and maintainable large-scale commercial off-the-shelf integrated software systems, in *Reliability and Maintainability Symposium. 1997 Proceedings, Annual*, pp. 361–367.

Wendland, M.F., Kranz, M., and Schieferdecker, I. (2012) A systematic approach to risk-based testing using risk-annotated requirements models, in *ICSEA 2012*, pp. 636–642.

Yahav, I., Kenett, R.S., and Bai, X. (2014a) Risk based testing of open source software (oss), in *Computer Software and Applications Conference Workshops (COMPSACW), 2014 IEEE 38th International*, IEEE, pp. 638–643.

Yahav, I., Kenett, R.S., and Bai, X. (2014b) Data driven testing of open source software, in *International Symposium On Leveraging Applications of Formal Methods, Verification and Validation*, Berlin: Springer, pp. 309–321.

Yoo, S. and Harman, M. (2012) Regression testing minimization, selection and prioritization: A survey. *Software Testing, Verification and Reliability,* 22(2), 67–120.

Yoon, H. and Choi, B. (2011) A test case prioritization based on degree of risk exposure and its empirical study. *International Journal of Software Engineering and Knowledge Engineering*, 21(02), 191–209.

Zech, P. (2011) Risk-based security testing in cloud computing environments, in *ICST 2011*, IEEE, pp. 411–414.

# 2

## Improving Software Testing with Causal Modeling

*Norman Fenton and Martin Neil*

## Synopsis

This chapter introduces the idea of causal modeling using Bayesian networks, and shows they can be used to improve software testing strategies and better understand and predict software reliability. The goal is to help software professionals (be they developers, testers, managers, or maintainers) make decisions under uncertainty. We are ultimately interested in predicting how much effort and time will be required to produce a system to a particular set of requirements, how much more testing to do before the system has tolerably few bugs, etc. All of these decisions involve uncertainty, risk, and trade-offs. The approach allows analysts to incorporate causal process factors as well as combine qualitative and quantitative measures, hence overcoming some of the well-known limitations of traditional software metrics and reliability methods. The approach has been used and reported on by major organizations, especially in the telecommunications industry.

## 2.1  Introduction

Several years ago, at a leading international software metrics conference, a keynote speaker recounted an interesting story about a company-wide metrics program that he had been instrumental in setting up. He said that one of the main objectives of the program was to achieve process improvement by learning, from metrics, which process activities worked and which did not. To do this, the company looked at those projects that, in metrics terms, were considered most successful. These were the projects with especially low rates of customer-reported defects, measured by defects per thousand lines of code (KLOC). The idea was to learn what processes characterized such successful projects. A number of such "star" projects were identified, including some that achieved the magical perfect quality target of zero defects per KLOC in the first six months post-release. But, it turned out that what they learned from this was very different from what they had expected. Few of the star projects were, in fact, at all successful from any commercial or subjective perspective:

> The main explanation for the very low number of defects reported by customers was that the products the projects developed were generally so poor that they never got properly completed or used.

*Analytic Methods in Systems and Software Testing*, First Edition.
Edited by Ron S. Kenett, Fabrizio Ruggeri, and Frederick W. Faltin.
© 2018 John Wiley & Sons Ltd. Published 2018 by John Wiley & Sons Ltd.

This story exposes the classic weakness of relying only on traditional static code metrics and defect counts: the omission of sometimes obvious and simple *causal* factors that can have a major explanatory effect on what is observed and learnt. If you are managing a software development project and you hear that very few defects were found during a critical testing phase, is that good news or bad news? Of course, it depends on the testing effort, just as it does if you heard that a large number of defects were discovered. The danger is to assume that defects found in testing correlates with defects found in operation, and that it is thus possible to build a, typically, regression style of model where *defects in operation* (the uncertain quantity we are trying to predict) is a function of the variable *defects found in test* (the known quantity we have observed and measured). Similarly, it is dangerous to assume that we can build regression models in which other metrics, such as size and complexity metrics, act as independent variables that can predict dependent variables like defects and maintenance effort. In Section 2.2 we will explain why such attempts at prediction based solely on linear correlation are not normally effective and cannot be used for risk assessment.

If you are interested in, say, predicting defects in operation from information gained during development and testing, then certainly metrics like defects found in test and other size and complexity metrics would be very useful pieces of evidence, but you need to use this information in a smart way and also include other sources of necessary data.

The "smart" and rational way to use this evidence to update your beliefs about defects in operation is to use Bayesian reasoning. When there are multiple, dependent metrics, as we have here, then we use Bayesian reasoning together with causal models, in the form of Bayesian networks (BNs). In this chapter we explain the basics of Bayesian reasoning and BNs in Section 2.3. In Section 2.4 we describe BNs that have been effectively used to support software testing and defects prediction, while in Section 2.5 we look at the special problems of predicting software reliability. In Section 2.6 we look at typical approaches to risk assessment used by software project managers and demonstrate how the BN approach can provide far more logic and insight.

There are numerous software tools available that make it very easy to build and run the BN models described in this chapter; Fenton and Neil (2012) provides a comprehensive overview. In particular, the AgenaRisk BN tool (Agena Ltd., 2018) can be downloaded free, and it contains all of the models described in the chapter.

## 2.2 From Correlation and Regression to Causal Models

Standard statistical approaches to risk assessment and prediction in all scientific disciplines seek to establish hypotheses from relationships discovered or postulated to be empirically observable in data. To take a non-software example, suppose we are interested in the risk of fatal automobile crashes. Table 2.1 gives the number of crashes resulting in fatalities in a recent year in the USA broken down by month (source: US National Highways Traffic Safety Administration). It also gives the average monthly temperature.

We plot the fatalities and temperature data in a scatterplot graph as shown in Figure 2.1.

**Table 2.1** Fatal automobile crashes per month.

| Month | Total fatal crashes | Average monthly temperature (°F) |
| --- | --- | --- |
| January | 297 | 17.0 |
| February | 280 | 18.0 |
| March | 267 | 29.0 |
| April | 350 | 43.0 |
| May | 328 | 55.0 |
| June | 386 | 65.0 |
| July | 419 | 70.0 |
| August | 410 | 68.0 |
| September | 331 | 59.0 |
| October | 356 | 48.0 |
| November | 326 | 37.0 |
| December | 311 | 22.0 |

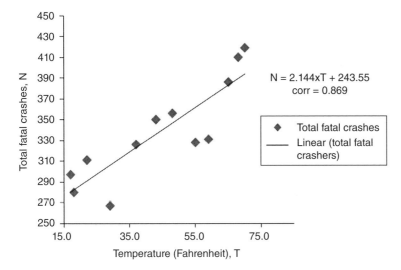

**Figure 2.1** Scatterplot of temperature against road fatalities (each dot represents a month).

There seems to be a clear relationship between temperature and fatalities – fatalities increase as the temperature increases. Indeed, using the standard statistical tools of correlation and $p$-values, statisticians would accept the hypothesis of a relationship as "highly significant" (the correlation coefficient here is approximately 0.869 and it comfortably passes the criteria for a $p$-value of 0.01).

However, in addition to serious concerns about the use of $p$-values generally, as described comprehensively in Ziliak and McCloskey (2008), there is an inevitable temptation arising from such results to infer causal links such as, in this case, higher temperatures cause more fatalities. Even though any introductory statistics course teaches that correlation is not causation, the regression equation might be used for prediction (e.g., in this case the equation relating $N$ to $T$ is used to predict that at 80 °F we might expect to see 415 fatal crashes per month).

But there is a grave danger of confusing prediction with risk assessment (and by implication with causation). For risk assessment and risk management the regression model is useless, because it provides no explanatory power at all. In fact, from a risk perspective this model would provide irrational, and potentially dangerous, information: it would suggest that if you want to minimize your chances of dying in an automobile crash you should do your driving when the highways are at their most dangerous, in winter.

One obvious improvement to the model, if the data were available, would be to factor in the number of miles traveled (i.e. journeys made). But there are other underlying causal and influential factors that might do much to explain the apparently strange statistical observations and provide better insights into risk. With some common sense and careful reflection we can easily recognize the following:

- Temperature influences the highway conditions (which will be made worse by decreasing temperature).
- Temperature also influences the number of journeys made; people generally make more journeys in spring and summer and will generally drive less when weather conditions are bad.
- When the highway conditions are bad, people tend to reduce their speed and drive more slowly, so highway conditions influence speed.
- The actual number of crashes is influenced not just by the number of journeys, but also the speed. If relatively few people are driving, and taking more care, we might expect fewer fatal crashes than we would otherwise experience.

The influence of these factors is shown in Figure 2.2.

The crucial message here is that the model no longer involves a simple single causal explanation; instead, it combines the statistical information available in a database (the "objective" factors) with other causal "subjective" factors derived from careful reflection and our experience. These factors now interact in a non-linear way that helps us to arrive at an explanation for the observed results. Behavior, such as our natural caution in driving more slowly when faced with poor road conditions, leads to lower accident rates (people are known to adapt to the perception of risk by tuning the risk to tolerable levels – this is formally referred to as risk homeostasis). Conversely, if we insist on driving fast in poor road conditions then, irrespective of the temperature, the risk of an accident increases, and so the model is able to capture our intuitive beliefs that were contradicted by the counterintuitive results from the simple regression model.

By accepting the statistical model we are asked to defy our senses and experience and actively ignore the role unobserved factors play. In fact, we cannot even explain the results without recourse to factors that do not appear in the database. This is a key point: with causal models we seek to dig deeper behind and underneath the data to explore richer relationships missing from over-simplistic statistical models. In doing so we gain insights into how best to control risk and uncertainty. The regression model, based on the idea that we can predict automobile crash fatalities based on temperature, fails to answer the substantial question: how can we control or influence behavior to reduce fatalities. This at least is achievable; control of the weather is not.

Statistical regression models have played a major role in various aspects of software prediction, with size (whether it be solution size, as in complexity metrics or LOC-based

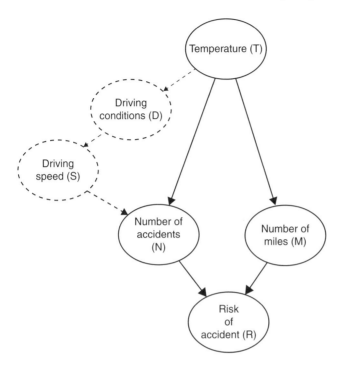

**Figure 2.2** Causal model for fatal crashes.

approaches, or problem size, as in function-point-based approaches) being the key driver. Thus, it is assumed that we can fit a function $f$ such that:

Defects $= f$(size, complexity).

Figures 2.3–2.6 show data from a major case study (Fenton and Ohlsson, 2000) that highlighted some of the problems with these approaches. In each figure the dots represent modules (which are typically self-contained software components of approximately 2000 LOC) sampled at random from a very large software system. Figure 2.3 (in which "faults" refers to all known faults discovered pre- and post-delivery) confirms what many studies have shown: module size is correlated with number of faults, but is not a good predictor of it. Figure 2.4 shows that complexity metrics, such as cyclomatic complexity, are not significantly better (and in any case are very strongly correlated to LOC). This is true even when we separate out pre- and post-delivery faults.

Figure 2.5 shows that there is no obvious empirical support for the widely held software engineering theory that smaller modules are less "fault-prone" than large ones. The evidence here shows a random relationship.

But the danger of relying on statistical models is most evident when we consider the relationship between the number of pre-release faults (i.e. those found during pre-release testing) and the number of post-release faults (the latter being the number you really are most interested in predicting). The data alone, as shown in Figure 2.6 (an empirical result that has been repeated in many other systems) shows a relationship that is contrary to widely perceived assumptions. Instead of the expected strong (positive) correlation between pre-release and post-release faults (i.e. the expectation

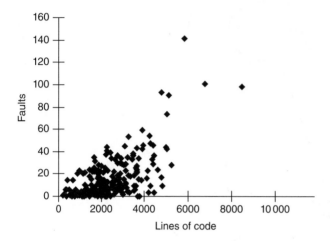

**Figure 2.3** Scatterplot of LOC against all faults for a major system (each dot represents a module).

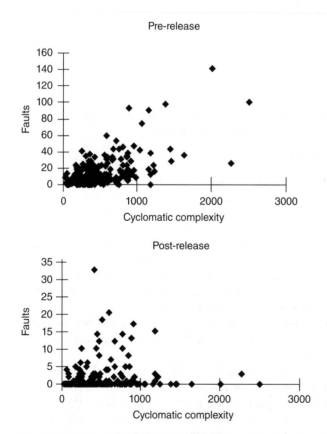

**Figure 2.4** Scatterplots of cyclomatic complexity against number of pre- and post-release faults for release *n* + 1 (each dot represents a module).

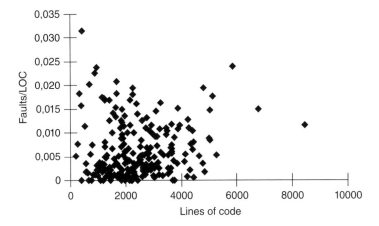

**Figure 2.5** Scatterplot of module fault density against size (each dot represents a module).

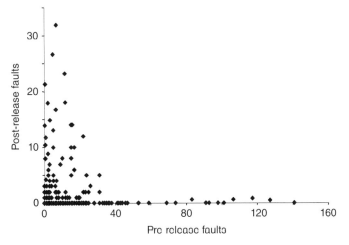

**Figure 2.6** Scatterplot of pre-release faults against post-release faults for a major system (each dot represents a module).

that modules which are most prone to faults pre-release will be the ones that are most fault prone post-release), there is strong evidence of a *negative correlation*. The modules that are very fault prone pre-release are likely to reveal very few faults post-release. Conversely, the truly "fault-prone" modules post-release are the ones that mostly revealed few or no faults pre-release.[1]

There are, of course, very simple causal explanations for the phenomenon observed in Figure 2.6. One possibility is that most of the modules that had many pre-release and few post-release faults were very well tested. The amount of testing is therefore a very simple explanatory factor that must be incorporated into any predictive model of defects. Similarly, a module that is simply never executed in operation will reveal

---

1 These results are potentially devastating for some regression-based fault prediction models, because many of those models were "validated" on the basis of using pre-release fault counts as a surrogate measure for operational quality.

no faults no matter how many are latent. Hence, operational usage is another obvious explanatory factor that must be incorporated.

The absence of any causal factors that explain variation is also a feature of the classic approach to software metrics resource prediction, where again the tendency has been to produce regression-based functions of the form:

$$\text{Effort} = f(\text{size, process quality, product quality}),$$

$$\text{Time} = f(\text{size, process quality, product quality}).$$

There are five fundamental problems with this approach:

- It is inevitably based on limited historical data of projects that just happened to have been completed and were available. Based on typical software projects these are likely to have produced products of variable quality, including many that are poor. It is therefore difficult to interpret what a figure for effort prediction based on such a model actually means.
- It fails to incorporate any true causal relationships, relying often on the fundamentally flawed assumption that somehow the solution size can influence the amount of resources required. This is contrary to the economic definition of a production model where output $= f(\text{input})$ rather than input $= f(\text{output})$.
- There is a flawed assumption that projects do not have prior resource constraints. In practice all projects do, but these cannot be accommodated in the models. Hence, the "prediction" is premised on impossible assumptions and provides little more than a negotiating tool.
- The models are effectively "black boxes" that hide or ignore crucial assumptions explaining the relationships and trade-offs between the various inputs and outputs.
- The models provide no information about the inevitable inherent uncertainty of the predicted outcome; for a set of "inputs" the models will return a single point value with no indication of the range or size of uncertainty.

To provide genuine risk assessment and decision support for managers we need to provide the following kinds of predictions:

- For a problem of this size, and given these limited resources, how likely am I to achieve a product of suitable quality?
- How much can I scale down the resources if I am prepared to put up with a product of specified poorer quality?
- The model predicts that I need four people over two years to build a system of this size. But I only have funding for three people over one year. If I cannot sacrifice quality, how good must the staff be to build the systems with the limited resources? Alternatively, if my staff are no better than average and I cannot change them, how much required functionality needs to be cut in order to deliver at the required level of quality?

Our aim is to show that causal models, using BNs, expert judgment, and incomplete information are also available in software engineering projects.

## 2.3 Bayes' Theorem and Bayesian Networks

While Section 2.2 provided the rationale for using causal models rather than purely statistically driven models, it provided no actual mechanism for doing so. The necessary mechanism is driven by Bayes' theorem, which provides us with a rational means

**Figure 2.7** Causal view of evidence.

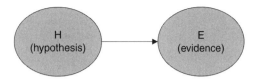

of updating our belief in some unknown hypothesis in the light of new or additional evidence (i.e. observed metrics).

At their core, all of the decision and prediction problems identified so far incorporate the basic causal structure shown in Figure 2.7.

There is some unknown hypothesis $H$ about which we wish to assess the uncertainty and make some decision. Does our system contain critical bugs? Does it contain sufficiently few bugs to release? Will it require more than three person-months of effort to complete the necessary functionality? Will the system fail within a given period of time?

Consciously or unconsciously we start with some (unconditional) prior belief about $H$. Taking a non-software example, suppose that we are in charge of a chest clinic and are interested in knowing whether a new patient has cancer; in this case, $H$ is the hypothesis that the patient has cancer. Suppose that 10% of previous patients who came to the clinic were ultimately diagnosed with cancer. Then a reasonable prior belief for the probability that $H$ is true, written $P(H)$, is 0.1. This also means that the prior $P(\text{not } H) = 0.9$.

One piece of evidence $E$ we might discover that could change our prior belief is whether or not the person is a smoker. Suppose that 50% of the people coming to the clinic are smokers, so $P(E) = 0.5$. If we discover that the person is indeed a smoker, to what extent do we revise our prior judgment about the probability $P(H)$ that the person has cancer? In other words we want to calculate the probability of $H$ given the evidence $E$. We write this as $P(H \mid E)$; it is called the *conditional probability* of $H$ given $E$, and because it represents a revised belief about $H$ once we have seen the evidence $E$ we also call it the *posterior probability* of $H$. To arrive at the correct answer for the posterior probability we use a type of reasoning called *Bayesian inference*, named after Thomas Bayes who determined the necessary calculations for it in 1763.

What Bayes recognized was that (as in this and many other cases) we might not have direct information about $P(H \mid E)$ but we do have *prior* information about $P(E \mid H)$. The probability $P(E \mid H)$ is called the *likelihood* of the evidence – it is the chance of seeing the evidence $E$ if $H$ is true[2] (generally we will also need to know $P(\text{not } E \mid H)$). Indeed, we can find out $P(E \mid H)$ simply by checking the proportion of people with cancer who are also smokers. Suppose, for example, that we know that $P(E \mid H) = 0.8$.

Now Bayes' theorem tells us how to compute $P(H \mid E)$ in terms of $P(E \mid H)$. It is simply:

$$P(H|E) = \frac{P(E|H) \times P(H)}{P(E)}.$$

In this example, since $P(H) = 0.1$, $P(E \mid H) = 0.8$, and $P(E) = 0.5$, it follows that $P(H \mid E) = 0.16$. Thus, if we get evidence that the person is a smoker we revise our probability of the person having cancer (it increases from 0.1 to 0.16).

---

2  For simplicity, we assume for the moment that the hypothesis and the evidence are simple Boolean propositions, i.e. they are simply either true or false. Later on we will see that the framework applies to arbitrary discrete and even numeric variables.

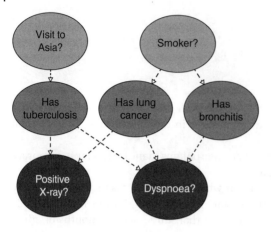

**Figure 2.8** Bayesian network for diagnosing disease.

| | |
|---|---|
| Yes | 0.01 |
| No | 0.99 |

Probability table for "Visit to Asia?"

| Smoker? | Yes | No |
|---|---|---|
| Yes | 0.6 | 0.3 |
| No | 0.4 | 0.7 |

Probability table for "Bronchitis?"

**Figure 2.9** Node probability table examples.

In general, we may not know $P(E)$ directly, so the more general version of Bayes' theorem is one in which $P(E)$ is also determined by the prior likelihoods:

$$P(H|E) = \frac{P(E|H) \times P(H)}{P(E)} = \frac{P(E|H) \times P(H)}{P(E|H) \times P(H) + P(E|\text{not } H) \times P(\text{not } H)}.$$

While Bayes' theorem is a rational way of revising beliefs in the light of observing new evidence, it is not easily understood by people without a statistical/mathematical background. But, if Bayes' theorem is difficult for lay people to compute and understand in the case of a single hypothesis and piece of evidence (as in Figure 2.7), the difficulties are obviously compounded when there are multiple related hypotheses and evidence as in the example of Figure 2.8.

As in Figure 2.7, the nodes in Figure 2.8 represent variables (which may be known or unknown), and the arcs[3] represent causal (or influential) relationships. In addition to the graphical structure, we specify, for each node, a *node probability table* (NPT), such as the examples shown in Figure 2.9. These tables capture the relationship between a node and its parents by specifying the probability of each outcome (state) given every combination of parent states. The resulting model is called a BN.

So, a BN is a directed graph together with a set of probability tables. The directed graph is referred to as the "qualitative" part of the BN, while the probability tables are referred to as the "quantitative" part.

The BN in Figure 2.8 is intended to model the problem of diagnosing diseases (tuberculosis, cancer, bronchitis) in patients attending a chest clinic. Patients may have symptoms (like dyspnoea – shortness of breath) and can be sent for diagnostic tests (X-ray);

---

3 Some of the arcs in Figure 2.8 are dotted. This simply means there are some "hidden" nodes through which the causal or influential relationship passes.

(a) Prior beliefs point to bronchitis
  as most likely

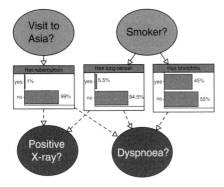

(b) Patient is '"non-smoker"' experiencing
  dyspnoea (shortness of breath): strengthens
  belief in bronchitis

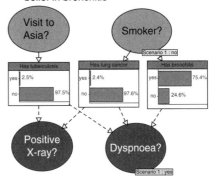

(c) Positive X-ray result increases
  probability of TB tuberculosis and
  cancer, but bronchitis still most likely

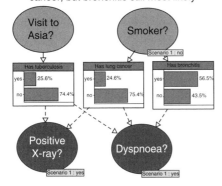

(d) Visit to Asia makes TB tuberculosis
  most likely now

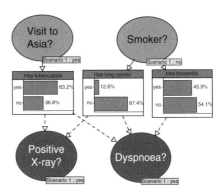

**Figure 2.10** Reasoning within the Bayesian network.

there may also be underlying causal factors that influence certain diseases more than
others (such as smoking, visiting Asia).

To use Bayesian inference properly in this type of network necessarily involves multiple applications of Bayes' theorem in which evidence is "propagated" throughout the
model. This process is complex and quickly becomes infeasible when there are many
nodes and/or nodes with multiple states. This complexity is the reason why, despite its
known benefits, there was for many years little appetite to use Bayesian inference to
solve real-world decision and risk problems. Fortunately, due to breakthroughs in the
late 1980s that produced efficient calculation algorithms, there are now widely available
software tools that enable anybody to do the Bayesian calculations without ever having
to understand, or even look at, a mathematical formula. These developments were the
catalyst for an explosion of interest in BNs. Using such a software tool we can do the
kind of powerful reasoning shown in Figure 2.10.

Specifically:

- With the prior assumptions alone (Figure 2.10a). Bayes' theorem computes what
  are called the prior marginal probabilities for the different disease nodes (note that
  we did not "specify" these probabilities – they are computed automatically; what we

specified were the conditional probabilities of these diseases given the various states of their parent nodes). So, before any evidence is entered the most likely disease is bronchitis (45%).

- When we enter evidence about a particular patient, the probabilities for all of the unknown variables get updated by the Bayesian inference. So, in Figure 2.10b, once we enter the evidence that the patient has dyspnoea and is a non-smoker, our belief in bronchitis being the most likely disease increases (75%).
- If a subsequent X-ray test is positive (Figure 2.10c) our belief in both tuberculosis (26%) and cancer (25%) are raised, but bronchitis is still the most likely (57%).
- However, if we now discover that the patient visited Asia (Figure 2.10d) we overturn our belief in bronchitis in favor of tuberculosis (63%).

Note that we can enter any number of observations anywhere in the BN and update the marginal probabilities of all the unobserved variables. As the above example demonstrates, this can yield some exceptionally powerful analyses that are simply not possible using other types of reasoning and classical statistical analysis methods.

In particular, BNs offer the following benefits:

- explicitly model causal factors
- reason from effect to cause and vice versa
- overturn previous beliefs in the light of new evidence (also called "explaining away")
- make predictions with incomplete data
- combine diverse types of evidence, including both subjective beliefs and objective data
- arrive at decisions based on visible auditable reasoning (unlike black-box modeling techniques there are no "hidden" variables and the inference mechanism is based on a long-established theorem).

With the advent of the BN algorithms and associated tools, it is therefore no surprise that BNs have been used in a range of applications that were not previously possible with Bayes' theorem alone.

## 2.4 Applying Bayesian Networks to Software Defects Prediction

Once we think in terms of causal models that relate unknown hypotheses and evidence, we are able to make much better use of software metrics and arrive at more rational predictions that not only avoid the kind of fallacies we introduced earlier, but are also able to explain apparent empirical anomalies. This is especially evident in the area of defect prediction.

### 2.4.1 A Very Simple BN for Understanding Defect Prediction

When we count *defects found in testing*, what we actually have is evidence about the unknown variable we are most interested in, namely the *number of defects present*. A very simple, but rational, causal model is shown in Figure 2.11.

Clearly, the number of defects present will influence the number of defects found, but the latter will also be influenced by the testing quality (as we will see later, this will turn out to be a fragment of a larger model; for example, the node representing defects

**Figure 2.11** Simple causal model for defects found.

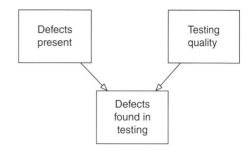

**Table 2.2** Node probability table for "defects found" node.

| Defects present | Low | | Medium | | High | |
|---|---|---|---|---|---|---|
| Testing quality | Poor | Good | Poor | Good | Poor | Good |
| Low | 1 | 1 | 0.9 | 0.1 | 0.7 | 0 |
| Medium | 0 | 0 | 0.1 | 0.9 | 0.2 | 0.2 |
| High | 0 | 0 | 0 | 0 | 0.1 | 0.8 |

present is a synthesis of a number of factors including process quality and problem complexity).

To keep things as simple as possible initially, we will assume that the number of defects is classified into just three states (low, medium, high) and that testing quality is classified into just two (poor, good). Then we might reasonably specify the NPT for the node *defects found in testing* as shown in Table 2.2. For example, this specifies that if testing quality is "poor" and there are a "medium" number of defects present then there is a 0.9 probability (90% chance) that the number of defects found is "low," 0.1 probability (10% chance) that the number of defects found is "medium," and 0 probability that the number of defects found is "high."

So, if testing quality is "poor," then the number of defects found is likely to be "low" even when there are a high number of defects present. Conversely, if testing quality is "good" we assume that most (but not all) defects present will be found. Assuming that the prior probabilities for the states of the nodes *defect present* and *testing quality* are all equal, then in its marginal state the model is shown in Figure 2.12.

One of the major benefits of the causal models that we are building is that they enable us to reason in both a "forward" and a "reverse" direction, using the notion of propagation introduced earlier. That is, we can identify the possible causes given the observation of some effect. In this case, if the number of defects found is observed to be "low," the model will tell us that low testing quality and a low number of defects present are *both* possible explanations (perhaps with an indication as to which one is the most likely explanation), as shown in Figure 2.13.

The power of the model becomes clearer as we discover more "evidence." Suppose, for example, that we also have independent evidence that the testing quality is good. Then, as shown in Figure 2.14a, we can deduce with some confidence that there were indeed a low number of defects present. However, if it turns out that the testing quality is poor, as shown in Figure 2.14b, we are little wiser than when we started; the evidence of a

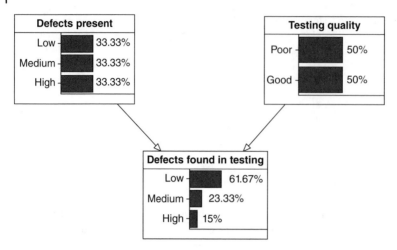

**Figure 2.12** Model in its initial (marginal) state.

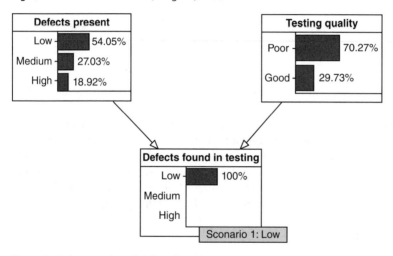

**Figure 2.13** Low number of defects found.

low number of defects found in testing has been "explained away" by the poor testing quality; we now have no real knowledge of the actual number of defects present, despite the low number found in testing.

### 2.4.2 More Complete Model for Software Testing and Defects Prediction

A more complete causal model for software testing and defects and prediction is shown in Figure 2.15; this is itself a simplified version of a model that has been extensively used in a number of real software development environments (Fenton et al., 2007).

In this case, the *number of defects found in operation*[4] (i.e. those found by customers) in a software module is what we are really interested in predicting. We know this is

---

4 Handling numeric nodes such as *the number of defects found in operation* in a BN was, until relatively recently, a major problem because it was necessary to "discretize" such nodes using some predefined range and intervals. This is cumbersome, error prone, and highly inaccurate – see Fenton et al. (2008) for a comprehensive explanation. Such inaccuracies, as well as the wasted effort over selecting and defining

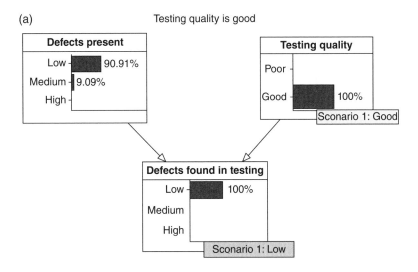

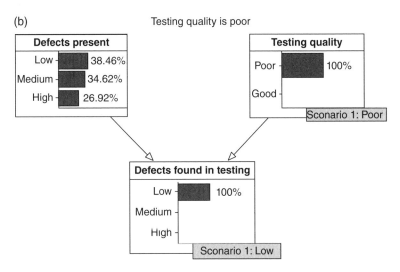

**Figure 2.14** Effect of different evidence about testing quality.

clearly dependent on the number of *residual defects*, but it is also critically dependent on the amount of *operational usage*. If you do not use the system you will find no defects, irrespective of the number there. The *number of residual defects* is determined by the *number you introduce during development* minus the *number you successfully find and fix*. Obviously, defects found and fixed are dependent on the *number introduced*. The number introduced is influenced by *problem complexity* and *design process quality*. The better the design the fewer the defects, and the less complex the problem the fewer the defects. Finally, how many defects you find is influenced not just by the number there to find but also by the amount of *testing effort*.

discretization intervals, can now be avoided by using *dynamic discretization* – in particular, the algorithm described in Neil et al. (2007) and implemented in AgenaRisk. Dynamic discretization allows users to simply define a numeric node by a single range (such as –infinity to infinity, 0 to 100, 0 to infinity, etc.).

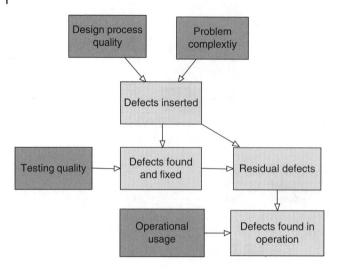

**Figure 2.15** Bayesian network model for software defects and reliability prediction.

**Table 2.3** Probability distributions for the nodes of the BN model in Figure 2.15.

| Node name | Probability distribution |
| --- | --- |
| Defects found in operation | $B(n, p)$, where $n$ = "residual defects" and $p$ = "operational usage" |
| Residual defects | defects inserted – defects found (and fixed) in testing |
| Defects found in testing | $B(n, p)$, where $n$ = "defects inserted" and $p$ = "testing quality" |
| Defects inserted | This is a distribution based on empirical data from a particular organization. For full details see Fenton et al. (2007). |

The task of defining the NPTs for each node in this model is clearly more challenging than the previous examples. In many situations we would need the NPT to be defined as a *function* rather than as an exhaustive table of all potential parent state combinations. Some of these functions are deterministic rather than probabilistic: for example, the "residual defects" is simply the numerical difference between the "defects inserted" and the "defects found and fixed." In other cases, we can use standard statistical functions. For example, in this version of the model we assume that "defects found and fixed" is a binomial distribution, $B(n, p)$, where $n$ is the number of defects inserted and $p$ is the probability of finding and fixing a defect (which in this case is derived from the "testing quality"); in more sophisticated versions of the model the $p$ variable is also conditioned on $n$ to reflect the increasing relative difficulty of finding defects as $n$ decreases. Table 2.3 lists the full set of conditional probability distributions for the nodes (that have parents) of the BN model of Figure 2.15.

The nodes "design quality," "complexity," "testing quality," and "operational usage" are all examples of what are called *ranked nodes* (Fenton, Neil, and Gallan, 2007); ranked nodes have labels like {very poor, poor, average, good, very good}, but they have an underlying [0, 1] scale that makes it very easy to define relevant probability tables for nodes which have them as parents (as in some of the functions described in Table 2.3).[5] The nodes without parents are all assumed to have a prior uniform distribution, i.e. one in which any state is equally as likely as any other state (in the "real" models the distributions for such nodes would normally not be defined as uniform but would reflect the historical distribution of the organization either from data or expert judgment).

We next illustrate the BN calculations (that are performed automatically), which show that the case for using BNs as causal models for software defects and reliability prediction is both simple and compelling.

Figure 2.16 shows the marginal distributions of the model before any evidence has been entered; this represents our uncertainty before we enter any specific information about this module. Since we assumed uniform distributions for nodes without parents, we see, for example, that the module is just as likely to have very high complexity as very low, and that the number of defects found and fixed in testing is in a wide range where the median value is about 18–20 (the prior distributions here were for a particular organization's modules).

Figure 2.17 shows the result of entering two observations about this module:

1) that it had zero defects found and fixed in testing; and
2) that the problem complexity is "high."

Note that all the other probability distributions updated. The model is doing both forward inference to predict defects in operation and backwards inference about causes, including, say, design process quality. Although the fewer than expected defects found does indeed lead to a belief that the post-release faults will drop, the model shows that the most likely explanation is inadequate testing.

So far we have made no observation about operational usage. If, in fact, the operational usage is "very high" (Figure 2.18) then what we have done is replicate the apparently counterintuitive empirical observations we discussed in Section 2.2 whereby a module with no defects found in testing has a high number of defects post-release.

But, suppose we find out that the test quality was "very high" (Figure 2.19).

Then we completely revise our beliefs. We are now confident that the module will be fault free in operation. Note also that the "explanation" is that the design process is likely to be very high quality. This type of reasoning is unique to BNs. It provides a means for decision makers (such as quality assurance managers in this case) to make decisions and interventions dynamically as new information is observed.

### 2.4.3 Commercial-Scale Versions of the Defect Prediction Models

The ability to do the kind of prediction and what-if analysis described in the model in Section 2.4.2 has proved to be very attractive to organizations who need to monitor and predict software defects and reliability, and who already collect defect-type metrics. Hence, organizations such as Motorola (Gras, 2004), Siemens (Wang et al., 2006), and Philips (Fenton et al., 2007) have exploited models and tools originally developed in Fenton, Neil, and Krause (2002) to build large-scale versions of the kind of model described in Section 2.4.2.

---

5 Again, note that AgenaRisk provides comprehensive support for ranked nodes and their associated NPTs.

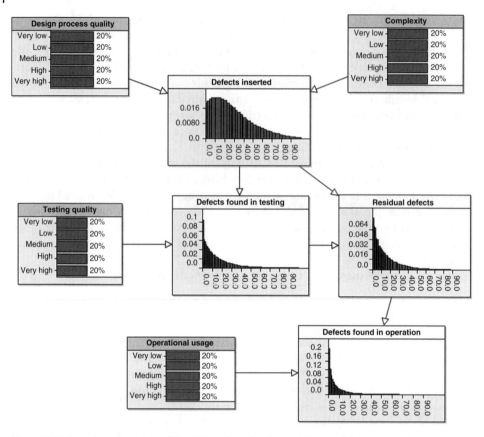

**Figure 2.16** Bayesian network model with marginal distributions for variables superimposed on nodes. All of the graphs are probability distributions, but there is a standard convention to represent discrete distributions (such as the node *testing quality*) with horizontal bars (i.e. the probability values are on the *x*-axis), whereas continuous/numeric distributions (such as the node *defects found in testing*) have vertical bars (i.e. the probability values are on the *y*-axis).

It is beyond the scope of this chapter to describe the details of these models and how they were constructed and validated, but what typifies the approaches is that they are based around a sequence of testing phases, by which we mean those testing activities such as system testing, integration testing, and acceptance testing that are defined as part of the companies' software processes (and, hence, for which relevant defect and effort data is formally recorded). In some cases a testing phase is one that does not involve code execution, such as design review. The final "testing" phase is generally assumed to be the software in operation. Corresponding to each phase is a "risk object" like that in Figure 2.20, where a risk object is a component of the BN with interface nodes to connect the component risk object to other parent and child risk objects. For the final "operational" phase, there is, of course, no need to include the nodes associated with defect fixing and insertion.

The distributions for nodes such as "probability of finding defect" derive from other risk objects such as that shown in Figure 2.21. The particular nodes and distributions will, of course, vary according to the type of testing phase.

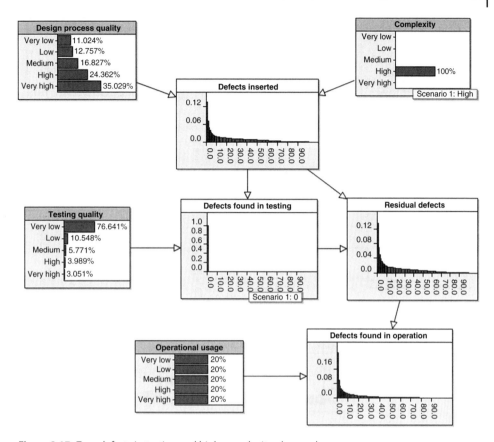

**Figure 2.17** Zero defects in testing and high complexity observed.

To give a feel for the kind of expert elicitation and data that was required to complete the NPTs in these kinds of models, we look at two examples, namely the nodes "probability of finding a defect" and "testing process overall effectiveness."

The "probability of finding a defect" node is a continuous node in the range [0,1] that has a single parent, "testing process overall effectiveness," which is a ranked node – in the sense of Fenton, Neil, and Gallan (2007) on a five-point scale from "very low" to "very high"). For a specific type of testing phase (such as integration testing) the organization had both data and expert judgment that enabled them to make the following kinds of assessments:

- "Typically (i.e. for our average level of test quality) this type of testing will find approximately 20% of the residual defects in the system."
- "At its best (i.e. when our level of testing is at its best) this type of testing will find 50% of the residual defects in the system; at its worst it will only find 1%."

Based on this kind of information, the NPT for the node "probability of finding a defect" is a partitioned expression like the one in Table 2.4. Thus, for example, when the overall testing process effectiveness is "average," the probability of finding a defect is a truncated normal distribution over the range [0, 1] with mean 0.2 and variance 0.001.

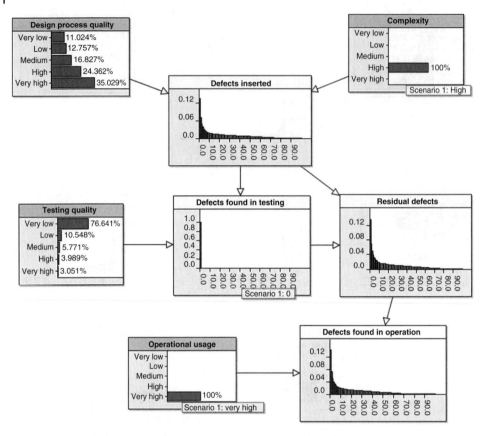

**Figure 2.18** Very high operational usage.

The "testing process overall effectiveness" node is a ranked node on a five-point ranked scale from "very low" to "very high." It has three parents, "testing process quality," "testing effort," and "quality of overall documentation," which are all also ranked nodes on the same five-point ranked scale from "very low" to "very high." Hence, the NPT in this case is a table of 625 entries. Such a table is essentially impossible to elicit manually, but the techniques described in Fenton, Neil, and Gallan (2007), in which ranked nodes are mapped on to an underlying [0, 1] scale, enabled experts to construct a sensible table in seconds using an appropriate "weighted expression" for the child node in terms of the parents. For example, the expression elicited in one case was a truncated normal distribution (on the range [0, 1]) with mean equal to the weighted minimum of the parent values (where the weights were 5.0 for "testing effort," 4.0 for "testing quality," and 1.0 for "documentation quality") and variance 0.001. Informally, this weighted minimum expression captured expert judgment like the following:

> Documentation quality cannot compensate for lack of testing effort, although a good testing process is important.

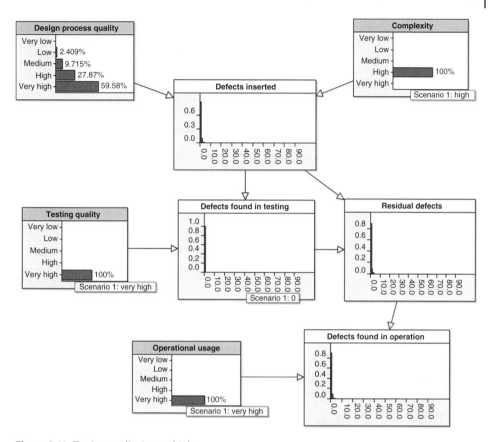

**Figure 2.19** Testing quality is very high.

As an illustration, Figure 2.22 shows the resulting distribution for overall testing process effectiveness when the testing process quality is "average," the quality of documentation is "very high," but the testing effort is "very low."

Using a BN tool such as AgenaRisk the various risk objects are joined according to the BN Object approach (Koller and Pfeffer, 1997), as shown in Figure 2.23. Here, each box represents a BN where only the "input" and "output" nodes are shown. For example, for the BN representing the defects in phase 2, the "input" node *residual defects pre* is defined by the marginal distribution of the output node *residual defects post* of the BN representing the defects in phase 1.

The general structure of the BN model proposed here is relevant for any software development organization whose level of maturity includes defined testing phases in which defect and effort data is recorded. However, it is important to note that a number of the key probability distributions will inevitably be organization/project specific. In particular, there is no way of producing a generic distribution for the "probability of finding a defect" in any given phase (and this is especially true of the operational testing phase); indeed, even within a single organization this distribution will be conditioned on many factors (such as ones that are unique to a particular

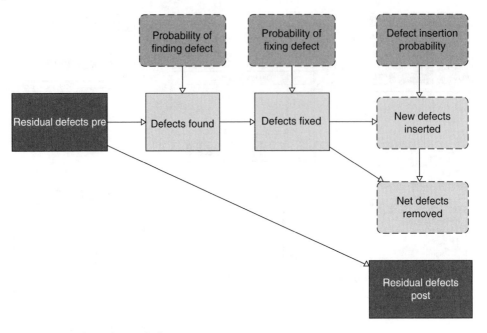

**Figure 2.20** Defects phase risk object.

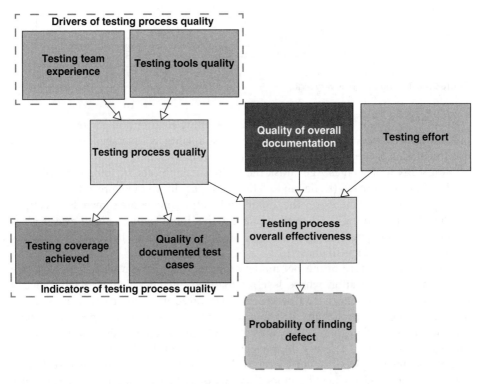

**Figure 2.21** Typical risk object for testing quality.

**Table 2.4** Node probability table for the node "probability of finding a defect."

| Parent state ("overall testing process effectiveness") | Probability of finding a defect |
|---|---|
| Very low | TNormal(0.01, 0.001, 0, 1) |
| Low | TNormal(0.1, 0.001, 0, 1) |
| Average | TNormal(0.2, 0.001, 0, 1) |
| High | TNormal(0.35, 0.001, 0, 1) |
| Very high | TNormal(0.5, 0.001, 0, 1) |

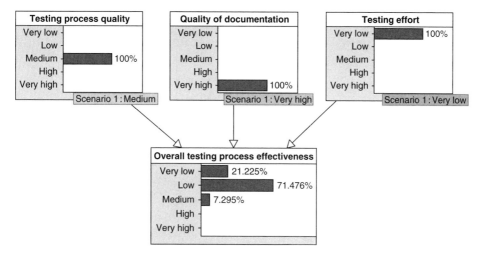

**Figure 2.22** Scenario for "overall testing effectiveness."

project) that may be beyond the scope of a workable BN model. At best we assume that there is sufficient maturity and knowledge within an organization to produce a "benchmark" distribution in a given phase. Where necessary, this distribution can then still be tailored to take account of specific factors that are not incorporated in the BN model. It is extremely unlikely that such tailoring will always be able to take account of extensive relevant empirical data; hence, as in most practically usable BN models, there will be a dependence on subjective judgments, but at least the subjective judgments and assumptions are made explicit and visible.

The assumptions about the "probability of finding a defect" are especially acute in the case of the operational testing phase because, for example, in this phase the various levels of "operational usage" will be much harder to standardize on. What we are doing here is effectively predicting reliability, and to do this accurately may require the operational usage node to be conditioned on a formally defined operational profile such as described in the literature on statistical testing (Musa, 1993).

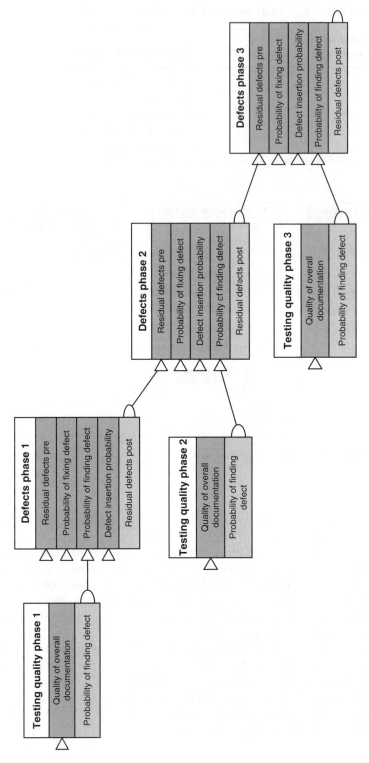

**Figure 2.23** Sequence of software testing phases as linked BN risk objects.

## 2.5 Using Software Failure Data to Determine Reliability

The BN models in Section 2.4 can be considered as fairly crude models of software reliability. In this section we consider more formal predictions of software reliability based on testing data, including models that enable us to determine when to stop testing.

### 2.5.1 Case 1: During the Testing and Fix Process

The idea behind this model is that, as before, there are a number of testing phases where defects are discovered and attempted fixes made before the next testing phase. The differences now are that:

1) Instead of the rather vague notion of "testing quality" being the driver of how many of the residual defects are discovered, the more formalized notion of "number of demands" is used. Thus, testing phase $T$ is made up of a number of "demands," where a "demand" is some appropriate unit of software testing – it could be a single execution of the software or a time period over which it is run such as an hour, a day, etc. All that matters is that the unit is used consistently throughout.
2) We are trying to learn the probability $p$ of triggering a defect per demand from continued periods of testing. We assume that all defects have the same probability of being triggered per demand.
3) We are only interested in the number of unique defects discovered, since we do not wish to "double count" defects. Thus, the same defect may be triggered 100 times during a testing period (since attempted fixes are only made at the end of the testing period), but – since it is just a single residual defect that we need to attempt to fix – it is logical that we only count it once.

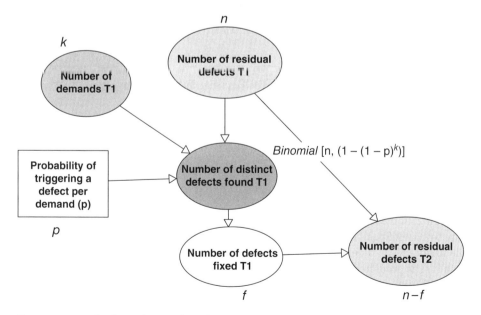

**Figure 2.24** Single-phase (demand-based) testing model.

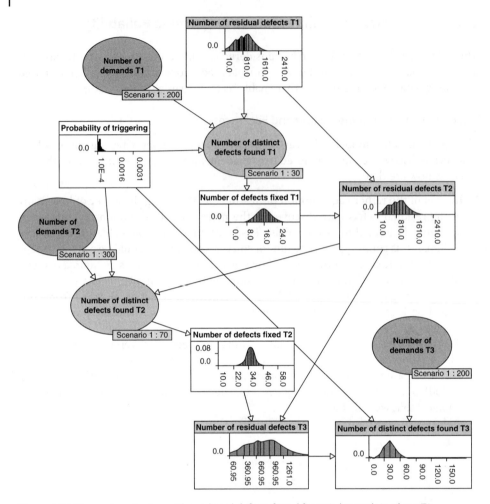

**Figure 2.25** Three-phase testing with predicted defects found for 200 demands in phase T3.

The single-phase version of this model is shown in Figure 2.24.

Here we assume some uncertain proportion of defects found are fixed (specifically, the number of defects fixed is a TNormal distribution with range 0 to D1 and mean $0.5 \times D1$, where D1 is the number of distinct defects found).

The reason for the particular binomial distribution for the node "number of defects found" is proved in Appendix 2.1.

Figure 2.25 shows a three-phase version of the model with observations entered for the first two phases. Specifically, in phase T1 30 distinct defects were found in 200 demands, while in phase T2 70 defects were found in 300 demands. The model shows the distribution for the predicted number of defects if the next phase, T3, involves 200 demands. This distribution has a mean of 38 with 25–75 percentile range 31 to 46.

A more comprehensive version of the model would incorporate at least one parent of the "number of defects fixed" node to take more explicit account of fix quality, and would also allow for the potential of introducing completely new defects during defect fixing.

However, the major weakness of the model is the assumption that each defect is equally likely to be detected per demand. The strong empirical evidence of Adams (1984) suggests that a very small proportion of defects have a very high probability of being triggered on demand (and hence these account for the vast majority of customer-reported defects), while the majority of defects have a very low probability of being triggered on demand (and hence may remain "dormant" for very longer periods of demand). A more comprehensive version of the model takes account of these differences.

### 2.5.2 Case 2: After Release

Once a software system is built and delivered we can no longer assume that there will be an attempt to fix all (or even any) defects found. Even if such attempts are made, there will be continued periods of use before doing so. In such situations we are interested in using information about failures (meaning defects triggered in operation) to determine measures of software reliability, such as *probability of failure on demand* and *mean time to next failure*. In cases of safety critical software – where few or no failures can be tolerated – we are also interested in how reliable the system is if it has never failed for a given period/number of demands. We are assuming that failures are independent.

A simple BN model to capture these aspects of software reliability is shown in Figure 2.26.

The idea behind this model is that we are trying to learn the probability of failure on demand (pfd) from continued periods of usage where, as before, a period of usage $T$ is made up of a number of demands.

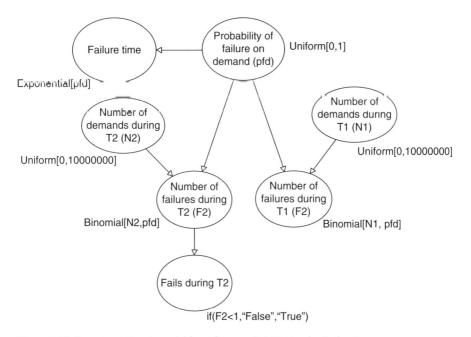

**Figure 2.26** Bayesian network model for software reliability (no faults fixed).

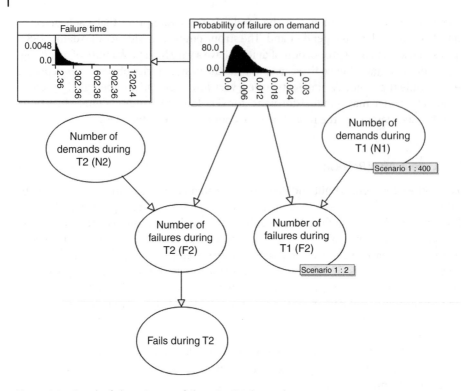

**Figure 2.27** Result of observing two failures in 400 demands.

The model shows two time periods, T1 and T2, but it can of course be extended to many. If we have any prior knowledge about the pfd then, of course, this can be captured in the NPT for the node. However, here we assume no prior knowledge and use a Uniform[0, 1] distribution.

Let us consider what we can learn from a single period of usage. Suppose, for example, that we have observed two failures in 400 demands. Then the BN mode will "learn" both the probability of failure on demand distribution and failure time distributions as shown in Figure 2.27. The pfd upper and lower quartiles fall between 0.0042 and 0.0074, with a mean value of 0.0059 and median value 0.0056.

Table 2.5 shows five different scenarios, all of which involve observing, on average, one failure per 200 demands. Clearly, the more demands observed the lower the variance.

Of most interest is what we can infer when we have observed *no* failures after many demands. This is the classic reliability assessment problem for high-integrity systems. Littlewood and Strigini (1993) considered how many failure-free demands need to be observed in order to be confident that the system meets a "high reliability requirement" such as $10^{-9}$ pfd. They showed analytically that, with the Uniform[0, 1] prior assumption for the pfd, if we observe $n$ failure-free demands then the probability of observing no failures in the next $n$ demands is 0.5.

Using the above model we not only easily replicate the analytical result but also provide more extensive information about uncertainty in different scenarios – see Table 2.6.

For there to be a probability of just below 10% that a system will be failure free in $10^n$ demands we will need to have seen a failure-free period of $10^{n+1}$ demands.

**Table 2.5** Different scenarios for time period T1.

| Number of demands | Number of failures observed | Lower–upper quartiles of pfd | Mean, median pfd | Time to failure (demands) mean, median |
|---|---|---|---|---|
| 400 | 2 | 0.0043–0.0098 | 0.0075, 0.0067 | 202, 105 |
| 1000 | 5 | 0.0042–0.0074 | 0.0059, 0.0056 | 202, 124 |
| 5000 | 25 | 0.0045–0.0058 | 0.0052, 0.0051 | 201, 136 |
| 10 000 | 50 | 0.0046–0.0056 | 0.0051, 0.005 | 201, 138 |
| 100 000 | 500 | 0.0049–0.0052 | 0.005, 0.005 | 201, 140 |

**Table 2.6** Different scenarios for failure-free periods of demands.

| Number of demands $n$ | Lower–upper pfd quartiles | Mean, median pfd | Time to failure (demands) mean, median | Probability fails in next $n$ demands | Probability fails in next $n/10$ demands |
|---|---|---|---|---|---|
| $10^2$ | 0.0029–0.0137 | 0.0028, 0.0136 | 1427, 101 | 0.4975 | 0.00901 |
| $10^3$ | $2.9 \times 10^{-2}$–$1.3 \times 10^{-3}$ | $10^{-3}, 6.9 \times 10^{-2}$ | 13 094, 1003 | 0.4998 | 0.00908 |
| $10^4$ | $2.9 \times 10^{-3}$–$1.4 \times 10^{-4}$ | $10^{-4}, 6.9 \times 10^{-3}$ | $1.5 \times 10^4, 10^4$ | 0.4999 | 0.0091 |
| $10^5$ | $2.9 \times 10^{-4}$–$1.4 \times 10^{-5}$ | $10^{-5}, 6.9 \times 10^{-4}$ | $1.2 \times 10^5, 10^5$ | 0.5001 | 0.00912 |
| $10^6$ | $2.9 \times 10^{-5}$–$1.39 \times 10^{-6}$ | $10^{-6}, 6.9 \times 10^{-5}$ | $7.8 \times 10^4, 10^5$ | 0.5001 | 0.009111 |
| $10^7$ | $2.9 \times 10^{-6}$–$1.39 \times 10^{-7}$ | $10^{-7}, 6.9 \times 10^{-6}$ | $7.9 \times 10^6, 10^7$ | 0.5001 | 0.00911 |
| $10^8$ | $2.88 \times 10^{-7}$–$1.39 \times 10^{-8}$ | $10^{-8}, 7 \times 10^{-7}$ | $7.9 \times 10^7, 10^8$ | 0.50009 | 0.00909 |
| $10^9$ | $2.88 \times 10^{-8}$–$1.38 \times 10^{-9}$ | $10^{-9}, 7 \times 10^{-8}$ | $7 \times 10^8, 10^9$ | 0.50001 | 0.00904 |

## 2.6 Bayesian Networks for Software Project Risk Assessment and Prediction

In Section 2.3 we suggested that effective metrics-driven software risk methods should be able to provide answers to straightforward resource, quality, and schedule questions, and more specifically questions regarding trade-offs between these.

We now describe a "project-level software risk" model that provides support to help answer these types of questions. The model is a very general-purpose quality and risk assessment model for large software projects, and was developed as part of a major international consortium, with key empirical and expert judgment provided by a range of senior software managers and developers. Although it is beyond the scope of this chapter to describe the full details of the model and its validation – these details are provided in Fenton et al. (2004) – we can show how the model is used to predict different aspects of resources and quality while monitoring and mitigating different types of risks. The full model is shown in Figure 2.28, but is too complex to understand all at once.

Figure 2.29 provides an easier to understand schematic view of the model. We can think of the model as comprising six risk objects (shown as square boxes).

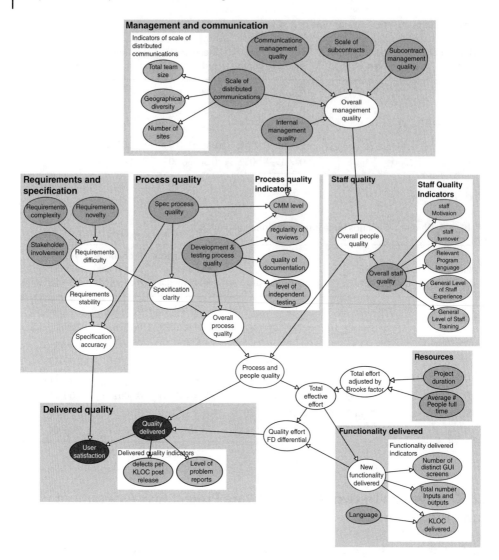

**Figure 2.28** Full BN model for software project risk.

The risk objects are:

- Distributed communications and management. Contains variables that capture the nature and scale of the distributed aspects of the project and the extent to which these are well managed.
- Requirements and specification. Contains variables relating to the extent to which the project is likely to produce accurate and clear requirements and specifications.
- Process quality. Contains variables relating to the quality of the development processes used in the project.
- People quality. Contains variables relating to the quality of people working on the project.

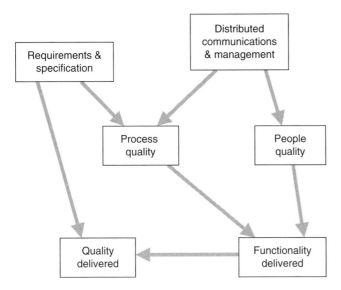

**Figure 2.29** Schematic for the project-level model.

- Functionality delivered. Contains all relevant variables relating to the amount of new functionality delivered on the project, including the effort assigned to the project.
- Quality delivered. Contains all relevant variables relating to both the final quality of the system delivered and the extent to which it provides user satisfaction (note the clear distinction between the two).

The full model enables us to cope with variables that cannot be observed directly. Instead of making direct observations of the process and people quality, the functionality delivered, and the quality delivered, the states of these variables are inferred from their causes and consequences. For example, the process quality is a synthesis of the quality of the different software development processes – requirements analysis, design, and testing.

The quality of these processes can be inferred from "indicators." Here, the causal link is from the "quality" to directly observable values like the results of project audits and process assessments, such as the Capability Maturity Model (CMM; Paulk, Weber, and Curtis, 1995). Of course, only some organizations have been assessed to a CMM level, but this need not be a stumbling block since there are many alternative indicators. An important and novel aspect of our approach is to allow the model to be adapted to use whichever indicators are available.

At its core, the model captures the classic trade-offs between:

- quality – where we distinguish and model both *user satisfaction* (the extent to which the system meets the user's true requirements) and *quality delivered* (the extent to which the final system works well)
- effort – represented by the *average number of full-time people* who work on the project
- time – represented by the *project duration*
- functionality – meaning *functionality delivered*.

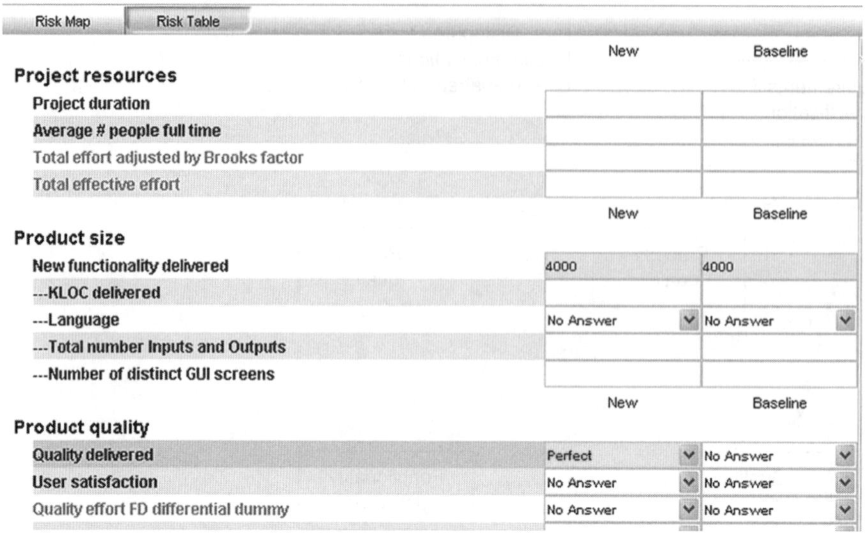

| Risk Map | Risk Table | | | |
|---|---|---|---|---|
| | | New | | Baseline |
| **Project resources** | | | | |
| Project duration | | | | |
| Average # people full time | | | | |
| Total effort adjusted by Brooks factor | | | | |
| Total effective effort | | | | |
| | | New | | Baseline |
| **Product size** | | | | |
| New functionality delivered | | 4000 | | 4000 |
| ---KLOC delivered | | | | |
| ---Language | | No Answer | ⌄ | No Answer ⌄ |
| ---Total number Inputs and Outputs | | | | |
| ---Number of distinct GUI screens | | | | |
| | | New | | Baseline |
| **Product quality** | | | | |
| Quality delivered | | Perfect | ⌄ | No Answer ⌄ |
| User satisfaction | | No Answer | ⌄ | No Answer ⌄ |
| Quality effort FD differential dummy | | No Answer | ⌄ | No Answer ⌄ |

**Figure 2.30** Two scenarios in Risk Table view.

So, for example, if you want a lot of functionality delivered with little effort in a short time then you should not expect high quality. If you need high quality then you will have to be more flexible on at least one of the other factors (i.e. use more effort, use more time, or deliver less functionality).

What makes the model powerful, when compared with traditional software cost models, is that we can enter observations anywhere in the model to perform not just predictions but also many types of trade-off analysis and risk assessment. So, we can enter requirements for quality and functionality and let the model show us the distributions for effort and time. Alternatively, we can specify the effort and time we have available and let the model predict the distributions for quality and functionality delivered – measured in function points (Symons, 1988). Thus, the model can be used like a spreadsheet – we can test the effects of different assumptions.

To explain how this works we consider two scenarios called "new" and "baseline" (Figure 2.30 shows how, in AgenaRisk, you can enter observations for the different scenarios into a table view of the model). Suppose the new project is to deliver a system of size 4000 function points (this is around 270 KLOC of Java, an estimate you can see for the node KLOC by entering the observation "java" for the question "language"). In the baseline scenario we enter no observations other than the one for functionality. We are going to compare the effect against this baseline of entering various observations into the new scenario.

We start with the observations shown in Figure 2.30, i.e. the only change from baseline in the new project is to assert that the quality delivered should be "perfect." Running the model produces the results shown in Figure 2.31 for the factors *process and people quality, project duration*, and *average number of people full time*. First note that the distributions for the latter factors have high variances (not unexpected given the minimal data entered) and that generally the new scenario will require a bit more effort for a bit longer. However, the factor *process and people quality* (which combines

(a) Process and people quality

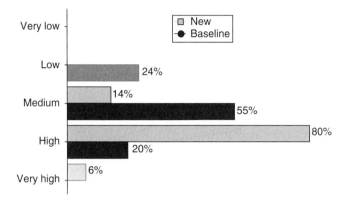

(b) Project duration (median 31, 39)

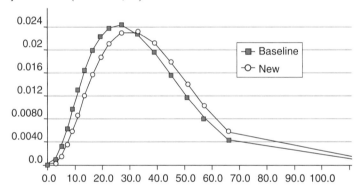

(c) Average number of people full time (19, 23)

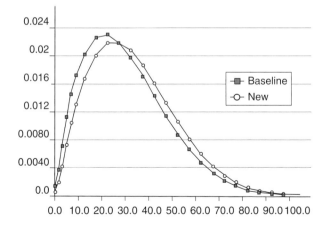

**Figure 2.31** Distributions when functionality delivered is set as "perfect" for the new project (compared with the baseline).

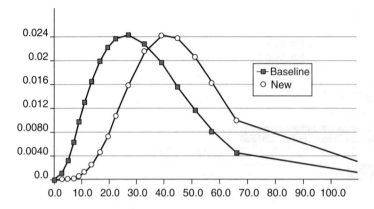

**Figure 2.32** When staff quality is "medium," project duration jumps to a median value of 54 months.

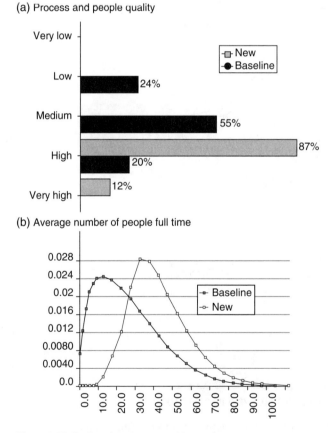

(a) Process and people quality

(b) Average number of people full time

**Figure 2.33** Project duration set to 18 months.

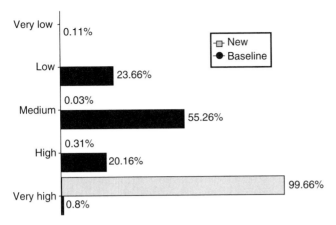

**Figure 2.34** Project duration = 12 months, people = 10.

all the process and people factors) shows a very big difference from the baseline. The prediction already suggests that it will be unlikely (a 14% chance) to deliver the system to the required level of quality unless the quality of staff is better than average.

Suppose, however, that we can only assume *process and people quality* is "medium." Then the predictions for project duration and effort increase significantly. For example, the median value for project duration is up from 31 months in the baseline case to around 54 months (Figure 2.32), with full-time staff increasing to 33.

Now we withdraw the observation of *process and people quality* and suppose, as is typical in software projects, that we have a hard schedule deadline of 18 months in which to complete (i.e. a target that is significantly lower than the one the model predicts). With this observation we get the distributions shown in Figure 2.33 for *process and people quality* and *average number of people full time*. Now, not only do we need much higher quality people, we also need a lot more of them compared with the baseline.

But typically, we will only have a fixed amount of effort. Suppose, for example, that additionally we enter the observation that we have only ten full-time people (so the project is really under-resourced compared with the predictions). Then the resulting distribution for *process and people quality* is shown in Figure 2.34.

What we see now is that the probability of the overall *process and people quality* being "very high" (compared to the industry average) is 0.9966. Put a different way, if there is even a tiny chance that your processes and people are not among the best in the industry then this project will not meet its quality and resource constraints. In fact, if we know that the *process and people quality* is just "average" and now remove the observation "perfect" for quality delivered, then Figure 2.35 shows the likely quality to be delivered: it is likely to be "abysmal" (with probability 0.69).

Nevertheless, suppose we insist on perfect quality and all the previous resource constraints. In this case the *only* thing left to "trade-off" is the functionality delivered. So, to represent this we remove the observation 4000 in the new scenario. Figure 2.36 shows the result when we run the model with these assumptions: we are likely to deliver only a tenth of the functionality originally planned. Armed with this information a project manager can make an informed decision about how much functionality needs to be dropped to meet the quality and resource constraints.

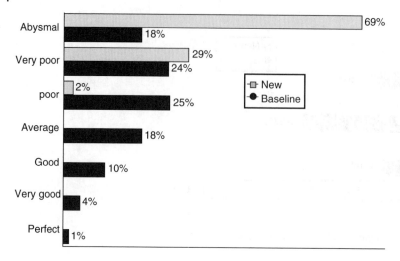

**Figure 2.35** Quality delivered if process and people quality = medium with resource constraints set.

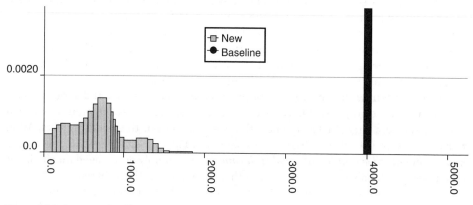

**Figure 2.36** Functionality (function points) delivered if process and people quality = medium with resource constraints set.

## 2.7 Summary

There have been many non-causal models for software quality and resource prediction, mainly in the form of regression and correlation. Some of these have achieved very good accuracy (for a detailed overview, see Fenton et al., 2008) and they provide us with an excellent empirical basis. However, in general these models are typically data-driven statistical models; they provide us with little insight when it comes to effective risk management and assessment. What we have shown is that, by complementing the empirical data with expert judgment, usually about the software engineering process, we are able to build causal Bayesian network models that enable us to address the kind of dynamic decision making that software professionals have to confront as a project develops.

The BN approach helps to identify, understand, and quantify the complex interrelationships (underlying even seemingly simple situations) and can help us make sense of how risks emerge and are connected, and how we might represent our control and mitigation of them. By thinking about the causal relations between events we can investigate alternative explanations, weigh up the consequences of our actions, and identify unintended or (un)desirable side effects. Above all else, the BN approach quantifies the uncertainty associated with every prediction.

We are not suggesting that building a useful BN model from scratch is simple. It requires an analytical mindset to decompose the problem into "classes" of events and relationships that are granular enough to be meaningful, but not too detailed that they are overwhelming. The states of variables need to be carefully defined, and probabilities need to be assigned that reflect our best knowledge. Fortunately, there are tools that help avoid much of the complexity of model building, and once built the tools provide dynamic and automated support for decision making. Also, we have presented some predefined models that can be tailored for different organizations.

## Appendix 2.1

The probability that $x$ distinct defects are chosen in $k$ demands is the binomial distribution

$$\binom{n}{n-x}(1-p)^{k(n-x)}(1-(1-p)^k)^x.$$

*Proof:* There are $\binom{n}{n-x}$ configurations in which $x$ distinct defects are chosen at least once in $k$ demands and the remaining $n-x$ defects are not chosen at all.

For each of these configurations we multiply the probability of the $x$ defects being selected at least once by the probability of the $n-x$ defects not being selected at all.

The probability that any given defect is selected at least once in $k$ demands is one minus the probability it is not selected at all in $k$ demands; this is:

$$1-(1-p)^k.$$

So, the probability that $x$ specific defects are selected at least once is

$$(1-(1-p)^k)^x. \tag{2.1}$$

The probability that any given defect is not selected at all in $k$ demands is $(1-p)^k$. So, the probability that $n-x$ specific defects are not selected at all is:

$$(1-p)^{k(n-x)}. \tag{2.2}$$

So, the probability of exactly $x$ distinct defects being chosen is

$$\binom{n}{n-x}(1-(1-p)^k)^x(1-p)^{k(n-x)},$$

which is the combination term multiplied by 2.1 and 2.2.

# References

Adams, E. (1984). Optimizing preventive service of software products. *IBM Research Journal*, 28(1), 2–14.

Agena Ltd., 2018. AgenaRisk. www.agenarisk.com [accessed 30 January, 2018].

Fenton, N.E. et al. (2004). Making resource decisions for software projects. *26th International Conference on Software Engineering (ICSE2004)*, pp. 397–406.

Fenton, N.E. et al. (2008). On the effectiveness of early life cycle defect prediction with Bayesian nets. *Empirical Software Engineering*, 13, 499–537.

Fenton, N.E., Neil, M., Marsh, D.W.R., et al. (2007). Predicting software defects in varying development lifecycles using Bayesian nets. *Information and Software Technology*, 49, 32–43.

Fenton, N.E. and Neil, M. (2012). *Risk Assessment and Decision Analysis with Bayesian Networks*. Boca Raton, FL: CRC Press. Available at: www.bayesianrisk.com.

Fenton, N.E., Neil, M. and Gallan, J. (2007). Using ranked nodes to model qualitative judgements in Bayesian networks. *IEEE Transactions on Knowledge and Data Engineering*, 19(10), 1420–1432.

Fenton, N.E., Neil, M. and Krause, P. (2002). Software measurement: Uncertainty and causal modelling. *IEEE Software*, 10(4), 116–122.

Fenton, N.E. and Ohlsson, N. (2000). Quantitative analysis of faults and failures in a complex software system. *IEEE Transactions on Software Engineering*, 26(8), 797–814.

Gras, J.-J. (2004). End-to-end defect modeling. *IEEE Software*, 21(5), 98–100.

Koller, D. and Pfeffer, A. (1997). Object-oriented Bayesian networks. *Proceedings of the 13th Annual Conference on Uncertainty in AI (UAI)*, pp. 302–313.

Littlewood, B. and Strigini, L. (1993). Validation of ultra-high dependability for software-based systems, in *Predictably Dependable Computing Systems*, New York: Springer, pp. 473–493.

Musa, J. (1993). Operational profiles in software reliability engineering. *IEEE Software*, 10(2), 14–32.

Neil, M., Tailor, M., and Marquez, D. (2007) Inference in hybrid Bayesian networks using dynamic discretization. *Statistics and Computing*, 17(3), 219–233.

Paulk, M., Weber, C. V. and Curtis, B. (1995). *The Capability Maturity Model for Software: Guidelines for Improving the Software Process*. Reading, MA: Addison-Wesley.

Symons, C.R. (1988). Function point analysis: Difficulties and improvements. *IEEE Transactions on Software Engineering*, 14(1), 2–11.

Wang, H. et al. (2006). Software project level estimation model framework based on Bayesian belief networks. *Sixth International Conference on Quality Software (QSIC'06)*, pp. 209–218.

Ziliak, S.T. and McCloskey, D.N. (2008). *The Cult of Statistical Significance*. Ann Arbor, MI: The University of Michigan Press.

# 3

# Optimal Software Testing across Version Releases

*Simon P. Wilson and Seán Ó Ríordáin*

## Synopsis

Much software is now updated with a regular schedule of version releases. This chapter looks at models for bug detection in software where this is the case, and illustrates the ideas through a well-known non-homogeneous Poisson process software reliability model due to Goel and Okumoto (1979). Model inference is developed from a Bayesian perspective and applied to data on bug reports for several versions of Mozilla Firefox. Finally, the model is applied to the decision problem of determining the optimal time between releases based on bug detection data. A decision theoretic solution is developed and compared to the actual 42-day version release cycle that was being used by Mozilla in the period when the data was collected.

## 3.1 Introduction

Many statistical models have been developed in the context of modeling the reliability of bespoke software or the process of pre-release testing by the software development team; see Chapter 9 on software reliability models. However, important software is increasingly maintained and updated as a series of releases. Rapid scheduled version releases help to maintain the competitiveness of the software, improve reliability and defend against the malicious exploitation of vulnerabilities (Humble and Farley, 2011). The general availability of internet access at a sufficient bandwidth has made this approach practical. It also suits the increasingly popular agile approach to software development (Larman, 2003). In this situation, bug detection is not only conducted by the software development team before release, but is also done by users who can report bugs, either after full release or during a pre-release stage where a set of users have the opportunity to test out the new version (so-called beta testing).

This chapter discusses several approaches that generalize existing software reliability models to the case of a sequence of releases. Some of those approaches are illustrated with the non-homogeneous Poisson process model of Goel and Okumoto (1979). One common feature of much rapid-release software is a fixed release schedule; for example, at the time of writing, the popular internet browser Mozilla Firefox has a new version release every 42 days and has held to that schedule with very little deviation for several years. A decision theoretic approach to determining the optimal release schedule

*Analytic Methods in Systems and Software Testing*, First Edition.
Edited by Ron S. Kenett, Fabrizio Ruggeri, and Frederick W. Faltin.
© 2018 John Wiley & Sons Ltd. Published 2018 by John Wiley & Sons Ltd.

is discussed and applied with the Goel–Okumoto model. Other chapters dealing with related topics include Chapter 14 on agile testing with user data and Chapter 16 on dynamic test case selection.

## 3.2 Mozilla Firefox: An Example of Bug Detection Data

Extensive data on bug discovery for the internet browser Mozilla Firefox is publicly available through its Bugzilla tracking system – see bugzilla.mozilla.org. The authors have downloaded from this source and formatted a dataset of the dates of bug discoveries, as well as other information, from 1 January 2011 to 18 July 2013. The data are a snapshot of what was in the database on 18 July 2013, after security- and personnel-related bugs were removed, as well as bugs that were considered as requests for enhancements, or were identified as a duplicate of an existing reported bug, or that the development team deemed to be unnecessary to fix. The data include bugs reported up to version 25. Version 5 had marked the beginning of the "rapid-release" cycle, where new versions were released every 42 days. All versions prior to the fifth are labeled "PreRapid." The formatted dataset is also publicly available (Ó Ríordáin, 2013).

Figure 3.1 shows the cumulative number of detected bugs for each of the releases of Firefox in this data set. The dashed vertical lines in the figure correspond to release dates. Figure 3.2 shows a smoothed proportion of bugs reported from each version over time. Both plots illustrate the dominant feature of the data as a function of time, namely the rise and fall in the detection rate of bugs for a particular version as the date for its release arrives and it is then used, followed by a decline once the next version is made available.

Each bug in the data has the following information recorded:

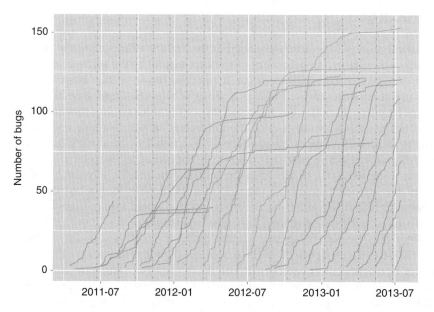

**Figure 3.1** Cumulative number of bugs logged to Bugzilla for each individual rapid-release version of Firefox as a function of time from release 5 (the earliest in the figure) to 25 (the latest).

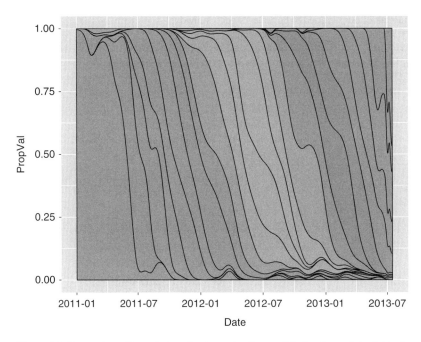

**Figure 3.2** Proportion of bugs belonging to each release of Firefox for each individual rapid-release version of Firefox as a function of time from release 5 (the earliest in the figure) to 25 (the latest).

**version:** The version number of Firefox associated with this bug, e.g. 14, or "PreRapid" to describe any version prior to 5, or "Unknown" for any bug that did not have a version number associated with it.

**bug_id:** The original bug number in the Bugzilla database. Further details on this bug can be queried at bugzilla.mozilla.org/.

**bug_severity:** One of: blocker, critical, major, normal, minor, trivial.

**bug_status:** One of: ASSIGNED, NEW, REOPENED, RESOLVED, VERIFIED.

**priority:** One of: --, P1, P2, P3, P4, P5, where P1 has the highest priority and "--" means that no priority has been assigned.

**creation_ts:** A character string POSIX formatted time stamp of when the bug was first inserted into the Bugzilla database, e.g. "2011-04-13 17:19:05".

**reporter:** The ID number of the person reporting the bug from the underlying Bugzilla database, used for privacy reasons.

**component_id:** The ID number of the component from the Bugzilla database. There are 45 components used in Firefox, some of which are shared with other Mozilla Corporation products like Thunderbird.

It is noted that fields such as bug_severity and priority may be changed after the bug is originally reported, typically by those in triage or a developer who sees a trivial bug marked as high severity and high priority.

Some information for each release is also available:

**Version:** The release version number, e.g. 5.

**Release:** The release version as text, e.g. release-15.0.

**FChanged:** The number of files changed since the previous version.

**Table 3.1** The number of bugs logged to Bugzilla for each individual rapid-release version of Firefox.

| Release | No. bugs detected | FChanged | LInserts | LDeletions | ReleaseDate |
|---|---|---|---|---|---|
| PreRapid | 102 | — | — | — | — |
| 5 | 42 | 3 618 | 72 367 | 62 832 | 2011-06-21 |
| 6 | 41 | 4 276 | 90 398 | 83 114 | 2011-08-16 |
| 7 | 41 | 4 493 | 92 537 | 78 987 | 2011-09-27 |
| 8 | 65 | 7 341 | 74 473 | 70 222 | 2011-11-08 |
| 9 | 65 | 4 377 | 109 381 | 73 509 | 2011-12-20 |
| 10 | 99 | 6 037 | 148 795 | 122 633 | 2012-01-31 |
| 11 | 78 | 4 805 | 108 274 | 76 724 | 2012-03-13 |
| 12 | 119 | 4 457 | 123 432 | 84 129 | 2012-04-24 |
| 13 | 125 | 4 688 | 150 233 | 106 764 | 2012-06-05 |
| 14 | 118 | 16 835 | 355 941 | 560 277 | 2012-07-17 |
| 15 | 127 | 15 728 | 257 432 | 501 298 | 2012-08-28 |
| 16 | 85 | 6 349 | 132 879 | 84 268 | 2012-10-09 |
| 17 | 135 | 1 051 | 28 867 | 22 179 | 2012-11-20 |
| 18 | 124 | 1 315 | 29 593 | 22 844 | 2013-01-08 |
| 19 | 126 | 6 133 | 129 714 | 149 368 | 2013-02-19 |
| 20 | 121 | 6 393 | 250 433 | 180 606 | 2013-04-02 |
| 21 | 108 | 6 840 | 146 689 | 108 744 | 2013-05-14 |
| 22 | 91 | 9 569 | 220 244 | 171 637 | 2013-06-25 |
| 23 | 76 | 11 149 | 166 918 | 121 491 | 2013-08-06 |
| 24 | 48 | 6 924 | 153 191 | 115 052 | 2013-09-17 |
| 25 | 23 | 10 933 | 248 838 | 201 130 | 2013-10-29 |
| Unknown | 8 461 | | | | |
| Total | 10 420 | | | | |

**LInserts:** The number of lines added since the previous version.
**LDeletions:** The number of lines deleted since the previous version.
**ReleaseDate:** The release date for this version.

Table 3.1 shows the count of each bug that is labeled with a version in the dataset, as well as the release version information. Over 10 000 bugs are recorded, of which only about 2000 have a version label. This missing data is one of the most important features for inference.

## 3.3 Modeling Bug Detection across Version Releases

In what follows we assume that there is a statistical model that defines the distribution of the times $t_1, t_2, \ldots$ between bug detections. As can been seen in Chapters 7 and 9, the distributions of these times may be specified directly, e.g. through specification of a failure rate for each inter-discovery time, or their distributions may be specified implicitly, e.g.

a counting process for the number of bugs at time $t$, $N(t)$, is specified which implies an inter-discovery time distribution. The non-homogeneous Poisson process, of which the model of Goel and Okumoto (1979) is an example, is in the latter category.

Let $T_i$ be the random variable that represents the time between the detection of the $(i-1)$th and $i$th bugs in a piece of software, and let $p_i(t \mid t_{1:i-1}, \theta)$ represent the probability density function of $T_i$ with parameters $\theta$, which is possibly dependent on the previous $i-1$ inter-discovery times $t_{1:i-1} = (t_1, \ldots, t_{i-1})$. It is assumed that the bug discovery process starts at time 0, and so $s_i = \sum_{j=1}^{i} t_j$ is the discovery time of the $i$th bug. If $n$ bugs are discovered by time $\mathcal{T}$, so that $s_n \leq \mathcal{T} < s_{n+1}$, then the joint distribution of the $T_1, \ldots, T_n$ is

$$p(t_1, \ldots, t_n \mid \mathcal{T}, \theta) = \left( \prod_{i=1}^{n} p_i(t_i \mid t_{1:i-1}, \theta) \right) P(T_{n+1} > \mathcal{T} - s_n \mid t_{1:n}, \theta), \quad (3.1)$$

where $P(T_{n+1} > \mathcal{T} - s_n \mid t_{1:n}, \theta)$ is the probability of the right-censored inter-discovery time of the $(n+1)$th bug. If the discovery times are modeled by a counting process $N(\mathcal{T})$ then it may be easier to write the likelihood in the form

$$p(t_1, \ldots, t_n \mid \mathcal{T}, \theta) = P(N(\mathcal{T}) = n \mid \theta) \, p(t_1, \ldots, t_n \mid \theta, N(\mathcal{T}) = n); \quad (3.2)$$

for example, in the case of a non-homogeneous Poisson process with mean value function $E(N(t) \mid \theta) = \Lambda(t; \theta)$, this is:

$$p(t_1, \ldots, t_n \mid \mathcal{T}, \theta) \propto \frac{\Lambda(\mathcal{T}; \theta)^n}{n!} e^{-\Lambda(\mathcal{T};\theta)} \times n! \prod_{i=1}^{n} \frac{\lambda(s_i; \theta)}{\Lambda(t; \theta)}$$

$$= e^{-\Lambda(\mathcal{T};\theta)} \prod_{i=1}^{n} \lambda(s_i; \theta), \quad s_1 < s_2 < \cdots < s_n, \quad (3.3)$$

where $\lambda(t; \theta) = \Lambda'(t; \theta)$ is the process rate function.

If the software has been released as a sequence of $K$ versions, then let $T_{ki}$ denote the $i$th time between bug detections for version $k$. Suppose that the bug discovery process has been happening to version $k$ for a time $\mathcal{T}_k$, and $n_k$ bugs have been discovered. A quite general model is to assume that each version is conditionally independent given the $\theta_k$, so that the joint distribution of all the $T_{ki}$ is then

$$p(t_{1,1:n_1}, \ldots, t_{K,1:n_K} \mid \mathcal{T}_{1:K}, \theta_{1:K}) = \prod_{k=1}^{K} p(t_{k,1:n_k} \mid \mathcal{T}_k, \theta_k), \quad (3.4)$$

with $p(t_{k,1:n_k} \mid \mathcal{T}_k, \theta_k)$ given by Equation (3.1).

Different dependencies between the versions can then be modeled by assuming different relationships between the $\theta_k$. Four possible structures are:

**Independent:** The $\theta_k$ are independent of each other. For inference purposes, each $\theta_k$ can be independently estimated using the data from release $k$ only. This is the simplest model but ignores the dependencies that there may be between the reliability of different versions due to the large amount of code in common, large similarities in the development team, and common users across versions.

**Random sample:** All versions have the same parameter $\theta$ and are considered to be a random sample from a population of releases. This is possibly realistic for well-established software with few changes between releases so that variation in the reliability of different releases is due to random differences in code changes, the addition of features, the skill of the development team, etc.

**Hierarchical:** The $\theta_k$ are exchangeable random effects, that is to say they are assumed to be a random sample from a population with distribution $p_\theta(\theta \mid \psi)$, for hyper-parameters $\psi$. This means that the $\theta_k$ are now dependent, so that discovery-time data from one version has some information about $\theta$ for another version. Exchangeability implies that changing the order of the sequence $\theta_{1:K}$ does not alter the probability of it occurring, and so the sequential nature of the releases is ignored (Chow and Teicher, 1997). This is a more realistic model for software that is in a reasonably steady state, where releases are primarily concerned with addressing bugs and security updates rather than new features.

**Autoregressive:** The ordering of the $\theta_k$ is taken into account, so that the reliability of successive releases is more correlated than the reliability of releases that are far apart. Completely generally, one might have a relationship of the form $\theta_k = g_k(\theta_{1:k-1}, \phi_k; \psi)$ for some function $g_k$, independent random peturbation $\phi_k$, and some other fixed parameters $\psi$. In other words, there is a conditional distribution $p(\theta_k \mid \theta_{1:k-1}, \psi)$, that may be explicitly or implicitly defined, that models the time dependence. More concretely, a typical example might be to use a Gaussian autoregressive process with a real-valued parameter $\theta_k$ evolving as:

$$\theta_1 \sim N(a, s_1^2),$$
$$\theta_k = a + r(\theta_{k-1} - a) + \phi_k,$$

where $\phi_k \sim N(0, s^2)$ for some variance $s^2$, $a$ is the mean level for the $\theta_k$, and $r \in (0, 1)$ is the correlation parameter, so that the $\theta_k$ are a stationary process with positive correlation and the set of parameters is $\psi = (a, s_1^2, r, s_s^2)$. This model is better suited to software that is still evolving and where releases show considerable development of features.

## 3.4  Bayesian Inference

The inference task is to take data on bug detection times and learn about model parameters. The Bayesian solution requires a prior distribution on those parameters, which also neatly encapsulates the dependence structures that are defined in the last section.

For the independence model, each $\theta_k$ has its own independent prior, and so:

$$p(\theta_1, \ldots, \theta_K) = \prod_{k=1}^{K} p(\theta_k).$$

For the random sample model, a prior $p(\theta)$ is all that is required. For the hierarchical model there are extra parameters $\psi$ that describe the distribution of the population of $\theta_k$s, and the prior has the form:

$$p(\theta_1, \ldots, \theta_K, \psi) = \left[ \prod_{k=1}^{K} p(\theta_k \mid \psi) \right] p(\psi),$$

and finally, for the autoregressive model,

$$p(\theta_1, \ldots, \theta_K, \psi) = p(\theta_1 \mid \psi) \left[ \prod_{k=2}^{K} p(\theta_k \mid \theta_{k-1}, \psi) \right] p(\psi).$$

Note that for the hierarchical and autoregressive model, inference is also conducted on the hyper-parameters $\psi$.

The likelihood term, Equation (3.4), is the same regardless of the parameter structure. Bayes' law then gives the state of knowledge about the model parameters following observation of bug detection for $K$ versions, with version $k$ having been available for bug detection for a time $\mathcal{T}_k$ where $n_k$ bug detection times $t_{k,1:n_k} = (t_{k1}, \ldots, t_{kn_k})$ are observed. For the independence model, the form is:

$$p(\theta_{1:K} \mid t_{1,1:n_1}, \ldots, t_{K,1:n_K}) \propto \prod_{k=1}^{K} p(\theta_k) p(t_{k,1:n_k} \mid \mathcal{T}_k, \theta_k), \tag{3.5}$$

while for the random sample model it is

$$p(\theta \mid t_{1,1:n_1}, \ldots, t_{K,1:n_K}) \propto p(\theta) \prod_{k=1}^{K} p(t_{k,1:n_k} \mid \mathcal{T}_k, \theta); \tag{3.6}$$

for the hierarchical model it is

$$p(\theta_{1:K} \, \psi \mid t_{1,1:n_1}, \ldots, t_{K,1:n_K}) \propto \left[ \prod_{k=1}^{K} p(\theta_k \mid \psi) p(t_{k,1:n_k} \mid \mathcal{T}_k, \theta_k) \right] p(\psi), \tag{3.7}$$

and finally for the autoregressive model it is

$$p(\theta_{1:K} \, \psi \mid t_{1,1:n_1}, \ldots, t_{K,1:n_K}) \propto p(\theta_1 \mid \psi) p(t_{1,1:n_1} \mid \mathcal{T}_1, \theta_1)$$
$$\times \left[ \prod_{k=1}^{K} p(\theta_k \mid \theta_{k-1}, \psi) \, p(t_{k,1:n_k} \mid \mathcal{T}_k, \theta_k) \right] p(\psi), \tag{3.8}$$

where $p(t_{k,1:n_k} \mid \mathcal{T}_k, \theta_k)$ is given by Equation (3.1).

Note that for the independence case, the posterior distribution of $\theta_1, \ldots, \theta_K$ is a product of independent posterior distributions for each $\theta_k$ given $t_{k,1:n_k}$. For the hierarchical and autoregressive structures, the $\theta_k$ are not independent in the posterior, implying that learning about $\theta_k$ can take place with data from other releases.

## 3.5 The Goel–Okumoto Model

Goel and Okumoto (1979) proposed a non-homogenous Poisson process model for bug detection. Letting $N(t)$ be the number of bugs detected by time $t$, the mean value function $\Lambda(t) = E(N(t))$ satisfies the relationship

$$\frac{d\Lambda(t)}{dt} \propto (b - \Lambda(t)),$$

where $b > 0$ is the expected number of bugs that will eventually be detected in the software. Hence, the rate of bug discovery is proportional to the expected number of remaining bugs. Assuming that $\Lambda(0) = 0$, then the solution to the above equation is

$$\Lambda(t) = a(1 - e^{-bt})$$

for a detection rate parameter $a$, and so the distribution of $N(t)$ is

$$P(N(t) = n \mid a, b) = \frac{[a(1 - e^{-bt})]^n}{n!} \exp(-a(1 - e^{-bt})), \; n = 0, 1, \ldots,$$

for model parameters $\theta = (a, b)$. The likelihood for detecting $n$ bugs over a time $\mathcal{T}$ with inter-discovery times $t_1, \ldots, t_n$ is, following Equation (3.3),

$$p(t_{1:n} \mid \mathcal{T}, a, b) = \exp(-a(1 - e^{-b\mathcal{T}})) \prod_{i=1}^{n} abe^{-bs_i}$$

$$= a^n \, b^n \, \exp\left(-a(1 - e^{-b\mathcal{T}}) - b \sum_{i=1}^{n} s_i\right), \tag{3.9}$$

where $s_i = \sum_{j=1}^{i} t_j$. The likelihood over $K$ release versions extends in the way that has been described via Equation (3.4).

The inference equations of Section 3.4 can be applied to the Goel–Okumoto model. As regards a prior on the parameters $\theta = (a, b)$, McDaid and Wilson (2001) use independent gamma prior distributions

$$p(a, b) = \frac{\alpha_a^{\beta_a}}{\Gamma(\beta_a)} a^{\beta_a - 1} e^{-\alpha_a a} \frac{\alpha_b^{\beta_b}}{\Gamma(\beta_b)} b^{\beta_b - 1} e^{-\alpha_b b},$$

which allows for independent location and scale for $a$ and $b$.

Using this prior, one obtains the following forms for the posterior distribution in the independent case:

$$p(a_{1:K}, b_{1:K} \mid t_{1,1:n_1}, \ldots, t_{K,1:n_K}) \propto \prod_{k=1}^{K} \left\{ a_k^{\beta_a + n_k - 1} \, b_k^{\beta_b + n_k - 1} \right.$$

$$\left. \times \exp\left(-a_k(\alpha_a + 1 - e^{-b_k \mathcal{T}_k}) - b_k \left(\alpha_b + \sum_{i=1}^{n_k} s_{ki}\right)\right) \right\},$$

which, as has been mentioned, is a product of independent posterior distributions for each $(a_k, b_k)$. For the random sample case, one obtains:

$$p(a, b \mid t_{1,1:n_1}, \ldots, t_{K,1:n_K}) \propto a^{\sum n_k + \beta_a - 1} \, b^{\sum n_k + \beta_b - 1}$$

$$\times \exp\left(-(\alpha_a + K)a - \left(\alpha_b + \sum_{k=1}^{K} \sum_{i=1}^{n_k} s_{ki}\right) b + a \sum_{k=1}^{K} e^{-b\mathcal{T}_k}\right).$$

For the hierarchical case, it is the hyper-parameters $\psi = (\alpha_a, \beta_a, \alpha_b, \beta_b)$ that have a prior assessed on them. Ó Ríordáin (2016) proposed various forms such as the lognormal or exponential, but for a general prior $p(\psi)$, the posterior distribution is of the form:

$$p(a_{1:K}, b_{1:K}, \alpha_a, \beta_a, \alpha_b, \beta_b \mid t_{1,1:n_1}, \ldots, t_{K,1:n_K}) \propto \prod_{k=1}^{K} \left\{ a_k^{\beta_a + n_k - 1} \, b_k^{\beta_b + n_k - 1} \right.$$

$$\left. \times \exp\left(-a_k(\alpha_a + 1 - e^{-b_k \mathcal{T}_k}) - b_k \left(\alpha_b + \sum_{i=1}^{n_k} s_{ki}\right)\right) \right\} p(\alpha_a, \beta_a, \alpha_b, \beta_b).$$

Finally, for the autoregressive model, one can take the specific case of separate order-one autoregressive Gaussian processes on $\log(a)$ and $\log(b)$:

$$\log(a_1) \sim N(a, s_{a,1}^2),$$
$$\log(b_1) \sim N(b, s_{b,1}^2),$$
$$\log(a_k) = a + r_a(\log(a_{k-1}) - a) + \phi_{a,k},$$
$$\log(b_k) = b + r_b(\log(b_{k-1}) - b) + \phi_{b,k},$$

where $\phi_{a,k} \sim N(0, s_a^2)$ and $\phi_{b,k} \sim N(0, s_b^2)$, and so the hyper-parameters are $\psi = (a, b, s_{a,1}^2, s_{b,1}^2, r_a, r_b, s_a^2, s_b^2)$. This implies that the $a_k$ and $b_k$ are lognormally distributed, with the log mean and log variance parameters given by the Gaussian process. The posterior distribution then has the form:

$$p(a_{1:K}, b_{1:K}, a, b, s_{a,1}^2, s_{b,1}^2, r_a, r_b, s_a^2, s_b^2 \mid t_{1,1:n_1}, \dots, t_{K,1:n_K})$$

$$\propto \prod_{k=1}^{K} \left\{ a_k^{n_k-1} b_k^{n_k-1} \exp\left( -a_k(1 - e^{-b_k T_k}) - b_k \sum_{i=1}^{n_k} s_{ki} \right) \right\}$$

$$\times \exp\left( -\frac{1}{2s_{a,1}^2}(\log(a_1) - a)^2 - \frac{1}{2s_{b,1}^2}(\log(b_1) - b)^2 \right)$$

$$\times \prod_{k=2}^{K} \exp\left( -\frac{1}{2s_a^2}(\log(a_k) - r_a \log(a_{k-1}) - (1 - r_a)a)^2 \right.$$

$$\left. -\frac{1}{2s_b^2}(\log(b_k) - r_b \log(b_{k-1}) - (1 - r_b)b)^2 \right)$$

$$\times p(a, b, s_{a,1}^2, s_{b,1}^2, r_a, r_b, s_a^2, s_b^2).$$

In terms of computational feasibility, the independence and random sample structures present few difficulties because the posterior still consists of two-dimensional distributions – on $(a, b)$ for the random sample model, and on $K$ independent pairs $(a_k, b_k)$ for the independence model – that can be computed exactly on a discrete grid. However, for the hierarchical and autoregressive structures there is no independence a posteriori, and the posterior dimension increases with $K$. Ó Ríordáin (2016) details a Markov chain Monte Carlo (MCMC) approach for the hierarchical model that uses imputation to account for the bugs whose release version is missing. The imputation approach can also be extended to the other three model structures.

The independent, random sample, and hierarchical models have been fitted to the Mozilla Firefox bug detection data, with the MCMC imputation method of Ó Ríordáin (2016). Figures 3.3, 3.4, and 3.5 summarize the results with box plots of the posterior samples of $a_k$ and $b_k$ for releases 5 to 25. In Figure 3.4, the summarized distributions for release $k$ are the marginals of the data up to that release, e.g. $p(a, b \mid t_{1,n_1}, \dots, t_{k,1:n_k})$. It is noted that, for the independent and hierarchical models, there is greater a posteriori uncertainty around later versions, corresponding to those that are still in use and for which the detection process has not been completed.

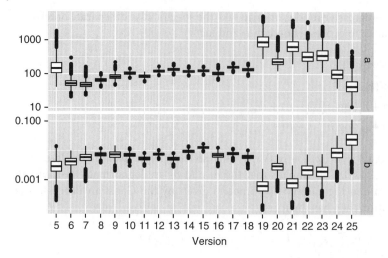

**Figure 3.3** Box plots of posterior samples of $a_k$ and $b_k$ for the Mozilla Firefox data under the independent model.

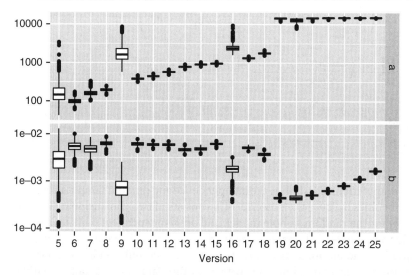

**Figure 3.4** Box plots of posterior samples of $a_k$ and $b_k$ for the Mozilla Firefox data under the random sample model.

## 3.6 Application to the Optimal Time Between Releases

McDaid and Wilson (2001) explored decision theoretic approaches to optimal testing times for a bespoke piece of software. They looked at a variety of different testing plans, from a simple one-off period of testing to sequential testing plans where further testing was based on data from previous testing stages. A similar problem exists in rapid-release software, namely to identify an optimal time between releases, assuming that this time is to remain unchanged from one release to the next.

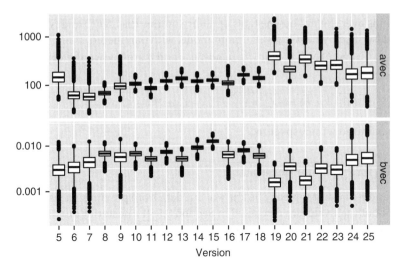

**Figure 3.5** Box plots of posterior samples of $a_k$ and $b_k$ for the Mozilla Firefox data under the hierarchical model.

### 3.6.1  Decision Theoretic Solution

The elements of a decision theory solution to this problem are:

**Decision variables:** In this case there is only one decision variable, an optimal time between releases $T^*$.

**Utility:** A function that describes the value of a particular decision about the value of the decision variable, so in this case a measure of the worth of releasing a new version every $T$ units of time.

**States of nature:** Unknown variables that affect the utility, described by probabilities; in this case, the times of detections of bugs before and after release and, if a model is being used to describe that process, then the parameters of that model.

A simple utility function could assign a fixed cost of $C_1$ for the detection and fixing of a bug during the $T$ days leading up to the release of a version, e.g. in the time from the release of the previous version. This makes a simplifying assumption that testing for a release begins at the release of the previous version, and ignores the detection of bugs that are found from a release prior to the previous release, as appears to happen in the Firefox data. Nevertheless, it is a guide to the bug detection cost before release. Then there is a cost $C_2 > C_1$ incurred with the detection of a bug post-release, which covers both any extra effort to fix it and other more intangible costs such as loss of reputation. Finally, a per-unit testing cost of $C_3$, again assumed to run for time $T$ from the prior version release. Thus, a simple utility of releasing every $T$ units of time, with $N(T)$ bugs detected before release and $\overline{N}(T)$ detected post-release, is:

$$U(T, N(T), \overline{N}(T)) = -C_1 N(T) - C_2 \overline{N}(T) - C_3 T.$$

The solution is to maximize the expected utility with respect to $T$:

$$T^* = \arg\max_T E(U(T, N(T), \overline{N}(T))),$$

where expectation is over the unknown states of nature, in this case $N(T)$ and $\overline{N}(T)$. If bug detection is described by the Goel–Okumoto model with parameters $a$ and $b$, then $E(N(T)) = a(1 - e^{-bT})$ and $E(\overline{N}(T)) = ae^{-bT}$; hence, if $a$ and $b$ are known the expected utility is:

$$E(U(T, N(T), \overline{N}(T)) \mid a, b) = -C_1 a(1 - e^{-bT}) - C_2 ae^{-bT} - C_3 T,$$

which is easily maximized with respect to $T$.

### 3.6.2 Accounting for Parameter Uncertainty

Unfortunately, $a$ and $b$ may not typically be known, but data from previous releases can be used to learn about their value, as discussed in Section 3.5. If a posterior distribution of $a$ and $b$ is available given data on $K$ previous releases, then expectation can be taken to obtain a posterior expected utility:

$$E(U(T, N(T), \overline{N}(T)) \mid t_{1,1:n_1}, \ldots, t_{K,1:n_K})$$
$$= -C_1 E(a(1 - e^{-bT}) \mid t_{1,1:n_1}, \ldots, t_{K,1:n_K})$$
$$- C_2 E(ae^{-bT} \mid t_{1,1:n_1}, \ldots, t_{K,1:n_K}) - C_3 T, \tag{3.10}$$

where expectation is over the posterior distribution of $a$ and $b$. With respect to our four approaches to modeling, this is easiest for the random sample model because the expectations in Equation (3.10) are with respect to the posterior $p(a, b \mid t_{1,1:n_1}, \ldots, t_{K,1:n_K})$ [Equation (3.6)]. The independent model is more problematic as inference can only be done on the release-specific parameters and these are releases that have already occurred; in other words, with the independence model, one cannot make predictions about the values of $a$ and $b$ for a future release except through the prior $p(a_{K+1}, b_{K+1})$. Alternatively, one could take expectation with respect to one of the $p(a_k, b_k \mid t_{k,1:n_k})$, but this only makes use of data from one release. However, for the hierarchical one has a distribution over the next $\theta_{K+1} = (a_{K+1}, b_{K+1})$ that can be used, making use of what has been learnt about the hyper-parameter $\psi$, which takes the form

$$p(a_{K+1}, b_{K+1} \mid t_{1,1:n_1}, \ldots, t_{K,1:n_K})$$
$$= \int p(a_{K+1}, b_{K+1} \mid \psi) \, p(\psi \mid t_{1,1:n_1}, \ldots, t_{K,1:n_K})) \, d\psi,$$

which is the expectation of $p(a_{K+1}, b_{K+1} \mid \psi)$ with respect to $p(\psi \mid t_{1,1:n_1}, \ldots, t_{K,1:n_K})$. Hence, the expectation in Equation (3.10) can be approximated in the usual Monte Carlo fashion as a sample average,

$$E(U(T, N(T), \overline{N}(T)) \mid t_{1,1:n_1}, \ldots, t_{K,1:n_K})$$
$$\approx - \left[ \frac{1}{M} \sum_{m=1}^{M} C_1 a^{(m)}(1 - e^{-b^{(m)}T}) + C_2 a^{(m)} e^{-b^{(m)}T} \right] - C_3 T, \tag{3.11}$$

where $(a^{(m)}, b^{(m)})$ is a sample from $p(a_{K+1}, b_{K+1} \mid \psi^{(m)})$ and $\psi^{(m)}$ are samples of $\psi$ from the posterior distribution obtained, for example, by MCMC.

Similarly, for the autoregressive model a posterior distribution for $\theta_{K+1} = (a_{K+1}, b_{K+1})$ is:

$$p(a_{K+1}, b_{K+1} \mid t_{1,1:n_1}, \ldots, t_{K,1:n_K})$$

$$= \int p(a_{K+1}, b_{K+1} \mid a_K, b_K, \psi) \, p(a_K, b_K, \psi \mid t_{1,1:n_1}, \ldots, t_{K,1:n_K})) \, da_K \, db_K \, d\psi,$$

which is the expectation of that autoregressive distribution $p(a_{K+1}, b_{K+1} \mid a_K, b_K, \psi)$, and so the expectation of Equation )(3.10)) can be approximated as a sample average

$$E(U(T, N(T), \overline{N}(T)) \mid t_{1,1:n_1}, \ldots, t_{K,1:n_K})$$

$$\approx - \left[ \frac{1}{M} \sum_{m=1}^{M} C_1 a^{(m)} (1 - e^{-b^{(m)}T}) + C_2 a^{(m)} e^{-b^{(m)}T} \right] - C_3 T, \tag{3.12}$$

where $(a^{(m)}, b^{(m)})$ is a sample from $p(a_{K+1}, b_{K+1} \mid a_K^{(m)}, b_K^{(m)}, \psi^{(m)})$, and $(a_K^{(m)}, b_K^{(m)}, \psi^{(m)})$ are samples from the posterior distribution obtained.

Equation (3.10), (3.11), or (3.12) can be then be maximized with respect to $T$ to find

$$T^* = \arg \max_{T \geq 0} E(U(T, N(T), \overline{N}(T)) \mid t_{1,1:n_1}, \ldots, t_{K,1:n_K}).$$

### 3.6.3 Application to the Firefox Data

This solution was applied to the Firefox data fit of the Goel–Okumoto model. The hierarchical model for the $(a_k, b_k)$ was used. Figure 3.6 shows the expected utility with costs $C_1 = 1$, $C_2 = 20$, and $C_3 = 12$; i.e., the cost of a bug post-release is 20 times that of pre-release, and the per-unit time cost of testing is 12 times that per day. The optimal time between releases is 50 days.

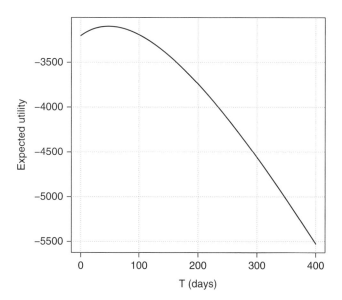

**Figure 3.6** The expected utility of testing to time $T$ for the Firefox data with $C_1 = 1, C_2 = 20$, and $C_3 = 12$.

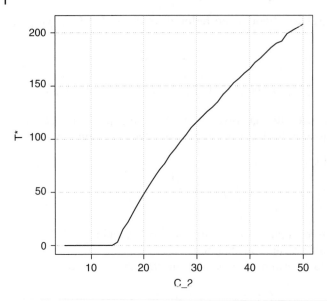

**Figure 3.7** The optimal time between releases as a function of $C_2$ for the Firefox data with $C_1 = 1$ and $C_3 = 12$.

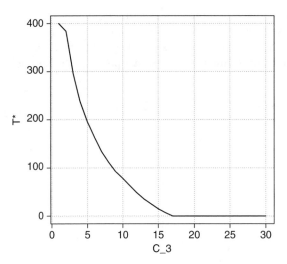

**Figure 3.8** The optimal time between releases as a function of $C_3$ for the Firefox data with $C_1 = 1$ and $C_2 = 20$.

The sensitivity of $T^*$ to changes in the utility coefficients is also explored. Figure 3.7 shows how $T^*$ varies as a function of $C_2$, the post-release cost of testing, with $C_1 = 1$ and $C_3 = 12$ fixed. Not surprisingly, it can be seen that $T^*$ increases with $C_2$ as it becomes preferable to identify more bugs before release takes place. Figure 3.8 shows how $T^*$ varies as a function of $C_3$, the unit-time cost of testing, with $C_1 = 1$ and $C_2 = 20$ fixed. Again, unsurprisingly, it can be seen that $T^*$ decreases with $C_3$ as it becomes preferable to release as soon as possible to reduce high unit-time testing costs. Indeed, for $C_3 > 17$ the release time becomes 0, indicating that the per-unit test cost is so high that in fact no testing should be done and bugs simply fixed as they arise.

**Figure 3.9** A set of values of $(C_2, C_3)$ that yield $T^* = 42$ days, the release cycle length that Firefox uses in the dataset, for $C_1 = 1$.

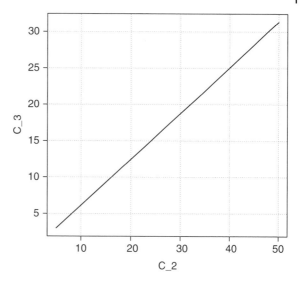

Finally, it is interesting to ask what combination of costs yields the 42-day release cycle that the Firefox versions in the data follow. Figure 3.9 shows the set of values of $C_2$ and $C_3$ that yield $T^* = 42$ for a fixed $C_1 = 1$. These lie roughly on a straight line with slope 0.63. This suggest that a 42-day release cycle is consistent with the cost of lengthening the release cycle by one day, is the equivalent of the occurrence of 0.63 bugs post-release, and with the principle that it is the cost of fixing bugs rather than per-unit time testing costs that drive the 42-day figure.

## 3.7 Summary

New versions of much consumer software are released on a rapid (order of a few months) schedule in order to maintain competitiveness and security. This paper illustrates some approaches to modeling the reliability of these versions in terms of the number of bugs reported, and how inference for model parameters can be implemented under the Bayesian paradigm. An application of this fitted model, to the question of how much time there should be between releases, has also been demonstrated for the case where this time is fixed across all releases with a decision theoretic solution.

It is worth pointing out that the hierarchical and autoregressive approaches to this problem also give a sequential solution; after each release, the posterior distribution for the next release is updated and the optimal release time could be re-evaluated. In this way the time between releases could vary according to the observed bug discovery times. Related methods that can complement this proposal are semi-parametric control methods that can help determine product release decisions – see, for example, Chapter 14 (Kenett and Pollak, 1986).

There is nothing particularly special about the choice of the Goel–Okumoto model for this work, other than that it is quite well known and studied. Other software reliability models, based on the non-homogeneous Poisson process or otherwise, can be used. For inference, the principal requirement is that it is possible to compute the likelihood so that schemes such as importance sampling or MCMC can be more easily implemented.

# References

Chow, Y.S. and Teicher, H. (1997) *Probability Theory: Independence, Interchangeability, Martingales*. New York: Springer, 3rd edn.

Goel, A.L. and Okumoto, K. (1979) Time-dependent error detection rate model for software reliability and other performance measures. *IEEE Transactions on Relativity*, R-28, 206–211.

Humble, J. and Farley, D. (2011) *Continuous Delivery: Reliable Software Releases through Build, Test and Deployment Automation*. Boston, MA: Pearson Education, Inc.

Kenett, R.S. and Pollak, M. (1986) A semi-parametric approach to testing for reliability growth, with application to software systems. *IEEE Transactions on Reliability*, R-35, 304–311.

Larman, C. (2003) *Agile and Iterative Development: A Manager's Guide*. New York: Addison-Wesley.

McDaid, K. and Wilson, S.P. (2001) Deciding how long to test software. *The Statistician*, 50, 117–134.

Ó Ríordáin, S. (2013) Firefox-2013. https://github.com/seanpor/Firefox-2013 [accessed 30 January, 2018].

Ó Ríordáin, S. (2016) Tracking the distribution of bugs across software release versions. Ph.D. thesis, Trinity College Dublin.

# 4

# Incremental Verification and Coverage Analysis of Strongly Distributed Systems

*Elena V. Ravve and Zeev Volkovich*

## Synopsis

In this chapter, we model both software and hardware hierarchical systems as logical structures. The main used formalism is the finite state machine; however, our method may be adopted to other presentations as well. We express the properties to be tested in different extensions of first-order logic. Coverage analysis is aimed to guarantee that the runs of the tests fully capture the functionality of the system. While functional verification must answer Boolean questions as to whether particular properties are true or false in the model of the system, the corresponding coverage analysis must provide quantitative information about the runs.

We propose a method to analyze quantitative coverage metrics, using the corresponding labeled weighted tree as a representation of the runs of the tests, as well as weighted monadic second-order logic to express the coverage metrics, and weighted tree automata as an effective tool to compute the value of the metric on the tree.

We introduce the notion of strongly distributed systems and present a uniform logical approach to incremental automated verification of such systems. The approach is based on systematic use of two logical reduction techniques: Feferman–Vaught reductions and syntactically defined translation schemes.

This chapter is completely theoretical and does not provide experimental results.

## 4.1  Verifying Modern Hardware and Software

Lately, verification has become the main bottleneck of the full design workflow, and may require more than 70% of the time to market. Simulation-based methods are still the main verification method for both software and hardware design. While the main purpose of a test is to activate some legal behavior of the design and to check that the results obtained are identical to expectation, the main purpose of coverage analysis is to objectively show what part of the overall behaviors of the design we have managed to exercise. In general, the result of a test run is Boolean: pass/fail; the result of coverage analysis is a quantitative value, related to the analysis. The particular interpretation of the value depends upon the analyzed coverage metric: it may be a counter of some event,

*Analytic Methods in Systems and Software Testing*, First Edition.
Edited by Ron S. Kenett, Fabrizio Ruggeri, and Frederick W. Faltin.
© 2018 John Wiley & Sons Ltd. Published 2018 by John Wiley & Sons Ltd.

some predefined (average) distance between tests, etc. Depending upon the coverage metric, we may want either to minimize or to maximize the value, we may want it either to fail into some interval or to be outside of some interval, and so on.

The increasing complexity of modern designs and intensive use of ready-made components in complex designs, like different libraries and systems on a chip, pose new challenges to verification teams. Moreover, while the design definition and specification are undertaken up–down, the design implementation and verification are done down–up. In fact (Wilcox and Wilcox, 2004):

> The only way to create testbenches that can be reused as the design develops from the unit to the block to the subsystem level is to plan ahead and know what will be required at higher subsystem levels. Similarly, even though the top–down flow develops the testbench from the top ... level down to the lowest unit-level the testing and integration are still performed in a bottom–up manner.

As a rule, modern hardware designs and well-structured programs are built from components (modules). The modules are the building blocks that create the design hierarchy. Classically, coverage properties are defined separately for each level of the design hierarchy. However, the fact that the corresponding properties are satisfied for a lower level does not guarantee anything about the coverage properties on the upper level, even partially. This leads to the well-known explosion of bug rates each time we pass in our design flow from lower to higher levels. In this chapter we propose an approach which guarantees that, given a coverage property in the upper level, it may be algorithmically translated to coverage properties in the lower levels. Moreover, if the derived coverage properties are satisfied in the lower level then the original property holds in the upper level.

We apply logical machinery to the verification of both Boolean and quantitative properties of hierarchical systems. Logical methods are widely used for functional verification of VLSI designs. We apply logical tools to incremental verification using runs of tests and coverage analysis. We present a hierarchical system (both hardware and software) as a logical structure. Assume we are given logical structure $\mathcal{A}$ and a formalized presentation of a property to be verified in the form of a formula $\phi$ (both Boolean and quantitative), possibly with free variables. Assume that $\mathcal{A}$ is built from well-defined components $\mathcal{A}_i$, where $i \in I$ is some index set or even structure. We systematically investigate how the fact that $\mathcal{A}$ is built from the components may be used in order to make an incremental evaluation of the property. Different special cases of the approach for the solution of particular problems are known as folklore. We provide a consistent generalized approach that may be easily used in a wide repertoire of particular applications. Thus, we propose a general way to reduce the answer on the given question of whether $\mathcal{A} \vDash \phi$ to computations on $\mathcal{A}_i$, where $i \in I$, and some combination of the local results, which gives the answer to the original question.

### 4.1.1 Verification of Structured Software

Structured programming often uses a *top–down design model*, where developers divide the full program into several subroutines. The extracted routines are coded separately. When a routine has been tested separately, it is then integrated into the full program.

Routines are tested using particular test suites. Code coverage is a measure that shows the quality of the applied suite, and seems to have first been introduced in Miller and Maloney (1963). High quality of the test suite should guarantee the absence of bugs in the routine. In order to measure the code coverage, different metrics may be applied. The main coverage criteria are (Myers and Sandler, 2004):

- Function coverage: Has each routine in the program been called?
- Statement coverage: Has each statement in the program been executed?
- Branch coverage: Has each branch of each control structure been executed?
- Condition coverage: Has each Boolean subexpression evaluated both to true and false?

A combination of function coverage and branch coverage is called *decision coverage*. In such a case, each entry and exit point in the routine must be activated at least once. Moreover, each decision in the program must have experienced each possible outcome at least once. *Condition/decision coverage* requires that both decision and condition coverage be satisfied. A detailed survey may be found in Paul and Lau (2014). Other coverage criteria are:

- State coverage: Has each state in a finite state machine been reached and explored?
- Linear code sequence and jump coverage: Has each linear code sequence and jump path been executed?
- Path coverage: Has every possible route through a given part of the code been executed?
- Entry/exit coverage: Has every possible call and return of the function been executed?
- Loop coverage: Has every possible loop been executed zero times, once, and more than once?

### 4.1.2 Verification of Hierarchical Hardware

In hierarchical VLSI designs, the communication between a module and its environment is executed using ports (or procedure calls). All but the top-level modules in a hierarchy have ports. Ports can be associated by order or by name and declared to be input, output, or inout. Modules connected by port order (implicit) require a correctly matched order. Such connections cause problems in debugging, when any port is added or deleted. Modules connected by name require name matching with the leaf module – the order is not important. Here is an example of port declaration in Verilog:

```
input   clk;                     // clock input
input   [15:0]  data\_in;      // 16-bit data input bus
output  [7:0]   count;          // 8-bit counter output
inout   data\_bi;              // bidirectional data bus
```

Functional verification of a VLSI design may be done either using formal verification methods or using massive runs of test patterns on the simulator of the design. If the second way is chosen, then we need to ensure completeness of the testing. Coverage analysis is aimed to guarantee that the runs of our tests fully capture the possible functionality of the design.

Coverage analysis in hardware verification has a long history (Wang and Tan, 1995; Drake and Cohen, 1998), and it is very rich (Jou and Liu, 1999a). It is an integral part of the verification task at each step of VLSI design. During VLSI design verification, we must take into account state coverage, transition coverage, toggle coverage, memory coverage, combinatorial logic coverage, assertion coverage, etc. We restrict ourselves to the step of functional verification only. One of the most popular categories of coverage metrics at this design step is related to the HDL code coverage. On the other hand, the design under test (DUT) at this step may be modeled as a finite state machine (FSM), effectively extractable from the HDL code of the design (Jou and Liu, 1998, 1999b), or as a Kripke model (Kripke, 1965). Finite state machine coverage is another category of coverage metrics at this design step. There is no single FSM coverage metric, but rather a family of coverage metrics related to the FSM and its behavior.

First of all, an FSM coverage metric must report how many times ($N_{J_{FSM}}$, $1 \leq J_{FSM} \leq N_{FSM}$, where $N_{FSM}$ is the number of states of the FSM) each state of the FSM was visited during the runs of the tests (*state visitations*). *State transitions* and *paired state visitations* are examples of other simple FSM coverage metrics (Jou and Liu, 1999a). More generally, all the state sequences generated during the runs of the tests may be analyzed in order to check how they satisfy more complicated coverage criteria. For example, we may be interested to know *how long* the runs of our tests were. We try to avoid both unpleasant situations where either only *very short* or only *very long* runs have been observed. Let us denote the value of the corresponding coverage metric by $N_{length}$. There may be several candidates for calculating the number; however, all of them are expected to take into account the total number of states visited during the runs, the number of final states observed during the runs, the length of the (longest) runs, etc.

### 4.1.3   Modeling Software and Hardware Using Finite State Machines

In this chapter we are mostly concentrating on FSMs as a way to model both software and hardware hierarchical systems. We use different extensions of first-order logic to express the tested properties of the systems and weighted monadic second-order logic to express the coverage properties. In general, different FSM coverage metrics capture different properties of the behavior of the design that was tested. Finite state machine coverage deals with lots of quantitative properties, which must be analyzed on (the sequences of) the states, toggled during the runs of the tests, and seems to be the most complex type of coverage analysis at this step of design and verification. On the other hand, based on FSM coverage we may guarantee that all bugs related to the FSM behavior have been found and fixed. However, in general, its computational complexity is very high.

If we consider the step of design verification, then the verification of the results of different simulation runs must answer a Boolean question as to whether a particular property is true or false in the model of the design. The corresponding theoretical basis has a long history and is well developed. The model of the design is either an FSM or a Kripke model, and the corresponding property to be checked is formulated in some fragment of a particular temporal logic. An extension of the Kripke model with a weighted transition relation and the corresponding generalization of computation tree logic was defined in Bouyer (2006), and Larsen, Fahrenberg, and Thrane (2009, 2012), and used in particular in the context of formal VLSI verification. However, coverage analysis should answer quantitative questions rather than Boolean ones. That is why, in this case, the contribution of classical (temporal) logics is very limited, in particular if we want to characterize the feasible computation of a particular FSM coverage metric.

We present an approach to effective coverage analysis of functional verification based on the concept of a *weighted* version of monadic second-order logic (WMSOL) and the corresponding weighted tree automata (WTA). Schützenberger (1961) investigated finite automata with weights, and characterized their behaviors as rational formal power series, which generalized Büchi's (1960) and Elgot's (1961) fundamental theorems. Droste and Gastin (2007) introduced WMSOL, and it was proven that, for commutative semirings, the behaviors of weighted automata are precisely the formal power series definable with this weighted logic. Mandrali and Rahonis (2009) considered WTA with discounting over commutative semirings. For them, Kleene's theorem and a WMSOL characterization were proven. Droste and Meinecke (2010) investigated automata operations like average, limit superior, limit inferior, limit average, or discounting. A connection between such new kinds of weighted automata and weighted logics was established. Droste et al. (2011a) investigated a more general cost model where the weight of a run may be determined by a global valuation function. It was shown that for bilocally finite bimonoids, WTA captures the expressive power of several semantics of full WMSOL. Droste et al. (2011b) considered weighted finite transition systems with weights from naturally ordered semirings. Such semirings comprise distributive lattices as well as the natural numbers with ordinary addition and multiplication, and the max-plus semiring. For these systems, the concepts of covering and cascade product were explored.

We apply the concepts of WMSOL and WTA to the field of effective quantitative coverage analysis during software and hardware verification. Weighted automata are classical finite automata in which the transitions are labeled by weights, using a set $L$ of labels. This allows us to define the general framework to express and to effectively check different FSM coverage metrics.

Assume we are given a design (software or hardware) and a set of tests. We use an effectively extractable FSM named $\mathcal{FSM}^{\text{Design}}$ as a model of the design. We run our tests and obtain the results, which may be presented as the tree $\mathcal{T}^{\text{Design}}$. Assume we are interested in analyzing a quantitative metric $\mathcal{P}^t_{\text{coverage}}$ on the results of the testing. Now, we are looking for a labeling $L$ of $\mathcal{T}^{\text{Design}}$ and a product tree valuation monoid ptv-$\mathcal{D}$ such that $\mathcal{P}^t_{\text{coverage}}$ may be expressed as a formula $\phi^t_{\text{coverage}}$ of a corresponding fragment of WMSOL. This means that we try to reduce the computation and analysis of $\mathcal{P}^t_{\text{coverage}}$ to a logical problem: whether $\mathcal{T}^{\text{Design}}_L \models \phi^t_{\text{coverage}}$. If we manage to find such an $L$ and ptv-$\mathcal{D}$, then we can effectively construct a variation $\mathcal{M}^{\phi^t_{\text{coverage}}}$ of WTA that computes the value of $\phi^t_{\text{coverage}}$ on $\mathcal{T}^{\text{Design}}_L$. We propose some candidates for such an $L$ and $\mathcal{D}$ for some examples of FSM coverage metrics. Our approach shows how analysis of different FSM coverage metrics may be reduced to computation of WMSOL-definable formulas on $\mathcal{T}^{\text{Design}}_L$.

## 4.2 Motivating Example: Cooperation of Three Computational Units

This example is taken mostly verbatim from Ravve, Volkovich, and Weber (2014). Assume that we are given a system with three computational units, which may communicate according to predefined rules. Assume that unit 1 may call both units 2 and 3. Moreover, unit 2 cannot call any unit, while unit 3 may call back to unit 1 – see

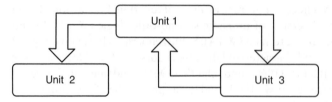

**Figure 4.1** A system with three units.

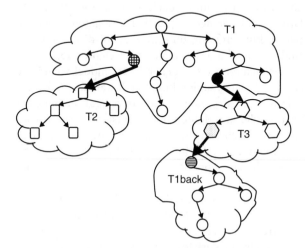

**Figure 4.2** Run tree $T$ of the system with three units.

Figure 4.1. We use the following formalization: The runs of the units are accumulated in weighted labeled trees; the weights are put on the edges of the trees. The vertices also may be labeled. A run is a path in the tree, as introduced in Definition 4.22.

Assume that run tree $T$ of the multi-unit system is presented in Figure 4.2. In this figure we omit the weights put on the edges. The meanings of the labels on the vertices are as follows:

- The gridded vertex in T1 corresponds to the call to unit 2 by unit 1 (the bold edge goes to the root of T2).
- The filled vertex in T1 corresponds to the call to unit 3 by unit 1 (the bold edge goes to the root of T3).
- The dotted vertex in T3 corresponds to the call to unit 1 by unit 3 (the bold edge does *not* go to the root of T1, but rather goes back to the striped vertex of T1).

Assume that we want to optimize (minimize) the runs in the tree. On the one hand, we may use one of the optimization algorithms on the complete tree $T$ that will give quantitative result $\mathcal{R}$. On the other hand, we observe that we may receive the optimal result in the following way:

1) Find the optimal run $\mathcal{R}_1$ in T1.
2) Find *all* labeled runs $\Lambda_{i_1}$ in T1: there are two such runs, $\Lambda_{1_1}$ and $\Lambda_{2_1}$.
3) Find the optimal run $\mathcal{R}_2$ in T2. For the optimization, we will use $\Lambda_{1_1} + \mathcal{R}_2$.
4) Find the optimal run $\mathcal{R}_3$ in T3. For the optimization, we will use $\Lambda_{2_1} + \mathcal{R}_3$.
5) Find *all* labeled runs $\Lambda_{i_3}$ in T3: there is one such run, $\Lambda_{1_3}$.

6) Find the optimal run $\mathcal{R}_{1_{back}}$ in T1back. We will use $\Lambda_{2_1} + \Lambda_{1_3} + \mathcal{R}_{1_{back}}$.
7) Finally, we find $\min\{\mathcal{R}_1, \Lambda_{1_1} + \mathcal{R}_2, \Lambda_{2_1} + \mathcal{R}_3, \Lambda_{2_1} + \Lambda_{1_3} + \mathcal{R}_{1_{back}}\}$.
8) We observe that $\mathcal{R} = \min\{\mathcal{R}_1, \Lambda_{1_1} + \mathcal{R}_2, \Lambda_{2_1} + \mathcal{R}_3, \Lambda_{2_1} + \Lambda_{1_3} + \mathcal{R}_{1_{back}}\}$.

In order to generalize the obtained observations of the example, we need:

- A precise definition of languages which describe different problems – see Section 4.3.
- A precise definition of the weighted labeled trees and computations on them – see Section 4.4.
- Tree $T$ is *not* a disjoint union of its subtrees. We need a formal framework to deal with such objects – see Section 4.9.
- Let $\mathfrak{T}_{old}(N)$ denote the time to solve the problem directly ($N$ stands for the size of the coding of $T$).
    - $\mathfrak{E}_I$ denotes the time to extract the index structure $I$ from $T$. We have four ordered numbers to distinguish the subtrees in our example.
    - $\mathfrak{E}_i$ denotes the time to extract each $T_i$ from $T$. We have four numbered subtrees in our example.
    - $\mathfrak{E}_i(n_i)$ denotes the time to compute the values $\mathcal{R}_1, \Lambda_{1_1}, \Lambda_{2_1}$ on T1; $\mathcal{R}_2$ on T2; $\mathcal{R}_3$, $\Lambda_{1_3}$ on T3; and $\mathcal{R}_{1_{back}}$ on T1back ($n_i$ is the size of the coding of $T_i$).
    - $\mathfrak{T}_F$ denotes the time to build a sentence like $\min\{\mathcal{R}_1, \Lambda_{1_1} + \mathcal{R}_2, \Lambda_{2_1} + \mathcal{R}_3, \Lambda_{2_1} + \Lambda_{1_3} + \mathcal{R}_{1_{back}}\}$.
    - $\mathfrak{T}_{comp}$ denotes the time to compute $\min\{\mathcal{R}_1, \Lambda_{1_1} + \mathcal{R}_2, \Lambda_{2_1} + \mathcal{R}_3, \Lambda_{2_1} + \Lambda_{1_3} + \mathcal{R}_{1_{back}}\}$.

The new computation time is

$$\mathfrak{T}_{new} = \mathfrak{E}_I + \sum_{i \in I} \mathfrak{E}_i + \sum_{i \in I} \mathfrak{E}_i + \mathfrak{T}_F + \mathfrak{T}_{comp}.$$

The question now is when is $\mathfrak{T}_{new} < \mathfrak{T}_{old}$? The analysis is provided in Section 4.10.

Moreover, we want the construction of the sentence like

$$\min\{\mathcal{R}_1, \Lambda_{1_1} + \mathcal{R}_2, \Lambda_{2_1} + \mathcal{R}_3, \Lambda_{2_1} + \Lambda_{1_3} + \mathcal{R}_{1_{back}}\}$$

to depend only upon the property to be optimized and the predefined "communication" rules, but *not* upon the given $T$. We will call them $F_{\Phi,\phi}$. The values $\mathcal{R}_1, \Lambda_{1_1} + \mathcal{R}_2, \Lambda_{2_1} + \mathcal{R}_3, \Lambda_{2_1} + \Lambda_{1_3} + \mathcal{R}_{1_{back}}$ will be referenced as evaluations of $\psi_{1,1}, \ldots, \psi_{1,j_1}, \ldots, \psi_{\beta,1}, \ldots, \psi_{\beta,j_\beta}$.

## 4.3 Extensions of First-Order Logic

In this section, we provide some logical and complexity background. First-order logic (FOL) is not powerful enough to express many useful properties. This obstacle can be overcome by adding different operators, as well as by richer quantification. Second-order logic (SOL) is like FOL but allows quantification over relations. If the arity of the relation is restricted to 1 then we are dealing with monadic second-order logic (MSOL).

For our purposes we will need some additional logical tools and notations. For all logics, we define:

**Definition 4.1    (Quantifier rank of formulae)**
*The quantifier rank of formula $\varphi$, rank($\varphi$), can be defined as follows:*

- *For $\varphi$ without quantifiers, rank($\varphi$) = 0.*
- *If $\varphi = \neg\varphi_1$ and rank($\varphi_1$) = $n_1$, then rank($\varphi$) = $n_1$.*
- *If $\varphi = \varphi_1 \cdot \varphi_2$, where $\cdot \in \{\vee, \wedge, \rightarrow\}$, and rank($\varphi_1$) = $n_1$, rank($\varphi_2$) = $n_2$, then rank($\varphi$) = $\max\{n_1, n_2\}$.*
- *if $\varphi = Q\varphi_1$, where Q is a quantifier, and rank($\varphi_1$) = $n_1$, then rank($\varphi$) = $n_1 + 1$.*

We use the following notation for arbitrary logics $\mathcal{L}$: $\mathcal{A} \equiv_{\mathcal{L}}^{n} \mathcal{B}$ means that all formulae of logic $\mathcal{L}$ with quantifier rank $n$ have in the structures $\mathcal{A}$ and $\mathcal{B}$ the same truth value.

It is well known that the expressive power of FOL is very limited. For example, transitive closure is not defined in this logic. The source of this defect is its lack of counting or recursion mechanism. Several attempts to augment the expressive power of FOL have been made in this direction. For example, Immerman (1982) introduced the *counting quantifier* $\exists i x$, which can be read as "there are at least $i$ elements $x$ such that …" On the other hand, these attempts were inspired by Mostowski (1957), which introduced the notion of *cardinality quantifiers* (for example, "there are infinitely many elements"), and Tarski (1961), who studied *infinitary languages*. The next development appeared in Lindström (1966, 1969), which introduced *generalized quantifiers*. In this chapter we mostly follow Kolaitis and Väänänen (1992). We use the notation $\mathcal{K}$ (or $\mathcal{Q}$) for an arbitrary class of structures. If $\tau$ is a vocabulary, $\mathcal{K}(\tau)$ is the class of structures over $\tau$ that are in $\mathcal{K}$.

**Definition 4.2    (Simple unary generalized quantifier)**
*A* simple unary generalized quantifier *is a class Q of structures over the vocabulary consisting of a unary relation symbol P, such that Q is closed under isomorphism, i.e. if $\mathcal{U} = \langle U, P^{U} \rangle$ is a structure in Q and $\mathcal{U}' = \langle U', P^{U'} \rangle$ is a structure that is isomorphic to $\mathcal{U}$, then $\mathcal{U}'$ is also in Q.*

The *existential* quantifier is the class of all structures $\mathcal{U} = \langle \mathcal{U}, P^{U} \rangle$ with $P^{U}$ being a non-empty subset of $U$, while the *universal* quantifier consists of all structures of the form $\mathcal{U} = \langle \mathcal{U}, \mathcal{U} \rangle$. Numerous natural examples of simple unary generalized quantifiers on classes of finite structures arise from properties that are not FOL definable on finite structures, such as "there is an even number of elements," "there are at least $\log(n)$ many elements," … In particular, the quantifier "there is an even number of elements" can be viewed as the class $Q_{\text{even}} = \{\langle U, P^{U} \rangle : U \text{ is a finite set}, P^{U} \subseteq U, \text{ and } |P^{U}| \text{ is even}\}$. We may expand Definition 4.2 to the *n-ary generalized quantifier*.

**Definition 4.3    (Lindström quantifiers)**
*Let $(n_1, n_2, \ldots, n_{\ell})$ be a sequence of positive integers. A* Lindström quantifier of type $(n_1, n_2, \ldots, n_{\ell})$ *is in a class Q of structures over the vocabulary consisting of relation symbols $(P_1, P_2, \ldots, P_{\ell})$ such that $P_i$ is $n_i$ − ary for $1 \leq i \leq \ell$ and Q is closed under isomorphisms.*

One of the best-known examples of non-simple quantifiers is the *equicardinality* or *Härtig quantifier I*. This is a Lindström quantifier of type (1, 1) which comprises all structures $\mathcal{U} = \langle \mathcal{U}, \mathcal{X}, \mathcal{Y} \rangle$ when $|X| = |Y|$. Another example is the *Rescher* quantifier whose mean is *more*.

Another way to extend FOL is to allow countable disjunctions and conjunctions.

### Definition 4.4 (Infinitary logics)

- $L_{\omega_1 \omega}$ is the logic that allows countable disjunctions and conjunctions.
- $L^k_{\omega_1 \omega}$ is the logic that allows countable disjunctions and conjunctions, but has a total of only $k$ distinct variables.
- $L^k_{\infty \omega}, k \geq 1$ is the logic that allows infinite disjunctions and conjunctions, but has a total of only $k$ distinct variables.
- $L^\omega_{\infty \omega} = \bigcup L^k_{\infty \omega}$.

We assume that the only variables involved are $v_0, \dots, v_{k-1}$.

Now we introduce the syntax and semantics of the logic $L^k_{\infty \omega}$ that contains simple unary generalized quantifiers.

### Definition 4.5
Let $Q = \{Q_i : i \in I\}$ be a family of simple unary generalized quantifiers, and let $k$ be a positive integer. The infinitary logic $L^k_{\infty \omega}(Q)$ with $k$ variables and the generalized quantifiers $Q$ have the following syntax (for any vocabulary $\tau$):

- the variables of $L^k_{\infty \omega}(Q)$ are $v_1, \dots, v_k$.
- $L^k_{\infty \omega}(Q)$ contains all FOL formulae over $\tau$ with variables among $v_1, \dots, v_k$.
- If $\varphi$ is a formula of $L^k_{\infty \omega}(Q)$, then so is $\neg \varphi$.
- If $\Psi$ is a set of formulae of $L^k_{\infty \omega}(Q)$, then $\vee \Psi$ and $\wedge \Psi$ are also formulae of $L^k_{\infty \omega}(Q)$.
- If $\varphi$ is a formula of $L^k_{\infty \omega}(Q)$, then each of the expressions $\exists v_j \varphi, \forall v_j \varphi, Q_i v_j \varphi$ is also a formula of $L^k_{\infty \omega}(Q)$ for every $j$ such that $1 \leq j \leq k$ and for every $i \in I$.

The semantic of $L^k_{\infty \omega}(Q)$ is defined by induction on the construction of the formulae. So, $\vee \Psi$ is interpreted as a disjunction over all formulae in $\Psi$, and $\wedge \Psi$ is interpreted as a conjunction. Finally, if $\mathcal{U}$ is a structure having $U$ as its universe and $\varphi(v_j, \bar{y})$ is a formula of $L^k_{\infty \omega}(Q)$ with free variables among the variables of $v_j$ and the variables in the sequence $\bar{y}$, and $\bar{u}$ is a sequence of elements from the universe of $\mathcal{U}$, then $U, \bar{u} \vDash Q_i v_j \varphi(v_j, \bar{y})$ iff the structure $\langle U, \{a : U, a, \bar{u} \vDash \varphi(v_j, \bar{y})\} \rangle$ is in the quantifier $Q_i$.

We may also enrich the expressive power of FOL by allowing quantification over relation symbols. Second-order logic is like FOL, but also allows variables and quantification over relation variables of various but fixed arities. Monadic second-order logic is the sublogic of SOL where relation variables are restricted to be unary. The meaning function of formulae is explained for arbitrary $\tau$-structures, where $\tau$ is the vocabulary, i.e. a finite set of relation and constant symbols. Fixed-point logic (LFP) can be viewed as a fragment of SOL where the second-order variables only occur positively and in the fixed point construction. Similarly, MLFP corresponds to the case where the arity of the relation variables is restricted to 1. The semantics of the fixed point is given by the least fixed point, which always exists because of the positivity assumption on the set variable. The logic LFP is defined similarly, with operators $k$-LFP for every $k \in N$ which bind $2k$ variables. On ordered structures, LFP expresses exactly the polynomially recognizable classes of finite structures. Without order, every formula in LFP has a polynomial model checker. For transition systems, MLFP corresponds exactly to $\mu$-calculus (Vardi, 1982; Emerson, 1990; Arnold and Niwiński, 1992).

The logic MTC (monadic transitive closure) is defined inductively, like FOL. For a thorough discussion of it, see Immerman (1987). Atomic formulae are as usual. The inductive clauses include closure under the Boolean operations, existential and universal quantification, and one more clause:

**Syntax** If $\phi(x, y, \overline{u})$ is an MTC formula with $x$, $y$, and $\overline{u} = u_1, \ldots, u_n$ its free variables, $s$, $t$ are terms, then $MTCx, y, s, t\phi(x, y, \overline{u})$ is an MTC formula with $x$, $y$ bound and $\overline{u}$ free.

**Semantics** The formula $MTCx, y, s, t\phi(x, y, \overline{u})$ holds in a structure $\mathcal{U}$ under an assignment of variables $z$ if $s_z, t_z \in \text{TrCl}(\phi^{\mathcal{U}})$.

The logic TC is defined similarly, with operators $k$-TC for every $k \in N$ which bind $2k$ variables. For a more detailed exposition, see Abiteboul, Hull, and Vianu (1995), Bosse (1993), Ebbinghaus, Flum, and Thomas (1994), Grädel (1992), and Grohe (1994).

### 4.3.1 Complexity of Computation for Extensions of First-Order Logic

Computation for MSOL sits fully in the polynomial hierarchy. For the complexity of FOL and MSOL, see Frick and Grohe (2004).

More precisely, the complexity of computation (in the size of the structure) of SOL-expressible properties can be described as follows. The class $NP$ of non-deterministic polynomial-time problems is the set of properties that are expressible by existential SOL on finite structures (Fagin, 1974). Computation for SOL-definable properties is in the polynomial hierarchy (Ebbinghaus and Flum, 1995). Moreover, for every level of the polynomial hierarchy there is a problem, expressible in SOL, that belongs to this class. The same fact holds for MSOL as well, as observed in Makowsky and Pnueli (1993a).

Computation for properties definable in LFP is polynomial (Vardi, 1982). CTL* is a superset of computational tree logic and linear temporal logic. All the problems that are expressible by CTL* can be computed in polynomial time (Emerson, 1990). The relation between FOL with generalized quantifiers and computations with oracles is investigated in Makowsky and Pnueli (1993b). Most properties that appear in real-life applications are stronger than FOL but weaker than MSOL, and their computational complexity is polynomial.

## 4.4 Weighted Monadic Second-Order Logic and Weighted Tree Automata

In this section, we mostly follow the style of Droste et al. (2011a).

**Definition 4.6 (Semiring)**
A semiring $\mathcal{K}$ is a structure $(K, +, \cdot, 0, 1)$, where $(K, +, 0)$ is a commutative monoid, $(K, \cdot, 1)$ is a monoid, multiplication distributes over addition, and $0 \cdot x = x \cdot 0 = 0$ for each $x \in K$.

If the multiplication is commutative then $\mathcal{K}$ is also *commutative*. If the addition is idempotent[1] then the semiring is called *idempotent*.

---

[1] *Idempotence* is the property of certain operations in mathematics and computer science that can be applied multiple times without changing the result beyond the initial application.

Let $\mathbf{N} = \{1, 2, \ldots\}$ be the set of natural numbers and let $\mathbf{N}_0 = \mathbf{N} \cup \{0\}$. A ranked alphabet $\Xi$ is a pair $\Xi = (\sum, \mathbf{rk}_{\sum})$ consisting of a finite alphabet $\sum$ and a mapping $\mathbf{rk}_{\sum}$ : $\sum \to \mathbf{N}_0$ that assigns to each symbol of $\sum$ its rank. $\sum^{(m)}$ denotes the set of all symbols with rank $m \in \mathbf{N}_0$, and $a^{(m)}$ denotes that $a \in \sum^{(m)}$. Let $\max_{\sum} = \max\{\mathbf{rk}_{\sum}(a)|a \in \sum\}$, the maximal rank of $\sum$. Let $\mathbf{N}^*$ be the set of all finite words over $\mathbf{N}$. A *tree domain* $B$ is a finite, non-empty subset of $\mathbf{N}^*$ such that for all $u \in \mathbf{N}^*$ and $i \in \mathbf{N}$, $u.i \in B$ implies $u.1, \ldots, u.(i-1) \in B$; $u.1, \ldots, u.(i-1)$ is called the immediate prefix of $u.i$. Note that the tree domain of $B$ is prefix closed. A *tree over a set $L$ (of labels)* is a mapping $t : B \to L$ such that $\mathrm{dom}(t) = B$ is a tree domain. The elements of $\mathrm{dom}(t)$ are called *positions* of $t$, and $t(u)$ is called the *label* of $t$ at $u \in \mathrm{dom}(t)$. The set of all trees over $L$ is denoted by $T_L$.

**Definition 4.7** *A tree valuation monoid (tv-monoid) is a quadruple $(D, +, \mathrm{Val}, \mathbf{0})$ such that $(D, +, \mathbf{0})$ is a commutative monoid, and $\mathrm{Val} : T_D \to D$ is a function with $\mathrm{Val}(d) = d$ for every tree $d \in T_D$ and $\mathrm{Val}(t) = \mathbf{0}$ whenever $\mathbf{0} \in \mathrm{im}(t)$ for $t \in T_D$. $\mathrm{Val}$ is called a (tree) valuation function.*

**Definition 4.8** *A product tree valuation monoid (a ptv-monoid) $(D, +, \mathrm{Val}, \Diamond, \mathbf{0}, \mathbf{1})$ consists of a valuation monoid $(D, +, \mathrm{Val}, \mathbf{0})$, a constant $\mathbf{1} \in D$ with $\mathrm{Val}(t) = \mathbf{1}$ whenever $\mathrm{im}(t) = \{\mathbf{1}\}$ for $t \in T_D$, and an operation $\Diamond : D^2 \to D$ with $\mathbf{0}\Diamond d = d\Diamond \mathbf{0} = \mathbf{0}$ and $\mathbf{1}\Diamond d = d\Diamond \mathbf{1} = \mathbf{1}$.*

Note that the operation $\Diamond$ has to be neither commutative nor associative.

### 4.4.1 Weighted Monadic Second-Order Logic

Given a ptv-monoid $D$, the syntax of WMSOL over $D$ is defined in the following way.

**Definition 4.9 (Syntax of WMSOL)**
*Boolean formulae:*

- $\mathrm{label}_a(x)$ *and* $\mathrm{edge}_i(x, y)$ *for* $a \in \sum$ *and* $1 \leq i \leq \max_{\sum}$;
- $x \in X$, $\neg\beta_1$, $\beta_1 \wedge \beta_2$, $\forall x\beta_1$, $\forall X\beta_1$ *for first-order variable $x$ and second-order variable $X$.*

*Weighted formulae:*

- $d$ *for* $d \in D$;
- $\beta$ *for Boolean formula $\beta$;*
- $\phi_1 \vee \phi_2$, $\phi_1 \wedge \phi_2$, $\exists x\phi_1$, $\forall x\phi_1$, $\forall X\phi_1$.

The set $\mathrm{free}(\phi)$ of free variables occurring in $\phi$ is defined as usual. Semantics of WMSOL valuates trees by elements of $D$. There is no change in the semantics of Boolean formulas. $\mathbf{0}$ defines the semantics of the truth value "false." $\mathbf{1}$ defines the semantics of the truth value "true." The monoid operation "+" is used to define the semantics of the disjunction and existential quantifier. The monoid $\mathrm{Val}$ function is used to define the semantics of the first-order universal quantification. If we use the max-plus semiring, for example, the semantic interpretation of $\forall x\phi$ is the sum of all weights (rewards or time) defined by $\phi$ for all different positions $x$. More precisely, for a $(\mathcal{V}, t)$ assignment that maps $\tilde{\sigma} : \mathcal{V} \to \text{dom}(t) \cup PS(\text{dom}(t))$, with $\tilde{\sigma}(x) \in \text{dom}(t)$ and $\tilde{\sigma}(X) \subseteq \text{dom}(t)$, and $s \in T_{\Sigma_v}$.

### Definition 4.10 (Semantics of WMSOL)

*Boolean formulae:*

$$[\text{label}_a(x)]_\mathcal{V}(s) = \begin{cases} 1 & \text{if } t(\tilde{\sigma}(x)) = a, \\ 0 & \text{otherwise;} \end{cases}$$

$$[\text{edge}_i(x, y)]_\mathcal{V}(s) = \begin{cases} 1 & \text{if } \tilde{\sigma}(y) = \tilde{\sigma(x)}.i, \\ 0 & \text{otherwise;} \end{cases}$$

$$[x \in X]_\mathcal{V}(s) = \begin{cases} 1 & \text{if } \tilde{\sigma}(x) \in \tilde{\sigma}(X), \\ 0 & \text{otherwise;} \end{cases}$$

$$[\neg\beta_1]_\mathcal{V}(s) = \begin{cases} 1 & \text{if } [\beta_1]_\mathcal{V}(s) = 0, \\ 0 & \text{otherwise;} \end{cases}$$

$$[\beta_1 \vee \beta_2]_\mathcal{V}(s) = [\beta_1]_\mathcal{V}(s) + [\beta_2]_\mathcal{V}(s);$$

$$[\beta_1 \wedge \beta_2]_\mathcal{V}(s) = [\beta_1]_\mathcal{V}(s)\Diamond[\beta_2]_\mathcal{V}(s);$$

$$[\forall x \beta_1]_\mathcal{V}(s) = \begin{cases} \text{Val}(s_D) \text{ for } s_D \in T_D \text{ with } \text{dom}(s_D) = \text{dom}(s) \text{ and} \\ s_D(u) = [\beta_1]_{\mathcal{V}\cup\{x\}}(s[x \to u]) \text{ for all } u \in \text{dom}(s); \end{cases}$$

$$[\forall X \beta_1]_\mathcal{V}(s) = \begin{cases} 1 & \text{if } [\beta_1]_{\mathcal{V}\cup\{X\}}(s[X \to I]) = 1 \text{ for all } I \subseteq \text{dom}(s), \\ 0 & \text{otherwise.} \end{cases}$$

*Weighted formulae:*

$$[d]_\mathcal{V}(s) = d;$$

$$[\phi_1 \vee \phi_2]_\mathcal{V}(s) = [\phi_1]_\mathcal{V}(s) + [\phi_2]_\mathcal{V}(s);$$

$$[\beta_1 \wedge \beta_2]_\mathcal{V}(s) = [\phi_1]_\mathcal{V}(s)\Diamond[\phi_2]_\mathcal{V}(s);$$

$$[\exists x \phi_1]_\mathcal{V}(s) = \sum_{u \in \text{dom}(s)} [\phi]_{\mathcal{V}\cup\{x\}}(s[x \to u]);$$

$$[\exists X \phi_1]_\mathcal{V}(s) = \sum_{I \subseteq \text{dom}(s)} [\phi]_{\mathcal{V}\cup\{X\}}(s[X \to I]);$$

$$[\forall x \phi_1]_\mathcal{V}(s) = \begin{cases} \text{Val}(s_D) \text{ for } s_D \in T_D \text{ with } \text{dom}(s_D) = \text{dom}(s) \text{ and} \\ s_D(u) = [\phi_1]_{\mathcal{V}\cup\{x\}}(s[x \to u]) \text{ for all } u \in \text{dom}(s); \end{cases}$$

$$[\forall X \phi_1]_\mathcal{V}(s) = \begin{cases} 1 & \text{if } [\phi_1]_{\mathcal{V}\cup\{X\}}(s[X \to I]) = 1 \text{ for all } I \subseteq \text{dom}(s), \\ 0 & \text{otherwise.} \end{cases}$$

### 4.4.2 Weighed Tree Automata

Let $\Sigma$ be a ranked alphabet and $(D, +, \text{Val}, \mathbf{0})$ a tv-monoid.

**Definition 4.11** *A weighted bottom-up tree automaton (WTA) over a tv-monoid D is a quadruple $M = \langle Q; \Sigma, \mu, F \rangle$ where Q is a non-empty finite set of states, $\Sigma$ is a ranked alphabet, $\mu = (\mu_m)_{0 \leq m \leq \max_\Sigma}$ is a family of transition mappings $\mu_m : \Sigma^{(m)} \to D^{Q^m \times Q}$, and $F \subseteq Q$ is a set of final states.*

The behavior of a WTA $M$ is defined by a run semantics. A run $r$ of $M$ on a tree $t \in T_{\tilde{\sigma}}$ is a mapping $r : \text{dom}(t) \to Q$. For all positions $u \in \text{dom}(t)$, labeled with $t(u) \in \Sigma^{(m)}$, we call $\mu_m(t(u))_{r(u.1)...r(u.m).r(u)}$ the weight of $r$ on $t$ at $u$. Since the domain of a run is a tree domain, each run $r$ on $t$ defines a tree $\mu(t, r) \in T_D$, where $\text{dom}(\mu(t, r)) = \text{dom}(t)$ and $\mu(t, r)(u) = \mu_m(t(u)^{(m)})_{r(u.1)...r(u.m).r(u)}$ for all $u \in \text{dom}(t)$. $r$ on $t$ is *valid* if $\mathbf{0} \notin \text{im}(\mu(t, r))$ and *successful* if $r(\epsilon) \in F$. Furthermore, $\text{succ}(M, t)$ denotes the set of all successful runs of $M$ on $t$. $\text{Val}(\mu(t, r))$ is the *weight* of $r$ on $t$. $\text{Val}(\mu(t, r)) = \mathbf{0}$ if $r$ is not valid.

**Definition 4.12** *The behavior of a WTA M is the function $\text{bhv}_M : T_\Sigma \to D$ defined by*

$$\text{bhv}_M(t) = \sum (\text{Val}(\mu(t, r)) | r \in \text{succ}(M, t))$$

*for all $t \in T_\Sigma$. If no successful run on t exists, then $\text{bhv}_M(t) = \mathbf{0}$.*

A *(formal) tree series* is a mapping $S : T_\Sigma \to D$. A tree series $S$ is called *recognizable* if $S = \text{bhv}_M$ for some WTA $M$.

#### 4.4.2.1 Fragments of WMSOL

In Droste and Meinecke (2010) and Droste et al. (2011a) the following fragments of the WMSOL were introduced.

**Definition 4.13** *An almost Boolean formula is a WMSOL formula consisting of finitely many conjunctions and disjunctions of Boolean formulas and elements of D.*
*A $\forall$-restricted WMSOL formula $\psi$ is almost Boolean for each subformula $\forall \psi$ occurring in $\phi$.*

Let $\text{const}(\phi)$ be the set of all $d \in D$ occurring in $\phi$. Two subsets $D_1, D_2 \subseteq D$ commute if $d_1 \Diamond d_2 = d_2 \Diamond d_1$ for all $d_1 \in D_1, d_2 \in D_2$.

**Definition 4.14** *Whenever a strongly $\wedge$-restricted WMSOL formula $\phi$ contains a subformula $\phi_1 \wedge \phi_2$, either both $\phi_1$ and $\phi_2$ are almost Boolean or $\phi_1$ or $\phi_2$ is Boolean.*
*Whenever a $\wedge$-restricted WMSOL formula $\phi$ contains a subformula $\phi_1 \wedge \phi_2$, $\phi_1$ is almost Boolean or $\phi_2$ is Boolean.*
*Whenever a commutatively $\wedge$-restricted WMSOL formula $\phi$ contains a subformula $\phi_1 \wedge \phi_2$, $\phi_1$ is almost Boolean or $\text{const}(\phi_1)$ and $\text{const}(\phi_2)$ commute.*

Obviously, each strongly $\wedge$-restricted WMSOL formula is $\wedge$-restricted. If subformula $\phi_2$ of $\phi_1 \wedge \phi_2$ is Boolean, then $\text{const}(\phi_2) = \emptyset$, so $\text{const}(\phi_1)$ and $\text{const}(\phi_2)$ commute. Hence, each $\wedge$-restricted WMSOL formula is *commutatively $\wedge$-restricted*.

#### 4.4.2.2 Restrictions of WTA

In Droste and Meinecke (2010) and Droste et al. (2011a) the following properties of ptv-monoids were defined.

**Definition 4.15** *A ptv-monoid D is called* regular *if, for all $d \in D$ and all ranked alphabets $\sum$, there exists WTA $M_d$ with behavior $\mathrm{bhv}_{M_d}$ such that $\mathrm{bhv}_{M_d} = d$ for each $t \in T_{\sum}$.*

**Definition 4.16** *A* left-$\Diamond$-distributive *ptv-monoid D satisfies $d\Diamond(d_1 + d_2) = d\Diamond d_1 + d\Diamond d_2$ for all $d, d_1, d_2 \in D$.*
   A left-multiplicative *ptv-monoid D satisfies $d\Diamond\mathrm{Val}(t) = \mathrm{Val}(t')$ for all $d \in D, t, t' \in T_D$ with $\mathrm{dom}(t) = \mathrm{dom}(t')$, $t'(\epsilon) = d\Diamond t(\epsilon)$, and $t'(u) = t(u)$ for every $u \in \mathrm{dom}(t)\backslash\{\epsilon\}$.*
   A left-Val-distributive *ptv-monoid D satisfies $d\Diamond\mathrm{Val}(t) = \mathrm{Val}(t')$ for all $d \in D, t, t' \in T_D$ with $\mathrm{dom}(t) = \mathrm{dom}(t')$, $t'(u) = d\Diamond t(u)$ for every $u \in \mathrm{dom}(t)$.*

Note that each left-multiplicative or left-Val-distributive ptv-monoid is regular.

**Definition 4.17** *A ptv-monoid D is called* left-distributive *if it is:*

- *left-multiplicative or left-Val-distributive,*
- *and, moreover, left-$\Diamond$-distributive.*

**Definition 4.18** *A* right-$\Diamond$-distributive *ptv-monoid D satisfies $(d_1 + d_2)\Diamond d = d_1\Diamond d + d_2\Diamond d$ for all $d, d_1, d_2 \in D$.*
   *A* distributive *ptv-monoid D is both left- and right-distributive.*
   *In an* associative *ptv-monoid D, $\Diamond$ is associative.*

Note that if $D$ is associative, then $(D, +, \Diamond, \mathbf{0}, \mathbf{1})$ is a strong bimonoid. If, moreoverm $D$ is $\Diamond$-distributive, $(D, +, \Diamond, \mathbf{0}, \mathbf{1})$ is a semiring, called a *tree valuation semiring (tv-semiring)*.

**Definition 4.19** *A ptv-monoid D is* conditionally commutative *if $\mathrm{Val}(t_1)\Diamond\mathrm{Val}(t_2) = \mathrm{Val}(t)$ for all $t_1, t_2, t \in T_D$ with $\mathrm{dom}(t_1)=\mathrm{dom}(t_2)=\mathrm{dom}(t)$ and $\mathrm{im}(t_1)$ and $\mathrm{im}(t_2)$ commute and $t(u) = t_1(u)\Diamond t_2(u)$ for all $u \in \mathrm{dom}(t)$.*

**Definition 4.20** *A* cctv-semiring *is a conditionally commutative tv-semiring which is, moreover, left-multiplicative or left-Val-distributive.*

Obviously, each cctv-semiring is left-distributive. The main theorem of Droste et al. (2011a) states:

**Theorem 4.1** *Let $S : T_{\sum} \to D$ be a tree series.*

- *If D is* regular, *then S is recognizable iff S is definable by a $\forall$-restricted and strongly $\wedge$-restricted WMSOL sentence $\phi$.*
- *If D is* left-distributive, *then S is recognizable iff S is definable by a $\forall$-restricted and $\wedge$-restricted WMSOL sentence $\phi$.*
- *If D is a* cctv-semiring, *then S is recognizable iff S is definable by a $\forall$-restricted and commutatively $\wedge$-restricted WMSOL sentence $\phi$.*

Note that $D$ must be regular. Otherwise, there is at least one $d \in D$ without a WTA recognizing $d$, and hence the semantics of the $\forall$-restricted and strongly $\forall$-restricted sentence $d$ is not recognizable.

### 4.4.3   Expressive Power of Weighted Monadic Second-Order Logic

Weighted monadic second-order logic and its fragments have considerable expressive power. Lots of optimization problems and counting problems are expressible in WMSOL – see multiple examples in Droste and Gastin (2007), Droste and Vogler (2009), Mandrali and Rahonis (2009), Bollig and Gastin (2009), and Droste and Meinecke (2010). However, the larger the particular fragment gets, the more restrictions on the underlying ptv-monoid we need.

## 4.5   Modeling Computational Units as Logical Structures

In this section, we describe how we use FSMs, as introduced by Courcelle and Walukiewicz (1995), in order to model (the behavior of) computational units (both hardware and software). In computer science, the most popular ways to model computational units are FSMs, Kripke models, Petri nets, etc. For hardware units, an FSM is even effectively extractable from the hardware descriptive language code of the design (Jou and Liu, 1999b).

We use the modeling in order to present computational systems as logical structures, $\mathcal{A}$s, built from components. Each component is also presented as a logical structure $\mathcal{A}_i$ (unit $i$). The properties, constraints, and desired behavior of computational system are presented as logical formulae, $\phi$s. The formal presentation of the problem that we want to solve is:

> Is it possible to evaluate $\phi$ on $\mathcal{A}$, using some effectively derived formulae on $\mathcal{A}_i$, and (maybe) some additional manipulations?

In order to show how computation may be expressed as logical formulae, we take an example from Immerman (1989), where addition was shown to be expressible in FOL in the following way. Let $\tau = \langle A, B, k \rangle$ be a vocabulary consisting of two unary relations and a constant $k$. In a structure $\mathfrak{M}$ of the vocabulary, $A$ and $B$ are binary strings of length $k = |\mathfrak{M}|$. $\mathfrak{M}$ satisfies the additional property that if the $k$th bit of the sum of $A$ and $B$ is one,

$$\text{CARRY}(x) = (\exists y < x)[A(y) \wedge B(y)(\forall zy < z < x)A(z) \vee B(z)],$$
$$\text{PLUS}(x) = A(x) \oplus B(x) \oplus \text{CARRY}(x).$$

The last line expresses the addition property.

### 4.5.1   The Behavior of a Finite State Machine as Logical Structures

Now, we describe in great detail how FSMs may be used as models of units in a computational system. However, our approach works for *any* suitable modeling of units as logical structures. We use logical formulae over the logical structures to describe computations, properties, constraints, and the maintenance and management of the computational system. The FSM may be defined as a set of states and transitions between them in the following way.

**Definition 4.21   (Finite state machine)**
*(Courcelle and Walukiewicz, 1995)* $\mathcal{M}_{FSM} = \langle S \cup T, P_S, P_T, \mathbf{init}, \mathbf{src}, \mathbf{tgt}, P_1, \ldots, P_k \rangle$, *where:*

- *S and T are two sets, called the set of* states *and the set of* transitions, *respectively.*
- $P_S$ *is a unary predicate for the states.*
- $P_T$ *is a unary predicate for the transitions.*
- **init** *is a state called the* initial state.
- **src** *and* **tgt** *are two mappings from T to S that define respectively the* source *and the* target *of a transition.*
- $P_1, \ldots, P_k$ *are subsets of either S or T that specify properties of the states or the transitions, respectively.*

A state predicate $P_\ell$ *may denote the* Idle *state, when there is nothing for the FSM to do,* Accepting *states,* Rejecting *states,* Error *states, etc. The transition predicate* $P_{Ack}$ *may stand for the perception of an input* Acknowledgment *signal, and so on.*

An FSM is not really interesting as an object under consideration. Given an FSM, we are instead interested in different properties of its runs.

**Definition 4.22   (Paths of runs of an FSM)**
A path of a run *of FSM* $\mathcal{M}$ *is a finite sequence of transitions* $(t_1, t_2, \ldots, t_n)$ *such that the source of* $t_1$ *is the initial state and for every* $i = 1, \ldots, n-1$, $t_i = (s_{i_1}, s_{i_2}) \in \mathbf{T}$.
   **Paths**($\mathcal{M}$) *is the set of paths of runs of FSM* $\mathcal{M}$, *and* $\prec$ *is the* prefix order *on* **Paths**($\mathcal{M}$).[2]

Now, we define the notion of the *behavior* of an FSM.

**Definition 4.23   (Behavior of an FSM)**
*The* behavior *of an FSM* $\mathcal{M}$ *is a logical structure*

$$\mathbf{bhv}(\mathcal{M}) = \langle \mathbf{Paths}(\mathcal{M}), \prec_1, P_1^*, \ldots, P_k^* \rangle,$$

*where each* $P_i^*$ *is the property of a path saying that the last state of this path satisfies* $P_i$.[3]

### 4.5.2   The Expressive Power of MSOL on Finite State Machines

One of the ways to express the properties, constraints, and desired behavior of computational units is the use of *logical* formulae. Monadic second-order logic is an extension of FOL where quantification on unary relations is allowed. This logic has considerable expressive power for FSMs and their behavior. Most of the logics used in different applications of computer science are sublogics of it. Among these we have:

**Monadic fixed-point logic (with inflationary fixed points)** This logic is MSOL expressible on the FSM. fixed-point logic, in general, is definable in SOL.
**Propositional dynamic logic** (Harel, 1984) This is definable in monadic fixed-point logic on the FSM.

---

2  The prefix order here means the prefix order on strings, which is a special kind of substring relation. $p \prec_1 p'$ iff $p \prec p'$ and $p'$ has exactly one more transition than $p$.
3  In the definition, the prefix order $\prec_1$ generalizes the intuitive concept of a tree by introducing the possibility of continuous progress and continuous branching.

**Fixed-point definable operations on the power set algebra of trees** (Arnold and Niwiński, 1992) These are MSOL expressible on the FSM.

**Linear temporal logic (LTL)** This is already expressible in FOL on the behavior of the FSM as was shown by Burgess (1984).

**Computation tree logic** (Lichtenstein, Pnueli, and Zuck, 1985) This is in monadic fixed-point logic on the behavior of the FSM.

**Fragments of $\mu$ and $\nu$ calculus** The operations that can be defined without alternation on $\mu$ and $\nu$ (Arnold and Niwiński, 1992). Actually, they are definable in WMSOL on the behavior of the FSM.

Moreover, using powerful tools based on ideas related to Rabin's theorem on the decidability of MSOL on infinite trees, Rabin (1969) and Courcelle (1993, 1995) proved:

**Theorem 4.2  (Courcelle, 1995)**
*Every MSOL-expressible property of behavior of the FSM is equivalent to some MSOL-expressible property of the FSM.*

### 4.5.3 Parallel Runs of Computational Units with Message Passing

Let us consider two FSMs, FSM$^1$ and FSM$^2$, which can communicate via two one-way channels (message-passing parallel computation) – see Figure 4.3. Both asynchronous and synchronous communication may be modeled in this way. For synchronous communication, after state $P_\text{Send}$, the FSM goes to a state $P_\text{Idle}$ until perception of an input Ack signal is observed.

We assume that the vocabularies of the FSM$^i$ ($i \in \{1, 2\}$) in the tildeplest case contain at least two unary predicates, which indicate $P_\text{Send}$ and $P_\text{Receive}$ for communication states. Communication with the external environment is omitted for tildeplicity, and

$$\mathcal{M}^i_\text{FSM} = \langle S^i \cup T^i, P^i_S, P^i_T, \textbf{init}^i, \textbf{src}^i, \textbf{tgt}^i, P^i_\text{Receive}, P^i_\text{Send} P^i_1, \ldots, P^i_{k^i} \rangle.$$

The resulting FSM may be described as

$$\mathcal{M}_\text{FSM} = \langle S^1 \cup S^2 \cup T^1 \cup T^2 \cup \{(P^1_\text{Send}, P^2_\text{Receive}), (P^2_\text{Send}, P^1_\text{Receive})\},$$
$$\cup \{(\textbf{init}, \textbf{init}^1), (\textbf{init}, \textbf{init}^2)\}, P^1_S, P^1_T, P^2_S, P^2_T, P_T,$$
$$\textbf{init}, \textbf{src}, \textbf{tgt}, \textbf{init}^1, \textbf{src}^1, \textbf{tgt}^1, \textbf{init}^2, \textbf{src}^2, \textbf{tgt}^2,$$
$$P^1_\text{Receive}, P^1_\text{Send}, \ldots, P^2_\text{Receive}, P^2_\text{Send} \rangle.$$

**Figure 4.3** Composition of two FSMs.

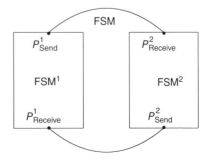

Here:

- The new added transitions are $(P_{Send}^1, P_{Receive}^2)$, $(P_{Send}^2, P_{Receive}^1)$ and $(\mathbf{init}, \mathbf{init}^1)$, $(\mathbf{init}, \mathbf{init}^2)$.
- The initial state of $\mathcal{M}_{FSM}$ is labeled by **init**.
- The unary predicate for the new added transitions is $P_T$.
- The new functions **src** and **tgt** map the new transitions to their sources and targets.

## 4.6 Parallel Distributed Solution of a Partial Differential Equation

We now give an example of the use of an FSM for modeling the behavior of computational systems. Let us consider a domain decomposition method for solving one of the most classical partial differential equations, presented in Ravve and Volkovich (2013):

$$\frac{\partial^2 u}{\partial t^2} - \frac{\partial^2 u}{\partial x^2} = 0$$

with the following conditions:

$$t \geq 0, \quad 0 \leq x \leq 1,$$

$$\begin{cases} u(0, t) = 0, \\ u(1, t) = 0, \\ u(x, 0) = f(x), \\ u_t(x, 0) = 0. \end{cases}$$

### 4.6.1 Decomposition of the Domain

In order to receive the numerical solution of the equation, we divide the domain into $N$ intervals and use final differences as an approximation of derivatives. The approximate solution in point number $j$ can be obtained using iterations ($n$ denotes the number of the iteration; see Figure 4.4):

$$U_j^n \approx u(j \cdot \Delta x, n \cdot \Delta t), \text{ where}$$

$$j = 1, \ldots, N - 1; \quad n = 1, 2, 3, \ldots; \quad \Delta x = \frac{1}{N}; \quad \mu = \frac{\Delta t}{\Delta x}.$$

We finish the computation when the solutions at iterations $n$ and $n + 1$ are almost the same, i.e. a *convergence* condition holds. The approximate solution $U_j^{n+1}$ for point $j$ at iteration $n + 1$ can be obtained with the help of the following formula that uses the values for three points, $j - 1$, $j$, and $j + 1$, from the previous iteration, and the value for point $n$ at iteration $n - 1$:

$$U_j^{n+1} = 2 \cdot (1 - \mu^2) \cdot U_j^n + \mu^2 \cdot (U_{j+1}^n + U_{j-1}^n) - U_j^{n-1}.$$

The boundary values and the initial values are defined in the following way:

$$\begin{cases} U_0^n = 0, \\ U_N^n = 0, \\ U_j^0 = f(j \cdot \Delta x), \\ U_j^{-1} = U_j^0. \end{cases}$$

**Figure 4.4** Domain decomposition.

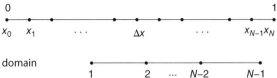

**Figure 4.5** Computation Scheme for the domain decomposition method.

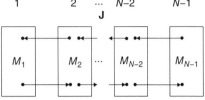

The computation domain can be divided among (up to) $N - 1$ elementary computation elements (FSMs) such that each one computes $U_j^{n+1}$ or some subset of the values (see Figure 4.5). Each FSM $\mathcal{M}_j$ ($j \neq 1$, $j \neq N - 1$) communicates with two neighbors, which process $U_{j+1}^n$ and $U_{j-1}^n$. If there are not enough FSMs then each one processes more than one point. Pseudocode for one iteration of an FSM that executes the computation of *MyValueNew* for point $j = MyTaskId$, using values from the right and left neighbors as well as *MyValue* from iteration $n$ and *MyValueOld* from iteration $n - 1$, looks like:

> Identify the left and the right neighbors
> if *MyTaskId* = first then use the boundary condition
> if *MyTaskId* = last then use the boundary condition
> *LeftNeighbor* = *MyTaskId* − 1
> *RightNeighbor* = *MyTaskId* + 1
> Send *MyValue* to the *LeftNeighbor* FSM
> Receive *LeftValue* from the *RightNeighbor* FSM
> Send *MyValue* to the *RightNeighbor*
> Receive *RightValue* from the *LeftNeighbor* FSM
> Calculate *MyValueNew* = $2 \cdot (1 - \mu^2) \cdot$ *MyValue* + $\mu^2 \cdot$ (*RightValue* + *LeftValue*) − *MyValueOld*
> Check and report the convergence condition

Assume that each $\mathcal{M}_j$ is a finite state machine FSM$_j$, as defined above, with the following additional predicates:

- $P_{\text{done}_j}$ defines the successful termination of the computation.
- $P_{\text{error}_j}$ defines the failure of the computation.
- $P_{\text{out-left}_j}, P_{\text{out-right}_j}, P_{\text{in-left}_j}, P_{\text{in-right}_j}$ define the communication ports to the neighbors from two sides.

$$\mathcal{M}_j = \langle S_j \cup T_j, P_{S_j}, P_{T_j}, \mathbf{init}_j, P_{S_j}, P_{T_j}, \mathbf{src}_j, \mathbf{tgt}_j, P_{\text{done}_j},$$
$$P_{\text{error}_j}, P_{\text{out-left}_j}, P_{\text{out-right}_j}, P_{\text{in-left}_j}, P_{\text{in-right}_j}, P_{1_j}, \dots, P_{k_j} \rangle.$$

In addition:

- In $\mathcal{M}_1, P_{\text{out-left}_1} = \emptyset$ and $P_{\text{in-left}_1} = \emptyset$.
- In $\mathcal{M}_{N-1}, P_{\text{out-right}_{N-1}} = \emptyset$ and $P_{\text{in-right}_{N-1}} = \emptyset$.

In such a case, the complete parallel computation system with communication via the channels is presented as:

$$\mathcal{M} = \langle \bigcup_{j\in J} S_j \bigcup_{j\in J} T_j \cup L_{\text{left-right}} \cup L_{\text{right-left}} \cup$$

$$(\textbf{init}, \textbf{init}_1) \cup \cdots \cup (\textbf{init}, \textbf{init}_{n-1}), \textbf{init}, \textbf{init}_1, \ldots, \textbf{init}_{N-1},$$

$$P_{S_1}, \ldots, P_{S_{N-1}}, P_{T_1}, \ldots, P_{T_{N-1}}, P_T,$$

$$\textbf{src}_1, \ldots, \textbf{src}_{N-1}, \textbf{src}, \textbf{tgt}_1, \ldots, \textbf{tgt}_{N-1}, \textbf{tgt},$$

$$P_{\text{done}_1}, \ldots, P_{\text{done}_{N-1}}, P_{\text{done}}, P_{\text{error}_1}, \ldots, P_{\text{error}_{N-1}}, P_{\text{error}},$$

$$P_{\text{out-left}_1}, \ldots, P_{\text{out-left}_{N-1}}, P_{\text{out-right}_1}, \ldots, P_{\text{out-right}_{N-1}},$$

$$P_{\text{in-left}_1}, \ldots, P_{\text{in-left}_{N-1}}, P_{\text{in-right}_1}, \ldots, P_{\text{in-right}_{N-1}}, \ldots \rangle,$$

where $J = \{1, 2, \ldots, N-1\}$ is an index set, and

- **init** is the initial state of the computation;
- $L_{\text{left-right}}(a, b) = 1$ iff $h(a) = h(b) + 1$ and $P_{\text{out-right}_{h(a)}}(a)$ and $P_{\text{in-left}_{h(b)}}(b)$ and $\neg(h(a) \approx (N-1))$;
- $L_{\text{right-left}}(a, b) = 1$ iff $h(a) = h(b) - 1$ and $P_{\text{out-right}_{h(b)}}(b)$ and $P_{\text{in-left}_{h(a)}}(a)$ and $\neg(h(b) \approx 1)$;
- **src** and **tgt** map new added transitions $L_{\text{left-right}}$, $L_{\text{right-left}}$ and $(\textbf{init}, \textbf{init}_1)$, ..., $(\textbf{init}, \textbf{init}_{n-1})$ to their sources or targets.

Here we use the function $h : \mathcal{M} \mapsto J$, $h(s) = j$ iff $s \in S_j$ or $h(t) = j$ iff $t \in T_j$.

### 4.6.2 Incremental Evaluation of Properties

Now we show a method of incremental evaluation of properties of behavior of FSMs.

#### 4.6.2.1 Verifying Convergence of the Iterative Computation
In order to determine the completion of our iterative computation we use:

**Convergence:** results of the iterations approach the analytical solution;
**Stability:** errors at one step of the iterations do not cause increasingly large errors as the computations are continued;
**Consistency:** the truncation error of the approximation tends to zero as the mesh length does.

It is sometimes possible to approximate a parabolic or hyperbolic equation with a finite difference scheme that is stable but which does not converge to the solution of the differential equation as the mesh lengths tend to zero. Such a scheme is called inconsistent or incompatible. Consistency alone also does not guarantee that the solution of the difference equations approximates the exact solution of the partial differential equation. However, consistency and stability imply convergence. More precisely: $O(\Delta x^{\ell})$ local truncation error and stability imply $O(\Delta x^{\ell})$ global error, where $\Delta x$ is the mesh width.

We finish the computation when the solutions at iterations $n$ and $n+1$ are almost the same, i.e. there exists a formula $\varphi_{\text{Convergence}}$ on $\textbf{bhv}(\mathcal{M})$ that expresses the convergence condition. The formula is written in a tildeilar way to $\varphi_{\text{PLUS}}(x)$ from Immerman (1989), where we use $\varphi_{\text{MINUS}}(x)$ rather than $\varphi_{\text{PLUS}}(x)$, $R_<$, and a constant $c_{\epsilon}$. Recall that the behavior of our $\mathcal{M}$ was defined as $\textbf{bhv}(\mathcal{M}) = \langle \textbf{Paths}(\mathcal{M}), \prec_1, P_1^*, \ldots, P_k^* \rangle$, where $P_i^*$ is the property of a path saying that the last state of this path satisfies $P_i$. However, in order to satisfy $\varphi_{\text{Convergence}}$ on $\mathbf{bhv}(\mathcal{M})$ we must satisfy it on all $\mathbf{bhv}(\mathcal{M}_j)$, $j \in J$. Indeed, $\mathbf{bhv}(\mathcal{M}) \vDash \varphi_{\text{Convergence}}$ iff $\mathbf{bhv}(\mathcal{M}_1) \vDash \varphi_{\text{Convergence}}$ and ... $\mathbf{bhv}(\mathcal{M}_j) \vDash \varphi_{\text{Convergence}}$ ... and $\mathbf{bhv}(\mathcal{M}_{N-1}) \vDash \varphi_{\text{Convergence}}$. We note that $\varphi_{\text{Convergence}} \in \text{FOL}(\tau_{\mathbf{bhv}(\mathcal{M})})$, and

- the corresponding Boolean function is defined as

$$F_{\Phi, \varphi_{\text{Convergence}}}(b_{\text{Convergence}_1}, \dots, b_{\text{Convergence}_{N-1}}) = \bigwedge_{j \in J} b_{\text{Convergence}_j};$$

- $b_{\text{Convergence}_j} = 1$ iff $\mathbf{bhv}(\mathcal{M}_j) \vDash \varphi_{\text{Convergence}}$.

Now we apply Theorem 4.2, and translate the FOL-expressible properties of behaviors of the FSMs ($\varphi_{\text{Convergence}}$) to some MSOL-expressible properties of the FSMs themselves. This means that we receive $\phi_{\text{Convergence}_j} \in \text{MSOL}(\tau_{\mathcal{M}_j})$ such that $\mathbf{bhv}(\mathcal{M}_j) \vDash \varphi_{\text{Convergence}}$ iff $\mathcal{M}_j \vDash \phi_{\text{Convergence}}$. In this way we have managed to reformulate the FOL-definable convergence condition $\mathbf{bhv}(\mathcal{M}) \vDash \varphi_{\text{Convergence}}$ to a set of MSOL formulas on $\mathcal{M}_j, j \in J$.

### 4.6.2.2 Verifying Safety and Liveness of the Iterative Computation
The two basic properties that describe the legal behavior of a computation unit are safety and liveness. The meanings of these properties are:

**Safety:** nothing bad happens, i.e. error/STOP state is not reachable.
**Liveness:** something good eventually happens, i.e. the computation is eventually completed (done).

The properties describe the *behavior* of our $\mathcal{M}$. In the vocabulary $\mathbf{bhv}(\mathcal{M})$ we emphasize each time the relevant predicates and hide with "..." others that are relevant to other cases. In our case,

$$\mathbf{bhv}(\mathcal{M}) = \langle \mathbf{Paths}(\mathcal{M}), P_{\text{init}}, P_{\text{init}_1}, \dots, P_{\text{init}_{N-1}},$$
$$P_{\text{done}}, P_{\text{done}_1}, \dots, P_{\text{done}_{N-1}},$$
$$P_{\text{error}}, P_{\text{error}_1}, \dots, P_{\text{error}_{N-1}},$$
$$P_{\text{out-left}_1}, \dots, P_{\text{out-left}_{N-1}}, P_{\text{out-right}_1}, \dots, P_{\text{out-right}_{N-1}},$$
$$P_{\text{in-left}_1}, \dots, P_{\text{in-left}_{N-1}}, P_{\text{in-right}_1}, \dots, P_{\text{in-right}_{N-1}} \rangle.$$

Our safety and liveness properties are definable on $\mathbf{bhv}(\mathcal{M})$ as follows:

$$\phi_{\text{Safety}} = \neg \exists p \bigvee_{j \in J} P_{\text{error}_j}(p);$$
$$\phi_{\text{Liveness}} = \exists p_1 \dots \exists p_{N-1} \bigwedge_{j \in J} P_{\text{done}_j}(p_j).$$

We note that both $\phi_{\text{Safety}}$ and $\phi_{\text{Liveness}}$ are FOL formulas on the vocabulary of $\mathbf{bhv}(\mathcal{M})$. Moreover, their evaluation may be easily reduced to evaluation of the derived formulas on the behaviors of the components of $\mathcal{M}$. Indeed, $\mathbf{bhv}(\mathcal{M}) \vDash \psi_{\text{Safety}}$ iff $\mathbf{bhv}(\mathcal{M}_1) \vDash \psi_{\text{Safety}_1} = \neg \exists p P_{\text{error}_1}(p)$ and $\mathbf{bhv}(\mathcal{M}_j) \vDash \psi_{\text{Safety}_j} = \neg \exists p P_{\text{error}_j}(p)$ and $\mathbf{bhv}(\mathcal{M}_{N-1}) \vDash \psi_{\text{Safety}_{N-1}} = \neg \exists p P_{\text{error}_{N-1}}(p)$. We note that in our case, $\phi_{\text{Safety}} \in \text{FOL}(\tau_{\mathbf{bhv}(\mathcal{M})})$, and

- the relevant Boolean function is defined as $F_{\Phi, \phi_{\text{Safety}}}(b_{\text{Safety}_1}, \dots, b_{\text{Safety}_{N-1}}) = \bigwedge_{j \in J} b_{\text{Safety}_j}$;
- $b_{\text{Safety}_j} = 1$ iff $\mathbf{bhv}(\mathcal{M}_j) \vDash \psi_{\text{Safety}_j}$;
- $\psi_{\text{Safety}_1}, \dots, \psi_{\text{Safety}_{N-1}}$ are FOL formulas on $\mathbf{bhv}(\mathcal{M}_j)$.

In the case of $\phi_{\text{Liveness}}(p_1, \ldots, p_{N-1})$, we have $\mathbf{bhv}(\mathcal{M}) \vDash \psi_{\text{Liveness}}$ iff $\mathbf{bhv}(\mathcal{M}_1) \vDash \psi_{\text{Liveness}_1} = \exists p P_{\text{done}_1}(p)$ and ... $\mathbf{bhv}(\mathcal{M}_j) \vDash \psi_{\text{Liveness}_j} = \exists p P_{\text{done}_j}(p)$ ... and $\mathbf{bhv}(\mathcal{M}_{N-1}) \vDash \psi_{\text{Liveness}_{N-1}} = \exists p P_{\text{done}_{N-1}}(p)$. In fact, $\phi_{\text{Liveness}} \in \text{FOL}(\tau_{\mathbf{bhv}(\mathcal{M})})$, and

- the relevant Boolean function is defined tildeilarly as
$$F_{\Phi, \phi_{\text{Liveness}}}(b_{\text{Liveness}_1}, \ldots, b_{\text{Liveness}_{N-1}}) = \wedge_{j \in J} b_{\text{Liveness}_j};$$
- $b_{\text{Liveness}_j} = 1$ iff $\mathbf{bhv}(\mathcal{M}_j) \vDash \psi_{\text{Liveness}_j}$;
- $\psi_{\text{Liveness}_1}, \ldots, \psi_{\text{Liveness}_{N-1}}$ are FOL formulas on $\mathbf{bhv}(\mathcal{M}_j)$.

Now, according to Theorem 4.2, the FOL-expressible properties of behaviors of the FSMs are equivalent to some MSOL-expressible properties of the FSMs themselves. This means that there exist $\zeta_{\text{Safety}_j} \in \text{MSOL}(\tau_{\mathcal{M}_j})$ and $\zeta_{\text{Liveness}_j} \in \text{MSOL}(\tau_{\mathcal{M}_j})$ such that $\mathbf{bhv}(\mathcal{M}_j) \vDash \psi_{\text{Safety}_j}$ iff $\mathcal{M}_j \vDash \zeta_{\text{Safety}_j}$, and $\mathbf{bhv}(\mathcal{M}_j) \vDash \psi_{\text{Liveness}_j}$ iff $\mathcal{M}_j \vDash \zeta_{\text{Liveness}_j}$.

In this way, we have managed to reduce the verification of the FOL-definable claims that $\mathbf{bhv}(\mathcal{M}) \vDash \psi_{\text{Safety}}$ and $\mathbf{bhv}(\mathcal{M}) \vDash \psi_{\text{Liveness}}$ to verification of a set of MSOL formulas on $\mathcal{M}_j$, $j \in J$. Note, that we treated our example manually.

Our approach guarantees that there exists an algorithm that leads to equivalent results.

However, the order of our actions would be a bit different.

## 4.7 Disjoint Union and Shuffling of Structures

The first reduction technique that we use is *Feferman–Vaught reductions*. A Feferman—Vaught reduction sequence (or tildeply, reduction) is a set of formulae such that each such formula can be evaluated locally in some component or index structure. Next, from the local values received from the components, and possibly some additional information about the components, we compute the value for the given global formula. In the logical context, the reductions are applied to relational structure $\mathcal{A}$ distributed over different components with structures $\mathcal{A}_i, i \in I$. The reductions allow the formulae over $\mathcal{A}$ to be computed from formulae over the $\mathcal{A}_i$s and formulae over the index structure $\mathcal{I}$.

In this section, we start by discussing different ways of obtaining structures from components. The *disjoint union* of a family of structures is the tildeplest example of juxtaposing structures over index structure $\mathcal{I}$ with universe $I$, where none of the components are linked to each other. In such a case, the index structure $\mathcal{I}$ may be replaced by index set $I$.

**Definition 4.24 (Disjoint union)**
*Let $\tau_i = \langle R_1^i, \ldots, R_{j^i}^i \rangle$ be a vocabulary of structure $\mathcal{A}_i$. In the general case, the resulting structure is $\mathcal{A} = \bigsqcup_{i \in I} \mathcal{A}_i$, where*

$$\mathcal{A} = \langle I \cup \bigcup_{i \in I} A_i, P(\iota, v), \text{index}(x), R_j^I (1 \leq j \leq j^I), R_{j^i}^i (i \in I, 1 \leq j^i \leq j^i)) \rangle$$

*for all $i \in I$, and $P(i, v)$ is true iff element a came from $\mathcal{A}_i$; index$(x)$ is true iff $x$ came from $I$.*

**Definition 4.25  (Partitioned index structure)**
*Let $\mathcal{I}$ be an index structure over $\tau_{\text{ind}}$. $\mathcal{I}$ is called* finitely partitioned *into $\ell$ parts if there are unary predicates $I_\alpha$, $\alpha < \ell$, in the vocabulary $\tau_{\text{ind}}$ of $\mathcal{I}$ such that their interpretation forms a partition of the universe of $\mathcal{I}$.*

Using *Ehrenfeucht–Fraïssé* games for MSOL (Ehrenfeucht, 1961), it is easy to see the following.

**Theorem 4.3**  *Let $\mathcal{I}, \mathcal{J}$ be two (not necessary finitely) partitioned index structures over the same vocabulary such that for $i, j \in I_\ell$ and $i', j' \in J_\ell$, $\mathfrak{A}_i$ and $\mathfrak{A}_j$ ($\mathfrak{B}_{i'}$ and $\mathfrak{B}_{j'}$) are isomorphic.*

- *If $\mathcal{I} \equiv_{\text{MSOL}}^{n} \mathcal{J}$ and $\mathfrak{A}_i \equiv_{\text{MSOL}}^{n} \mathfrak{B}_i$, then $\bigcup_{i \in I} \mathfrak{A}_i \equiv_{\text{MSOL}}^{n} \bigcup_{j \in J} \mathfrak{B}_j$.*
- *If $\mathcal{I} \equiv_{\text{MSOL}}^{n} \mathcal{J}$ and $\mathfrak{A}_i \equiv_{\text{FOL}}^{n} \mathfrak{B}_i$, then $\bigcup_{i \in I} \mathfrak{A}_i \equiv_{\text{FOL}}^{n} \bigcup_{j \in J} \mathfrak{B}_j$.*

If, as in most of our applications, there are only finitely many different components, we can prove a stronger statement, dealing with formulae rather than theories. The following holds:

**Theorem 4.4**  *Let $\mathcal{I}$ be a finitely partitioned index structure. Let $\mathcal{A} = \bigsqcup_{i \in I} \mathcal{A}_i$ be a $\tau$-structure, where each $\mathcal{A}_i$ is isomorphic to some $\mathcal{B}_1, \ldots, \mathcal{B}_\ell$ over the vocabularies $\tau_1, \ldots, \tau_\ell$, in accordance with the partition ($\ell$ is the number of classes). For every $\phi \in \text{MSOL}(\tau)$, there are:*

- *a Boolean function $F_\phi(b_{1,1}, \ldots, b_{1,j_1}, \ldots, b_{\ell,1}, \ldots, b_{\ell,j_\ell}, b_{I,1}, \ldots, b_{I,j_I})$;*
- *MSOL formulae $\psi_{1,1}, \ldots, \psi_{1,j_1}, \ldots, \psi_{\ell,1}, \ldots, \psi_{\ell,j_\ell}$;*
- *MSOL formulae $\psi_{I,1}, \ldots, \psi_{I,j_I}$*

*such that for every $\mathcal{A}$, $\mathcal{I}$, and $\mathcal{B}_i$ as above with $\mathcal{B}_i \vDash \psi_{i,j}$ iff $b_{i,j} = 1$ and $\mathcal{B}_I \vDash \psi_{I,j}$ iff $b_{I,j} = 1$, we have*

$$\mathcal{A} \vDash \phi \text{iff } F_\phi(b_{1,1}, \ldots, b_{1,j_1}, \ldots, b_{\ell,1}, \ldots, b_{\ell,j_\ell}, b_{I,1}, \ldots, b_{I,j_I}) = 1.$$

*Moreover, $F_\phi$ and the $\psi_{i,j}$ are computable from $\phi$, $\ell$, and the vocabularies alone, but the number of $\psi_{i,j}$ is tower exponential in the quantifier rank of $\phi$.*[4]

*Proof:* The proof is classical; see, in particular, Chang and Keisler (1990).

We now introduce an abstract preservation property of the *XX*-combination of logics $\mathcal{L}_1, \mathcal{L}_2$, denoted by *XX*-PP$(\mathcal{L}_1, \mathcal{L}_2)$. *XX* may mean, for example, disjoint union. The property says roughly that if two *XX*-combinations of structures $\mathfrak{A}_1, \mathfrak{A}_2$ and $\mathfrak{B}_1, \mathfrak{B}_2$ satisfy the same sentences of $\mathcal{L}_1$ then the disjoint unions $\mathfrak{A}_1 \bigsqcup \mathfrak{A}_2$ and $\mathfrak{B}_1 \bigsqcup \mathfrak{B}_2$ satisfy the same sentences of $\mathcal{L}_2$.

The reason we look at this abstract property is that it can be proven for various logics using their associated pebble games. The proofs usually depend in detail on the

---

4  However, in practical applications, the complexity is, as a rule, tildeply exponential.

particular pebble games. However, the property $XX$-PP$(\mathcal{L}_1, \mathcal{L}_2)$ and its variants play an important role in our development of the Feferman–Vaught-style theorems. This abstract approach was initiated by Feferman and Vaught (1959) and further developed by Makowsky (1978, 1985).

We now spell out various ways in which the theory of a disjoint union depends on the theory of the components. We first look at the case where the index structure is fixed.

**Definition 4.26**    (**Preservation properties with a fixed index set**)
*For two logics $\mathcal{L}_1$ and $\mathcal{L}_2$ we define:*
Disjoint pair:

**Input of operation:** *Two structures.*
**Preservation property:** *If two pairs of structures $\mathfrak{A}_1$, $\mathfrak{A}_2$ and $\mathfrak{B}_1$, $\mathfrak{B}_2$ satisfy the same sentences of $\mathcal{L}_1$ then the disjoint unions $\mathfrak{A}_1 \bigsqcup \mathfrak{A}_2$ and $\mathfrak{B}_1 \bigsqcup \mathfrak{B}_2$ satisfy the same sentences of $\mathcal{L}_2$.*
**Notation:** P-PP$(\mathcal{L}_1, \mathcal{L}_2)$.

Disjoint union:

**Input of operation:** *Indexed set of structures.*
**Preservation property:** *If, for each $i \in I$ (index set), $\mathfrak{A}_i$ and $\mathfrak{B}_i$ satisfy the same sentences of $\mathcal{L}_1$, then the disjoint unions $\bigsqcup_{i \in I} \mathfrak{A}_i$ and $\bigsqcup_{i \in I} \mathfrak{B}_i$ satisfy the same sentences of $\mathcal{L}_2$.*
**Notation:** DJ-PP$(\mathcal{L}_1, \mathcal{L}_2)$.

The *disjoint union* of a family of structures is the tildeplest example of juxtaposing structures where none of the components are linked to each other. Another way of producing a new structure from several given structures is by mixing (shuffling) structures according to a (definable) prescribed way along the index structure.

**Definition 4.27**    (**Shuffle over partitioned index structure**)
*Let $\mathcal{I}$ be an index structure partitioned into $\beta$ parts using unary predicates $I_\alpha$, $\alpha < \beta$. Let $\mathcal{A}_i, i \in I$ be a family of structures such that, for each $i \in I_\alpha$, $\mathcal{A}_i \cong B_\alpha$ holds, according to the partition. In this case, we say that $\bigsqcup_{i \in I} \mathcal{A}_i$ is the shuffle of $B_\alpha$ along the partitioned index structure $\mathcal{I}$, and denote it by $\biguplus_{\alpha < \beta}^{\mathcal{I}} B_\alpha$.*

Note that the shuffle operation, as defined here, is a special case of the disjoint union, and that the disjoint pair is a special case of the finite shuffle.

In the case of variable index structures and FOL, Feferman and Vaught observed that it is not enough to look at the FOL theory of the index structures, but one has to look at the FOL theories of expansions of the Boolean algebras PS$(\mathcal{I})$ and PS$(\mathcal{J})$, respectively, where PS$(X)$ denotes the power set of $X$. Gurevich (1979) suggested another approach, by looking at the MSOL theories of the structures $\mathcal{I}$ and $\mathcal{J}$. This is really the same, but more in the spirit of the problem, as the passage from $I$ to an expansion of PS$(\mathcal{I})$ remains on the semantic level, whereas the comparison of theories is syntactic. There is not much freedom in choosing the logic in which to compare the index structures, so we assume it is always MSOL.

**Definition 4.28**    (**Preservation properties with variable index structures**)
*For two logics $\mathcal{L}_1$ and $\mathcal{L}_2$ we define:*
Disjoint multiples:

**Input of operation:** *Structure and index structure.*

**Preservation property:** *Given two pairs of structures $\mathfrak{A}, \mathfrak{B}$ and $\mathcal{I}, \mathcal{J}$ such that $\mathfrak{A}, \mathfrak{B}$ satisfy the same sentences of $\mathcal{L}_1$ and $\mathcal{I}, \mathcal{J}$ satisfy the same MSOL sentences, then the disjoint unions $\bigsqcup_{i \in I} \mathfrak{A}$ and $\bigsqcup_{j \in J} \mathfrak{B}$ satisfy the same sentences of $\mathcal{L}_2$.*

**Notation:** Mult-PP$(\mathcal{L}_1, \mathcal{L}_2)$.

Shuffles:

**Input of operation:** *A family of structures $\mathfrak{B}_\alpha : \alpha < \beta$ and a (finitely) partitioned index structure $\mathcal{I}$ with $I_\alpha$ a partition.*

**Preservation property:** *Assume that for each $\alpha < \beta$ the pair of structures $\mathfrak{A}_\alpha, \mathfrak{B}_\alpha$ satisfy the same sentences of $\mathcal{L}_1$, and $\mathcal{I}, \mathcal{J}$ satisfy the same MSOL sentences; then the shuffles $\biguplus_{\alpha < \beta}^{I} \mathfrak{A}_\alpha$ and $\biguplus_{\alpha < \beta}^{J} \mathfrak{B}_\alpha$ satisfy the same sentences of $\mathcal{L}_2$.*

**Notation:** Shu-PP$(\mathcal{L}_1, \mathcal{L}_2)$ (FShu-PP$(\mathcal{L}_1, \mathcal{L}_2)$).

**Observation 4.1** *Assume that for two logics $\mathcal{L}_1$ and $\mathcal{L}_2$ we have the preservation property XX-PP$(\mathcal{L}_1, \mathcal{L}_2)$, and that $\mathcal{L}_1'$ is an extension of $\mathcal{L}_1$ and $\mathcal{L}_2'$ is a sublogic of $\mathcal{L}_2$; then XX-PP$(\mathcal{L}_1', \mathcal{L}_2')$ holds as well.*

**Observation 4.2** *For two logics $\mathcal{L}_1, \mathcal{L}_2$ the following implications between preservation properties hold:*

- *DJ-PP$(\mathcal{L}_1, \mathcal{L}_2)$ implies P-PP$(\mathcal{L}_1, \mathcal{L}_2)$, and, for fixed index structures, Mult-PP$(\mathcal{L}_1, \mathcal{L}_2)$, Shu-PP$(\mathcal{L}_1, \mathcal{L}_2)$, and FShu-PP$(\mathcal{L}_1, \mathcal{L}_2)$.*
- *For variable index structures we have that Shu-PP$(\mathcal{L}_1, \mathcal{L}_2)$ implies FShu-PP$(\mathcal{L}_1, \mathcal{L}_2)$ and Mult-PP$(\mathcal{L}_1, \mathcal{L}_2)$.*

**Definition 4.29** (Reduction sequence for shuffling)
*Let $\mathcal{I}$ be a finitely partitioned $\tau_{\mathrm{ind}}$-index structure, and $\mathcal{L}$ be logic. Let $\mathcal{A} = \biguplus_{\alpha < \beta}^{I} B_\alpha$ be the $\tau$-structure which is the finite shuffle of the $\tau_\alpha$-structures $B_\alpha$ over $\mathcal{I}$. The $\mathcal{L}_1$-reduction sequence for shuffling for $\phi \in \mathcal{L}_2(\tau_{\mathrm{shuffle}})$ is given by*

- *a Boolean function $F_\phi(b_{1,1}, \dots, b_{1,j_1}, \dots, b_{\beta,1}, \dots, b_{\beta,j_\beta}, b_{I,1}, \dots, b_{I,j_I})$,*
- *a set $\Upsilon$ of $\mathcal{L}_1$ formulae $\Upsilon = \{\psi_{1,1}, \dots, \psi_{1,j_1}, \dots, \psi_{\beta,1}, \dots, \psi_{\beta,j_\beta}\}$,*
- *and MSOL formulae $\psi_{I,1}, \dots, \psi_{I,j_I}$,*

*and has the property that for every $\mathcal{A}, \mathcal{I},$ and $B_\alpha$ as above with $B_\alpha \vDash \psi_{\alpha,j}$ iff $b_{\alpha,j} = 1$ and $B_I \vDash \psi_{I,j}$ iff $b_{I,j} = 1$, we have*

$$\mathcal{A} \vDash \phi \text{ iff } F_\phi(b_{1,1}, \dots, b_{1,j_1}, \dots, b_{\beta,1}, \dots, b_{\beta,j_\beta}, b_{I,1}, \dots, b_{I,j_I}) = 1.$$

*Note that we require that $F_\phi$ and the formulae $\psi_{\alpha,j}$ depend only on $\phi, \beta,$ and $\tau_1, \dots, \tau_\beta$, but not on the structures involved.*

**Remark 4.1** *If $\mathcal{I}$ is finite and fixed, then the MSOL formulae $\psi_{I,1}, \dots, \psi_{I,j_I}$ in the reduction sequences can be hidden in the function $F$.*

In many applications, we want to keep the quantifier rank $m = \mathrm{rank}(\phi)$ or the number of variables $k = var(\phi)$ of the formulae $\phi$ under control. Let $\mathcal{L}^{m,k}(\tau) \subseteq \mathcal{L}(\tau)$ be the set of formulae of quantifier rank $\leq m$ and the total number of variables be $\leq k$. Frequently, a property $XX$-PP$(\mathcal{L}_1, \mathcal{L}_2)$ also holds uniformly for subsets of formulae of bounded quantifier rank and total number of variables, i.e. $XX$-PP$(\mathcal{L}_1^{m_1,k_1}, \mathcal{L}_2^{m_2,k_2})$ holds, where $m_1$ is a tildeple function of $m_2$ and $k_1$ is a tildeple function of $k_2$. A strong form of such a uniform version of $XX$-PP$(\mathcal{L}_1^{m_1,k_1}, \mathcal{L}_2^{m_2,k_2})$ is given by Theorem 4.5.

For our various notions of combinations, preservation theorems can be proven by suitable pebble games, which are generalizations of Ehrenfeucht–Fraïssé games. This usually gives a version of $XX$-PP$(\mathcal{L}_1^{m_1,k_1}, \mathcal{L}_2^{m_2,k_2})$ for suitable chosen definitions of quantifier rank and counting of variables.

Usually, a close examination of the winning strategies for pebble games gives more: we can get an algorithm which, for each $\phi$, produces the corresponding reduction sequence. However, for several logics the preservation theorems can be proven directly, and the proofs give a transparent way to build the corresponding reduction sequence (Chang and Keisler, 1990). Such a proof for Theorem 4.3 is given in Ravve (2016); see also Monk (1976) and Chang and Keisler (1990). We now list which preservation properties hold for which logics.

**Theorem 4.5** *Let $\mathcal{I}$ be an index structure and $\mathcal{L}$ be any of FOL, $FOL^{m,k}$, $L_{\omega_1,\omega}^{\omega}$, $L_{\omega_1,\omega}^{k}$, $MSOL^m$, $MTC^m$, $MLFP^m$, or $FOL[Q]^{m,k}$ ($L_{\omega_1,\omega}[Q]^k$) with unary generalized quantifiers. Then DJ-PP$(\mathcal{L}, \mathcal{L})$ and FShu-PP$(\mathcal{L}, \mathcal{L})$ hold. This includes DJ-PP$(FOL^{m,k}, FOL^{m,k})$ and FShu-PP$(FOL^{m,k}, FOL^{m,k})$ with the same bounds for both arguments, and tildeilarly for the other logics.*

*Proof:*

**FOL and FOL$^{m,k}$:** The proofs for FOL and MSOL are classical – see, in particular, Chang and Keisler (1990). Extension for $FOL^{m,k}$ can be done directly from the proof for FOL.
**MLFP and MLFP$^m$:** The proof for MLFP was given in Bosse (1995).
$L_{\omega_1,\omega}(Q)^k$: The proof was given in Dawar and Hellat (1993).
**MTC$^m$:** The proof was given in Ravve, Volkovich, and Weber (2015a).

**Theorem 4.6** *Let $\mathcal{L}$ be any of FOL, $FOL^{m,k}$, $L_{\omega_1,\omega}^{\omega}$, $L_{\omega_1,\omega}^{k}$ $MSOL^m$, $MTC^m$, $MLFP^m$, or $FOL[Q]^{m,k}$ with unary generalized quantifiers. There is an algorithm which, for given $\mathcal{L}$, $\tau_{\text{ind}}$, $\tau_{\alpha}$, $\alpha < \beta$, $\tau_{\text{shuffle}}$, and $\phi \in \mathcal{L}(\tau_{\text{shuffle}})$ produces a reduction sequence for $\phi$ for $(\tau_{\text{ind}}, \tau_{\text{shuffle}})$ shuffling. However, $F_\phi$ and the $\psi_{\alpha,j}$ are tower exponential in the quantifier rank of $\phi$, and $F$ depends on the MSOL theory of the index structure restricted to the same quantifier rank as $\phi$.[5]*

*Proof:* By analyzing the proof of Theorem 4.5. A special case was analyzed in Gurevich (1979).

We finally show that our restriction to unary generalized quantifiers (MTC and MLFP) is necessary.

**Proposition 4.1** *Theorem 4.5 does not hold for 2-TC or 2-LFP.*

---

5 Note that this includes reduction sequences for disjoint pairs and disjoint multiples.

*Proof:* Let $I = \{0, 1\}$, and let the components be finite linear orders. Using a counting argument, it is easy to produce arbitrary large pairs of linear orders $\mathcal{A}_0$, $\mathcal{A}_1$ that are 2-TC-$m$ equivalent, but of different cardinalities. Now consider the structure $\mathcal{B}_0 = \mathcal{A}_0 \bigsqcup \mathcal{A}_0$ and $\mathcal{B}_1 = \mathcal{A}_0 \bigsqcup \mathcal{A}_1$. The 2-TC formula that distinguishes $\mathcal{B}_0$ from $\mathcal{B}_1$ is the formula $\theta$, which asserts that the two components have the same cardinality. $\theta$ can be written as

$$\text{2-TC} x_0, x_1, y_0, y_1; \text{first}_0, \text{first}_1, \text{last}_0, \text{last}_1 (\text{succ}_0(x_0, y_0) \wedge \text{succ}_1(x_1, y_1)),$$

where $\text{succ}_i$ is the FOL formula expressing the successor in the $i$th component, and $\text{first}_i$, $\text{last}_i$ are the constant symbols that are interpreted by the first last element, respectively, in the $i$th component.

We now discuss various ways of obtaining weighted labeled trees from components, as in Ravve, Volkovich, and Weber (2014).

**Definition 4.30    (Finite disjoint union of weighted labeled trees)**
*Let $\tau_i = \langle \text{label}^\tau_{a i}, \text{edge}^\tau_{i\,i} \rangle$ be a vocabulary of a weighted labeled tree $\mathcal{T}_i$ over $\sum$. In the general case, the tree over $\sum \cup I$ is*

$$\mathcal{T} = \bigsqcup_{i \in I} \mathcal{T}_i = \langle \bigcup_{i \in I} \mathcal{B}_i, D; \text{label}_i(u)(i \in I), \text{label}^\tau_{ai}(i \in I), \text{edge}^\tau_{i\,i}(\iota \in I) \rangle$$

*for all $i \in I$, where $\text{label}_i(u)$ is true iff $u$ came from $\mathcal{B}_i$, $I$ is finite, and each element in $I$ is of rank 1.*

The following theorem can be stated (Makowsky and Ravve, 1995; Ravve, Volkovich, and Weber, 2014):

**Theorem 4.7**    *Let $I$ be a finite index set with $\ell$ elements. Let $\mathcal{T} = \bigsqcup_{i \in I} \mathcal{T}_i$ be a weighted labeled tree. Then for every $\varphi \in \text{WMSOL}(\tau)$ over Boolean semirings, there are:*

- *a computation over weighted WMSOL formulae $F_\varphi(\varpi_{1,1}, \dots, \varpi_{1,j_1}, \dots, \varpi_{\ell,1}, \dots, \varpi_{\ell,j_\ell})$,*
- *WMSOL formulae $\psi_{1,1}, \dots, \psi_{1,j_1}, \dots, \psi_{\ell,1}, \dots, \psi_{\ell,j_\ell}$*

  *such that for every $\mathcal{T}_i$ and $I$ as above with $\varpi_{i,j} = \varrho_{i,j}$ iff $[\psi_{i,j}] = \varrho_{i,j}$, we have*

  $$[\varphi] = \varrho \text{ iff } F_\varphi(\varpi_{1,1}, \dots, \varpi_{1,j_1}, \dots, \varpi_{\ell,1}, \dots, \varpi_{\ell,j_\ell}) = \varrho.$$

*Moreover, $F_\varphi$ and the $\psi_{i,j}$ are computable from $\varphi$, $\ell$, and the vocabularies alone, but are tower exponential in the quantifier rank of $\varphi$.*

We also list some other options of commutative semirings to choose.

**Theorem 4.8**    *In addition, the following semirings satisfy Theorem 4.7:*

**Subset semiring:** $(PS(A), \cap, \cup, \emptyset, A)$. *Proof by analyzing and extension of the proof in Feferman and Vaught (1959).*
**Fuzzy semiring:** $([0, 1], \vee, \wedge, 0, 1)$. *Proof by analyzing and extension of the proof in Makowsky and Ravve (1995).*
**Extended natural number:** $(\mathbf{N} \cup \{\infty\}, +, \cdot, 0, 1)$. *Proof by analyzing and extension of the proof in Makowsky and Ravve (1995).*

**Tropical semiring:** $(\mathbf{R}_+ \cup \{+\infty\}, \min, +, +\infty, 0)$. *See Ravve, Volkovich, and Weber (2014).*

**Arctic semiring:** $(\mathbf{R}_+ \cup \{-\infty\}, \max, +, -\infty, 0)$. *Proof by analyzing and extension of the proof in Ravve, Volkovich, and Weber (2014).*

## 4.8 Syntactically Defined Translation Schemes

The second logical reduction technique that we use is *syntactically defined translation schemes*, which describe transformations of logical structures. The notion of abstract translation schemes goes back to Rabin (1965). They give rise to two induced maps, translations and transductions. Transductions describe the induced transformation of logical structures, and translations describe the induced transformations of logical formulae.

**Definition 4.31 (Translation schemes $\Phi$)**
*Let $\imath_1$ and $\imath_2$ be two vocabularies and $\mathcal{L}$ be a logic. Let $\tau_2 = \{R_1, \ldots, R_m\}$, and let $\rho(R_i)$ be the arity of $R_i$. Let $\Phi = \langle \varphi, \psi_1, \ldots, \psi_m \rangle$ be formulae of $\mathcal{L}(\tau_1)$. Then $\Phi$ is named $\kappa$-feasible for $\tau_2$ over $\tau_1$ if $\varphi$ has exactly $\kappa$ distinct free variables and each $\psi_i$ has $\kappa \rho(R_i)$ distinct free variables. Such a $\Phi = \langle \varphi, \psi_1, \ldots, \psi_m \rangle$ is also called a $\kappa - \tau_1 - \tau_2$ translation scheme or tildeply a translation scheme if the parameters are clear in the context. If $\kappa = 1$ we speak of scalar or non-vectorized translation schemes.*

The above definition as a rule assumes *one-sorted* logical structures. However, if we deal with weighted logics like WMSOL this is not the case. In general, if $\mathcal{L}$ is defined over particular kinds of $\mathcal{L}$ objects (like graphs, words, etc.) then the exact logical presentation of the objects must be explicitly provided. Weighted monadic second-order logic is defined over the weighted labeled trees, and we present them as logical structures in the following way.

**Definition 4.32 (Weighted labeled tree over a ptv-monoid $D$)**
*Given a ptv-monoid $D$, the weighted labeled tree over $D$ is the logical many-sorted structure $\mathcal{T} = \langle B, D; \text{label}_a, \text{edge}_i \rangle$, where:*

- *The universe of $\mathcal{T}$ is many sorted: $B$ is the tree domain and $D$ comes from the monoid.*
- *Relations of $\mathcal{T}$:*
  - *$\text{label}_a$ is a unary relation, which for each $a \in \sum$ means that $u \in B$ is labeled by $a$.*
  - *$\text{edge}_i$ is a binary relation, which for each $1 \le i \le \max_{\sum}$ and two $u_1, u_2 \in B$ means that $u_2$ is an immediate prefix of $u_1$.*

In the context of WMSOL, Definition 4.31 may be paraphrased as follows:

**Definition 4.33 (Translation schemes $\Phi_D$ on weighted labeled trees)**
*Let $\tau_1$ and $\tau_2$ be two vocabularies of weighted labeled trees. Let $\tau_1 = \langle \text{label}_a^{\tau_1}, \text{edge}_i^{\tau_1} \rangle$ over ptv-monoid $D$. Let $\Phi = \langle \phi_B, \phi_D; \psi_{\text{label}_a}, \psi_{\text{edge}_i} \rangle$ be almost Boolean WMSOL formulae (for each $a \in \sum$ and $1 \le i \le \max_{\sum}$). We say that $\Phi_D$ is feasible for $\tau_2$ over $\tau_1$ if:*

- $\phi_B$ has exactly one distinct free first-order variable over $B$;
- $\phi_D$ is a tautology with exactly one free variable over $D$;
- each $\psi_{\text{label}_a}$ has exactly one distinct free first-order variable over $B$;
- each $\psi_{\text{edge}_i}$ has exactly two distinct free first-order variables over $B$.

In general, Definition 4.31 must be adopted to the given logic $\mathcal{L}$, if it is not straightforward.

For $\mathcal{L}$ like FOL, MSOL, MTC, MLFP, or FOL with unary generalized quantifiers, with a translation scheme $\Phi$ we can *naturally* associate a (partial) function $\Phi^*$ from $\tau_1$ structures to $\tau_2$ structures.

### Definition 4.34 (The induced map $\Phi^*$)

Let $\mathcal{A}$ be a $\tau_1$ structure with universe $A$ and $\Phi$ be $\kappa$-feasible for $\tau_2$ over $\tau_1$. The structure $\mathcal{A}_\Phi$ is defined as follows:

- The universe of $\mathcal{A}_\Phi$ is the set $A_\Phi = \{\bar{a} \in A^\kappa : \mathcal{A} \vDash \varphi(\bar{a})\}$.
- The interpretation of $R_i$ in $\mathcal{A}_\Phi$ is the set

$$A_\Phi(R_i) = \{\bar{a} \in A_\Phi^{\rho(R_i)\cdot\kappa} : \mathcal{A} \vDash \psi_i(\bar{a})\}.$$

  Note that $\mathcal{A}_\Phi$ is a $\tau_2$ structure of cardinality at most $|A|^\kappa$.
- The partial function $\Phi^* : \text{Str}(\tau_1) \to \text{Str}(\tau_2)$ is defined by $\Phi^*(\mathcal{A}) = \mathcal{A}_\Phi$. Note that $\Phi^*(\mathcal{A})$ is defined iff $\mathcal{A} \vDash \exists \bar{x}\varphi$.

The case of WMSOL is a bit more tricky. The (partial) function $\Phi_D^*$ from $\tau_1$ trees to $\tau_2$ trees is defined as follows:

### Definition 4.35 (The induced map $\Phi_D^*$)

Let $\mathcal{T}^{\tau_1}$ be a $\tau_1$ tree and $\Phi_D$ be feasible for $\tau_2$ over $\tau_1$. The structure $\mathcal{T}^{\tau_2}_{\Phi_D}$ is defined as follows:

- The many-sorted universe $B, D$ are the sets:
  - $B_{\Phi_D} = \{u \in B^{\tau_1} : \mathcal{T}^{\tau_1} \vDash \phi_B(u)\}$;
  - $D_{\Phi_D} = D$.
- Relations $\text{label}_a^{\tau_2}$, $\text{edge}_i^{\tau_2}$: For each $a \in \Sigma$ and $1 \leq i \leq \max_\Sigma$,
  - the interpretation of each $\text{label}_a$ in $\mathcal{T}^{\tau_2}_{\Phi_D}$ is the set

$$\mathcal{T}^{\tau_2}_{\Phi_D}(\text{label}_a) = \{u \in B^{\tau_1} : \mathcal{T}^{\tau_1} \vDash \psi_{\text{label}_a}(u)\};$$

  - the interpretation of each $\text{edge}_i$ is the set of pairs

$$\mathcal{T}^{\tau_2}_{\Phi_D}(\text{edge}_i) = \{(u_1, u_2) \in {B^{\tau_1}}^2 : \mathcal{T}^{\tau_1} \vDash \psi_{\text{edge}_i}(u_1, u_2)\};$$

- $\Phi_D^*$: The partial function $\Phi_D^* : \text{Trees}(\tau_1) \to \text{Trees}(\tau_2)$ is defined by

$$\Phi_D^*(\mathcal{T}^{\tau_1}) = \mathcal{T}^{\tau_2}_{\Phi_D}.$$

Note that $\Phi_D^*(\mathcal{T}^{\tau_1})$ is defined iff $\mathcal{T}^{\tau_1} \vDash \phi_B(u)$.

Again, for $\mathcal{L}$ like FOL, MSOL, MTC, MLFP, or FOL with unary generalized quantifiers, with a translation scheme $\Phi$ we can also *naturally* associate a function $\Phi^\#$ from $\mathcal{L}(\tau_2)$ formulae to $\mathcal{L}(\tau_1)$ formulae.

**Definition 4.36** (**The induced map** $\Phi^\#$)
*Let $\theta$ be a $\tau_2$ formula and $\Phi$ be $\kappa$-feasible for $\tau_2$ over $\tau_1$. The formula $\theta_\Phi$ is defined inductively as follows:*

- *For $R_i \in \tau_2$ and $\theta = R(x_1, \dots, x_m)$ let $x_{j,h}$ be new variables with $i \leq m$ and $h \leq \kappa$, and define $\bar{x}_i = \langle x_{i,1}, \dots, x_{i,\kappa} \rangle$. We put $\theta_\Phi = \psi_i(\bar{x}_1, \dots, \bar{x}_m)$.*
- *For the Boolean connectives the translation distributes, i.e., if $\theta = (\theta_1 \vee \theta_2)$ then $\theta_\Phi = (\theta_{1\Phi} \vee \theta_{1\Phi})$, and if $\theta = \neg\theta_1$ then $\theta_\Phi = \neg\theta_{1\Phi}$; tildeilarly for $\wedge$.*
- *For the existential quantifier we use relativization, i.e., if $\theta = \exists y\theta_1$, let $\bar{y} = \langle y_1, \dots, y_\kappa \rangle$ be new variables. We put $\theta_\Phi = \exists\bar{y}(\varphi(\bar{y}) \wedge \theta_{1\Phi})$.*
- *For (monadic) second-order variables $U$ of arity $\ell$ ($\ell = 1$ for MSOL) and $\bar{v}$ a vector of length $\ell$ of first-order variables or constants, we translate $U(\bar{v})$ by treating $U$ like a relation symbol above and put*

$$\theta_\Phi = \exists V(\forall\bar{v}(V(\bar{v}) \rightarrow (\phi(\bar{v_1}) \wedge \dots \phi(\bar{v_\ell}) \wedge (\theta_1)_\Phi))).$$

- *For generalized quantifiers, if $\theta = Q_i v^1, v^2, \dots, v^m \theta_1(v^1, v^2, \dots, v^m, \dots)$, then let $\vec{v}^j = \langle v_1^j, \dots, v_k^j \rangle$ be new variables for $v^j$. We make*

$$\theta_\Phi = Q_i\bar{v}^1, \bar{v}^2, \dots, \bar{v}^m (\theta_1(\bar{v}^1, \bar{v}^2, \dots, \bar{v}^m, \dots)_\Phi).$$

- *For infinitary logics, if $\theta = \wedge\Psi$ then $\theta_\Phi = \wedge\Psi_\Phi$.*
- *For LFP, if $\theta = n\text{-}LFP\bar{x}, \bar{y}, \bar{u}, \bar{v}\theta_1$ then $\theta_\Phi = (n \cdot \kappa)\text{-}LFP\bar{x}, \bar{y}, \bar{u}, \bar{v}\theta_{1\Phi}$.*
- *For TC, if $\theta = n\text{-}TC\bar{x}, \bar{y}, \bar{u}, \bar{v}\theta_1$ then $\theta_\Phi = (n \cdot \kappa)\text{-}TC\bar{x}, \bar{y}, \bar{u}, \bar{v}\theta_{1\Phi}$.*
- *For weighted formulae over ptv-monoid $D$:*
  - *for $d$ we do nothing;*
  - *for a Boolean formula $\beta$ we put $\zeta_{\Phi_D} = \beta$;*
  - *for Boolean connectives and quantifiers the translation distributes.*
- *The function $\Phi^\# : \mathcal{L}(\tau_2) \rightarrow \mathcal{L}(\tau_1)$ is defined by $\Phi^\#(\theta) = \theta_\Phi$.*

Note that the case of weighted formulae over ptv-monoid $D$ is the most complicated in the definition. In general, given $\mathcal{L}$, Definition 4.36 must be adopted to all well-formed formulae of the logic.

**Observation 4.3** *If we use MSOL, and $\Phi^*$ is over MSOL too, and it is vectorized, then we do not obtain MSOL for $\mathcal{A}_\Phi$. However, in most feasible applications, we have that $\Phi^*$ is not vectorized, but not necessarily (Benedikt and Koch, 2009; Makowsky and Ravve, 2012).*

**Observation 4.4**
- $\Phi^\#(\theta) \in$ FOL *(SOL, TC, LFP) provided $\theta \in$ FOL (SOL, TC, LFP), even for vectorized $\Phi$.*
- $\Phi^\#(\theta) \in$ MSOL *provided $\theta \in$ MSOL, but only for scalar (non-vectorized) $\Phi$.*
- $\Phi^\#(\theta) \in nk\text{-}TC$ ($nk\text{-}LFP$) *provided $\theta \in n\text{-}TC$ ($n\text{-}LFP$) and $\Phi$ is a $k$-feasible.*
- $\Phi^\#(\theta) \in TC^{kn}(LFP^{kn}, L_{\infty\omega}^{kn})$ *provided $\theta \in TC^n(LFP^n, L_{\infty\omega}^n)$ and $\Phi$ is $k$-feasible.*

**Figure 4.6** Translation scheme and its components.

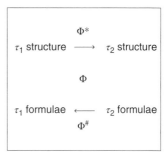

The following fundamental theorem is easily verified for correctly defined $\mathcal{L}$ translation schemes – see Figure 4.6. Its origins go back at least to the early years of modern logic (Hilbert and Bernays, 1970, pp. 277ff). See also Ebbinghaus, Flum, and Thomas (1994).

**Theorem 4.9** *Let* $\Phi = \langle \varphi, \psi_1, \ldots, \psi_m \rangle$ *be a* $\kappa - \tau_1 - \tau_2$ *translation scheme,* $\mathcal{A}$ *a* $\tau_1$ *structure, and* $\theta$ *an* $\mathcal{L}(\tau_2)$ *formula. Then* $\mathcal{A} \vDash \Phi^\#(\theta)$ *iff* $\Phi^*(\mathcal{A}) \vDash \theta$.

## 4.9 Strongly Distributed Systems

The disjoint union and shuffles are not, in themselves, very interesting. However, combining them with translation schemes gives a rich repertoire of composition techniques. Let $\tau_0, \tau_1, \tau$ be finite vocabularies. For a $\tau_0$ model $\mathcal{I}$ (serving as index model), $\tau_1$ structures $\mathcal{A}_i$   $(i \in I)$ are pairwise disjoint for tildeplicity, and a $\tau$ structure $\mathcal{A}$ is the disjoint union of $\langle \mathcal{A}_i : i \in I \rangle$ with $\mathcal{A} = \bigsqcup_{i \in I} \mathcal{A}_i$. Now, we generalize the disjoint union or shuffling of structures to *strongly distributed systems* in the following way.

**Definition 4.37** **(Strongly distributed systems)**
*Let* $\mathcal{I}$ *be a finitely partitioned index structure, and* $\mathcal{L}$ *be any of FOL, MSOL, WMSOL, MTC, MLFP, or FOL with unary generalized quantifiers. Let* $\mathcal{A} = \bigsqcup_{i \in I} \mathcal{A}_i$ *be a* $\tau$ *structure, where each* $\mathcal{A}_i$ *is isomorphic to some* $\mathcal{B}_1, \ldots, \mathcal{B}_\beta$ *over the vocabularies* $\tau_1, \ldots, \tau_\beta$, *in accordance with the partition. For* $\Phi$ *a scalar (non-vectorized)* $\tau_1 - \tau_2 - \mathcal{L}$ *translation scheme, the* $\Phi$ *-strongly distributed system composed from* $\mathcal{B}_1, \ldots, \mathcal{B}_\beta$ *over* $\mathcal{I}$ *is the structure* $S = \Phi^*(\mathcal{A})$, *or rather any structure isomorphic to it.*

The above definition is pretty general. If $\mathcal{A} = \bigsqcup_{i \in I} \mathcal{A}_i$ models a computational system then we are talking about a *strongly distributed computational system*. Now, our main theorem can be formulated as follows:

**Theorem 4.10** *Let* $\mathcal{I}$ *be a finitely partitioned index structure,* $\mathcal{L}$ *be any of FOL, MSOL, MTC, MLFP, MSOL, or FOL with unary generalized quantifiers. Let* $S$ *be a* $\Phi$ *-strongly distributed system composed from* $\mathcal{B}_1, \ldots, \mathcal{B}_\beta$ *over* $\mathcal{I}$, *as above. For every* $\phi \in \mathcal{L}(\tau)$ *there are:*

- *a Boolean function* $F_{\Phi,\phi}(b_{1,1}, \ldots, b_{1,j_1}, \ldots, b_{\beta,1}, \ldots, b_{\beta,j_\beta}, b_{I,1}, \ldots, b_{I,j_I})$;
- $\mathcal{L}$ *formulae* $\psi_{1,1}, \ldots, \psi_{1,j_1}, \ldots, \psi_{\beta,1}, \ldots, \psi_{\beta,j_\beta}$;
- *MSOL formulae* $\psi_{I,1}, \ldots, \psi_{I,j_I}$

*such that for every $S$, $\mathcal{I}$, and $\mathcal{B}_i$ as above with $\mathcal{B}_i \vDash \psi_{i,j}$ iff $b_{i,j} = 1$ and $\mathcal{I} \vDash \psi_{I,j}$ iff $b_{I,j} = 1$, we have*

$$S \vDash \phi \text{iff } F_{\Phi,\phi}(b_{1,1}, \ldots, b_{1,j_1}, \ldots, b_{\beta,1}, \ldots, b_{\beta,j_\beta}, b_{I,1}, \ldots, b_{I,j_I}) = 1.$$

*Moreover, $F_{\Phi,\phi}$ and $\psi_{i,j}$ are computable from $\Phi^\#$ and $\phi$, but are tower exponential in the quantifier rank of $\phi$.*

*Proof:* By analyzing the proof of Theorem 4.6 and using Theorem 4.9.

Moreover, Ravve, Volkovich, and Weber (2014) proved the following for WMSOL.

**Theorem 4.11** *Let $\mathcal{I}$ be a finite index structure and let $\mathcal{T}$ be $\Phi$-strongly distributed over $\mathcal{T}_1, \ldots, \mathcal{T}_\ell$ over $I$, as above. For every $\varphi \in$ WMSOL($\tau$) that satisfies Theorem 4.7 there are:*

- *a computation over weighted formulae $F_{\Phi,\varphi}(\varpi_{1,1}, \ldots, \varpi_{1,j_1}, \ldots, \varpi_{\ell,1}, \ldots, \varpi_{\ell,j_\ell})$;*
- *WMSOL formulae $\psi_{1,1}, \ldots, \psi_{1,j_1}, \ldots, \psi_{\ell,1}, \ldots, \psi_{\ell,j_\ell}$*

*such that for every $\mathcal{T}_i$ and $\mathcal{T}$ as above with $\varpi_{i,j} = \varrho_{i,j}$ iff $[\psi_{i,j}] = \varrho_{i,j}$ we have*

$$[\varphi] = \varrho \text{iff } F_{\Phi,\varphi}(\varpi_{1,1}, \ldots, \varpi_{1,j_1}, \ldots, \varpi_{\ell,1}, \ldots, \varpi_{\ell,j_\ell}) = \varrho.$$

*Moreover, $F_{\Phi,\varphi}$ and $\psi_{i,j}$ are computable from $\Phi^\#$ and $\varphi$, but are tower exponential in the quantifier rank of $\varphi$.*

## 4.10 Complexity Analysis

In this section, we discuss under what conditions our approach improves the complexity of computations, when measured by the size of the composed structures only. Our scenarios are as follows: A strongly distributed system is now submitted to a computation unit, and we want to know how long it takes to check whether $\phi$ is true on it. We give the general complexity analysis of the computation on strongly distributed systems.

Assume that $\mathcal{A}$ is a strongly distributed system. Its components are $\mathcal{A}_\iota$ with index structure $\mathcal{I}$, and we want to check whether $\phi$ is true in $\mathcal{A}$. Assume that:

- $\mathfrak{T}(N)$ or $\mathfrak{T}_{\text{old}}(N)$ denotes the time to solve the problem by the traditional sequential way ($N$ denotes the size of the coding of $\mathcal{A}$);
- $\mathfrak{C}_I$ denotes the time to extract index structure $\mathcal{I}$ from $\mathcal{A}$;
- $\mathfrak{C}_\iota$ denotes the time to extract each $\mathcal{A}_\iota$ from $\mathcal{A}$;
- $\mathfrak{C}_I(n_I)$ denotes the time to compute all values of $b_{I,j}$, where $n_I$ is the size of $I$;
- $\mathfrak{C}_\iota(n_\iota)$ denotes the time to compute all values of $b_{\iota,j}$, where $n_\iota$ is the size of $A_\iota$;
- $\mathfrak{T}_{F_{\Phi,\phi}}$ denotes the time to build $F_{\Phi,\phi}$;
- $\mathfrak{T}_S$ denotes the time to achieve one result of $F_{\Phi,\phi}$.

According to these symbols, the new computation time is:

$$\mathfrak{T}_{\text{new}} = \mathfrak{C}_I + \sum_{\iota \in I} \mathfrak{C}_\iota + \mathfrak{C}_I + \sum_{\iota \in I} \mathfrak{C}_\iota + \mathfrak{T}_{F_{\Phi,\phi}} + \mathfrak{T}_S, \tag{4.1}$$

and the question to answer is: When does it hold that $\mathfrak{T}_{\text{old}} > \mathfrak{T}_{\text{new}}$?

### 4.10.1 Scenario A: Single Computation on Repetitive Structures

The design of software and hardware widely uses hierarchical structures and repetition of blocks. Many basic elements of VLSI design such as adders (Figure 4.7) or registers (Figure 4.8) are built in this manner. Repetition of modules is also explored in control logic and other kinds of hardware design. Memory is another example of a repetitive structure in VLSI design. Inside a CPU, cache memory allows the minimization of the latency of memory access, and the CPU also has several independent caches: an instruction cache, a data cache, and a translation lookahead table. The data cache, as a rule, has

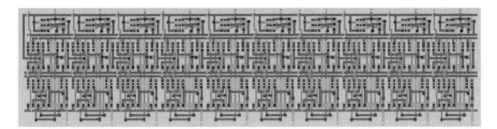

**Figure 4.7** Layout of a full ten-bit adder.

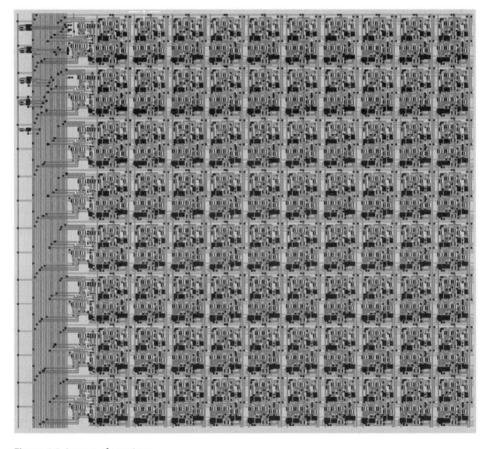

**Figure 4.8** Layout of a register.

a hierarchical structure of several levels: L1, L2, etc. In software, multiple calls to the same procedure is an example.

Another example is the *cell* microprocessor architecture jointly developed by Sony, Sony Computer Entertainment, Toshiba, and IBM, an alliance known as "STI" (Hennessy and Patterson, 2011). Cell is shorthand for "Cell Broadband Engine Architecture," commonly abbreviated to CBEA or just Cell BE (Figure 4.9).

A multicore processor is a single computing component with two or more independent actual processors (called "cores"), which are the units that read and execute program instructions (Yan, 2016). A dual-core processor has two cores (e.g. AMD Phenom II X2, Intel Core Duo), a quad-core processor contains four cores (Figure 4.10; e.g. AMD Phenom II X4, the Intel 2010 core line that includes three levels of quad-core processors – i3, i5, and i7), a hexa-core processor contains six cores (e.g. AMD Phenom II X6, Intel Core i7 Extreme Edition 980X), an octa-core processor contains eight cores (e.g. AMD FX-8150). A multicore processor implements multiprocessing in a single physical package. Multicore processors are widely used across many application domains, including general purpose, embedded, network, digital signal processing, and graphics.

In this section, we consider the complexity gain of our approach in such cases.

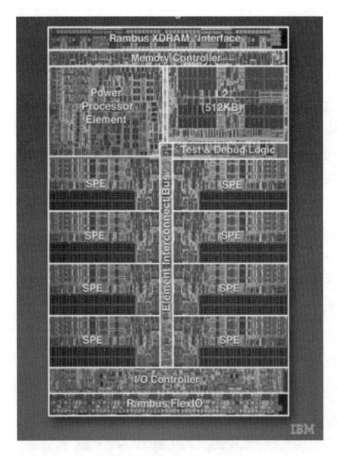

**Figure 4.9** IBM Cell Broadband Engine processor.

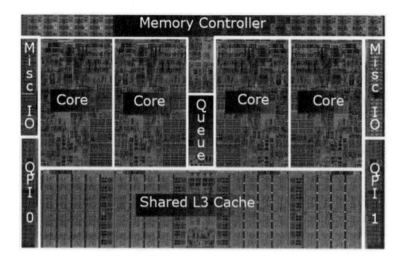

**Figure 4.10** Intel multicore processor.

#### 4.10.1.1 Complexity Gain for MSOL

Assume that our system is presented as an FSM such that:

- $N$ is the size of $\mathcal{A}$, $n$ is the size of $\tilde{\mathcal{A}}$, and $l$ is the size of the index structure $I$.
- The decomposition is given as $\mathcal{E}_I = \mathcal{E}_t = 0$.
- The computation is exponential in the form $\mathfrak{T} = e^{g(x)}$.

In this case,

$$\mathfrak{T}_{\text{new}} = P^p(\mathfrak{T}(n), \mathfrak{T}(l)),$$

where $P^p$ denotes polynomial of degree $p$, and

$$\mathfrak{T}_{\text{old}} = \mathfrak{T}(l \cdot n).$$

The question to answer is: When is $f(n \cdot l) > P^p(f(n), f(l))$? According to our assumptions, we find that the comparison of the computation times in Equation (4.1) looks like:

$$e^{g(n \cdot l)} > a_p(e^{p \cdot g(n)} + e^{p \cdot g(l)}).$$

Assume that $n = l$. Then $g(n^2) > p \cdot g(n) + \ln 2 + \ln(a_p)$. Assume that $g(x) = \ln^2(x)$; then $f(x) = x^{\ln(x)}$. In this case we obtain that Equation (4.1) is transformed to:

$$\ln^2(n^2) > p \cdot \ln^2(n) + \ln 2 + \ln(a_p)$$

or

$$\ln^2(n) > \frac{\ln(2 \cdot a_p)}{2 - p}.$$

#### 4.10.1.2 Complexity Gain for Polynomial Checker

If we have a logic where the computational procedure is polynomial in the sizes of $A$, $I$, and each $A_i$ too, then we do not obtain an analogous result.

### 4.10.2 Scenario B: Incremental Recomputations

Assume that we change several times (let us denote the number of the times by $\varsigma$) some fixed component $\check{A}$ of $A$. We check each time whether $A \vDash \phi$.

#### 4.10.2.1 Complexity Gain for Polynomial Checker

Let $\mathfrak{I}_{\text{old}}$ be the time to solve the given problem in the traditional manner. It should be clear that $\mathfrak{I}_{\text{old}} = \varsigma \cdot \mathfrak{I}(N)$. Let $\mathfrak{I}_{\text{new}}$ be the time to solve the same problem, when structure $A$ is viewed as a strongly distributed system. It is easy to see that

$$\mathfrak{I}_{\text{new}}(N, n) = \mathfrak{I}(N - n) + \varsigma \cdot \mathfrak{I}(n) + \mathfrak{I}_{F_{\Phi,\phi}} + \varsigma \cdot \mathfrak{I}_S.$$

The question to answer is: Which value of $n$ ensures that $\mathfrak{I}_{\text{old}} > \mathfrak{I}_{\text{new}}$? Assume that $\mathfrak{I}(x) = x^2$; then Equation (4.1) becomes

$$\varsigma \cdot N^2 > (N - n)^2 + \varsigma \cdot n^2 + \mathfrak{I}_{F_{\Phi,\phi}} + \varsigma \cdot \mathfrak{I}_S,$$

$$N^2 - 2 \cdot n \cdot N + n^2(\varsigma + 1) + \mathfrak{I}_{F_{\Phi,\phi}} + \varsigma \cdot \mathfrak{I}_S - \varsigma \cdot N^2 < 0,$$

$$n_{1,2} = \frac{N \pm \sqrt{N^2 + (\varsigma + 1)(N^2(\varsigma - 1) - \mathfrak{I}_{F_{\Phi,\phi}} - \varsigma \cdot \mathfrak{I}_S)}}{\varsigma + 1}.$$

If $n_1 \le n \le n_2$, then $\mathfrak{I}_{\text{old}} > \mathfrak{I}_{\text{new}}$.

$$n_2 = \frac{N + \sqrt{\varsigma^2(N^2 - \mathfrak{I}_S) - \varsigma(\mathfrak{I}_S + \mathfrak{I}F_{\Phi,\phi}) - \mathfrak{I}_{F_{\Phi,\phi}}}}{\varsigma + 1},$$

$$\lim_{\varsigma \to \infty} n_2 = \sqrt{N^2 - \mathfrak{I}_S}.$$

#### 4.10.2.2 Complexity Gain for Other Logics

Let $\mathcal{L}$ be any proper sublogic of MSOL stronger that FOL. Our theorems do not hold in the following: if we apply it, then $\psi_{i,j}$ are not necessary in $\mathcal{L}$.

### 4.10.3 Scenario C: Parallel Computations

In Equation (4.1), the new computation time is calculated as

$$\mathfrak{T}_{\text{new}} = \mathfrak{E}_I + \sum_{i \in I} \mathfrak{E}_i + \mathfrak{C}_I + \sum_{i \in I} \mathfrak{C}_i + \mathfrak{T}_{F_{\Phi,\phi}} + \mathfrak{T}_S$$

for the case when all the computations are done sequentially on a single computational unit. In fact, now even personal computers and smartphones have several cores. In this case, the computation may be done in the following way (we assume that there exist enough computational units for total parallelism):

**Extraction super step** The extraction of the index structure $I$ from $A$ and each $A_i$ from $A$ may be done in parallel, as well as the building of $F_{\Phi,\phi}$. We denote the extraction time by

$$\mathfrak{C} = \max\{\mathfrak{C}_I, \max_{i\in I}\{\mathfrak{C}_i\}, \mathfrak{T}_{F_{\Phi,\phi}}\}.$$

**Computational super step** The computation of all values of $b_{I,J}$ and $b_{i,J}$ may also be done in parallel. In this case,

$$\mathfrak{C} = \max\{\mathfrak{C}_I(n_I), \max_{i\in I}\{\mathfrak{C}_i(n_i)\}\}.$$

In fact, at this step, even more parallelism may be reached if we compute all $b_{i,J}$ in parallel.

**Final processing** $\mathfrak{T}_S$ still denotes the time to search for one result of $F_{\Phi,\phi}$.

The new computation time for the case of full parallelism is:

$$\mathfrak{T}_{new}^{BSP} = \mathfrak{C} + \mathfrak{C} + \mathfrak{T}_S.$$

The computation model fails in the general framework of BSP (Valiant, 1990).

#### 4.10.3.1  Complexity Gain for MSOL

The computation is exponential in the form $\mathfrak{T} = e^{g(x)}$. In this case,

$$\mathfrak{T}_{old} = \mathfrak{T} = f(N) = e^{g(N)} \text{and } \mathfrak{T}_{new}^{BSP} = \mathfrak{C} + P^p(e^{g(\frac{N}{k})}) + \mathfrak{T}_S,$$

and the question to answer is: When is $f(n\cdot k) > P^p(f(n))$? According to our assumptions, we obtain

$$e^{g(n\cdot k)} > \mathfrak{C} + a_p \cdot e^{p\cdot g(n)} + \mathfrak{T}_S.$$

Assume that $k = n$, which means that there exist enough computation units for full parallelization. In this case, the condition for effective computation looks like:

$$e^{g(n^2)} > \mathfrak{C} + a_p \cdot e^{p\cdot g(n)} + \mathfrak{T}_S.$$

#### 4.10.3.2  Complexity Gain for Polynomial Checker

Assume again that $\mathfrak{T}(x) = x^2$, and each $A_i$ is of the same size $\frac{N}{k}$. Then:

$$\mathfrak{T}_{old} = N^2 \text{and } \mathfrak{T}_{new}^{BSP} = \mathfrak{C} + (\frac{N}{k})^2 + \mathfrak{T}_S;$$

$$N^2 > \mathfrak{C} + \left(\frac{N}{k}\right)^2 + \mathfrak{T}_S; N^2 - \left(\frac{N}{k}\right)^2 > \mathfrak{C} + \mathfrak{T}_S.$$

Now, the condition of the effective computation looks like:

$$N^2 \cdot \frac{(k^2 - 1)}{k^2} > \mathfrak{C} + \mathfrak{T}_S.$$

The complexity considerations for other logics $\mathcal{L}$ that are proper sublogics of MSOL stronger than FOL are tildeilar to those given in Section 4.10.2.2.

### 4.10.4 Scenario D: Parallel Activities on Distributed Environments

In this section, we consider the underlying structure, the formula, and the modularity, as well as possible applications of our method for parallel activities on distributed environments.

#### 4.10.4.1 Complexity Gain

The full computation process is now composed of the following steps (the above $\mathfrak{E} = 0$).

**Computational super step** The computation of all values of $b_{I,J}$ and $b_{i,J}$ is done in parallel in the corresponding sites. We still have

$$\mathfrak{C} = \max\{\mathfrak{C}_I(n_I), \max_{\iota \in I}\{\mathfrak{C}_\iota(n_\iota)\}\}.$$

Recall that in each site, the $b_{i,J}$ still may be computed in parallel if the corresponding computer has several cores.

**Communication super step** The results $b_{I,J}$ and $b_{i,J}$ must be sent for final processing. We denote by $\mathfrak{T}_I$ the time to transfer all values of $b_{I,J}$, and by $\mathfrak{T}_\iota$ the time to transfer all values of $b_{i,J}$. The communication time now is

$$\mathfrak{T} = \max\{\mathfrak{T}_I, \max_{\iota \in I}\{\mathfrak{T}_\iota)\}\}.$$

**Final processing** $\mathfrak{T}_S$ still denotes the time to search for one result of $F_{\Phi,\phi}$.

The new computation time of Equation (4.1) for the case of distributed storage and computation is

$$\mathfrak{T}_{new}^{distr} = \mathfrak{C} + \mathfrak{T} + \mathfrak{T}_S.$$

If the computations and the data transfer in each site may be done in parallel then we may combine the first two super steps in the above model into one step that, in fact, leads to some variation of the LogP model introduced in Culler et al. (1993). We define

$$\mathfrak{D} = \max\{(\mathfrak{C}_I(n_I) + \mathfrak{T}_I), \max_{\iota \in I}\{(\mathfrak{C}_\iota(n_\iota) + \mathfrak{T}_\iota)\}\}.$$

Now, the corresponding computation time is:

$$\mathfrak{T}_{new}^{LogP} = \mathfrak{D} + \mathfrak{T}_S.$$

**Observation 4.5**
**Communication load** *Note that the only values transferred between different computational sites are $b_{I,J}$ and $b_{i,J}$, which are binary.*
**Confidentiality** *All meaningful information is still stored in the corresponding locations in a secure way and is not transferred.*

## 4.11 Description of the Method

Our general scenario is as follows: given logic $\mathcal{L}$, structure $\mathcal{A}$ as a composition of structures $\mathcal{A}_i, i \in I$, index structure $\mathcal{I}$, and formula $\phi$ of the logic to be evaluated on $\mathcal{A}$, the question is: What is the reduction sequence of $\phi$, if any? We propose a general approach to try to answer the question and to investigate the computation gain of the incremental evaluations. The general template is defined as follows.

- **Prove preservation theorems**
  Given logic $\mathcal{L}$:
  - **Define disjoint union of $\mathcal{L}$ structures** The logic may be defined for arbitrary structures, or rather for a class of structures like graphs, (directed) acyclic graphs, trees, words, (Mazurkiewicz) traces (Diekert and Rozenberg, 1995), (lossy) message sequence charts, etc. In the general case, we use Definition 4.24, which provides a logical definition of the *disjoint union* of the components: $\mathcal{A} = \bigsqcup_{i \in I} \mathcal{A}_i$. Adaptation of Definition 4.24 to the case of WMSOL, which is introduced over trees, is presented in Definition 4.30. If logic $\mathcal{L}$ is introduced over another class of structures then Definition 4.24 must be aligned accordingly.
  - **Define preservation property** *XX*-PP **for** $\mathcal{L}$ After we have defined the appropriate disjoint union of structures, we define the notion of an *(XX) preservation property (PP)* for logics; see Definitions 4.26 and 4.27.
  - **Prove the preservation property** *XX*-PP **for** $\mathcal{L}$ Now we try to prove the corresponding preservation property for $\mathcal{L}$. As a rule, such a preservation theorem can be proven by suitable pebble games, which are generalizations of Ehrenfeucht–Fraïssé games. This usually gives a version of $XX\text{-PP}(\mathcal{L}_1^{m_1,k_1}, \mathcal{L}_2^{m_2,k_2})$ for suitable chosen definitions of quantifier rank $(m_j)$ and counting of variables $(k_j)$. Theorem 4.5 shows that the preservation theorems hold for lots of extensions of FOL. However, theorems like Theorem 4.5 are not always true. In Proposition 4.1, we show that our restriction to unary generalized quantifiers (MTC and MLFP) is necessary. Moreover, Theorem 4.8 shows for which semirings WMSOL possesses the preservation property.
- **Define translation schemes**
  Given logic $\mathcal{L}$:
  Definition 4.31 gives the classical syntactically defined translation schemes. Definition 4.32 is an adaption of Definitions 4.31 to the case of many-sorted structures. In general, Definitions 4.31 must be adopted in a tildeilar way to the given logic $\mathcal{L}$. Definitions 4.31 give rise to two induced maps, translations and transductions. The transductions describe the induced transformation of $\mathcal{L}$ structures, and the translations describe the induced transformations of $\mathcal{L}$ formulae; see Definitions 4.35 and 4.36. Again, the presented adaptation of the definitions to the case of WMSOL is a bit tricky. If the $\mathcal{L}$-translation scheme is defined correctly then the proof of the corresponding variation of Theorem 4.9 is easily verified.
- **Strongly distributed system**
  Given $\mathcal{L}$ structure $\mathcal{A}$:
  If the given $\mathcal{L}$ structure $\mathcal{A}$ is a $\Phi$-strongly distributed composition of its components then we may apply $\mathcal{L}$ variation of our main Theorem 4.10 to it. Theorem 4.10 shows how to effectively compute the reduction sequences for the different logics under investigation for the strongly distributed systems.
- **Complexity gain**
  Detailed complexity analysis of the method for different combinations of logics and strongly distributed systems is provided in Section 4.10.

Finally, we derive a method for evaluating $\mathcal{L}$ formula $\phi$ on $\mathcal{A}$, which is a $\Phi$-strongly distributed composition of its components. The method proceeds as follows:

**Preprocessing:** Given $\phi$ and $\Phi$, but not $\mathcal{A}$, we construct a sequence of formulas $\psi_{i,j}$ and an evaluation function $F_{\Phi,\phi}$ as in Theorem 4.10.

**Incremental computation:** We compute the local values $b_{i,j}$ for each component of $\mathcal{A}$.

**Solution:** Theorem 4.10 now states that $\phi$, expressible in the corresponding logic $\mathcal{L}$, on $\mathcal{A}$ may be effectively computed from $b_{i,j}$, using $F_{\Phi,\phi}$.

## 4.12 Application of WMSOL and WTA to Coverage Analysis

Assume we are given a strongly distributed system. The model of the system is the extracted FSM. We name the model $\mathcal{FSM}^{\text{System}}$. We run our tests and obtain the tree $\mathcal{T}^{\text{System}}$, which represents runs of the tests on $\mathcal{FSM}^{\text{System}}$. The root of the tree is the initial state of $\mathcal{FSM}^{\text{System}}$, and each path in $\mathcal{T}^{\text{System}}$ is a sequence of states visited during the run of a particular test – see Figure 4.11.

Assume that we are interested in effectively computing and analyzing a quantitative coverage metric $P^{l}_{\text{coverage}}$ on the results of our testing. If we want to use Theorem 4.1, then we are looking for a labeling $L$ of $\mathcal{T}^{\text{System}}$ and a ptv-monoid $D$ such that $P^{l}_{\text{coverage}}$ may be expressed as a formula $\phi^{l}_{\text{coverage}}$ of a fragment of the corresponding WMSOL from the theorem. If we manage to find such an $L$ and ptv-$D$, then (according to Theorem 4.1) there may effectively be constructed a corresponding variation $\mathcal{M}^{\phi_{\text{coverage}}}$ of WTA that computes the value of $\phi^{l}_{\text{coverage}}$ on $\mathcal{T}^{\text{System}}_{L}$ – see Figure 4.12.

### 4.12.1 Examples of FSM Coverage Metrics Expressible in WMSOL

Assume we are given a DUT and the extracted FSM, such that all possible runs of the FSM fail in the framework of a labeled tree over a ranked alphabet $\Xi$, as introduced in Section 4.4. $\Xi$ is defined by the FSM. Each run of a test on the FSM produces labeled word $u$ in the tree domain $B$. The particular labeling $L$ and ptv-monoid $D$ depend upon the particular metric of the coverage analysis we are interested in. Let us consider some possible coverage analysis metrics, which may be expressed as WMSOL formulae for particular choices of $L$ and $D$.

#### 4.12.1.1 $(\mathbb{N}, +, \cdot, 0, 1)$ (Droste and Gastin, 2007)

Let $P_a$ denote a unary predicate symbol for each $a \in \Sigma$. The formula $\exists x P_a$ counts how often $a$ occurs in the tree.

Note that each choice of the unary predicate, $L$, as well as $D$ (see some options of commutative semirings in Section 4.12.2), affects the meaning of "how often" and generates lots of variations of the coverage metrics. This means that the formula is a good candidate for state visitation and state transition coverage metrics, like $N_{J_{\text{FSM}}}$ from Section 4.1.

#### 4.12.1.2 $(\mathbb{R}_{+} \cup \{+\infty\}, \min, +, +\infty, 0)$ (Droste and Vogler, 2009)

Let $\Sigma = \{\alpha, \beta\}$. Let $\phi_{\alpha\alpha}$ be defined as follows:

$$\phi_{\alpha\alpha} = \forall x \forall y_1 \forall y_2 ((\phi_{\text{Child}}(x, y_1) \wedge \phi_{\text{Next}}(y_1, y_2) \wedge$$
$$\text{label}_{\alpha}(y_1) \wedge \text{label}_{\alpha}(y_2)) \rightarrow 1),$$

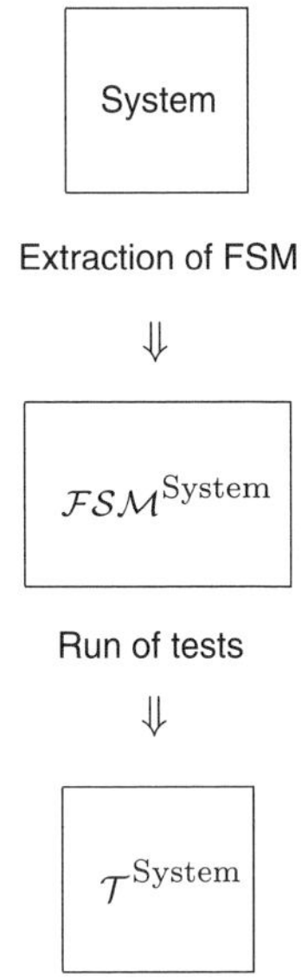

**Figure 4.11** From System to $\mathcal{T}^{\text{System}}$.

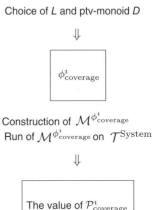

**Figure 4.12** Computation of an FSM coverage metric for functional verification.

where

- $\phi_{\text{Child}}(x,y) = \phi_{\text{Desc}}(x,y) \wedge \forall z(\phi_{\text{Desc}}(x,z) \rightarrow^+ \phi_{\leq}(y,z))$ defines $y$ as the first child of $x$;
- $\phi_{\text{Next}}(y_1,y_2) = \phi_{\leq}(y_1,y_2) \wedge \neg\phi_{\leq}(y_2,y_1) \wedge \forall z(\phi_{\leq}(y_1,y_2) \rightarrow^+ (\phi_{\leq}(z,y_1) \vee \phi_{\leq}(y_2,z)))$.

For precise definitions of $\phi_{\text{Desc}}$, $\phi_{\leq}$, and $\rightarrow^+$, see Droste and Vogler (2009). $\phi_{\alpha\alpha}$ counts the number of positions for which the first and second descendants are labeled by $\alpha$. A very wide repertoire of coverage metrics related to the observation of different sequences of states and transitions may be defined in a tildeilar way.

### 4.12.1.3 $Q_{max}$ (Droste et al., 2011a)

Let us consider

$$Q_{max} = (\mathbf{Q} \cup \{-\infty, +\infty\}, \max, \text{avg}, \min, -\infty, +\infty)$$

with

$$\text{avg}(t) = \frac{\sum_{u \in \text{dom}(t)} t(u)}{|\text{dom}(t)|},$$

which calculates the average of all weights of a tree. Let us consider the WMSOL formula $\phi_{\text{Leaf}} = \bigvee_{a \in \Sigma^{(0)}} \text{label}_a(x)$, which is equal to $\infty$ if $\sigma(x)$ is a leaf and $-\infty$ otherwise. Now, the semantics of the formula

$$\phi_{\text{Leaves-to-size}} = \forall x((\phi_{\text{Leaf}}(x) \wedge 1) \vee (\neg\phi_{\text{Leaf}}(x) \wedge 0))$$

is equal for every tree $t$ to its *leaves-to-size* ratio, where the size of a tree is the total number of nodes of the tree. The corresponding weighted automaton, which calculates the ratio, looks like $\mathcal{M}_{\text{Leaves-to-size}} = (\{q\}, \Sigma, \mu, \{q\})$, with $\mu_0(a)_{.q} = 1$ for all $a \in \Sigma^{(0)}$ and $\mu_m(b)q \dots q.q = 0$ for all $b \in \Sigma^{(m)}$ and $m \geq 1$.

Moreover, it is easy to show that there exists a Boolean formula $\phi_{\text{Path}}(X)$ whose semantics is $\infty$ if $\sigma(X)$ is a path of a tree and $-\infty$ otherwise. Now, the semantics of the formula

$$\phi_{\text{Height-to-size}} = \exists X(\phi_{\text{Path}}(X) \wedge \forall x((x \in X \wedge 1) \vee (x \notin X \wedge 0)))$$

is equal for every tree $t$ to its *height-to-size* ratio, where the height of a tree is the number of nodes on the longest paths decremented by one. The corresponding weighted automaton, which calculates the ratio, looks like $\mathcal{M}_{\text{Height-to-size}} = (\{p, n\}, \Sigma, \mu', \{p\})$, with $\mu_0'(a)_{.p} = \mu_0'(a)_{.n} = 0$ for all $a \in \Sigma^{(0)}$. For more details, see Droste et al. (2011a). Both $\mathcal{M}_{\text{Leaves-to-size}}$ and $\mathcal{M}_{\text{Height-to-size}}$ are good candidates for $N_{\text{Length}}$, introduced in Section 4.1.

While the automata $\mathcal{M}_{\text{Leaves-to-size}}$ and $\mathcal{M}_{\text{Height-to-size}}$ are constructed manually, Theorem 4.1 states that if a quantitative coverage metric is expressible in a certain fragment of WMSOL then there exists some variant of weighted tree automata that computes the value of the metric. This means that, given a DUT, we are looking for a labeling $L$ and product tree valuation monoid $\mathcal{D}$ such that the desired coverage metric may be expressed as a WMSOL formula from Theorem 4.1. In such a case, the metric is effectively evaluated by the corresponding WTA.

### 4.12.2  Hunting for Labelings and Monoids

Weighted monadic second-order logic and its fragments have considerable expressive power. However, the larger the particular fragment gets, the more restrictions on the underlying ptv-monoid we need. Theorem 4.8 gives some possible options.

One of the promising techniques to enrich the repertoire of ptv-monoids is *discounting* (de Alfaro, Henzinger, and Majumdar, 2003; Mandrali and Rahonis, 2009). Let us consider two semirings $K_1$ and $K_2$.

**Definition 4.38  (A semiring endomorphism)**
*A mapping $f : K_1 \to K_2$ is called a semiring homomorphism if $f(a + b) = f(a) + f(b)$ and $f(a \cdot b) = f(a) \cdot f(b)$ for every $a, b \in K_1$, and $f(0) = 0$ and $f(1) = 1$. A homomorphism $f : K \to K$ is an endomorphism of $K$.*

For the arctic semiring and every $p \in \mathbf{R}_+$, we put $p \cdot (-\infty) = -\infty$. The mapping $p : x \to p \cdot x$ is an endomorphism of the semiring.

The set $\mathrm{End}(K)$ of all endomorphisms of $K$ is a monoid itself with the operation of the usual composition mapping $\circ$ and unit element the identity mapping id on $K$. Let $\Xi$ be a ranked alphabet and $K$ be a semiring.

**Definition 4.39** (A $\Phi$ **discounting**)
*A* $\Phi$ *discounting over* $\Xi$ *and* $K$ *is a family* $\Phi = (\Phi_k)_{k \geq 0}$ *of mappings* $\Phi_k : \Sigma^{(k)} \to (\mathrm{End}(K))^k$ *for* $k \geq 1$, *and* $\Phi_0 : \Sigma^{(0)} \to K$.

For every $\Xi$-ranked tree $t$ and every $u \in \mathrm{dom}(t)$, the endomorphism $\Phi_u^t$ of $K$ is defined as follows:

$$\Phi_u^t = \begin{cases} \mathrm{id} & \text{if } u = \epsilon, \\ \Phi_{t(\epsilon)}^{\iota_1} \circ \Phi_{t(\iota_1)}^{\iota_2} \circ \cdots \circ \Phi_{t(\iota_1 \dots \iota_{n-1})}^{\iota_n} & \text{if } u = \iota_1 \dots \iota_n; \iota_1, \dots, \iota_n \in \mathbf{N}_+. \end{cases}$$

In Droste et al. (2011a), the following ptv-monoid with discounting was considered:

$$(\mathbf{R} \cup \{-\emptyset\}, \max, \mathrm{disc}_\Lambda, -\emptyset),$$

where the tree valuation function $\mathrm{disc}_\Lambda$ models a discounting on trees as defined above, $\Lambda = (\lambda_i)_{i \in \mathbf{N}}$ with $\lambda_i > 0$ and $i \in \mathbf{N}$. The weight of some $t \in T_{\mathrm{disc}}$ on $u$ of length $m$ is defined as

$$\mathrm{wgt}_u(t) = \begin{cases} t(u) & \text{if } u = \epsilon, \\ (\Pi_{i=1}^m \lambda_{k_i}) \cdot t(u) & \text{otherwise.} \end{cases}$$

Then $\mathrm{disc}_\Lambda(t) = \sum_{u \in \mathrm{dom}(t)} \mathrm{wgt}_u(t)$. In this case, the discounting depends on the distance of the node from the root. More examples may be found in Mandrali and Rahonis (2009). In particular, it was shown how to count the number of occurrences of $\varsigma \in \Sigma$ in the greatest initial $\varsigma$ subtree of $t$, which extends $N_{J_{\mathrm{FSM}}}$ from Section 4.1.

## 4.13  Summary

In this chapter we introduced the notion of strongly distributed systems and presented a uniform logical approach to incremental automated verification of such systems. The approach is based on systematic use of two logical reduction techniques: Feferman–Vaught reductions and syntactically defined translation schemes.

Feferman–Vaught reductions are applied in situations of strongly distributed system. The reduction describes how queries over a strongly distributed system can be computed from queries over the components and queries over the index set. Feferman–Vaught reductions were first introduced in model theory by Feferman and Vaught (1959). There exist several variants of the composition theorem, depending on the type of operations and the logic. The first result of this type is due to Beth (Skolem, 1958), and concerns the operation of juxtaposition of labeled linear orders; another proof was given independently by Fraïssé (1955a,b).Apreliminary version of the Feferman–Vaught theorem, basically dealing with a fixed indexing structure, was given by Mostowski (1952). There are composition theorems for MSOL under generalized sums, such as Gurevich (1979) and Gurevich and Shelah (1982). A summary of the composition theorem for a generalized sum may be found in Rabinovich (2007).

Use of Feferman–Vaught reductions in computer science was seemingly first suggested in Makowsky and Ravve (1995) and Ravve (1995) in the context of formal verification and model checking, in Ravve (2014a)(@) for coverage analysis, and in Courcelle, Makowsky, and Rotics (1998) in the context of graph algorithms for graphs of bounded clique width. Their use in operational research and data mining was introduced in Ravve, Volkovich, and Weber (2014, 2015a,b),and Ravve and Volkovich (2016). For recent applications in the field of database theory, see Ravve (2014b, 2016). The approach of Suciu (2000) and Amer-Yahia, Srivastava, and Suciu (2004) deals with distributed evaluation of network directory queries in terms of modal languages; however, they do not explicitly appeal to Feferman–Vaught reductions. For algorithmic uses of the Feferman–Vaught theorem see also Makowsky (2004).

Syntactically defined translation schemes are also known in model theory as interpretations (Immerman, 1999). They describe transformations of data and queries. They give rise to two induced maps, translations and transductions. Transductions describe the induced transformation of data instances, and translations describe the induced transformations of queries. The fundamental property of translation schemes describes how to compute transformed queries in the same way that Leibniz' theorem describes how to compute transformed integrals. The fundamental property of translation schemes has a long history, but was first succinctly stated by Rabin (1965). Recent uses of interpretations in the field of database theory may be found in particular in Benedikt and Koch (2009) and Grädel and Siebertz (2012).

Our general scenario is as follows: given a logic $\mathcal{L}$, a structure $\mathcal{A}$ as a composition of structures $\mathcal{A}_i, i \in I$, an index structure $\mathcal{I}$, and a formula $\varphi$ of the logic to be evaluated on $\mathcal{A}$, the question is: What is the reduction sequence for $\varphi$, if any?

We have shown that if we can prove preservation theorems for $\mathcal{L}$ as well as if $\mathcal{A}$ were a strongly distributed composition of its components then the corresponding reduction sequence for $\mathcal{A}$ may be algorithmically computed. In such a case, we derive a method for evaluating an $\mathcal{L}$ formula $\varphi$ on $\mathcal{A}$, which is a $\Phi$-strongly distributed composition of its components. First of all, given $\varphi$ and $\Phi$, but not $\mathcal{A}$, we construct a sequence of formulas $\psi_{i,j}$ and an evaluation function $F_{\Phi,\varphi}$. Next, we compute the local values $\varpi_{i,j}$ for each component of $\mathcal{A}$. Finally, our main theorems state that $\varphi$, expressible in the corresponding logic $\mathcal{L}$, on $\mathcal{A}$ may be effectively computed from $\varpi_{i,j}$ using $F_{\Phi,\varphi}$.

We have shown that the approach works for many extensions of FOL, but not all. The considered extensions of FOL are suitable candidates for expression of verification and coverage properties.

From our main results, Theorems 4.10 and 4.11, we derive a method for computing $\varphi$ on strongly distributed (weighted) systems which proceeds as follows:

**Preprocessing:** Given $\varphi$ and $\Phi$, but not the strongly distributed (weighted) system, we construct a sequence of formulae $\psi_{i,j}$ for each $j$th component to be locally evaluated, and the final evaluation function $F_{\Phi,\varphi}$.

**Incremental computation:** Given the strongly distributed (weighted) system, we compute the local values $\varpi_{i,j}$ of $\psi_{i,j}$ on each $j$th component.

**Final exact solution:** Theorems 4.10 and 4.11 now state that $\varphi$ on the strongly distributed (weighted) system may be effectively computed from $\varpi_{i,j}$, using $F_{\Phi,\varphi}$.

We emphasize the following:

**Scalability:** The formulae $\psi_{i,j}$ and $F_{\Phi,\varphi}$ do not depend upon the the strongly distributed (weighted) system itself, but rather upon $\varphi$ and $\Phi$.

**Communication load:** The only values that are transferred between different computational components are $\varpi_{i,j}$, which significantly reduces the communication load.

**Confidentiality:** All meaningful information is still stored in the corresponding locations in a secure way and is not transferred. The transferred values $\varpi_{i,j}$, as a rule, are meaningless without knowledge of the final processing.

We plan to apply the proposed methodology to incremental reasoning based on the promising variations of WMSOL introduced recently by Kreutzer and Riveros (2013), Labai and Makowsky (2013), and Monmege (2013) (see also Labai and Makowsky, 2015).

# References

Abiteboul, S., Hull, R., and Vianu, V. (1995) *Foundations of Databases*. Reading, MA: Addison-Wesley.

Amer-Yahia, S., Srivastava, D., and Suciu, D. (2004) Distributed evaluation of network directory queries. *IEEE Transactions on Knowledge and Data Engineering*, 16(4), 474–486.

Arnold, A. and Niwiński, D. (1992) Fixed point characterization of weak monadic logic definable sets of trees. In A. Podelski, M. Nivat (eds.) *Tree Automata and Languages*. Amsterdam: Elsevier Science Publishers B.V., pp. 159–188.

Benedikt, M. and Koch, C. (2009) From XQuery to relational logics. *ACM Transactions on Database Systems*, 34(4), 25:1–25:48.

Bollig, B. and Gastin, P. (2009) Weighted versus probabilistic logics. In V. Diekert and D. Nowotka (eds.) *DLT 2009*, LNCS R5583, pp. 18–38. Berlin: Springer-Verlag.

Bosse, U. (1993) An Ehrenfeucht–Fraïssé game for fixed point logic and stratified fixed point logic. In *CSL'92, volume 702 of Lecture Notes in Computer Science*, pp. 100–114. New York: Springer.

Bosse, U. (1995) *Ehrenfeucht–Fraïssé Games for Fixed Point Logic*. PhD thesis, Department of Mathematics, University of Freiburg, Germany.

Bouyer, P. Weighted timed automata: Model-checking and games. *Electronic Notes in Theoretical Computer Science*, 158, 3–17.

Büchi, J.R. (1960) Weak second-order arithmetic and finite automata. *Z. Math. Logik Grundlagen Math.*, 6, 66–92.

Burgess, J. Basic tense logic. In D. Gabbay and F. Günthner (eds.) *Handbook of Philosophical Logic*, Vol. 2. Dordrecht: D. Reidel Publishing Company.

Chang, C.C. and Keisler, H.J. (1990) *Model Theory*. Studies in Logic, vol 73. Amsterdam: North–Holland, 3rd edn.

Courcelle, B. (1993) The monadic second-order logic of graphs IX: Machines and their behaviours. Preprint.

Courcelle, B. (1995) The monadic second-order logic of graphs VIII: Orientations. *Annals of Pure and Applied Logic*, 72, 103–143.

Courcelle, B., Makowsky, J.A., and Rotics, U. (1998) Linear time solvable optimization problems on graph of bounded clique width, extended abstract. In J. Hromkovic and

O. Sykora (eds.) *Graph Theoretic Concepts in Computer Science,* volume 1517 of *Lecture Notes in Computer Science,* pp. 1–16. Berlin: Springer-Verlag.

Courcelle, B. and Walukiewicz, I. (1995) Monadic second-order logic, graphs and unfoldings of transition systems. *Annals of Pure and Applied Logic,* 92, 35–62.

Culler, D., Karp, R., Patterson, D., Sahay, A., Schauser, K.E., Santos, E., Subramonian, R., and von Eicken, T. (1993) LogP: Towards a realistic model of parallel computation. In *PPOPP '93 Proceedings of the fourth ACM SIGPLAN Symposium on Principles and Practice of Parallel Programming,* Vol. 28(7), pp. 1–12.

Dawar, A. and Hellat, L. (1993) The expressive power of finitely many genaralized quantifiers. Technical Report CSR 24–93, Computer Science Department, University of Wales, University College of Swansea.

de Alfaro, L. Henzinger, T.A., and Majumdar, R. (2003) Discounting the future in systems theory. In ICALP, volume 2719 of *Lecture Notes in Computer Science.* New York: Springer.

Diekert, V. and Rozenberg, G. (1995) *The Book of Traces.* Singapore: World Scientific.

Drake, D. and Cohen, P. (1998) HDL verification coverage. *Integrated Systems Design Magazine,* June.

Droste, M. and Gastin, P. (2007) Weighted automata and weighted logics. *Theoretical Computer Science,* 380, 69–86.

Droste, M., Götze, D., Märcker, S., and Meinecke, I. (2011a) Weighted tree automata over valuation monoids and their characterization by weighted logics. In W. Kuich and G. Rahonis (eds.) *Bozapalidis Festschrift,* LNCS 7020, pp. 30–55, Berlin: Springer-Verlag.

Droste, M. and Meinecke, I. (2010) Describing average- and longtime-behavior by weighted MSO logics. In P. Hliněný and A. Kučera (eds.) *MFCS 2010,* LNCS 6281, pp. 537–548, Berlin: Springer-Verlag.

Droste, M., Meinecke, I., Šešelja, B., and Tepavčević, A. (2011b) A cascade decomposition of weighted finite transition systems. In G. Mauri and A. Leporati (eds.) *DLT 2011,* LNCS 6795, pp. 472–473, Berlin: Springer-Verlag.

Droste, M. and Vogler, H. (2009) Weighted logics for unranked tree automata. *Theory of Computing Systems,* 48(1), 23–47.

Ebbinghaus, H.D. and Flum, J. (1995) *Finite Model Theory.* Perspectives in Mathematical Logic. New York: Springer.

Ebbinghaus, H.D., Flum, J., and Thomas, W. (1994) *Mathematical Logic,* 2nd edn, Undergraduate Texts in Mathematics. Berlin: Springer-Verlag.

Ehrenfeucht, A. (1961) An application of games to the completeness problem for formalized theories. *Fundamenta Mathematicae,* 49, 129–141.

Elgot, C.C. (1961) Decision problems of finite automata design and related arithmetics. *Transactions of the American Mathematical Society,* 98, 21–52.

Emerson, E.A. (1990) Temporal and modal logic. In J. van Leeuwen (ed.) *Handbook of Theoretical Computer Science,* Vol. 2. Amsterdam: Elsevier Science Publishers.

Fagin, R. (1974) Generalized first-order spectra and polynomial time recognizable sets. In R. Karp (ed.) *Complexity of Computation,* vol. 7 of American Mathematical Society Proc., pp. 27–41.

Feferman, S. and Vaught, R. (1959) The first order properties of products of algebraic systems. *Fundamenta Mathematicae,* 47, 57–103.

Fraïssé, R. (1955a) Sur quelques classifications des relations, basées sur les isomorphisms restraints, I: Études générale. *Publications Scientifiques de l'Université d'Alger, Serie A*, 2, 15–60.

Fraïssé, R. (1955b) Sur quelques classifications des relations, basées sur les isomorphisms restraints, II: Applications aux relations d'ordre, et constructions d'exemples montrant que les classifications sont distinctes. *Publications Scientifiques de l'Université d'Alger, Serie A*, 2, 273–295.

Frick, M. and Grohe, M. (2004) The complexity of first-order and monadic second-order logic revisited. *Annals of Pure and Applied Logic*, 130(1), 3–31.

Grädel, E. (1992) On transitive closure logic. In E. Börger, G. Jäger, H. Kleine Büning, and M.M. Richter (eds.) *Computer Science Logic*, volume 626 of *Lecture Notes in Computer Science*, pp. 149–163. Berlin: Springer-Verlag.

Grädel, E. and Siebertz, S. (2012) Dynamic definability. In *15th International Conference on Database Theory, ICDT '12*, Berlin, Germany, March 26–29, pp. 236–248.

Grohe, M. (1994) *The Structure of Fixed Point Logics*. PhD thesis, Department of Mathematics, University of Freiburg.

Gurevich, Y. (1979) Modest theory of short chains, I. *Journal of Symbolic Logic*, 44, 481–490.

Gurevich, Y. and Shelah, S. (1982) Monadic theory of order and topology in ZFC. *Annals of Mathematical Logic*, 23(2), 179–198.

Harel, D. (1984) Dynamic logic. In D. Gabbay and F. Günthner (eds.) *Handbook of Philosophical Logic*, Vol. 2. Dordrecht: D. Reidel Publishing Company.

Hennessy, J.L. and Patterson, D.A. (2011) *Computer Architecture: A Quantitative Approach*. San Francisco, CA: Morgan Kaufmann Publishers Inc., 5th edn.

Hilbert, D. and Bernays, P. (1970) *Grundlagen der Mathematik, I*, volume 40 of *Die Grundlehren der mathematischen Wissenschaften in Einzeldarstellungn*. Heidelberg: Springer-Verlag, 2nd edn.

Immerman, N. (1982) Relational queries computable in polynomial time. In *STOC'82*, pp. 147–152. New York: ACM.

Immerman, N. (1987) Languages that capture complexity classes. *SIAM Journal on Computing*, 16(4), 760–778.

Immerman, N. (1989) Expressibility and parallel complexity. *SIAM Journal on Computing*, 18, 625–638.

Immerman, N. (1999) *Descriptive Complexity*. New York: Springer.

Jou, J.-Y. and Liu, C.-N.J. (1998) An FSM extractor for HDL description at RTL level. In *Proceedings of the Fifth Asia-Pacific Conference on Hardware Description Languages*.

Jou, J.-Y. and Liu, C.-N.J. (1999a) Coverage analysis techniques for HDL design validation. In *Proceedings of the Sixth Asia-Pacific Conference on Hardware Description Languages*, pp. 3–10.

Jou, J.-Y. and Liu, C.-N.J. (1999b) An efficient functional coverage test for HDL descriptions at RTL. In *International Conference on Computer Design*, pp. 325–327.

Kolaitis, P.G. and Väänänen, J.A. (1992) Generalized quantifiers and pebble games on finite structures. In *LiCS'92*, pp. 348–359. New York: IEEE.

Kreutzer, S. and Riveros, C. (2013) Quantitative monadic second-order logic. In *28th Annual ACM/IEEE Symposium on Logic in Computer Science, LICS 2013*, New Orleans, LA, USA, June 25–28, pp. 113–122.

Kripke, S.A. (1965) Semantical analysis of intuitionistic logic I. In J.N. Crossley and M.A.E. Dummett (eds.) *Formal Systems and Recursive Functions*, volume 40 of *Studies in Logic and the Foundations of Mathematics*, pp. 92–130. Amsterdam: Elsevier.

Labai, N. and Makowsky, J.A. (2013) Weighted automata and monadic second order logic. In *Proceedings Fourth International Symposium on Games, Automata, Logics and Formal Verification, GandALF 2013*, Borca di Cadore, Dolomites, Italy, 29–31 August, pp. 122–135.

Labai, N. and Makowsky, J.A. (2015) Logics of finite Hankel rank. In *Fields of Logic and Computation II: Essays Dedicated to Yuri Gurevich on the Occasion of His 75th Birthday*, pp. 237–252.

Larsen, F.G., Fahrenberg, U., and Thrane, C. (2009) A quantitative characterization of weighted Kripke structures in temporal logic. In P. Hlinený, V. Matyáš, and T. Vojnar (eds.) *Annual Doctoral Workshop on Mathematical and Engineering Methods in Computer Science*, volume 13 of *Open Access Series in Informatics*, Dagstuhl, Germany, 2009. Schloss Dagstuhl–Leibniz-Zentrum fuer Informatik.

Larsen, K.G., Fahrenberg, U., and Thrane, C. (2012) A quantitative characterization of weighted Kripke structures in temporal logic. *Computing and Informatics*, 29(6), 2012.

Lichtenstein, O., Pnueli, A., and Zuck, L. (1985) Logics of program. In *Lecture Notes in Computer Science*, Vol. 193, pp. 196–218. New York: Springer.

Lindström, P. (1966) First order predicate logic with generalized quantifiers. *Theoria*, 32, 186–195.

Lindström, P. (1969) On extensions of elementary logic. *Theoria*, 35, 1–11.

Makowsky, J.A. (1978) Some observations on uniform reduction for properties invariant on the range of definable relations. *Fundamenta Mathematicae*, 99, 199–203.

Makowsky, J.A. (1985) Compactness, embeddings and definability. In J. Barwise and S. Feferman (eds.) *Model-Theoretic Logics*, Perspectives in Mathematical Logic. Berlin: Springer-Verlag.

Makowsky, J.A. (2004) Algorithmic uses of the Feferman–Vaught theorem. *Annals of Pure and Applied Logic*, 126(1–3), 159–213.

Makowsky, J.A. and Pnueli, Y.B. (1993a) Arity vs. alternation in second order definability. In *LFCS'94*, volume 813 of *Lecture Notes in Computer Science*, pp. 240–252. New York: Springer.

Makowsky, J.A. and Pnueli, Y.B. (1993b) Oracles and quantifiers. In *Computer Science Logic, 7th Workshop, CSL '93*, Swansea, United Kingdom, September 13–17, pp. 189–222.

Makowsky, J.A. and Ravve, E.V. (1995) Incremental model checking for decomposable structures. In *Mathematical Foundations of Computer Science (MFCS'95)*, volume 969 of *Lecture Notes in Computer Science*, pp. 540–551. Berlin: Springer-Verlag.

Makowsky, J.A. and Ravve, E.V. (2012) BCNF via attribute splitting. In A. Düsterhöft, M. Klettke, and K.-D. Schewe (eds.) *Conceptual Modelling and Its Theoretical Foundations: Essays Dedicated to Bernhard Thalheim on the Occasion of His 60th Birthday*, volume 7260 of *Lecture Notes in Computer Science*, pp. 73–84. New York: Springer.

Mandrali, E. and Rahonis, G. (2009) Recognizable tree series with discounting. *Acta Cybernetica*, 19(2), 411–439.

Miller, J.C. and Maloney, C.J. (1963) Systematic mistake analysis of digital computer programs. *Communications of the ACM*, 6(2), 58–63.

Monk, J.D. (1976) *Mathematical Logic*. Graduate Texts in Mathematics. Berlin: Springer Verlag.

Monmege, B. (2013) *Spécification et Vérification de Propriétés Quantitatives: Expressions, Logiques et Automates*. PhD thesis, Laboratoire Spécification et Vérification, École Normale Supérieure de Cachan.

Mostowski, A. (1952) On direct products of theories. *Journal of Symbolic Logic*, 17, 1–31.

Mostowski, A. (1957) On a generalization of quantifiers. *Fundamenta Mathematicae*, 44, 12–36.

Myers, G.J. and Sandler, C. (2004) *The Art of Software Testing*. Chichester: Wiley.

Paul, T.K. and Lau, M.F. (2014) A systematic literature review on modified condition and decision coverage. In *Proceedings of the 29th Annual ACM Symposium on Applied Computing*, SAC '14, pp. 1301–1308.

Rabin, M.O. (1965) A tildeple method for undecidability proofs and some applications. In Y. Bar Hillel (ed.) Logic, *Methodology and Philosophy of Science II, Studies in Logic*, pp. 58–68. Amsterdam: North Holland.

Rabin, M. (1969) Decidability of second order theories and automata on infinite trees. *Transactions of the American Mathematical Society*, 141, 1–35.

Rabinovich, A. (2007) Composition theorem for generalized sum. *Fundamenta Informaticae*, 79(1–2), 137–167.

Ravve, E.V. (1995) Model checking for various notions of products. Master's thesis, Department of Computer Science, Technion–Israel Institute of Technology.

Ravve, E.V. (2014a) Analyzing WMSOL definable properties on sum-like weighted labeled trees. In *16th International Symposium on Symbolic and Numeric Algorithms for Scientific Computing, SYNASC 2014*, Timisoara, Romania, September 22–25, pp. 375–382.

Ravve, E.V. (2014b) Views and updates over distributed databases. In *16th International Symposium on Symbolic and Numeric Algorithms for Scientific Computing, SYNASC 2014*, Timisoara, Romania, September 22–25, pp. 341–348.

Ravve, E.V. (2016) Incremental computations over strongly distributed databases. *Concurrency and Computation: Practice and Experience*, 28(11). 3061–3076.

Ravve, E.V. and Volkovich, Z. (2013) A systematic approach to computations on decomposable graphs. In *15th International Symposium on Symbolic and Numeric Algorithms for Scientific Computing, SYNASC 2013*, Timisoara, Romania, September 23–26, pp. 398–405.

Ravve, E.V. and Volkovich, Z. (2016) Incremental reasoning on fuzzy strongly distributed systems. *Proc. 11th International Multi-Conference on Computing in the Global Information Technology*.

Ravve, E.V., Volkovich, Z., and Weber, G.-W. (2014) Effective optimization with weighted automata on decomposable trees. *Optimization Journal, Special Issue on Recent Advances in Continuous Optimization on the Occasion of the 25th European Conference on Operational Research (EURO XXV 2012)*, 63, 109–127.

Ravve, E.V., Volkovich, Z., and Weber, G.-W. (2015a) A uniform approach to incremental automated reasoning on strongly distributed structures. In G. Gottlob, G. Sutcliffe, and A. Voronkov (eds.) *GCAI 2015. Global Conference on Artificial Intelligence*, volume 36 of *EasyChair Proceedings in Computing*, pp. 229–251.

Ravve, E.V., Volkovich, Z., and Weber, G.-W. (2015b) Reasoning on strongly distributed multi-agent systems. In *Proceedings of the 17th International Symposium on Symbolic and Numeric Algorithms for Scientific Computing*, pp. 251–256.

Schützenberger, M.P. (1961) On the definition of a family of automata. *Information and Control*, 4(2–3), 245–270.

Skolem, T. (1958) Review: E.W. Beth, observations metamathematiques sur les structures tildeplement ordonnées. *Journal of Symbolic Logic*, 23(1), 34–35.

Suciu, D. (2000) Query evaluation on distributed semi-structured data. US Patent 6076087.

Tarski, A. (1961) A model-theoretical result concerning infinitary logics. *Notices of the American Mathematical Society*, 8, 260–280.

Valiant, L.G. (1990) A bridging model for parallel computation. *Communications of the ACM*, 33(B), 103–111.

Vardi, M. (1982) The complexity of relational query languages. In *STOC'82*, pp. 137–146. New York: ACM.

Wang, T.-H. and Tan, C.G. (1995) Practical code coverage for Verilog. In *Proceedings of the 4th International Verilog HDL Conference*, pp. 99–104.

Wilcox, P. and Wilcox, P. (2004) *Professional Verification: A Guide to Advanced Functional Verification*. Ifip Series. New York: Springer.

Yan, W.Q. (2016) *Introduction to Intelligent Surveillance*. New York: Springer.

# 5

# Combinatorial Testing: An Approach to Systems and Software Testing Based on Covering Arrays

*Joseph Morgan*

## Synopsis

Software and system test engineers are often faced with the challenge of selecting test cases that maximize the chance of discovering faults while working with a limited budget. The input space of the system under test (SUT) may be large, or the cost of executing the SUT for each input may be high. In any event, exhaustive testing is usually impossible and so the engineer is faced with the challenge of sampling from the input space in such a way that the budget is not exceeded and the chance of discovering faults is high. As it turns out, empirical evidence indicates that system failures are typically due to the interaction of just a few inputs. In fact, for a wide variety of systems, almost all observed failures are due to the interaction of six or fewer inputs. This means that exhaustive testing is usually not necessary. However, even with this evidence, test engineers are still faced with the challenge of selecting a suitable set of test cases. Fortunately, combinatorial testing is an approach to selecting test cases that seeks to address this challenge. This chapter introduces readers to the testing approach known as combinatorial testing. We will illustrate this approach by using examples derived from real software systems. The chapter is intended to provide readers with enough details to allow them to use this approach in their own organizations.

## 5.1 Introduction

### 5.1.1 Systems and Software Testing

In 1972, William Hetzel organized, at the University of North Carolina, the first formal conference devoted to software testing. On reflection, this conference could be considered as a critical juncture in the emergence of testing as a distinct activity in the software development lifecycle. In 1979, *The Art of Software Testing* (Myers, 1979) was published. It is in this book that Myers proposed his now established definition of software testing:

> Testing is the process of executing a program or system with the intent of finding errors.

*Analytic Methods in Systems and Software Testing*, First Edition.
Edited by Ron S. Kenett, Fabrizio Ruggeri, and Frederick W. Faltin.
© 2018 John Wiley & Sons Ltd. Published 2018 by John Wiley & Sons Ltd.

**Table 5.1** Hourly software outage cost.

| | |
|---|---|
| Brokerage operations | $6 450 000 |
| Credit card authorization | $2 600 000 |
| Ebay (one outage 22 hours) | $ 225 000 |
| Amazon | $ 180 000 |
| Home shopping channel | $ 113 000 |
| Airline reservation center | $ 89 000 |
| Cellular service activation | $ 41 000 |

Myers' intention was clearly aimed at reorienting the prevailing view of testing. Myers went further by pointing out that

> ... if our goal is to demonstrate that a program has no errors, then we are subconsciously steered towards that goal.

Since those early formative years, substantial progress has been made in developing software testing methods (Orso and Rothermel, 2014). Despite this progress, a National Institute of Standards and Technology (NIST) report (Tassey, 2002) investigating the economic impact of inadequate software validation and verification methods found:

> ... the national annual costs of an inadequate infrastructure for software testing is estimated to range from $22.2 billion to $59.5 billion.

The NIST report concludes that the current state of software validation and verification practice is woefully lacking. Hartman (2006) also contributes to this dire outlook. At a 2006 Fields Institute workshop on covering arrays, he presented a table of hourly software outage costs to illustrate this point. Table 5.1 contains some of the outage costs presented at the workshop.

Hartman used his table to illustrate the high cost of system/software failures when these failures are discovered in the field. His point is that, although software testing is expensive, inadequate software testing is substantially more expensive. Yet, debugging, testing, and verification activities in typical commercial software development organizations consume 50% to 75% of total software development costs (Hailpern and Santhanam, 2002). A question that naturally arises is

> Can software testing effectiveness be improved without further increasing software development costs?

In order to answer this question, it is useful to examine how test engineers construct test cases. We will use the phrase "test case selection" to refer to this activity.

### 5.1.2 The Test Case Selection Challenge

Imagine a system that has 20 switches that control various system behaviors. The objective is to come up with a set of test cases that will allow a test engineer to determine if the system is ready to be deployed to the field. Let us refer to this system as the system under test (SUT).

Since each switch can either be on or off then we can think of the SUT as having 20 inputs each of which has two settings. As a result, exhaustive testing of the SUT would require $2^{20}$ or 1 048 576 test cases. Let us say that a test engineer needs five seconds to set up, execute, and check actual against expected behavior for each test case. This means that 60½ days would be needed to complete exhaustive testing of this SUT. It is probably unlikely that this timeline would be acceptable and, as a result, exhaustive testing is perhaps not a feasible option. In fact, exhaustive testing is rarely a feasible option in practice. As a result, the primary challenge that test engineers face is the challenge of selecting a subset of cases from the set of test cases that constitutes the input space of the SUT that will allow them to assert with confidence that the SUT is ready to be deployed to the field. You can think of this as a sample selection problem, where the sampling strategy needs to ensure that each additional test case chosen for the sample is as effective as possible at uncovering faults.

There are a number of selection strategies that could be used but, in particular, there are two classes of strategies that are noteworthy, in part because they have received some level of acceptance amongst practitioners.

1) **All values testing**: For this strategy, the idea is that the selected subset of test cases must ensure that, for any input, all possible settings of that input exist at least once. Inputs with a large number of settings are usually partitioned into mutually exclusive categories in order to reduce the number of possible settings (Myers, 1979). There are three reasons why this strategy is appealing to many test engineers. First, it is easy to understand and is easy to implement. Second, it can be a low cost strategy, since the lower bound on the number of test cases required is determined by the input with the largest number of settings, and partitioning can be an effective way of keeping the number of settings low. Third, if for each test case, actual and expected behavior agree then the test engineer can confidently assert that the SUT is free of faults due to any setting of any input. Unfortunately, the test engineer is unable to make such an assertion about faults due to combinations of settings from two or more inputs. To illustrate, for our hypothetical SUT the test engineer would need at least two test cases to satisfy this strategy. The first test case could have all switches set to *on* and the second would then have all switches set to *off*.

2) **Random testing**: For this strategy, a particular test set size budget is usually the determinative factor. For example, for our hypothetical SUT, the test engineer may have the time and resources to execute 100 test cases. The idea is, given the test set size budget and some distribution, randomly select test cases from the input space until the desired number of test cases is achieved (Hamlet, 1994). Usually, a uniform distribution is used but, when available, a distribution based on usage profiles is sometimes used (Thévenod-Fosse and Waeselynck, 1991). Often, this selection strategy also incorporates an all values testing requirement. The appeal of this strategy is similar to that for all values testing but with the added benefit that the selected test set is considered to be somewhat representative of usage patterns in the field, especially when a usage profile distribution is used for test case selection. Unfortunately, as for all values testing, with this basic form of random testing the test engineer is unable to make definitive assertions about faults due to combinations of settings from two or more inputs. Arcuri, Iqbal and Briand (2010) present a formal analysis of random

testing that provides some insight into how this issue may be investigated, but they do not directly address the issue. On the other hand, Nie et al. (2015) directly address the issue and, in addition, provide a comparative analysis of random testing and the test selection strategy proposed by this chapter.

The inherent limitation of these two selection strategies is that they do not take into account a particularly insidious characteristic of software faults. Boris Beizer (1983) eloquently captures this characteristic in his now famous quote:

> Bugs lurk in corners and congregate at boundaries.

The phrase "… lurk in corners …" alludes to the fact that software faults are often due to the incorrect implementation of a requirement that involves two or more inputs. The implication is that, in selecting test cases, it is not sufficient to ensure that all settings of each input are covered, but that the selection strategy should ensure that combinations of settings from two or more inputs are also covered. As it turns out, we need not be too concerned about covering combinations of settings from more than five inputs: Kuhn, Wallace, and Gallo (2004), in examining failures for several classes of software systems, discovered that software failures are rarely due to faults that involve more than five inputs. This finding suggests two very important guiding principles. First, not only is exhaustive testing usually infeasible, it is perhaps not necessary, except for cases where the input space is small enough to be accommodated by the testing budget. Second, a test selection strategy that seeks to cover combinations of input settings is likely to be highly effective at discovering faults even if it is limited to covering just five inputs.

*Combinatorial testing* seeks to address the limitations described above. The term refers to the family of test case selection strategies that use the "covering combinations of input settings" idea as a test case selection criterion. The utility of these ideas was recognized as early as 1985, but it took another ten years for the early advocates to identify the mathematical constructs that are now used. The following section presents a brief timeline of major developments in the emergence of combinatorial testing as a test case selection strategy, and then revisits our hypothetical SUT as a way of motivating the overview of the underlying mathematical constructs that follow.

### 5.1.3 The Emergence of Combinatorial Testing as a Test Case Selection Strategy

There have been several key milestones in the development of combinatorial testing as a test case selection strategy.

1) **Orthogonal latin squares**: Mandl (1985) proposed orthogonal latin squares as the basis of a test case selection strategy. He demonstrated the utility of this approach by using it to derive a set of test cases to validate the Ada compiler.
2) **Orthogonal arrays**: Tatsumi et al. (1987) proposed orthogonal arrays as the basis of a test case selection strategy.
3) **Covering arrays**: Cohen et al. (1997) proposed covering arrays as the basis of a test case selection strategy.

Although all three of these constructs encompass the "covering combinations of inputs" principle described in the section above, covering arrays accomplish

**Table 5.2** Input coverage vs. test set size.

| Input coverage | Test set size |
| --- | --- |
| 6 | 300 |
| 5 | 99 |
| 4 | 39 |
| 3 | 18 |
| 2 | 8 |

this more efficiently than the others and, as a result, current usage of the phrase "combinatorial testing" almost always implies a covering array as the underlying mathematical construct. Consequently, the remainder of this chapter will focus on covering arrays only.

Let us revisit our hypothetical SUT and see if we can get some insight into how covering arrays may be beneficial in addressing this test case selection challenge. We discovered that covering combinations of inputs from more than six inputs is probably unnecessary, so let us examine the expected test set size for covering six through two inputs.

If we use a covering array as a test selection strategy, Table 5.2 shows that we can ensure that all combinations of settings, for *any* six inputs, will be covered by 300 test cases. If we choose to cover all combinations of settings for *any* five inputs then we reduce the test set size to 99. Furthermore, if we are satisfied with covering all combinations of settings for just two inputs, then we only need eight test cases. These test set sizes are the sizes of the smallest known covering arrays (see Colbourn, 2005) for the associated coverage and for 20 inputs.

Remember that our hypothetical SUT problem has 20 inputs, each of which has two settings. If we consider the case where we want to cover six inputs, there are 38 760 ways of selecting six inputs from 20 and, for each of these six input selections, there are $2^6$ or 64 combinations of settings. Remarkably, with only 300 test cases, all 64 combinations of settings will be covered for *any* of the 38 760 possible six input selections. If we consider the case where we want to cover two inputs, there are 190 ways of selecting two inputs from 20 and, for each of these two input selections, there are $2^2$ or 4 combinations of settings. For this case, we only need eight test cases to cover all four setting combinations for any of the 190 possible two input selections. It is instructive to compare these test set sizes to the 1 048 576 test cases that would be needed if we attempted exhaustive testing. Rather than 60½ days to complete exhaustive testing, only 25 minutes would be needed to complete 300 test cases, and only 40 seconds for eight test cases.

Kuhn, Wallace, and Gallo (2004) make the case that covering combinations of inputs from more than six inputs is rarely necessary. For our hypothetical SUT problem, we see that we can select a test set of only 300 test cases that will allow us to discover the typical faults found in the field and, if none are found, we can assert with confidence that the SUT is ready to be deployed. Furthermore, and perhaps just as importantly, if the testing budget does not allow for 300 test cases, we can determine which covering criterion aligns with the testing budget, while at the same time having some sense of the risk involved in the trade-off between coverage and budget. It is perhaps not surprising that Kuhn and Okum (2006) coined the phrase "pseudo-exhaustive testing" as a way to capture the beneficial cost as well as effectiveness of the combinatorial testing approach to test case selection.

The following section presents an overview of covering arrays. We will try to keep the mathematical details to a minimum, and will present simple examples where necessary to illustrate the ideas.

## 5.2 Covering Arrays

### 5.2.1 Basic Concepts

**Definition 5.1** A covering array **CA**$(N; t, k, v)$ is an $N \times k$ array such that the $i$th column contains $v$ distinct symbols. If a **CA**$(N; t, k, v)$ has the property that for any $t$ column projection, all $v^t$ combinations of symbols exist at least once, then it is a $t$-covering array (or strength $t$ covering array). A $t$-covering array is optimal if $N$ is minimal for fixed $t$, $k$, and $v$.

The following two figures illustrate this idea. Figure 5.1 is an example of a five-column 2-covering array in two symbols. If you examine any two columns from this array you will notice that all $2^2 = 4$ combinations of symbols exist at least once. For example, for columns 2 and 5, the combination $(1, 1)$ and $(2, 2)$ occur once but the combinations $(1, 2)$ and $(2, 2)$ each occur twice. Also, notice that this array contains six rows. As it turns out, a five-column, two-symbol, 2-covering array cannot be constructed with fewer than six rows (Renyi, 1971) and so, in that sense, this is an optimal construction. Figure 5.2 is an example of a 3-covering array, also in two symbols and five columns. If you examine any three columns of this array you will notice that all $2^3 = 8$ combinations of symbols exist at least once. This is not an optimal construction though, since we know that such an array can be constructed in ten rows (Johnson and Entringer, 1989).

**Definition 5.2** Consider the covering array $\mathcal{A} = \mathbf{CA}(N; t, k, v)$. If for any $t$-column projection of $\mathcal{A}$ all $v^t$ combinations of symbols exist exactly $\lambda$ times then $\mathcal{A}$ is a $t$-covering orthogonal array of index $\lambda$.

From this definition we see that orthogonal arrays may be thought of as a particular class of covering arrays.

**Definition 5.3** The size of a covering array is the covering array number **CAN**$(t, k, v)$:

$$\mathbf{CAN}(t, k, v) = \min\{N : \exists \mathbf{CA}(N; t, k, v)\}.$$

A related size measure, denoted **CAK**$(N, t, v)$, is the maximum $k$ when $N$, $t$, and $v$ are fixed:

$$\mathbf{CAK}(N, t, v) = \max\{k : \exists \mathbf{CA}(N; t, k, v)\}.$$

These two size measures are related. That is,

$$\mathbf{CAN}(t, k, v) = \min\{N : \mathbf{CAK}(N, t, v) \geq k\}.$$

**Figure 5.1** An optimal **CA**(6; 2, 5, 2).

| | | | | |
|---|---|---|---|---|
| 1 | 1 | 1 | 1 | 1 |
| 2 | 2 | 2 | 2 | 2 |
| 2 | 2 | 2 | 1 | 1 |
| 2 | 1 | 1 | 2 | 2 |
| 1 | 2 | 1 | 2 | 1 |
| 1 | 1 | 2 | 1 | 2 |

**Figure 5.2** **CA**(12; 3, 5, 2).

| | | | | |
|---|---|---|---|---|
| 1 | 1 | 1 | 1 | 1 |
| 1 | 1 | 2 | 1 | 1 |
| 1 | 2 | 1 | 1 | 2 |
| 1 | 2 | 2 | 1 | 2 |
| 2 | 1 | 1 | 2 | 1 |
| 2 | 1 | 2 | 2 | 1 |
| 2 | 2 | 1 | 2 | 2 |
| 2 | 2 | 2 | 2 | 2 |
| 1 | 1 | 1 | 2 | 2 |
| 2 | 2 | 1 | 1 | 1 |
| 2 | 1 | 2 | 1 | 2 |
| 1 | 2 | 2 | 2 | 1 |

**Table 5.3** Size of **CA**(*N*; 2, *k*, 2), *k* ≤ 3003.

| *N* | 4 | 5 | 6 | 7 | 8 | 9 | 10 | 11 | 12 | 13 | 14 | 15 |
|---|---|---|---|---|---|---|---|---|---|---|---|---|
| *k* | 3 | 4 | 10 | 15 | 35 | 56 | 126 | 210 | 462 | 792 | 1716 | 3003 |

As it turns out, **CAN**(2, *k*, 2) is known for all values of *k* (see Katona, 1973; Kleitman and Spencer, 1973). Table 5.3 presents *N* for *k* ≤ 3003.

However, when $t \neq 2$ and $v \neq 2$, **CAN**(*t*, *k*, *v*) is not generally known. Colbourn (2005) maintains tables for the smallest known value of *N* for $2 \leq t \leq 6$, $2 \leq v \leq 25$, and $k \leq 20\,000$, when $t = 2$, and $k \leq 10000$, when $3 \leq t \leq 6$. Also, it is worthwhile to keep in mind the obvious lower bound, **CAN**$(t, k, v) \geq v^t$.

It is useful to consider the case where the cardinality of the symbol set for each column is allowed to vary. For the sake of convenience we will assume that columns are ordered by cardinality from largest to smallest. The covering property of covering arrays is not affected by permuting the columns, and so this assumption is not limiting in any way.

**Definition 5.4** A mixed symbol covering array **MCA**$(N; t, (v_1 \cdot v_2 \cdot \ldots \cdot v_k))$ is an $N \times k$ array such that the *i*th column contains $v_i$ distinct symbols. If, for any *t*-column projection, all $\Pi_{k=1}^{t} v_{f(k)}$ combinations of symbols exist, where $f(k)$ is the *k*th column of the

$t$-column projection, then the array is a $t$-covering array and is optimal if $N$ is minimal for fixed $t$, $k$, and $(v_1 \cdot v_2 \cdot \ldots \cdot v_k)$.

It is usually more convenient to use exponential notation to more compactly express the symbol specification. That is, if $A = \mathbf{MCA}(N; t, (v_1 \cdot v_2 \cdot \ldots \cdot v_k))$ then we express A as:

$$A = \mathbf{MCA}(N; t, (v_1^{k_1} \cdot \ldots \cdot v_m^{k_m})), k = \Sigma_1^m k_m.$$

Figures 5.3 and 5.4 illustrate this generalization. Figure 5.3 is an example of a five-column 2-covering array where the first two columns each contain three symbols and the remaining three columns each contain two symbols. For this example there are three possible outcomes for a two-column projection. First, there is one outcome where each column contains three symbols. In that case there are $3^2 = 9$ possible combinations of symbols. Second, there are six outcomes where one column contains three symbols and the other column contains two symbols. In those cases there are $3 \times 2 = 6$ possible combinations of symbols. Finally there are three outcomes where each column contains two symbols. In those cases there are $2^2 = 4$ possible combinations of symbols. If you examine any two columns from Figure 5.3 you will notice that all possible combinations of symbols exist at least once. Furthermore, since the array contains nine rows we know that it is an optimal construction. Figure 5.4 has the same symbol specification as the array in Figure 5.3, but it is a 3-covering array. If you examine any three columns from Figure 5.4 you will notice that all possible combinations of symbols exist at least once. Note, though, that the array in Figure 5.4 is not an optimal construction since we know that such an array can be constructed in 18 rows (Colbourn et al., 2011).

**Definition 5.5**  The size of a mixed symbol covering array, denoted $\mathbf{MCAN}(t, (v_1^{k_1} \cdot \ldots \cdot v_m^{k_m}))$, is

$$\mathbf{MCAN}(t, (v_1^{k_1} \cdot \ldots \cdot v_m^{k_m})) = \min\{N: \exists \mathbf{MCA}(N; t, (v_1^{k_1} \cdot \ldots \cdot v_m^{k_m}))\}.$$

One way of defining the mixed symbol covering array counterpart to $\mathbf{CAK}(N, t, v)$ is

$$\mathbf{MCAK}_j(N, t, (v_1^{k_1} \cdot \ldots \cdot v_j^{k_j} \cdot \ldots \cdot v_m^{k_m})) = \max\{k_j: \exists \mathbf{MCA}(N; t, (v_1^{k_1} \cdot \ldots \cdot v_j^{k_j} \cdot \ldots \cdot v_m^{k_m}))\}.$$

That is, the maximum $k_j$ when $N$, $t$, and $v_i^{k_i}$, $i \in \{1, \ldots, j-1, j+1, \ldots, m\}$ are fixed. To illustrate, consider the $\mathbf{MCA}(9; 2, (3^2 \cdot 2^k))$ example from Figure 5.3, where for

| | | | | |
|---|---|---|---|---|
| 1 | 1 | 2 | 2 | 2 |
| 1 | 2 | 2 | 2 | 1 |
| 1 | 3 | 1 | 1 | 1 |
| 2 | 1 | 1 | 2 | 1 |
| 2 | 2 | 1 | 1 | 2 |
| 2 | 3 | 2 | 1 | 1 |
| 3 | 1 | 1 | 1 | 2 |
| 3 | 2 | 2 | 1 | 1 |
| 3 | 3 | 2 | 2 | 2 |

**Figure 5.3** An optimal $\mathbf{MCA}(9; 2, (3^2 \cdot 2^3))$.

**Figure 5.4** MCA(21; 3, ($3^2 \cdot 2^3$)).

| | | | | |
|---|---|---|---|---|
| 1 | 1 | 1 | 2 | 1 |
| 1 | 1 | 2 | 1 | 2 |
| 1 | 2 | 1 | 2 | 2 |
| 1 | 2 | 2 | 1 | 1 |
| 1 | 3 | 1 | 1 | 1 |
| 1 | 3 | 2 | 2 | 2 |
| 2 | 1 | 1 | 2 | 2 |
| 2 | 1 | 2 | 1 | 1 |
| 2 | 2 | 1 | 1 | 2 |
| 2 | 2 | 1 | 2 | 1 |
| 2 | 2 | 2 | 2 | 2 |
| 2 | 3 | 1 | 2 | 2 |
| 2 | 3 | 2 | 1 | 2 |
| 2 | 3 | 2 | 2 | 1 |
| 3 | 1 | 1 | 1 | 1 |
| 3 | 1 | 2 | 2 | 2 |
| 3 | 2 | 1 | 2 | 2 |
| 3 | 2 | 2 | 1 | 2 |
| 3 | 2 | 2 | 2 | 1 |
| 3 | 3 | 1 | 1 | 1 |
| 3 | 3 | 2 | 2 | 2 |

that example $k = 3$. The author has discovered by construction that **MCAK**$_2$(9, 2, ($3^2 \cdot 2^k$)) $\geq 20$. This finding indicates that, for this mixed symbol covering array, optimal constructions are possible for $1 \leq k \leq 20$.

Bounds on the size of mixed symbol covering arrays are in general not well explored except when $k$ is small (see Colbourn et al., 2011). It is worth keeping in mind the obvious lower bound, **MCAN**$(t, (v_1 \cdot v_2 \cdot \ldots \cdot v_k)) \geq \Pi_{k=1}^t v_k$, where columns are assumed ordered by cardinality from largest to smallest.

It is useful to further generalize covering arrays to account for potential constraints on combinations of symbols. There are several possible constraint types that we could con-template (for a complete discussion see Cohen et al., 2008), but in this chapter we will only address constraints that restrict the possible symbol combinations for one or more $p$-column projections (where $2 \leq p \leq k$). Such constraints are typically referred to as "dis-allowed combinations" or "forbidden configurations." We will use the term "disallowed combinations" subsequently.

**Definition 5.6** A constrained covering array **CCA**$(N; t, (v_1 \cdot v_2 \cdot \ldots \cdot v_k), \phi)$ is an $N \times k$ array such that the $i$th column contains $v_i$ distinct symbols and $\phi$ is a set of $p$-tuples $(2 \leq p \leq k)$ such that each tuple is a set of two or more column/symbol pairs identifying a disallowed combination. If for any $t$-column projection, all *possible* combinations of symbols exist at least once, then it is a $t$-covering array and is optimal if $N$ is minimal for fixed $t$, $k$, $(v_1 \cdot v_2 \cdot \ldots \cdot v_k)$, and $\phi$.

Figure 5.5 illustrates this idea. Notice that the exponential notation is used to express the symbol specification. As for the examples in Figures 5.3 and 5.4, Figure 5.5 is a five-column 2-covering array where the first two columns each contain three symbols and the remaining three columns each contain two symbols. However, in this case, there is a constraint on the possible symbol combinations for columns $c_1$ and $c_3$. That is, the symbol combinations (1, 2), (2, 1), (2, 2), (3, 1), and (3, 2) are allowed, but the symbol combination (1, 1) is not allowed. We will denote disallowed combinations as a tuple of ordered pairs where each ordered pair represents a column and its associated symbol from the disallowed combination. In this case, there are two columns involved, so we have the 2-tuple $\{(c_1, 1), (c_3, 1)\}$.

If you examine any two columns from this array you will notice that all possible symbol combinations exist at least once. Since columns $c_1$ and $c_2$ each contain three symbols and they are not involved in the constraint, we know that the construction must be optimal since the array contains nine rows.

Let us consider a slightly more complicated variant of the example above. As before, we have a constraint on the possible symbol combinations for columns $c_1$ and $c_3$. However, for this example the allowed symbol combinations are (2, 1), (2, 2), (3, 1), and (3, 2), and the disallowed symbol combinations are (1, 1) and (1, 2). That is, when column $c_1$ is set to 1 there are no allowable settings for column $c_3$. This poses a couple of dilemmas because, as we can see in Figure 5.6, these constraints lead to an array where the cells in column $c_3$ that correspond to cells in column $c_1$ that are set to 1 cannot be assigned a symbol from the allowed symbol set of column $c_3$. The first dilemma is a representational one, in that we need a way to indicate that the constraint set $\phi$ results in cells where an assignment cannot be made. In Morgan, 2009, the symbol "." is used to indicate an unassigned cell and so we will adopt that convention. The second dilemma has to do with the meaning of "covering" when a $t$ column projection contains tuples with unassigned cells. In Morgan (2009) such tuples are excluded when evaluating the $t$-covering of $t$-column projections, and so we will also adopt that convention. With this background, we see that the array in Figure 5.6 is in fact a 2-covering array since all possible symbol combinations exist at least once for any two-column projection.

| $c_1$ | $c_2$ | $c_3$ | $c_4$ | $c_5$ |
|-------|-------|-------|-------|-------|
| 1 | 1 | 2 | 2 | 1 |
| 1 | 2 | 2 | 1 | 2 |
| 1 | 3 | 2 | 2 | 2 |
| 2 | 1 | 1 | 2 | 2 |
| 2 | 2 | 2 | 2 | 2 |
| 2 | 3 | 2 | 1 | 1 |
| 3 | 1 | 2 | 1 | 1 |
| 3 | 2 | 1 | 2 | 1 |
| 3 | 3 | 1 | 1 | 2 |

**Figure 5.5** An optimal **CCA**$(9; 2, (3^2 \cdot 2^3), \phi = \{\{(c_1, 1), (c_3, 1)\}\})$.

**Figure 5.6** An optimal **CCA**$(9; 2, (3^2 \cdot 2^3), \phi = \{\{(c_1,1), (c_3,1)\}, \{(c_1,1), (c_3,2)\}\})$.

| $c_1$ | $c_2$ | $c_3$ | $c_4$ | $c_5$ |
|---|---|---|---|---|
| 1 | 1 | . | 1 | 1 |
| 1 | 2 | . | 2 | 2 |
| 1 | 3 | . | 2 | 1 |
| 2 | 1 | 1 | 2 | 1 |
| 2 | 2 | 2 | 1 | 1 |
| 2 | 3 | 2 | 1 | 2 |
| 3 | 1 | 2 | 2 | 2 |
| 3 | 2 | 1 | 1 | 1 |
| 3 | 3 | 1 | 2 | 2 |

The covering array in Figure 5.6 is a special subclass of constrained covering arrays in that unassigned cells are allowed. Such covering arrays are referred to as *unsatisfiable constrained covering arrays* (Morgan, 2009).

The size measures defined in Definition 5.5 extend naturally to constrained covering arrays. We will use **CCAN**$(t, (v_1^{k_1} \cdot \ldots \cdot v_m^{k_m}), \phi)$ and **CCAK**$_j(N, t, (v_1^{k_1} \cdot \ldots \cdot v_j^{k_j} \cdot \ldots \cdot v_m^{k_m}), \phi)$ to denote the corresponding constrained covering array size measures.

## 5.2.2 Metrics

Given a $t$-covering array, a natural question that arises is: To what extent does a $t$-covering array cover symbol combinations for projections involving $t+1$, $t+2$, ..., $t+i$ columns (where $t+i \leq k$)? Dalal and Mallows (1998) propose two metrics, $t$-Coverage and $t$-Diversity, to address this question. We present reformulated versions of these metrics below, and then generalize them to accommodate constrained covering arrays. We use the terms "*adjusted t-Coverage*" and "*adjusted t-Diversity*" (Morgan, 2014) to distinguish our generalized metrics from the metrics proposed by Dalal and Mallows.

The following notation will be used:

- $_uC_v$ is the number of combinations of $u$ things taken $v$ at a time.
- $t$ is the strength of the covering array.
- $K$ is the number of columns.
- $M = _KC_t$.
- $v_i$ is the number of symbols for the $i$th column.
- $n_i$ is the number of distinct $t$ tuples in the covering array for the $i$th projection.
- $p_i = \Pi_{k=1}^{t} v_{f(k)}$, where $f(k)$ is the $k$th column of the $i$th projection.
- $r$ is the number of rows of the covering array.

**Definition 5.7** For an $N \times k$ array where the $i$th column contains $v_i$ distinct symbols, its $t$-Coverage $(1 \leq t \leq k)$ is the ratio of the number of *distinct* $t$-tuples to the number of *possible* $t$-tuples, averaged over all $t$-column projections. That is, using the notation above,

$$t\text{-Coverage} = \frac{1}{M} \sum_{i=1}^{M} n_i/p_i.$$

If $t$-Coverage is 100% for the array then the array is a $t$-covering array.

**Definition 5.8** For an $N \times k$ array where the $i$th column contains $v_i$ distinct symbols, its $t$-Diversity $(1 \leq t \leq k)$ is the ratio of the number of *distinct* $t$-tuples to the *total* number of $t$-tuples in the array, averaged over all $t$-column projections. That is, using the notation above,

$$t\text{-Diversity} = \frac{1}{M} \sum_{i=1}^{M} n_i/r.$$

If $t$-Diversity is 100% then all $t$-tuples in the array are different, whereas a $t$-Diversity of 50% indicates that all $t$-tuples, on average, occur twice. The $t$-Diversity is therefore a measure of how well the covering array avoids replication. Dalal and Mallows (1998) argue that for a $t$-covering array where $t$-Diversity is less than 100%, higher coverage may be obtained by increasing $t$-Diversity. Colbourn and Syrotiuk (2016) and Nie et al. (2015) also explore this issue.

Table 5.4 presents $t$-Coverage and $t$-Diversity for the covering arrays in Figures 5.1–5.4. The intent is to provide some sense of how those $t$-covering arrays cover symbol combinations for projections involving $t+1$ and $t+2$ column projections. Notice that the 2-covering arrays cover 70% and 64% respectively of the symbol combinations when three columns are projected, and when four columns are projected, they are able to cover 38% and 30% of the symbol combinations. We see a similar pattern for the 3-covering arrays. Although these values only hold for these examples, they nevertheless indicate that a $t$-covering array could provide substantial coverage at the $t+1$ level. Table 5.4 also demonstrates how $t$-Coverage may be used to assess the trade-off between covering array size and its covering ability.

These coverage and diversity definitions cannot be applied to constrained designs without making adjustments to account for how constraints affect the number of *possible* $t$-column symbol combinations. Furthermore, to accommodate unsatisfiable covering arrays, rows that contain missing values (i.e. unassignable cells) must be excluded from the computation of $n_i$.

Some more notation:

- $a_i$ is the number of invalid $t$-tuples arising from columns in the $i$th projection.
- $m$ is the number of projections where there are no valid $t$-tuples.
- $q_i$ is the number of rows in the covering array with missing values for any column in the $i$th projection.
- $r_i = r - q_i$.
- $M' = M - m$.

**Definition 5.9** For an $N \times k$ array where the $i$th column contains $v_i$ distinct symbols, its *adjusted* $t$-Coverage $(1 \leq t \leq k)$ is the ratio of the number of *distinct* $t$-tuples to the

**Table 5.4** *t*-Coverage (%) and *t*-Diversity (%) for Figures 5.1–5.4.

| $t$ | CA(6;2,5,2) | | CA(12;3,5,2) | | MCA(9;2,($3^2 \cdot 2^3$)) | | MCA(21;3,($3^2 \cdot 2^3$)) | |
|---|---|---|---|---|---|---|---|---|
| | *t*-Cov | *t*-Div | *t*-Cov | *t*-Div | *t*-Cov | *t*-Div | *t*-Cov | *t*-Div |
| 2 | 100 | 67 | 100 | 33 | 100 | 63 | 100 | 27 |
| 3 | 70 | 93 | 100 | 67 | 64 | 92 | 100 | 64 |
| 4 | 38 | 100 | 70 | 93 | 30 | 100 | 65 | 94 |
| 5 | — | — | 38 | 100 | — | — | 29 | 100 |

number of *adjusted possible t*-tuples, averaged over all *t*-column projections that contain valid tuples. That is, using the additional notation above,

$$adjusted\ t\text{-Coverage} = \frac{1}{M'} \sum_{i=1}^{M'} n_i / (p_i - a_i).$$

**Definition 5.10** For an $N \times k$ array where the *i*th column contains $v_i$ distinct symbols, its *adjusted t*-Diversity ($1 \le t \le k$) is the ratio of the number of *distinct t*-tuples, to the *adjusted total* number of *t*-tuples in the array, averaged over all *t*-column projections that contain valid tuples. That is, using the notation above,

$$adjusted\ t\text{-Diversity} = \frac{1}{M'} \sum_{i=1}^{M'} n_i / r_i.$$

Note that if $M' = M$ and, $\forall i \in \{1, \ldots, M\}$, $r_i = r$, then *adjusted t*-Coverage reduces to *t*-Coverage and *adjusted t*-Diversity reduces to *t*-Diversity.

Table 5.5 presents *adjusted t*-Coverage and *adjusted t*-Diversity for the covering arrays in Figures 5.5 and 5.6. As for Table 5.4, the intent is to provide some sense of how those covering arrays cover $t + 1$ and $t + 2$ column projections. Notice that exponent notation is used in Table 5.5 for the constraint set (Cohen et al., 2008). This shorthand notation indicates the number of *p*-tuples in the constraint set (i.e. $p^k$ indicates $k$ *p*-tuples).

The examples we have used to illustrate these metrics are covering arrays. However, as defined in this section, coverage and diversity can be computed for any $N \times k$ array. If for some $N \times k$ array it turns out that there is no setting of $t$ that results in 100% *t*-Coverage, then the array is not a covering array. Conversely, if there are several values of $t$ where *t*-Coverage is 100%, and $t'$ is the largest of these values, then the array is a $t'$-covering array.

We have focused on *t*-Coverage and *t*-Diversity as well as their *adjusted* variants in this section. This is not intended to be a comprehensive overview of the topic. In fact, a variety of additional metrics have been proposed since Dalal and Mallows (1998). We direct the interested reader to Kuhn et al. (2013) for additional discussion of the topic.

### 5.2.3 Tools for Constructing Covering Arrays

There are currently several available tools that may be used to construct covering arrays. Some of these tools are commercial products offered at a wide range of prices, some are

**Table 5.5** *Adjusted t*-Coverage (%) and *adjusted t*-Diversity (%) for Figures 5.5 and 5.6.

| t | CCA(9;2,($3^2 \cdot 2^3$),$\phi = \{2^1\}$) | | CCA(9;2,($3^2 \cdot 2^3$),$\phi = \{2^2\}$) | |
|---|---|---|---|---|
|   | *t*-Cov | *t*-Div | *t*-Cov | *t*-Div |
| 2 | 100 | 62 | 100 | 71 |
| 3 | 68 | 92 | 60 | 96 |
| 4 | 32 | 100 | 28 | 100 |

**Table 5.6** Covering array construction tools.

| Tool | Notes |
|---|---|
| ACTS | Free, graphical user interface (GUI). Available from NIST. See Yu et al. (2013). |
| AETG | Commercial, web-based tool. Available from Applied Communication Sciences (formerly Telcordia). See Cohen et al. (1997). |
| AllPairs | Free tool. Available from Satisfice Inc. |
| DDA | See Colbourn, Cohen, and Turban (2004). |
| EXACT | See Yan and Zhang (2006). |
| Hexawise | Commercial, web-based tool. Free for some customers. |
| IPO-s | See Calvagna and Gargantini (2009). |
| Jenny | Free, command-line tool. |
| JMP | Commercial, GUI. "Covering Arrays" is one of several options under the DOE menu. Available from SAS Institute. |
| OATS | See Brownlie, Prowse, and Phadke (1992). |
| PICT | Free, open source, command-line tool developed at Microsoft. See Czerwonka (2006). |
| Pro-Test | Commercial, GUI. Available from SigmaZone. |
| rdExpert | Available from Phadke Associates Inc. |
| TConfig | See Williams (2000). |
| TestCover | Commercial, web-based tool. |
| WHITCH | Available from IBM. See Hartman, Klinger, and Raskin (2011). |

commercial products offered under a service model, and some are free. Table 5.6 gives an annotated list, sorted alphabetically, of 16 of these tools. It is not meant to be a comprehensive list but instead reflects tools that either appear frequently in the literature or that the author has personal experience with. Hopefully the list provides sufficient variety, in terms of pricing and capability, that the interested reader will be able to find a tool that they can use to further investigate these constructs.

These tools employ a variety of methods to construct covering arrays, the details of which are beyond the scope of this chapter. Furthermore, the construction methods used by some of these tools are not known. However, Colbourn (2004) provides a comprehensive and accessible overview of several classes of algorithms that are often used to construct covering arrays.

## 5.3 Systems and Software Testing in Practice

### 5.3.1 Testing Fundamentals

The process of systems and software testing involves several activities that can usually be organized into three broad categories:

1) Planning and test preparation.
2) Test case selection.
3) Test set execution and analysis.

This basic framework also applies to combinatorial testing, except for test case selection where, as was discussed earlier, a "covering combinations of inputs" test case selection criterion is used. In the previous section we saw how a covering array satisfies this covering criterion, and we will see in subsequent sections how this applies in practice. Before getting to our case studies, let us briefly review a few techniques and activities that are involved in the process of systems and software testing.

Determining the input space of an SUT is probably the most important test preparation task. First, inputs are identified and then, for each input, the set of allowable values is determined. It is often the case that the input space is very large; for those situations where the type of one or more inputs is continuous, the input space will be infinitely large. When faced with large input spaces, test engineers usually resort to a technique known as *equivalence partitioning* (Myers, 1979). The idea is to examine each input and, where necessary, divide the range of the input into a number of mutually exclusive categories known as equivalence classes. The expectation is that, from the standpoint of the SUT, any individual member of an equivalence class is representative of the entire class. In practice, test engineers select between one and three members from each equivalence class. Sometimes, *boundary value analysis* is used to choose the members from each equivalence class. This technique is inspired by the "bugs … congregate at boundaries" phrase from the Boris Beizer (1983) quote in our opening section. Software faults are sometimes due to the incorrect implementation of requirements that specify different behaviors for different input values and, as a result, the idea is that members selected from the boundaries of equivalence classes are more likely to identify such faults. In addition, as this activity unfolds, the test engineer will determine if there are constraints that bound the input space.

There are two distinct sets of activities that constitute test case selection. There are those activities that are concerned with selecting, or creating, individual test cases, and then there are those activities that are concerned with determining the *expected result* for each test case. For a covering array based approach, creating individual test cases first involves mapping inputs, input values, and constraints identified during test preparation to a covering array specification, deciding on the appropriate covering criterion (or strength *t*), and then finally construction of the covering array. The mechanics of how this is done will be determined by the tool used for construction (see Table 5.6). As it turns out, for any testing method, determining expected results for test cases is perhaps the most important of all testing activities. A test engineer must have some way of knowing the expected outcome of a test case in order to be able to assess if the outcome is, or is not, a failure. This is known as the *test oracle* problem (for an in-depth discussion, see Barr et al., 2015).

Test set execution results in two sets of outcomes: the set of test cases that result in failure and the set of test cases that do not result in failure. We will use the phrases *failure set* and *success set* respectively to refer to these outcomes. If the failure set turns out to be non-empty then analysis will be required. In this context, analysis is the task of identifying inputs and associated settings that precipitate failure outcomes. This is known as the *fault location* problem. We direct the reader to Colbourn and Syrotiuk (2016) and Ghandehari et al. (2013) for additional discussion of this topic.

### 5.3.2 Case Studies

#### 5.3.2.1 Case Study 1: The JMP Multivariate Preferences Dialog

This case study is intended to illustrate how a combinatorial testing approach may be used for deriving test cases for a graphical user interface (GUI).

Figure 5.7 is a screenshot of the preferences dialog for the Multivariate platform from the JMP statistical analysis tool. There are 120 such preferences dialogs for the variety of platforms that the tool offers. This dialog is an example of a moderately complex preferences dialog. The SUT is the preferences subsystem that manages preferences for the Multivariate platform. The relevant controls are therefore those in the Options panel of the dialog. The objective is to derive a set of test cases that would allow a test engineer to

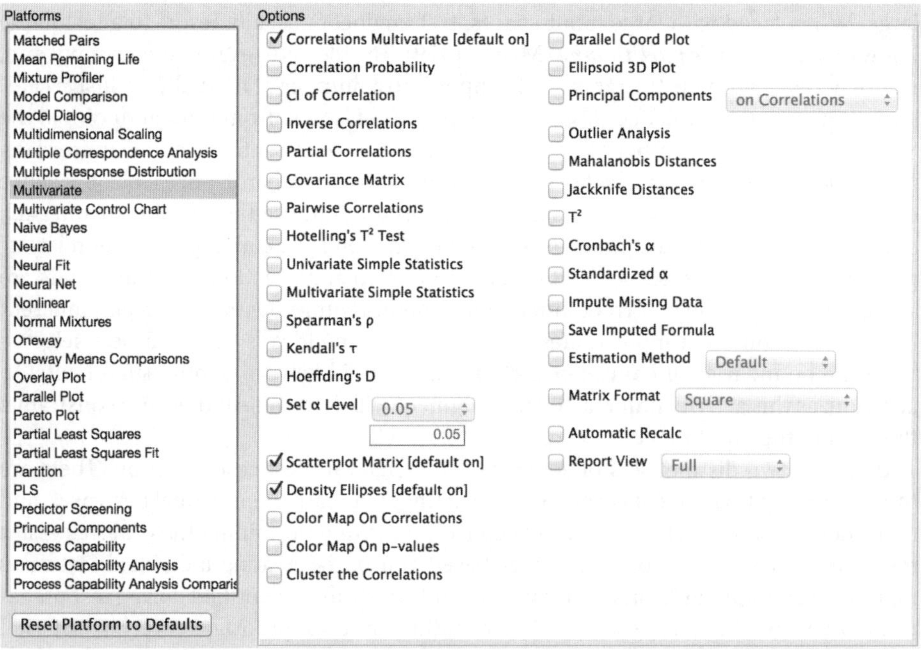

**Figure 5.7** JMP v13 Multivariate platform preferences dialog.

determine if this SUT is fit for use. As we will see in the test preparation section below, there are two ways in which we could approach this problem.

1) **Planning and test preparation**: The inputs of a GUI are determined by the controls that the GUI presents to the user. Usually each control is an input, and so allowable values for each input are determined by examining the options that the corresponding control presents. It is not always a one-to-one mapping, though. In this case there are three types of controls, 34 check boxes, five combo boxes, and a single edit box. However, the combo boxes are paired with the adjacent check boxes. They are in a disabled state until the checkbox is checked, so each check box / combo box pair is a single meta-control. Furthermore, the edit box is associated with its adjacent check box / combo box pair, so they constitute a triplet that is also a meta-control. There are therefore 34 inputs to consider: 29 that map to individual check boxes and five that map to meta-controls. Let us determine values for these controls.

   Note that the meta-controls could have been alternatively handled as the individual constituent controls with constraints restricting allowable level combinations. Case Study 3 presents a problem where constraints are used to restrict allowable level combinations.

   a) **Check box controls**: Check boxes can be in one of two states, checked or unchecked, so there are two possible values for each individual check box.

   b) **Meta-controls**: Figure 5.8 contains screenshots of the combo box menus for our meta-controls. The "Other" option in the "Set $\alpha$ Level" menu enables the corresponding edit box, thus allowing a user to enter a real number other than the four listed in the menu. To determine possible values we should therefore use equivalence partitioning to partition the range of possible values. Let us assume that three equivalence classes are sufficient, one for the four listed menu values and the other two for values below and above the listed values. Let us select two values from the *below* equivalence class, two from the *above* class, and the four listed values from the remaining class. This means that the input corresponding to this control will have eight possible values. Possible values for inputs corresponding to the remaining meta-controls will simply be the number of menu options for that control.

2) **Test case selection**: We determined that there are 29 two-level inputs from check boxes, one two-level input from the "Report View" meta control, one two-level input from the "Report View" meta control, one three-level input from the "Matrix Format" meta control, one six-level input from the "Estimation Method" meta control, one four-level input from the "Principal Components" meta control, and one eight-level input from the "Set $\alpha$ Level" meta control. Since there are no constraints, and if we assume that a strength two covering array is sufficient, then the **MCA**($N$; 2, ($8^1 \cdot 6^1 \cdot 4^1 \cdot 3^1 \cdot 2^{30}$)) array will be the basis for our test case selection strategy. As it turns out, this covering array can be optimally constructed in 48 rows. Table 5.7 presents the $t$-Coverage and $t$-Diversity for one such construction, indicating how the array covers $t = 2$, as well as $t + 1$ through $t + 4$ column projections.

   This 48-row covering array represents 48 test cases sampled from the 618 475 290 624 element input space for this SUT. In addition, these test cases cover all level combinations of any two of the 34 inputs of the SUT. From Table 5.7 we see that coverage remains high even for five input level combinations, and only falls below 50% for

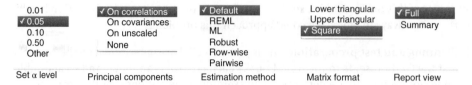

**Figure 5.8** Multivariate preferences meta-control menus.

**Table 5.7** *t*-Coverage (%) and *t*-Diversity (%): Multivariate preferences.

| | MCA(48;2,($8^1 \cdot 6^1 \cdot 4^1 \cdot 3^1 \cdot 2^{30}$)) | |
|---|---|---|
| *t* | *t*-Cov | *t*-Div |
| 2 | 100 | 12 |
| 3 | 97 | 25 |
| 4 | 85 | 45 |
| 5 | 62 | 66 |
| 6 | 37 | 81 |

six input level combinations. As impressive as these efficiencies are, a test engineer would probably choose the covering criterion that best aligns with the testing budget, so it is worth examining the covering array number for higher strengths.

Table 5.8 shows that the 3-covering array is four times the size of the 2-covering array. The incremental cost in size to achieve the extra 3% in coverage is therefore substantial. Nevertheless, this is the type of tradeoff that test engineers routinely make.

Note that we can also see from Table 5.8 that, as for $t = 2$, there is an optimal construction for $t = 3$ and $t = 4$ since $N$ is equal to the lower bound. However, the reported size for $t = 5$ is not known to be optimal. For an **MCA**($N$; 5, ($8^1 \cdot 6^1 \cdot 4^1 \cdot 3^1 \cdot 2^{30}$)), 2130 rows is an upper bound on $N$ and is the smallest $N$ known to the author.

3) **Test set execution and analysis**: One of the analysis tasks that test engineers are sometimes asked to perform is the assessment of *early release* risk (for details, see Yang et al., 2008). The term *early release* means releasing a software product for operational use before testing is complete. One of the benefits of the *t*-Coverage metric is that it can be computed for a subset of rows of a covering array. Imagine a scenario where, for this SUT, the test engineer has been asked to assess early release risk but has only executed 75% of the test cases from the 48-row 2-covering design. One way the engineer could provide an informed judgment about early release risk is to construct a *t*-Coverage report, as in Table 5.7, for the subset of the 36 executed test cases that result in success, and use that report to aid in making the risk assessment.

### 5.3.2.2 Case Study 2: The Traffic Collision Avoidance System (TCAS)

This case study presents an example of combinatorial testing applied to a real-time software system.

TCAS is an airborne collision avoidance system that warns pilots of nearby aircraft that pose a mid-air collision threat. The International Civil Aviation Organization mandates that aircraft of a certain mass, or a particular passenger capacity, must be fitted

**Table 5.8** Covering array number: Multivariate.

| $t$ | $MCAN(t,(8^1 \cdot 6^1 \cdot 4^1 \cdot 3^1 \cdot 2^{30}))$ |
|---|---|
| 2 | 48 |
| 3 | 192 |
| 4 | 576 |
| 5 | 2130 |

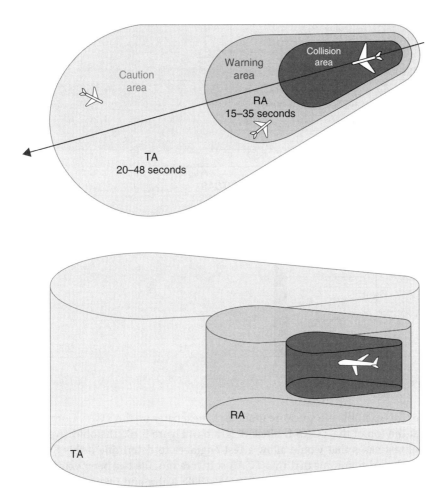

**Figure 5.9** TCAS protection envelope.

with TCAS. Figure 5.9 is an illustration of the protection envelope that surrounds each TCAS-equipped aircraft (see Federal Aviation Administration, 2011).

There are currently two versions of TCAS in use: TCAS I is intended for aircraft with a capacity of 10 to 30 passengers, and TCAS II is for aircraft with more than 30 passengers. Figure 5.10 is a block diagram of the TCAS II system (see Federal Aviation Administration, 2011).

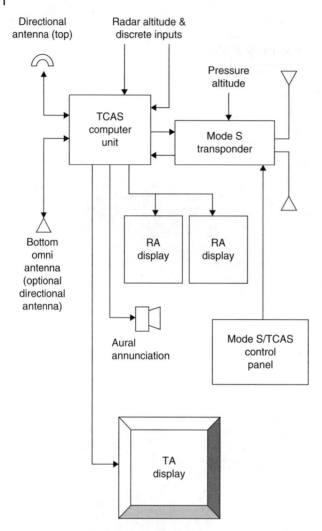

**Figure 5.10** TCAS II block diagram.

The SUT for this example is a software module that executes on the TCAS Computer Unit shown at the top of the TCAS II block diagram in Figure 5.10. The objective is to derive a set of test cases that would allow a test engineer to determine if this SUT is fit for use. We should point out that this TCAS software module has been extensively studied in the literature (Rothermel and Harrold, 1998; Kuhn and Okum, 2006). We chose TCAS for this case study partly because the interested reader will readily find examples of alternative approaches to validating TCAS that will allow them to assess the relative merits of the combinatorial testing approach.

1) **Planning and test preparation**: The SUT has 12 inputs that supply values of *own* aircraft and *other* aircraft. The term *own* is used to refer to the aircraft on which the SUT is executing, and *other* refers to the threat aircraft. The 12 inputs are of varying types: four are continuous, three are Boolean, and the remainder are discrete. Kuhn

and Okum (2006) use equivalence partitioning to partition the inputs and choose one representative value from each class. We will adopt their partitioning scheme; in Table 5.9 we reproduce from Kuhn and Okum (2006, Figure 5.2) the values they choose for each input.

2) **Test case selection**: As for Case Study 1 there are no constraints, and we will again start with a strength two covering array. The **MCA**$(N; 2, (10^2 \cdot 4^1 \cdot 3^2 \cdot 2^7))$ array will therefore be the basis for our test case selection strategy. As it turns out, this covering array can be optimally constructed in 100 rows (Moura et al., 2003, Theorem 5.1). Table 5.10 presents $t$-Coverage and $t$-Diversity for one such construction, indicating how that construction covers $t = 2$, as well as $t + 1$ and $t + 2$ column projections.

This 100-row covering array represents 100 test cases sampled from the 460 800-element input space for this SUT. In addition, these test cases cover all level combinations of any two of the 12 inputs of the SUT. From Table 5.10 we see that coverage is 71% for three input level combinations and falls below 50% for four input level combinations. Let us examine the covering array number for higher strengths.

From Table 5.11 we see that the 3-covering array is four times the size of the 2-covering array. The incremental cost in size to achieve the extra 29% in coverage is substantial, but less so than the example from Case Study 1.

Note that we can also see from Table 5.11 that, as for $t = 2$, there are optimal constructions for $t = 3, t = 4$, and $t = 5$, since $N$ is equal to the lower bound for these cases. However, the reported size for $t = 6$ is not known to be optimal. For an **MCA**$(N; 6, (10^2 \cdot 4^1 \cdot 3^2 \cdot 2^7))$, 9600 rows is an upper bound on $N$ and is the smallest $N$ known to the author.

3) **Test set execution and analysis**: One of the benefits of using the TCAS software module as a case study is that we have evidence of the effectiveness of using covering arrays as a test case selection method for this module. The TCAS software module, along with other software artifacts, was released to the public for research in 2005 (Do et al., 2005). At the time, 41 versions of the module were released. The variants were seeded with the types of faults typically found in real software applications. The intent was to create a set of variants that would allow researchers to evaluate the effectiveness of various testing methods. Kuhn and Okum (2006) used these variants to assess the effectiveness of using test cases derived from covering arrays to detect the seeded faults. They discovered that test cases derived from a strength 5 covering array were able to identify all faulty versions. We reproduce a summary of their findings in Table 5.12.

### 5.3.2.3 Case Study 3: The GNU Compiler Collection (GCC) Optimizer

This case study presents an example of combinatorial testing applied to a software system that has constraints on the input space.

GCC is a compiler infrastructure with support for multiple languages (e.g. C, C++, Ada) and many target machine architectures. Cohen, Dwyer, and Shi (2008) examine the GCC 4.1 documentation and, for the optimizer module, they identify 199 configuration options. Of these options, 189 are binary choices and the remaining ten each provide three choices. They also discover 40 constraints, involving 35 options, which limit the configuration space. Three of these constraints involve a 3-tuple of option/setting pairs, and the remaining 37 constraints are 2-tuples.

**Table 5.9** Inputs and values for TCAS software module.

| Input | Values |
|---|---|
| Cur_Vertical_Sep | 299, 300, 601 |
| High_Confidence | Boolean (i.e. T, F) |
| Two_of_Three_Reports_Valid | Boolean (i.e. T, F) |
| Own_Tracked_Alt | 1, 2 |
| Other_Tracked_Alt | 1, 2 |
| Own_Tracked_Alt_Rate | 600, 601 |
| Alt_Layer_Value | 0, 1, 2, 3 |
| Up_Separation | 0, 399, 400, 499, 500, 639, 640, 739, 740, 840 |
| Down_Separation | 0, 399, 400, 499, 500, 639, 640, 739, 740, 840 |
| Other_RAC | NO INTENT, DO_NOT_CLIMB, DO_NOT_DESCEND |
| Other_Capability | TCAS_TA, OTHER |
| Climb_Inhibit | Boolean (i.e. T, F) |

**Table 5.10** *t*-Coverage (%) and *t*-Diversity (%): TCAS.

| | MCA(100;2,($10^2 \cdot 4^1 \cdot 3^2 \cdot 2^7$)) | |
|---|---|---|
| *t* | *t*-Cov | *t*-Div |
| 2 | 100 | 13 |
| 3 | 71 | 28 |
| 4 | 39 | 42 |

**Table 5.11** Covering array number: TCAS.

| *t* | MCAN(*t*, ($10^2 \cdot 4^1 \cdot 3^2 \cdot 2^7$)) |
|---|---|
| 2 | 100 |
| 3 | 400 |
| 4 | 1200 |
| 5 | 3600 |
| 6 | 9600 |

The SUT for this example is the infrastructure that supports the configuration options. Notice that this example is conceptually similar to the example presented in Case Study 1, but in this case constraints limit the input space. The objective is to derive a set of test cases that would allow a test engineer to determine if this SUT is fit for use.

1) **Planning and test preparation**: For this example, each option may be thought of as an input and so there are 189 two level and 10 three level inputs. Furthermore, there are 40 constraints that limit the allowable level combinations for 35 of these inputs.

**Table 5.12** Faulty variants detected: TCAS.

| $t$ | % detected variants |
| --- | --- |
| 2 | 54 |
| 3 | 74 |
| 4 | 89 |
| 5 | 100 |

We cannot present all 40 of these disallowed combinations, but present instead the 3-tuples and six of the 2-tuples, as they would be expressed in the JMP tool (see Table 5.6).

Figure 5.11 expresses the set of disallowed combinations as a Boolean expression where the symbol "∧" denotes conjunction and "∨" denotes disjunction. Also, generic names are used for options (i.e. X$k$ represents the $k$th option) and for levels (i.e. L$k$ represents the $k$th level of an option). The ellipses represent the 31 tuples omitted from this expression. You may think of using this expression in the following way. Imagine building the constrained covering array one row at a time; if the expression is true for a candidate row then that row violates at least one of the constraints and so it would be discarded.

2) **Test case selection**: We will begin with a strength two covering array and, since we have constraints, we will need a **CCA**$(N; 2, (3^{10} \cdot 2^{189}), \phi = \{3^3 \cdot 2^{37}\})$ for our test case selection strategy. As it turns out, this constrained covering array can be constructed in 19 rows (Jia et al., 2015). However, we do not have metrics for that construction and so we present metrics for a 20-row construction in Table 5.13.

Note that, for the sake of brevity, we use the exponential notation for constraints (see Section 5.2.2).

The 20-row array in Table 5.13 represents 20 test cases sampled from the configuration space for this SUT. In addition, these test cases cover all level combinations of any two of the 199 inputs of the SUT. From Table 5.13 we see that coverage is 94% for three input level combinations and remains high up to four input level combinations; it only falls below 50% for five input level combinations. Let us examine the covering array number for higher strengths.

Note that the unconstrained configuration space has $4.6 \times 10^{61}$ possible configurations.

Table 5.14 presents the sizes for the best constructions as reported by Jia et al. (2015), as well as the sizes of constructions obtained by the author using JMP (see Table 5.6). For both constructions we see that the 3-covering array is almost five times the size of the 2-covering array. For the JMP construction, Table 5.13 shows that the 2-covering array covers 94% of three input level combinations, and so the incremental cost in size to achieve the extra 6% in coverage is substantial.

Note that the covering array numbers reported in Table 5.14 are not known to be optimal. These are upper bounds on $N$ and are the smallest $N$ known to the author.

3) **Test set execution and analysis**: In the preamble to the case study section we mentioned *fault location* as one of the activities that would be conducted if the *failure* set were non-empty. There are essentially two approaches to the fault location problem:

( X1 == L2 ∧ X2 == L1 ) ∨
( X2 == L1 ∧ X3 == L2 ) ∨

...

( X22 == L1 ∧ X36 == L3 ) ∨
( X23 == L2 ∧ X24 == L1 ∧ X25 == L1 ) ∨
( X23 == L2 ∧ X24 == L1 ∧ X25 == L2 ) ∨
( X23 == L2 ∧ X24 == L2 ∧ X25 == L1 ) ∨
( X25 == L1 ∧ X26 == L2 ) ∨

...

( X24 == L1 ∧ X35 == L2 ) ∨
( X24 == L1 ∧ X35 == L3 )

**Figure 5.11** GCC constraints as a Boolean expression.

**Table 5.13** *Adjusted t-*Coverage (%) and *adjusted t-*Diversity (%): GCC.

| | CCA(20;2,($3^{10} \cdot 2^{189}$),$\phi = \{3^3 \cdot 2^{37}\}$) | |
| --- | --- | --- |
| *t* | *t*-Cov | *t*-Div |
| 2 | 100 | 22 |
| 3 | 94 | 42 |
| 4 | 71 | 64 |
| 5 | 44 | 81 |

**Table 5.14** Covering array number: GCC.

| | CCAN(t,($3^{10} \cdot 2^{189}$),$\phi = \{3^3 \cdot 2^{37}\}$) | |
| --- | --- | --- |
| *t* | Best | JMP |
| 2 | 19 | 20 |
| 3 | 94 | 97 |
| 4 | NA | 562 |

an exact approach (see Colbourn and Syrotiuk, 2016) and a probabilistic approach (see Dumlu et al., 2011). Probabilistic approaches exhibit low computational cost but can be inaccurate, whereas exact approaches are accurate but are computationally costly. This is particularly relevant here, since for this case study, where there are 199 inputs, the difference in computational time between these two approaches will be distinctly more noticeable than for the other case studies.

### 5.3.3 Covering Arrays and Fault Detection Effectiveness

The topic of the effectiveness of various testing approaches at detecting faults has been an ongoing topic of investigation for several years, and there is empirical evidence that suggests that a covering array based approach can be very effective (Nie and Leung, 2011; Kuhn et al., 2004). In fact, Kuhn, Wallace, and Gallo (2004) examine failures from several categories of software, and they discover that it is rarely the case that a covering

**Table 5.15** Fault detection effectiveness.

| | % faults detected for *t*-covering array | | | |
| | $t = 2$ | $t = 3$ | $t = 4$ | $t = 5$ |
| --- | --- | --- | --- | --- |
| Medical device software | 97 | 99 | 100 | 100 |
| Browser application | 76 | 95 | 97 | 99 |
| Server software | 70 | 89 | 96 | 96 |
| NASA distributed system | 93 | 98 | 100 | 100 |

array of greater than strength 5 is needed. We present in Table 5.15 a summary of some of their findings.

Given a software system with $k$ inputs, a strength $t$ covering array may be used to generate a set of test cases that ensures that all $t$ input level combinations are covered by at least one test case. If $t \ll k$ then the cost savings when compared to exhaustive testing is dramatic. Covering arrays can therefore be an effective, and efficient, tool for deriving test cases to validate software systems.

## 5.4 Challenges

Although there is good reason to be encouraged by the current state of the combinatorial testing approach to validating software systems, there are challenges that need to be further addressed to ensure wider acceptance amongst practitioners. There appears to be some consensus among practitioners and academics that the following are important challenges.

1) **Variable-strength covering arrays**: It is often the case that subsets of inputs are more important than others. As a result, test engineers may be interested in specifying different strengths for different subsets of inputs, while ensuring that overall the covering array exhibits some minimum strength. This covering array variant is known as a variable-strength covering array. Some existing tools currently provide this capability. There appears to be agreement amongst practitioners that this capability should be commonplace.

2) **Test set prioritization**: Test engineers often have a keen sense of which test cases are most important. Bryce and Colbourn (2006) propose a covering array variant that they term *biased* covering arrays to address this issue. These covering arrays use a set of weights, provided by the test engineer, to ensure that certain level combinations occur early in the covering array, thus allowing for prioritization. Practitioners in particular seem to view this capability as important.

3) **Test set augmentation**: Software systems are not static entities, and so there is some agreement that covering array tools should provide a way to augment an existing covering array to ensure that as a software system evolves, and its inputs change, extant testing artifacts derived from the covering array can be updated rather than replaced (Colbourn, 2015).

4) **Constraint support**: Although many tools support constraints, there is some consensus that more research needs to be done to develop better construction algorithms as well as to develop techniques to provide automatic support for the *implied* constraint problem identified by Cohen, Dwyer, and Shi (2008).

5) **Fault location**: We mentioned the fault location problem in Case Study 3 and pointed out that there are essentially two approaches, an exact approach (Colbourn and Syrotiuk, 2016; and Lekivetz and Morgan, 2016) and a probabilistic approach (Dumlu et al., 2011). There seems to be some consensus that additional research is needed.

We have focused on just a handful of the challenges that need to be further addressed. There has been no attempt to rank them in any way; these happen to be a sample of several important challenges that seem to crop up in the literature with some frequency.

## References

Arcuri, A., Iqbal, M., and Briand, L. (2010) Formal analysis of the effectiveness and predictability of random testing, in *Proceedings of the 19th International Symposium on Software Testing and Analysis (ISSTA)*, pp. 219–229.

Barr, E., Harman, M., McMinn, P., Shahbaz, M., and Yoo, S. (2015) The oracle problem in software testing, *IEEE Transactions on Software Engineering*, 41(5), 507–525.

Beizer, B. (1983) *Software Testing Techniques*, New York: Van Nostrand Reinhold.

Brownlie, R., Prowse, J., and Phadke, M. (1992) Robust testing of AT&T PMX/StarMAIL using OATS, *AT&T Technical Journal*, 71(3), 41–47.

Bryce, R., and Colbourn, C. (2006) Prioritized interaction testing for pair-wise coverage with seeding and constraints, *Information & Software Technology*, 48(10), 960–970.

Calvagna, A., and Gargantini A. (2009) IPO-s: Incremental generation of combinatorial interaction test data based on symmetries of covering arrays, in *Software Testing Verification and Validation Workshop (ICSTW'09)*, IEEE, pp. 10–18.

Cohen, D., Dalal, S., Fredman, M., and Patton, G. (1997) The AETG system: An approach to testing based on combinatorial design, *IEEE Transactions on Software Engineering*, 23(7), 437–444.

Cohen, M., Dwyer, M., and Shi, J. (2008) Constructing interaction test suites for highly configurable systems in the presence of constraints: A greedy approach, *IEEE Transactions on Software Engineering*, 34(5), 633–650.

Colbourn, C. (2004) Combinatorial aspects of covering arrays, *Le Matematiche (Catania)*, 58, 121–167.

Colbourn, C. (2016) *Covering Array Tables*. http://www.public.asu.edu/~ccolbou/src/tabby/catable.html [accessed 30 January, 2018].

Colbourn, C. (2015) Augmentation of covering arrays of strength two, *Graphs and Combinatorics*, 31(6), 2137–2147.

Colbourn, C., Cohen, M., and Turban R. (2004) A deterministic density algorithm for pairwise interaction coverage, in *Proceedings of the IASTED International Conference on Software Engineering*.

Colbourn, C., Shi, C., Wang, C., and Yan, J. (2011) Mixed covering arrays of strength three with few factors, *Journal of Statistical Planning and Inference*, 1(11), 3640–3647.

Colbourn, C., and Syrotiuk, V. (2016) Coverage, location, detection, and measurement, in *Proceedings of the 9th International Conference on Software Testing Workshop (ICSTW)*, pp. 19–25.

Czerwonka, J. (2006) Pairwise testing in the real world: Practical extensions to test-case scenarios, in *Proceedings 24th Pacific Northwest Software Quality Conference*.

Dalal, S., and Mallows, C. (1998) Factor-covering designs for testing software, *Technometrics*, 40(3), 234–243.

Dumlu, E., Yilmaz, C., Cohen, C., and Porter A. (2011) Feedback adaptive combinatorial testing, in *Proceedings of the 20th International Symposium on Software Testing and Analysis (ISSTA)*, pp. 243–253.

Do, H, Elbaum, S., and Rothermel, G. (2005) Supporting controlled experimentation with testing techniques: An infrastructure and its potential impact, *Empirical Software Engineering*, 10(4), 405–435.

Federal Aviation Administration (2011) *Introduction to TCAS II v7.1*, Tech. Rep. HQ-111358.

Ghandehari, L., Lei, Y., Kung, D., Kacker, R., and Kuhn, R. (2013) Fault localization based on failure inducing combinations, in *Proceedings of the 24th International Symposium on Software Reliability Engineering (ISSRE)*, pp. 168–177.

Hailpern, B., and Santhanam, P. (2002) Software debugging, testing, and verification, *IBM Systems Journal*, 41(1), 4–12.

Hamlet, R. (1994) Random testing, in *Encyclopedia of Software Engineering*, Chichester: Wiley, pp. 970–978.

Hartman, A. (2006) Covering arrays: Mathematical engineering and scientific perspectives, in *Fields Institute Workshop on Covering Arrays*, May 2006.

Hartman, A., Klinger, T., and Raskin, L. (2011) IBM intelligent test case handler, *Discrete Mathematics*, 284(1), 149–156.

Jia, Y., Cohen, M., Harman, M., and Petke, J. (2015) Learning combinatorial interaction test generation strategies using hyperheuristic search, in *Proceedings of the 37th International Conference on Software Engineering (ICSE)*, pp. 540–550.

Johnson, K., and Entringer, R. (1989) Largest induced subgraphs of the *n*-cube that contain no 4-cycles, *Journal of Combinatorial Theory, Series B*, 46(3), 346–355.

Katona, G. (1973) Two applications (for search theory and truth functions) of Sperner type theorems, *Periodica Mathematica Hungarica*, 3(1–2), 19–26.

Kleitman D., and Spencer, J. (1973) Families of *k*-independent sets, *Discrete Mathematics*, 6(3), 255–262.

Kuhn, D., Mendoza I., Kacker, R., and Lei, Y. (2013) Combinatorial coverage measurement concepts and applications, in *Software Testing Verification and Validation Workshop*, ICSTW'13, IEEE, 352–361.

Kuhn, D., and Okum, V. (2006) Pseudo-exhaustive testing for software, in *30th Annual IEEE/NASA Software Engineering Workshop*, IEEE.

Kuhn, D., Wallace, D., and Gallo, A.M. (2004) Software fault interactions and implications for software testing, *IEEE Transactions on Software Engineering*, 30(6), 418–421.

Lekivetz, R., and Morgan, J. (2016) *Analysis of Covering Arrays Using Prior Information*, SAS Technical Report – 2016.1.

Mandl, R. (1985) Orthogonal latin squares: An application of experiment design to compiler testing, *Communications of the ACM*, 28(10), 1054–1058.

Morgan, J. (2009) *Unsatisfiable Constrained Covering Arrays*, SAS Technical Report – 2009.2.

Morgan, J. (2014) *Computing Coverage and Diversity Metrics for Constrained Covering Arrays*, SAS Technical Report – 2014.2.

Moura, L., Stardom, J., Stevens, B., and Williams, A. (2003) Covering arrays with mixed alphabet sizes, *Journal of Combinatorial Design*, 11(6), 413–432.

Myers, G. (1979) *The Art of Software Testing*, Chichester: Wiley.

Nie, C., and Leung, H. (2011) A survey of combinatorial testing, *ACM Computing Surveys*, 43(11), 1–29.

Nie, C., Wu, H., Niu, X., Kuo, F., Leung, H., and Colbourn, C. (2015) Combinatorial testing, random testing, and adaptive random testing for detecting interaction triggered failures, *Information & Software Technology*, 62, 198–213.

Orso, A., and Rothermel, G. (2014) Software testing: A research travelogue (2000–2014), in *Proceedings on the Future of Software Engineering*, ACM, May 2014, pp. 117–132.

Renyi, A. (1971) *Foundations of Probability*, Chichester: Wiley.

Rothermel, G., and Harrold, M. (1998) Empirical studies of a safe regression test selection technique, *IEEE Transactions on Software Engineering*, 24(6), 401–419.

Tassey, G. (2002) *The Economic Impacts of Inadequate Infrastructure for Software Testing*, National Institute of Standards and Technology.

Tatsumi, K., Watanabe, S., Takeuchi, Y., and Shimokawa, H. (1987) Conceptual support for test case design, in *Proceedings 11th Computer Software and Applications Conference (COMPSAC'87)*, pp. 285–290.

Thévenod-Fosse, P., and Waeselynck, H. (1991) An investigation of statistical software testing, *Software Testing, Verification and Reliability*, 1(2), 5–25.

Williams, A. (2000) Determination of test configurations for pair-wise interaction coverage, in *IFIP TC6/WG6.1, 13th International Conference on Testing Communicating Systems*, pp. 59–74.

Yan, J., and Zhang J. (2006) Backtracking algorithms and search heuristics to generate test suites for combinatorial testing, in *Proceedings 30th Annual International Computer Software and Applications Conference (COMPSAC'06)*, pp. 385–394.

Yang, B., Hu, H., and Jia, L. (2008) A study of uncertainty in software cost and its impact on optimal software release time, *IEEE Transactions on Software Engineering*, 34(6), 813–825.

Yu, L., Lei, Y., Kacker, R., and Kuhn, D. (2013) ACTS: A combinatorial test generation tool, in *Proceedings IEEE 6th International Conference on Software Testing, Verification and Validation*, 370–375.

# 6

# Conceptual Aspects in Development and Teaching of System and Software Test Engineering

*Dani Almog, Ron S. Kenett, Uri Shafrir, and Hadas Chasidim*

## Synopsis

System and software testing are important disciplines combining technical aspects with conceptual domains. While the reliability of systems and software is critical, many organizations involved in system and software development do not consider testing as an engineering discipline. Moreover, most testing team members do not have formal training, and university curricula, generally, do not prepare graduates with an understanding of test design and analysis methods. So, even though proper testing has become more critical to software development because of the increased complexity of systems and software, not much innovation in testing education or testing design is being reported. This chapter presents a novel approach focused on improving the teaching of system and software testing in academic and lifelong learning programs by introducing methodologies for conducting and assessing the process of education and training. The approach builds on meaning equivalence reusable learning objects (MERLO), which provide significant didactic advantages when one is interested in developing and assessing conceptual understanding of any knowledge domain. The scoring of MERLO items provides a statistical assessment of the level of conceptual understanding of students, and permits deep learning analysis of what is taught and how it is taught. Experience gained by teaching undergraduate and graduate students in a leading Israeli college, within a program dedicated to system and software testing, is presented. Some implications for test engineering methods and system operational performance signatures are also discussed.

## 6.1 Introduction

Software engineering, and especially software testing and software quality as subdisciplines, are relatively new. In software engineering (SE) education, designed to prepare software engineers for industry, it is obvious that quality and testing are important required software engineering skills. Software quality contains many elements, methods, and concepts that need to be integrated in an SE syllabus (Kenett and Baker, 2010; Ardis et al., 2015). Because of natural inertia, academic education often struggles to remain up to date and relevant from the perspective of industry. This challenge is addressed by two major questions: first, is the current research in SE relevant and innovative, and does it contribute to the direction industry is taking? Second, do we

*Analytic Methods in Systems and Software Testing*, First Edition.
Edited by Ron S. Kenett, Fabrizio Ruggeri, and Frederick W. Faltin.

educate and foster students to be holistic thinkers and leaders within industry? This chapter addresses the latter aspect. We first map a standard SE curriculum by looking at two sources of domain knowledge, SWEBOK and SE2014. We then suggest means and practical ways to prepare an integrated approach to SE education.

The remainder of the chapter consists of the following sections: Section 6.2 consists of an analysis of the current teaching curricula and a mapping of knowledge related to software quality and testing. This information is then compared to future predicted needs while considering the growing gap between what is being taught and what is needed by industry. This includes knowledge and conceptual understanding of "soft skills" aspects in future SE. Section 6.3 describes a systematic top-down information collection method creating an orderly database of items ready to be used as teaching resources. Section 6.4 gives an introduction to formative assessment in education and concept science with MERLO. The section includes examples of project-oriented learning for software engineering using gaming techniques. Section 6.5 presents methods for exercise and assessing student achievement. Different tools for achieving deeper understanding and appreciation of the student's level are presented. Section 6.6 is dedicated to online educational systems, with a focus on how MERLO techniques can be integrated in massive open online courses (MOOCs). Section 6.7 gives some examples of formative assessment scores with MERLO in SE courses.

Throughout the chapter we present examples from relevant courses taught in recent years. Every example demonstrates a different and complementary approach. The chapter is concluded with a summary section highlighting its main points.

The IEEE (2010) definition states that software engineering is

> The application of a systematic, disciplined, quantifiable approach to the development, operation and maintenance of software; that is, the application of engineering to software.

The IEEE *Guide to the Software Engineering Body of Knowledge* (SWEBOK; Bourque and Fairley, 2014) and software engineering curriculum guidelines (SE2014; IEEE, 2015) suggest that students participating in SE programs should gain:

> [**Professional Knowledge**] *Show mastery of software engineering knowledge and skills and of the professional standards necessary to begin practice as a software engineer.*
>
> [**Technical Knowledge**] *Demonstrate an understanding of and apply appropriate theories, models, and techniques that provide a basis for problem identification and analysis, software design, development, implementation, verification, and documentation.*
>
> [**Teamwork**] *Work both individually and as part of a team to develop and deliver quality software artifacts.*
>
> [**End-User Awareness**] *Demonstrate an understanding and appreciation of the importance of negotiation, effective work habits, leadership, and good communication with stakeholders in a typical software development environment.*
>
> [**Design Solutions in Context**] *Design appropriate solutions in one or more application domains using software engineering approaches that integrate ethical, social, legal, and economic concerns.*

[**Perform Trade-Offs**] *Reconcile conflicting project objectives, finding acceptable compromises within the limitations of cost, time, knowledge, existing systems, and organizations.*

[**Continuing Professional Development**] *Learn new models, techniques, and technologies as they emerge and appreciate the necessity of such continuing professional development.*

Figure 6.1 shows the current knowledge areas described in the final draft of *Software Engineering Education Knowledge* (SEEK; Sobel, 2013). Software quality (QUA), which is 2% of the SEEK curricula, and verification and validation (VEV, a.k.a. testing), 7%, are knowledge areas that cover the content we are addressing in this chapter. Today, less than 10% of SEEK directly addresses quality and testing. A taxonomy of software defects is available from Felderer and Beer (2013). A detailed taxonomy of risk-based testing is presented in Felderer and Schieferdecker (2014). This chapter is about introducing quality and testing into SE curricula.

Bloom's *Taxonomy* (Bloom et al., 1956) identifies two groups of cognitive skills: lower-order cognitive skills (LOCS), including knowledge, comprehension, and application; and higher-order cognitive skills (HOCS), including analysis, synthesis, and evaluation. Engineering education should expand beyond these two groups and offer a third one: proficiency order of skills (POK). Industry needs engineers to be experienced and skilled in practical implementations and not only in expressing analytics and abstract understanding. This is the reason why test engineering courses require hands-on exercises and practical activities during the educational program.

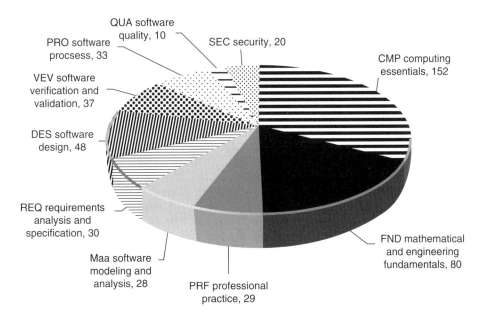

**Figure 6.1** Software Engineering Education Knowledge.

## 6.2 Determining What to Teach

The setting of an SE curriculum involves addressing several challenges, as listed below.

### 6.2.1 Closing the Gap Between Industry Needs and Academic Teaching

The academic ecosystem consists of systematic knowledge, composed of existing research, theories, and tools that are known and validated. In this context, the development and approval of curricula in academia is compatible with this approach. However, in the SE discipline, the pace of technology updates and changes in the development processes are accelerated. Therefore, in this particular area there is a built-in gap between academia and industry that is relatively difficult to close. Adopting a more dynamic approach that is fed by continuous feedback derived by both academia and industry can close this gap (see Figure 6.2). We suggest here a dynamic approach to address this gap by the integration of two feedback cycles driven by academia and industry. A common feedback session with representatives from industry and academia can help formulate a mechanism of mutual coordination and adaptation for the two participating parties. Figure 6.2 presents a detailed view of the feedback cycles affecting industry and academia.

The two cycles work differently. On the left, industry, which is driven by market and innovation, creates a professional demand for certain skills and knowledge. This dictates the job description and market demand of specific skills. This industry cycle is updated with a new cycle by a *feedback* step. The academic curriculum, on the other hand (the right cycle), has prepared courseware that slowly adapts to new research and future needs.

### 6.2.2 "Need to Know" Scope Explosion

It is not so clear how to precisely determine the scope of SE teaching. One might relate to each of the following lists as relevant to software quality and testing. The context of a specific course may dictate a different selection. It will be easy to relate first to the lower level of the pyramid in Figure 6.3. All of the pyramid topics are subject to reconsideration facing the rapid changes experienced in areas such as:

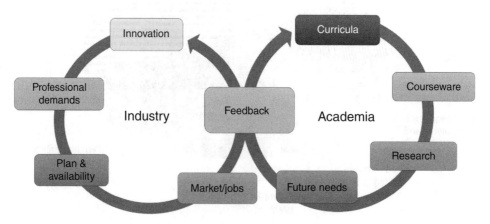

**Figure 6.2** Academia/industry feedback cycles.

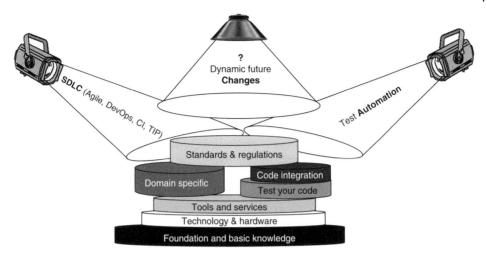

**Figure 6.3** Testing scope explosion.

- Software development life cycle (SDLC) – moving to Agile, DevOps, CI, TIP
- New testing techniques and test automation
- Dynamic future changes.

The following is an attempt to address the required SE knowledge. A list of relevant items includes:

1) Foundation and basic knowledge
2) Testing on different software development levels
   - Code testing (static and dynamic)
   - Unit test
   - Component test
   - Integration test
   - System test
   - User acceptance tests (UAT)
   - Alpha and beta test
3) Technology and hardware
4) Tools and services
5) Domain-specific testing
   - Real time
   - Web
   - Communication
   - NFT
   - IOT
   - Service orientation
   - Security
   - Cyber
   - Business line differentiation
6) Standards and regulation

The timing and the order for teaching these different topics is crucial in an SE program. Consider, for example, the following dilemma: Do we need to teach basic testing techniques, such as equivalent classes and boundary value testing, before presenting unit testing? This dilemma presents itself even more intensely when teaching test-driven development (TDD), which has become an industry standard software development paradigm where the development of the test precedes the actual coding.

Another technical issue is raised when considering mobile devices – it is obvious that the most common user interface (UI) currently is with mobile devices. However, software development platforms are usually on desktop technologies. One may argue that teaching testing should aim at mobile technology. These dilemmas, and others, accompany us throughout the chapter. The following example (Example 6.1) is a course curriculum aimed at undergraduate SE students – it covers what we consider the minimal knowledge required by all software engineers and serves as a prerequisite for entering the market as a certified software engineer.

---

**Example 6.1  The Basic Course**

**Course name:** Fundamentals of Software Quality Engineering
**Target students:** Second-year undergraduate software engineering students.
**Description:** Software quality testing is a major part of the development process that requires substantial resources. Software quality engineering is a leading area in the Israeli high-tech industry. This introductory course covers the main ideas and principles of software quality engineering. Students will learn how to deal with increasingly strict quality requirements by studying theoretical and practical software testing tools that are adapted to the rapidly changing, highly demanding, and impatient market.

**By successfully completing the course, the student will:**

- Become familiar with international quality standards and common software quality metrics.
- Be introduced to the importance of the software quality assurance in the system life cycle.
- Become familiar with the different methods and techniques for software testing.
- Be able to implement principles and methods of integration and regression testing.
- Gain practical knowledge and experience in unit testing, black-box testing, structural testing, UI testing, and system testing.
- Know how to use and benefit from test automation tools.

**Topics and subtopics of the courses:**

The quality of the code
- What is quality and software quality?
- Quality metrics for code
- The art of unit testing
- Object orientation and code quality
- Service orientation quality standards
- Testing levels

Software testing techniques
- What is a test case?
- Test oracle – the verification and validation issue
- Functional testing
- Domain testing
- Requirement-based testing
- Risk-based testing

Quality and testing projects
- Software testing during different SDLCs
- Test artifact management and control
- Defect management
- Traceability matrix and control
- Testing teams – qualification and improvement
- Test automation principles and tools

### 6.2.3 Software Test Engineers and Soft Skills

In SE, soft skills are as important as technical skills. These skills, or emotional intelligence, are often not taught in SE programs. They are needed in order to enable professionals to navigate smoothly and effectively through a wide variety of social and professional situations with a wide variety of people. Such skills include communication, cooperation, creativity, leadership, and organization. Addressing computer science (CS) teaching, Hazzan and Har-Shai (2013) reported on a special soft skills course developed at Technion's computer science faculty in Israel. The course description lists *students'* characterizations of soft skills as follows:

Cannot be measured:

- Time and pressure management
- Sensitivity to the environment
- Creativity.

Depends on the individual's personality:

- Problem-solving skills
- Communication.

Can be implemented in different ways:

- Abstraction
- Creativity capability.

Should be implemented differently in different situations according to the specific circumstances:

- Adaptation to teamwork.

Is developed gradually:

- Feedback
- The ability to examine known subjects from different perspectives.

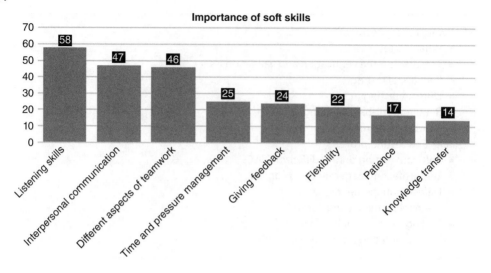

**Figure 6.4** Students' ranking of soft skills.

**Table 6.1** Top five soft skills for software testing / quality assurance.

| | |
|---|---|
| Oral/written – presentation skills | 65.12% |
| Teamwork | 62.79% |
| Initiative, proactive | 46.51% |
| Analytical, problem-solving | 27.91% |
| Methodical | 23.26% |

Figure 6.4 is what was rated by the students in the program as being important for teamwork. The students summarized their understanding of soft skills as follows:

- Characterized by interaction with the environment: listening ability, team management skills, interpersonal communication, giving feedback, teamwork, tolerance.
- Developed on the basis of experience in different life situations: presentation skills, conflict-solving skills, ability to cope with stress, flexibility, expression of diversity, abstraction skills.

Schawbel (2012) showed that soft skills topped the list of "must haves" for employers, with 98% of them saying communication skills are essential and 92% teamwork skills. The data was collected from job descriptions in advertisements in Uruguay (Matturro, 2013) related to the abovementioned activities of software testing and also software quality assurance. The five most mentioned soft skills for these process areas are shown in Table 6.1.

For test engineers specifically, an internal survey of five software quality assurance managers from Motorola Israel regarding a software test engineer position prior to a major recruiting campaign (Porat, 2009) provides the following recommendations:

- Fast learning abilities and intelligence – the ability to understand client processes, enabling applicable simulations of them. We expect the engineer to represent us at a design review and contribute to the debate.
- Highly developed communication and cooperation skills – no need of lonely wolves.
- Ability to take part in the important e informal communication within teams, and sometimes with customer as well.
- Assertiveness – a tester must be able to convey their results (regardless of a developer's aggressive disputation).
- Managerial skills (resources, goals, priorities).
- Discipline and order – accurate and precise in all their activities.
- Readiness to learn and adjust to new methods and techniques.
- Good presentation skills – clear and articulate (both written and oral).
- Prior knowledge and experience in the application (real time, RF, data processing, billing, …).
- Language proficiency (English and local).
- Programming skills (for scripting and code reviewing).
- Prior test automation experience an advantage.
- Imaginative/creative – in many cases you need an unconventional approach in order to reveal faults.
- Formal qualifications (formal education and certification).

None of these required skills are dealt with properly by most formal engineering degrees. We propose to address some of this during courses via quizzes, exercises, and assignments. In addition to the above, we must consider other aspects of software testing, such as curiosity, persistence, skepticism, fast learning, passion, boundary rider, and others – each of these deserves attention and consideration.

## 6.3 Systematic Mapping of Software Test Engineering Knowledge

This section discusses a top-down approach for collecting teaching material for a specific course, considering the dynamic need for curriculum adjustment (Hazzan et al., 2015). We propose an approach to collecting and updating teaching material for SE courses that includes:

- A top-down approach to selecting the main topics of a course – enabling a mapping of its scope within the full picture.
- An orderly preparation of teaching materials that serves as a knowledge sharing mechanism between the different teachers.

In order to properly convey the actual objectives to the students, the list of course topics should be discussed internally. The approach is presented graphically in Figure 6.5.

The next step is to collect these items in a datasheet. Table 6.2 demonstrates how to collect teaching material in worksheets with all the required knowledge hierarchies.

The preparation process for a specific item within a subtopic indicates that this breakdown structure is not rigid and might require additional subdivision levels. For example, when teaching domain testing, one might realize the need for another subdivision so

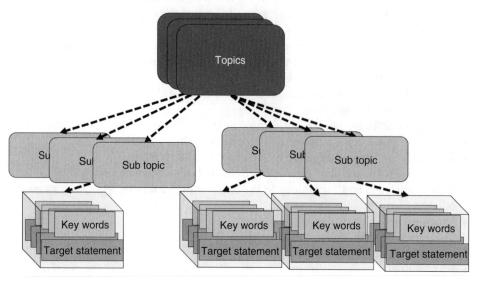

**Figure 6.5** Knowledge collection method.

**Table 6.2** Teaching material collection template. Topic: Software testing techniques. Subtopic: Test oracle.

| Topics | Subtopics | Key words | Target statements | Reference |
|---|---|---|---|---|
| Software testing techniques | What is a test case? | | | |
| | Test oracle | test oracle, input data, output data, logical rules, test automation | Answer the question: How will you know the outcome of the test? | |
| | | | The mechanism that determines success or failure of a test. | Hoffman (1999) |
| | | | One of the most humanly dependent properties during the testing work. | Hoffman (1999) |
| | | | Our ability to automate testing is fundamentally constrained by our ability to create and use oracles. | Hoffman (2003) |
| | Functional testing | | | |
| | Domain testing | | | |
| | Requirement-based testing | | | |
| | Risk-based testing | | | |

that terms such as equivalent classes and boundary values get the proper attention. This mapping is essential for determining the scope of a course, and provides the basis for assessing the level of conceptual understanding reached by the students.

## 6.4 Teaching for Conceptual Learning

This section is about a formative assessment measurement approach used in education to assess conceptual understanding, sometimes also labeled "deep" understanding. Such assessments are used during training or education sessions to provide feedback to the instructor, and directly contribute to the learning of students. It is also used for the improvement of material and delivery style. We first introduce the topic of formative assessment and the elements of concept science in education, and then introduce meaning equivalence reusable learning objects (MERLO), with an example on teaching quantitative literacy.

Listening to conversations among content experts reveals a common trend to flexibly reformulate the issue under discussion by introducing alternative points of view. This is often encoded in alternative representations in different sign systems. For example, a conversation that originated in a purely spoken exchange may progress to include written statements, images, diagrams, equations, etc., each with its own running (spoken) commentary. The term "meaning equivalence" designates a commonality of meaning across several representations. It signifies the ability to transcode meaning in a polymorphous (one to many) transformation of the meaning of a particular conceptual situation through multiple representations within and across sign systems. Listening to conversations among content experts also reveals a common trend to identify patterns of associations among important ideas, relations, and underlying issues. These experts engage in creative discovery and exploration of hidden, but potentially viable, relations that test and extend such patterns of associations that may not be obviously or easily identified. The term "conceptual thinking" is used to describe such ways of considering an issue; it requires the ability, knowledge, and experience to communicate novel ideas through alternative representations of shared meaning, and to create lexical labels and practical procedures for their nurturing and further development (Shafrir and Kenett, 2010; Shafrir and Etkind, 2010). The application of MERLO in education programs for statistics and quantitative literacy was introduced by Etkind, Kenett, and Shafrir (2010). A structured ladder matching conceptual aspects of management style and industrial statistics methods is provided in Kenett and Zacks (2014). For an application of MERLO in the curriculum of mathematics in Italian high schools, see Arzarello et al. (2015a, b). For an application of MERLO and concept mapping to new technologies and e-learning environments, including MOOCs, see Shafrir and Kenett (2015).

The pivotal element in conceptual thinking is the application of MERLO, which is a multidimensional database that allows the sorting and mapping of important concepts through exemplary target statements of particular conceptual situations, and relevant statements of shared meaning. Each node of MERLO is an item family, anchored by a target statement that describes a conceptual situation and encodes different features of an important concept, and also includes other statements that may, or may not, share equivalence of meaning with the target. Collectively, these item families encode the complete conceptual mapping that covers the full content of a course (a particular

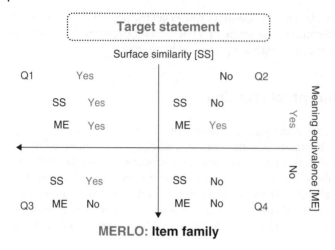

**Figure 6.6** Template for constructing an item family in MERLO.

content area within a discipline). Figure 6.6 is a template for constructing an item family anchored in a single target statement. Statements in the four quadrants of the template – Q1, Q2, Q3, and Q4 – are thematically sorted by their relation to the target statement that anchors the particular node (item family). They are classified by two sorting criteria: surface similarity to the target, and equivalence of meaning with the target. For example, if the statements contain text in natural language, then by "surface similarity" we mean same/similar words appearing in the same/similar order as in the target statement; by "meaning equivalence" we mean that a majority in a community that shares a sublanguage with a controlled vocabulary (e.g., statistics) would likely agree that the meaning of the statement being sorted is equivalent to the meaning of the target statement.

MERLO pedagogy guides sequential teaching/learning episodes in a course by focusing learners' attention on meaning. The format of MERLO items allows the instructor to assess deep comprehension of conceptual content by eliciting responses that signal learners' ability to recognize and produce multiple representations that share equivalence of meaning. A typical MERLO item contains five unmarked statements: a target statement plus four additional statements from quadrants Q2, Q3, and, sometimes, also Q4. Task instructions for a MERLO test are:

> At least two out of these five statements – but possibly more than two – share equivalence of meaning.
>
> 1) Mark all statements – but only those – that share equivalence of meaning.
> 2) Write down briefly the concept that guided you in making these decisions.

For example, the MERLO item in Figure 6.7 (mathematics/functions) contains five representations (A–E) that include text, equations, tables, and diagrams; at least two of these representations share equivalence of meaning. Thus, the learner is first asked to carry out a recognition task in situations where the particular target statement is not marked; namely, features of the concept to be compared are not made explicit. In order

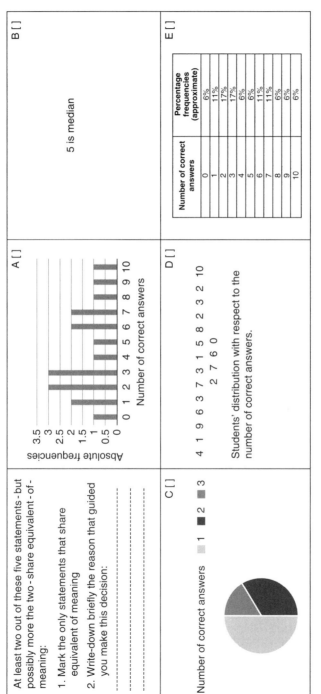

| | A [ ] | | B [ ] |

**A [ ]**

Absolute frequencies: 0, 0.5, 1, 1.5, 2, 2.5, 3, 3.5

Number of correct answers: 0 1 2 3 4 5 6 7 8 9 10

**B [ ]**

5 is median

**C [ ]**

Number of correct answers ▨ 1 ■ 2 ▧ 3

**D [ ]**

4 1 9 6 3 7 3 1 5 8 2 3 2 10
2 7 6 0

Students' distribution with respect to the number of correct answers.

**E [ ]**

| Number of correct answers | Percentage frequencies (approximate) |
|---|---|
| 0 | 6% |
| 1 | 11% |
| 2 | 17% |
| 3 | 17% |
| 4 | 6% |
| 5 | 6% |
| 6 | 11% |
| 7 | 11% |
| 8 | 6% |
| 9 | 6% |
| 10 | 6% |

At least two out of these five statements - but possibly more the two - share equivalent - of - meaning:

1. Mark the only statements that share equivalent of meaning

2. Write-down briefly the reason that guided you make this decision:

_____

_____

_____

**Figure 6.7** Example of a MERLO item (mathematics/functions).

to perform this task, a learner needs to begin by decoding and recognizing the meaning of each statement in the set. This decoding process is carried out by analyzing concepts that define the "meaning" of each statement. Successful analysis of all the statements in a given five-statement set (item) requires deep understanding of the conceptual content of the specific domain. The MERLO item format requires both rule inference and rule application in a similar way to the solution of analogical reasoning items. Once the learner marks those statements that in their opinion share equivalence of meaning, they formulate and briefly describe the concept they had in mind when making these decisions. Figure 6.8 is a MERLO item from the course described in Example 6.1.

A learner's response to a MERLO item combines a multiple-choice/multiple-response (also called recognition) question, and a short answer (called production). Subsequently, there are two main scores for each MERLO item: recognition score and production score. Specific comprehension deficits can be traced as low recognition scores on quadrants Q2 and Q3, due to the mismatch between the valence of surface similarity and meaning equivalence (Figure 6.6). The production score of MERLO test items is based on the clarity of the learner's description of the conceptual situation anchoring the item, and the explicit inclusion in that description of lexical labels of relevant and important concepts and relations. Classroom implementation of MERLO pedagogy includes interactive MERLO quizzes, as well as inclusion of MERLO items as part of mid-term tests and final exams. A MERLO interactive quiz is an in-class procedure that provides learners with opportunities to discuss a PowerPoint display of a MERLO item in small groups, and send their individual responses to the instructor's computer via mobile text messaging, or by using a clicker (classroom response system: CRS). Such a quiz takes 20–30 minutes, and includes the following four steps: small group discussion; individual response; feedback on production response; feedback on recognition response and class discussion.

The implementation of MERLO has been documented to enhance learning outcomes. Such implementations were carried out at different instructional situations (Shafrir and Etkind, 2006). Figure 6.8 is another MERLO item sample addressing the test oracle subtopic.

The software development environment in industry is complex and affects the life cycle of production as well as the evolution of the software. It might be challenging to imitate these conditions in academia, even though these may be crucial tools for qualification, and particularly for work in industry. For these reasons, teachers attempt to use creative and effective ways to enable students to experience these environments and to achieve significant learning. Project-based learning (PBL) is a worthy candidate as a teaching methodology for software engineering courses.

Project-based learning is a dynamic classroom approach in which students actively explore real-world problems and challenges and acquire deeper knowledge (Terenzini et al., 2001; Fosnot, 2013). It emphasizes learning activities that are long term, interdisciplinary, and student centered. Unlike traditional, teacher-led classroom activities, students often must organize their own work and manage their own time in a project-based class. Project-based instruction differs from traditional inquiry in its emphasis on students' collaborative or individual artifact construction to represent what is being learned. PBL also gives students the opportunity to explore problems and challenges that have real-world applications, increasing the possibility of long-term retention of skills and concepts. In PBL, students merge previous knowledge with new

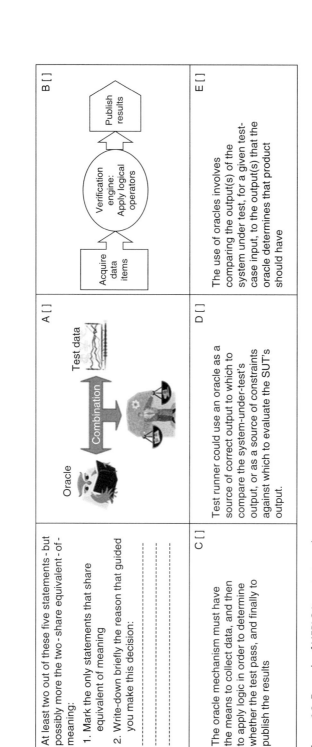

At least two out of these five statements - but possibly more the two - share equivalent - of - meaning:

1. Mark the only statements that share equivalent of meaning
2. Write-down briefly the reason that guided you make this decision:

_____
_____
_____

A [ ]

Oracle    Combination    Test data

B [ ]

Acquire data items → Verification engine: Apply logical operators → Publish results

C [ ]

The oracle mechanism must have the means to collect data, and then to apply logic in order to determine whether the test pass, and finally to publish the results

D [ ]

Test runner could use an oracle as a source of correct output to which to compare the system-under-test's output, or as a source of constraints against which to evaluate the SUT's output.

E [ ]

The use of oracles involves comparing the output(s) of the system under test, for a given test-case input, to the output(s) that the oracle determines that product should have

**Figure 6.8** Example of MERLO item: test oracle.

methodologies and concepts (e.g., Agile) to solve problems that reflect the complexity of the real world, while working in an integrated environment. The teacher plays the role of facilitator, working with students to frame worthwhile questions, structuring meaningful tasks, coaching both knowledge development and social skills, and carefully assessing what students have learned from the experience. Example 6.2 demonstrates how PBL can be implemented in a course in the software engineering department.

---

### Example 6.2  Learning Agile Development in a Project-Oriented Ecosystem

**Course name:** Software Development and Management in Agile Approach

**General course information**: About 220 students on average participate in our course in two campuses. The students have periodic meetings in order to discuss the theories they have learned and how these are implemented in their projects. At the meetings, students present how they have progressed and are encouraged to ask questions. This project-oriented learning within an Agile atmosphere format compensates for the lack of teacher presence that is typical of online learning environments. At the same time, students work on their projects in groups, and function like a real scrum team. Figure 6.9 provides an overview of the program. In each of the three sprints the students are required to:

- Plan the product backlog and sprint backlog (SBL) by using Microsoft Team Foundation Server (TFS).
- Receive task assignments from the scrum master using Microsoft TFS.
- Conduct and report on the daily meetings.

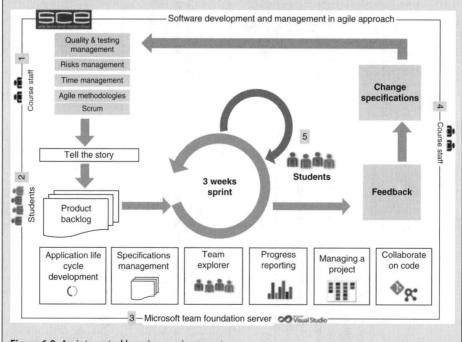

**Figure 6.9** An integrated learning environment.

- Report on the sprint review.
- Deliver a working product according to the SBL via Microsoft TFS and the connected account in Visual Studio.
- Additionally, each team is required to present at the end of the semester in a special meeting called "Integration and discussion," a short session that includes one of the course theoretical topics, how it was implemented in the project, and a related academic paper (that they should find) that added something to their knowledge base.

**Detailed course information:**

- Who: Students in their second year learn about Agile methodologies and related management theories and processes in SE projects.
- Scope: Students focus on integration, compliance with constraints, and applying the object-oriented approach. In addition to the theoretical stratum, students experience a pseudo-live environment. In order to integrate applied methodologies and theories into the practice of software development in an academic course, Agile methodologies, concepts, and processes are introduced prior to a particular assignment. Students are asked to study and implement a wide range of theories and principles in their project using the scrum methodology and the object-oriented approach.
- The approach: A course is project-oriented when students have to accomplish tasks independently, and cope with problems that encourage them to develop their knowledge based on their experience, with course staff support. A given project consists of the whole software development life cycle including: telling the story and defining the specification, planning (e.g., resources and risk management), design, programming, testing, validation and verification processes, as well as specification changes according to course staff instructions.
- How: The work is done in small teams of 4–5 students who are directed to manage their work according to scrum definitions. The roles in each team are alternately assigned among the team members, including the role of the scrum master and the customer. The course staff sometimes takes part in the project as a customer that changes specifications or as a managerial entity that changes the priorities or determines constraints. Students are familiar with a programming language, which is a compulsory course. However, the chosen programming language simulates real-world situations where a developer is required to learn new things by themselves. The teams collaborate and manage software processes via commercial platforms such as Microsoft TFS. The unique format of our course exposes our students to independent work, but also provides adequate support.
- Progress: Students have periodic meetings in order to discuss the theories they have learned and how these are implemented in their projects. At the meetings, students present how they have progressed, what the problems and challenges were that they experienced, and how they solved the problems. Students are encouraged to ask questions. This format provides adequate support to the student, despite of the lack of the traditional presence of the teacher.

**Course summary:**

In this project-oriented course, students integrate applied methodologies and theories into the practice of software development to solve problems that reflect the complexity

*(Continued)*

---

**Example 6.2 (Continued)**

of the real world while working in an integrated environment. The students work in small teams according to the scrum method, and manage their tasks according to the course guidelines and staff mentoring (Kenett, 2013). This course provides an important opportunity to develop problem-solving and soft skills to prepare them for problems they may encounter as professional engineers. The contribution of soft skills for software engineers is discussed in Section 6.2.3. Figure 6.9 illustrates the difference processes and principles that are implemented in the course.

Currently, students perform integration at team level. However, we think that subdividing one comprehensive project into separate projects that are handled by different teams, as is customary in industry, can challenge and benefit our students. Extending the project size and complexity will lead to a higher level of integration, and will emphasize the importance of testing to the integration process. As part of our mission we encourage industry collaboration by linking real-world problems to the learning environment.

---

### 6.4.1 Aspects To Be Considered in Project-Oriented Teaching

The following is a recommended list of aspects to be aware of in the preparation of a project-oriented course (McLoone et al., 2013).

- **Deliverables and assessment:** There should be limited number of deliverables required from each team (3–5). A student's grade may be determined based on the team's report and their individual ability to answer questions during interview or presentation.
- **Timeline and workshops:** The workshops typically consist of some lecture time and discussion time, and will be front loaded at the start of the semester. All of this information should be placed in electronic format onto the Moodle system (or equivalent) for students to access at any stage. The Moodle system is also where teams submit their various deliverables. The students will be given a detailed timeline incorporating all of the various deliverables; this is important, to allow the students to see an overall picture of key milestones.
- **Team selection:** Group formation should preferably be done in the first week. The project tasks will be given to each of the groups at the end of week 1, and they started the actual project task at the start of week 2. Group selection techniques should be considered. These include random selection, a combination of self- and staff-assigned selection, and selection based on project preference
- **Project specifications:** Specifications should be short and relatively open ended to give the teams as much scope as possible for research and exploration – in effect, we want the students to drive their project, to take ownership of it. The only conditions imposed on all projects are that it has to involve using the instructed material prepared for the course.
- **Facilitation:** The role of the staff is to act as facilitators to each of the teams, with the aim of encouraging and supporting the students in their work but without directly

involving themselves in that work. In this model, the teams are entirely responsible for all aspects of the project, including organizing meetings with the facilitator, booking suitable meeting rooms, writing agendas, etc. In cases where this does not happen, the facilitator will not, in general, intervene or try to arrange a meeting for the team.

- **Dissemination of information:** The main practical concern is to prevent a lack of information available upfront to the students. Everything must be presented on an as-needed basis. Facilitators and/or students identify needs on an ongoing basis, and subsequently react to those needs shortly thereafter. However, most of these needs relate to material that could, and should, be available to the students at the start of their project. Examples included templates for the reflective journals, templates for the reports, and, more importantly, a student handbook outlining the project concept, how facilitation works, information on teamwork, and general good-practice tips.

## 6.5 Formative Assessment: Student Assessment Methods

Software engineering is also about doing – hands-on activities during the course are necessary. It is suggested to prepare a quiz or case study for each topic (or subtopic), though the preparation of these items may prove to be a tiresome mission. We do believe in involving the student in the material collection process; for example, participation in the MERLO item buildup, starting with research on the internet using keywords from IEEE standards, SWEBOK, the ISTQB glossary, articles, and tutorials. This research aims at the collection, development, and organization of MERLO items. The following are some suggestions to the instructor for the preparation of teaching materials. It is possible to have graduate student assignments include the collection and preparation of MERLO items. A sample assignment is:

1) Map and arrange learning material (scope into topics) – suggested to be done as part of the curriculum preparation.
2) Divide into major subtopics (learning chapters) – these major subtopics will eventually become the actual learning goals and targets of the course.
3) Break down each chapter within the subtopics into target statements (TSs). Student participation in the preparation of a TS could be done as a group assignment, so active discussions and debates on the content will become part of the learning procedures.
4) Prepare MERLO items for each TS.

Student participation in MERLO item preparation may assist in understanding the meaning of a particular learning item. Key-word-driven searches may be performed, so that at the end every subtopic will be covered with content card files – see Figure 6.10.

In order to encourage diversity and innovative aspects of education, it is recommended to collect many alternative target statements for each subtopic. This provides a good way to derive a large variety of tests, quizzes, exercises, and examination materials. This collection may also serve as an addition to the software engineering public book of knowledge.

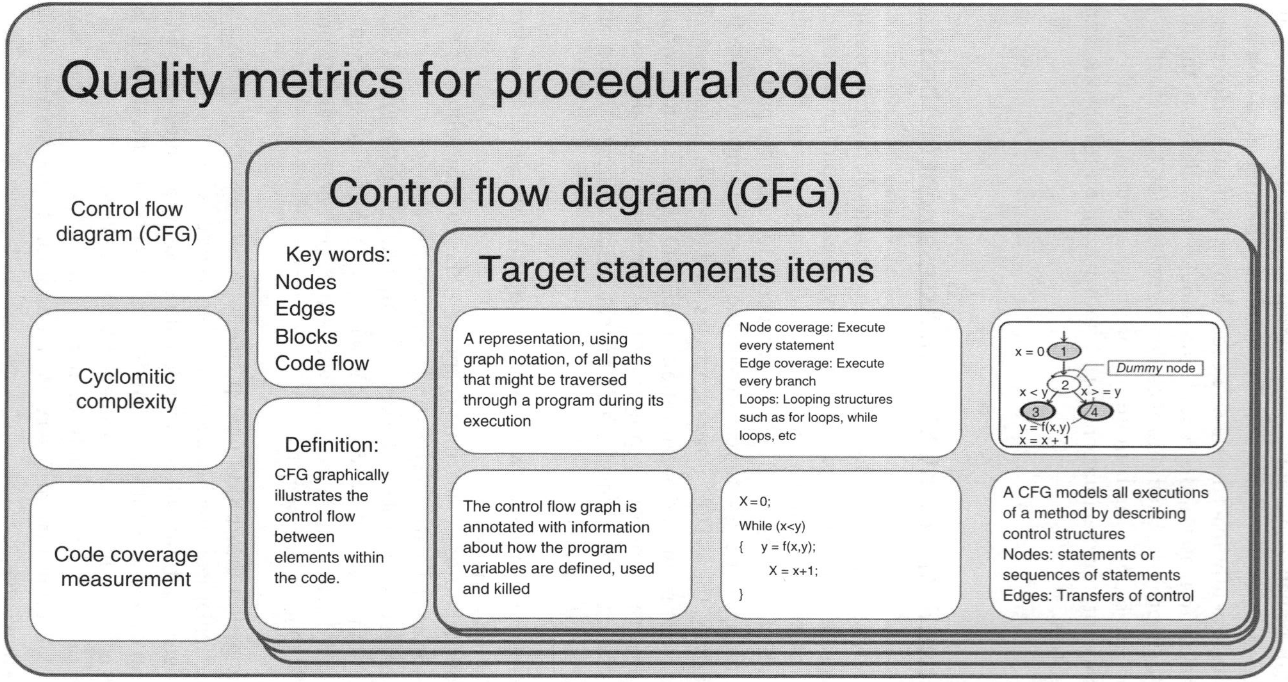
Quality metrics for procedural code
Control flow diagram (CFG)
Cyclomitic complexity
Code coverage measurement
Control flow diagram (CFG)
Key words:
Nodes
Edges
Blocks
Code flow
Definition:
CFG graphically illustrates the control flow between elements within the code.
Target statements items
A representation, using graph notation, of all paths that might be traversed through a program during its execution
Node coverage: Execute every statement
Edge coverage: Execute every branch
Loops: Looping structures such as for loops, while loops, etc
x = 0
1
Dummy node
2
x < y
x > = y
3
4
y = f(x,y)
x = x + 1
The control flow graph is annotated with information about how the program variables are defined, used and killed
X = 0;
While (x<y)
{ y = f(x,y);
X = x+1;
}
A CFG models all executions of a method by describing control structures
Nodes: statements or sequences of statements
Edges: Transfers of control

**Figure 6.10** Preparation of subtopic CFG.

Example 6.3 serves as a homework assignment, enforcing self-learning by the student using a text book (Kaner et al., 2013).

---

**Example 6.3 An Exercise in Domain Testing**

Domain testing and equivalence classes (Kaner et al., 2013) based on frequent flyer miles (Black, 2007, p. 217).

An airline rewards frequent flyers with credit according to Table 6.3. Your status is determined before you fly. The number of miles flown in this trip does not affect the status on this trip.

**Table 6.3** Frequent flyer points table.

| Status level | None | Silver | Gold | Premium |
|---|---|---|---|---|
| Trip bonus | 0 | 25% | 50% | 100% |
| Distance traveled | $d$ | $d$ | $d$ | $d$ |
| Points awarded | $d$ | $1.25 \times d$ | $1.5 \times d$ | $2 \times d$ |
| Miles required to reach this level | 0 | 25 000 | 50 000 | 100 000 |

**Exercise Mission:**
Determine whether the system awards the correct number of points for a given trip. Following the *A schema for domain testing* (Kaner et al., 2013, pp. 50–88) the student should respond to the following items.

Characterize the variables:

1) Identify the potentially interesting variables.
2) Identify the variable(s).
3) Determine the primary dimensions of the variable of interest.
4) Determine the type and scale of the variable's primary dimension and what values it can take.
5) Determine whether you can order the variable's values (from smallest to largest).
6) Determine whether this is an input variable or a result.
7) Determine how the program uses this variable.
8) Determine whether other variables are related to this one.

Analyze the variable and create tests:

9) Partition the variable (its primary dimension):
   - If the dimension is ordered, determine its subranges and transition points.
   - If the dimension is not ordered, base the partitioning on similarity.
10) Lay out the analysis in a classical boundary/equivalent table. Identify the best representatives.
11) Create a test for the consequences of the data entered, not just an input filter.
12) Identify secondary dimensions. Analyze them in the classical way.
13) Summarize your analysis with a risk/equivalence table.

*(Continued)*

---

**Example 6.3 (Continued)**

Generalize to multidimensional variables:

14) Analyze independent variables that should be tested together.
15) Analyze variables that hold results.
16) Analyze non-independent variables. Deal with relationships and constraints.

Prepare for additional testing:

17) Identify and list unanalyzed variables. Gather information for later analysis.
18) Imagine and document risks that don't necessarily map to an obvious dimension.

---

Game-based learning (GBL; Dhiraj, 2013) is a type of game that has defined learning outcomes. Game-based learning is designed to balance knowledge with gameplay and the ability of the player to retain and apply a target statement to the real world. Game-based learning describes an approach to teaching where students explore relevant aspect of games in a learning context designed by teachers. Teachers and students collaborate in order to add depth and perspective to the experience of playing the game.

Good game-based learning applications can draw us into virtual environments that look and feel familiar and relevant. Within an effective game-based learning environment we work toward a goal, choosing actions and experiencing the consequences of those actions along the way. We make mistakes in a risk-free setting, and through experimentation we actively learn and practice the right way to do things. This keeps us highly engaged in practicing behaviors and thought processes that we can easily transfer from the simulated environment to real life. The environments of gamification also use traditional learning techniques such as grading assignments, presenting corrective feedbacks, and encouraging collaborative projects. The research community broadly accepts the potential benefits of educational gaming in general, and videogames in particular, with online teaching like eLearning (Barata et al., 2013) and MOOC (Torrente et al., 2009) – MOOC will be addressed in the next section. Example 6.4 is a demonstration of a classroom game addressing the requirements engineering process – an important lesson for software development, including testing, qualifications.

Requirements engineering has become a crucial ingredient in the software development process; it defines what the system should do (Sommerville, 2011). This process consists of several subprocesses such as requirements elicitation, analysis, specification, and validation. However, these subprocesses are considered complex and are less technical, requiring the merging of information from different sources, including industrial standards, customer-provided requirements, and procedures from internal quality management systems. As such, software developers need to deal with complex processes of data collection, analysis, and translation to the specific context of the application. The exercise in Example 6.4 was given as part of the requirements engineering lesson, more specifically to demonstrate the processes of requirements elicitation and analysis, requirements specification, and validation.

## Example 6.4  A Classroom Requirement Engineering Game

In the following class example, the game mechanism is used for learning. The students are presented with the creative original game North Pole Camouflage, which is a development of a children's logical mind-challenging game. In this game the main principle is to arrange the six transparent puzzle pieces so that the animals are camouflaged in their correct environments: fish in the water and bears on the snow (see Figure 6.11).

**Figure 6.11** North Pole Camouflage game. Left: Solution for one level. Right: Basic principle. See color section.

Step 1: Figuring out the rules of the game. Students ask different questions in order to figure out the rules of the game/system:

- Can the fish be near the bear? (Here the answer was "Yes.")
- Can a bear stay on the blue area? (Here the answer is "No.")

After students ask enough questions, meaning that they feel that they can figure out the rules of the system, they should define the "system requirements." Additionally, as soon the students succeed in defining the general game/system rules (the instructor should decide on this), the students make the definition of the requirements more specific and accurate. Sometimes the board includes an implicit marking of a piece's shape (i.e., rectangular or other). This means that there are constraints embedded in the game.

Step 2: A new board is selected. The students are asked to solve the new board together, according to the requirements they defined during Step 1. Usually it takes several iterations to solve this task. Now the student can better understand how to play. In several cases the decision of placing the pieces becomes easier if, for example, the shapes of the pieces are marked – the dimensionality of the problem decreases and therefore the problem solving is easier.

Step 3: Validate the outcome by reviewing the requirements against the solved puzzle and summarize the learning process. A possible outcome may be a requirement collection/validation flowchart.

Example 6.4 shows how the teaching of a theoretical topic can become interactive and applicable. By applying a children's game, using an out-of-the-box approach, it is possible to encourage the student to implement the theoretical aspects and to practice real-world problem solving, in this case to develop the system requirements elicitation, specification, and validation.

---

**Example 6.5  Learn with a Free Online Game-Based Platform (Socrative)**

The concept of gamification has become part of e-learning and project-oriented learning (Barata et al., 2013). However, it is also possible to use it during classroom meetings, to encourage student engagement, active participation, and out-of-the-box thinking. From the teacher's point of view, it requires rethinking learning design and the implementation of mobile technologies. Today, there are many commercial tools for designing and implementing interactive learning in the classroom (e.g., Socrative, Kahoot). By using these apps, a teacher can have a clear visualization of the class understanding. The following example is taken from the course *Software Engineering Fundamental*. One of the basic topics studied in software development courses is the stages of testing, and the important differences between component, system, and release testing. After discussing these

*Quiz during class*

**Figure 6.12** Virtualization of results.

issues in class, in the last 15 minutes of the lesson the following questions were presented to the students via the Socrative website to verify their understanding. Students could fill in the survey, using their mobile, as anonymous or identified and get immediate feedback. When the user is identified, they can gain credit for correct answers. The left side of Figure 6.12 illustrates an example question presented to students using the mobile app, and on the right side of Figure 6.12 is the real-time visualization of the results as presented to the students in the classroom. Based on this method, students get feedback on their answers and the teacher can see the effectiveness of the learning process and whether there are any difficulties.

With minimal preparation, the teacher can prepare MERLO quizzes using the same technology and tools for generating live interactions during the class.

With the demand for fewer conventional classes, technology enables us to prepare class sessions in advance and record and publish lectures so that students can access them in their own time. This concept is very straightforward, and there are tools which record, publish, and index the lectures so that students can selectively access a specific topic.

## 6.6 Teaching Software Quality with a Massive Open Online Course

MOOCs have become a global innovative movement in education, and are considered as the latest effort to harness information technology for higher education (Altbach, 2014). Although the internet has offered open courses for many years, the quality and the quantity of the courses has changed. This change raises new challenges for education and has led many top universities to develop MOOCs using web platforms such as Coursera, edX, and Udacity. From the campus point of view, a MOOC is an additional learning channel. It enhances the educational experience of its on-campus students and it is also accessible by millions of external virtual learners around the world. Compared to networked online courses, MOOCs are more than just "transferring the information." The use of MOOCs enables, in addition to the presentation of information and the completion of assignments, the ability to perform peer-to-peer learning among the students, to use asynchronous forums, auto/peer grading, and collaborative activities among the learners and the course mentors (Day and Mulligan, 2015; Cheng-lin and Jian-wei, 2016). The development and the delivery of MOOCs involve financial considerations due to their extremely high costs. However, there have been many attempts to facilitate the rapid development of MOOCs at relatively low costs (Day and Mulligan, 2015) and to propose the most efficient methods to use in MOOCs to gain significant benefits from this approach (Lederman, 2013; Mulligan, 2013).

Universities and colleges have followed the global innovative trend and now develop MOOCs for hundreds of courses. MOOCs in different disciplines are varied in the construction of the course and the skills required from the students. Currently there are two main types: xMOOCs and cMOOCs. The former is more common, and structured similarly to traditional classes that usually split training into modules and tests for assessing student progression and learning (Vizoso, 2013). cMOOCs emphasize the connectivity among participants; they are based on discussions and contributions generated in a social learning network (Downes, 2006). When students are more familiar with innovative technologies and online learning culture, their experience can make the learning process simpler. It could be expected that researchers in educational software engineering would be among the key players in the education domain and in the coming age of MOOCs (Xie et al., 2013). However, the development of MOOCs in software engineering (e.g., coding, testing) includes hands-on assignments, collaboration, and the simulation of complex development environments, and therefore this kind of course still requires the support and the mentoring of a massive numbers of students (Leon Urrutia et al., 2015). An open source environment is one of the proposed solutions to teaching programming languages such as C#, Visual Basic, and F#, using an open platform in which anyone around the world can participate (Xie et al., 2013). The literature presents various attempts to develop MOOC environments for learning programming languages and software engineering related subjects (e.g., software testing and analysis, software analytics). For example, free/libre/open source software (FLOSS) projects are considered as early MOOCs. They are open to anyone to participate, and are driven mainly by the participant's willingness to contribute and to guide the direction of the project (Robles et al., 2014).

FLOSS projects require new developers to sustain them, and the controls over joining such projects are not strict. Another example of participant contribution to online learning is Pex4Fun (pex4fun.com) presented by Xie, Tillmann, and De Halleux (2014), an interactive gaming-based teaching and learning platform for.NET programming languages that was developed for software engineering education (Tillmann et al., 2013). The game is based on a coding duel where the player has to solve a particular programming problem, and during the game the player can get some feedback. However, both FLOSS and MOOCs suffer from several issues such as authentication, certification, and long-term participation (Robles et al., 2014).

The question of how to better teach and train software engineering skills by using a MOOC platform is still relevant. The target audience of software engineering courses may be familiar with online environments and programming languages to a certain extent. As such, it is possible to assume that they will be good candidates to contribute and learn more as a community. However, the role of the community in learning spaces is still questionable, so relying only on the contribution of the participant might be insufficient (Jones et al., 2016).

Education in the software engineering field requires more than just the technical aspects of the programming languages. Teaching software development processes includes a wide range of topics such as quality and testing, risk management, teamwork, soft skills, and related methodologies. These issues seem to be less well known and therefore need to be studied in a more guided way, as an integral part of the development process. Therefore, combining both xMOOC and cMOOC approaches can encourage the involvement of participants, and in addition provide them with a more instructive and well-structured environment.

---

**Example 6.6  A MOOC**

The following are the main ingredients of the MOOC *Software Quality Fundamental*. The course focuses on the fundamentals of software quality, including standards of quality and various testing techniques as part of the standard software development processes. This course puts quality and testing as integral parts of the development process. It covers managerial, technological, and process aspects. Facing the dynamic and constantly changing field of software engineering, and especially software quality processes and testing, in addition to the theoretical foundation, the course combines practical hands-on assignments using contemporary tools and technology as an interactive development environment using different languages (e.g., C, C#, Java, Python).

**Goals:**

- The student will become familiar with the principles and theories of software quality engineering as part of the programming.
- The student will experience applicable testing and software improvement techniques.
- The student will develop creative thinking for problem solving taken from real-life scenarios.
- The student will experience team collaboration as an integral part of the final project.

---

In this course, the participants work as a community to define the project specifications. However, the quality processes (e.g., requirement engineering processes, unit testing, component testing) are defined by the course instructors. By taking this approach it is possible to balance the potential contributions of the users and the need to complete the missing information using structured learning.

### 6.6.1  MOOC Challenges

The following are some of the challenges inherent in MOOC-based instruction:

- Distance between teacher and students. The growth of MOOC teaching methods is making direct impressions and acquaintance harder – the teacher is not facing a "real" student during classes.
- Lack of interaction. By enabling the student to pick their own time and pace we are decreasing the very necessary interaction and direct exchange needed in the personal assessment processes.
- Group learning utilization. Group activities and tasks lower our ability to differentiate between the different participants of the group.

## 6.7  Formative Assessment with MERLO

The application of formative assessment measurement typically involves more than one sample MERLO item for each subtopic taught in the course. Figures 6.13 and 6.14 are two examples of MERLO items developed for a course on risk-based testing.

| | |
|---|---|
| At list two out of these five statements, but possibly more then two, share equivalence of meaning<br>1. mark the statement that share equivalent of meaning<br>2. describe breathy the reasons that guided you to make these decisions | **A [ ]**<br><br>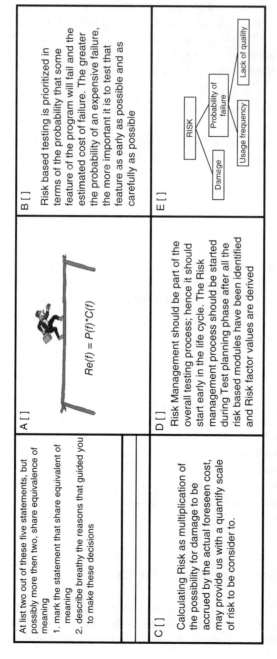<br><br>$Re(f) = P(f)*C(f)$ | **B [ ]**<br><br>Risk based testing is prioritized in terms of the probability that some feature of the program will fail and the estimated cost of failure. The greater the probability of an expensive failure, the more important it is to test that feature as early as possible and as carefully as possible |
| **C [ ]**<br><br>Calculating Risk as multiplication of the possibility for damage to be accrued by the actual foreseen cost, may provide us with a quantify scale of risk to be consider to. | **D [ ]**<br><br>Risk Management should be part of the overall testing process; hence it should start early in the life cycle. The Risk management process should be started during Test planning phase after all the risk based modules have been identified and Risk factor values are derived | **E [ ]**<br><br>RISK<br>Probability of failure<br>Damage<br>Usage frequency<br>Lack of quality |

**Figure 6.13** MERLO exam on risk-based testing (a).

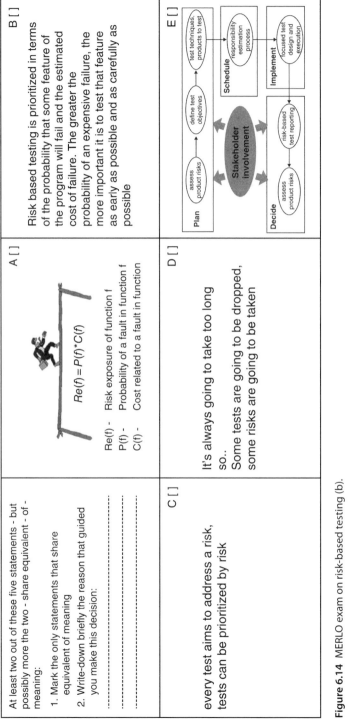

A [ ]

$Re(f) = P(f)*C(f)$

Re(f) - Risk exposure of function f
P(f) - Probability of a fault in function f
C(f) - Cost related to a fault in function

B [ ]

Risk based testing is prioritized in terms of the probability that some feature of the program will fail and the estimated cost of failure. The greater the probability of an expensive failure, the more important it is to test that feature as early as possible and as carefully as possible

C [ ]

every test aims to address a risk, tests can be prioritized by risk

D [ ]

It's always going to take too long so...
Some tests are going to be dropped, some risks are going to be taken

At least two out of these five statements - but possibly more the two - share equivalent - of - meaning:

1. Mark the only statements that share equivalent of meaning
2. Write-down briefly the reason that guided you make this decision:

----------------------------------------
----------------------------------------
----------------------------------------

E [ ]

Plan — assess product risks — define test objectives — test techniques, products to test

Schedule — responsibility estimation process

Implement — focused test design and execution

Stakeholder involvement

risk-based test reporting

Decide — assess product risks

**Figure 6.14** MERLO exam on risk-based testing (b).

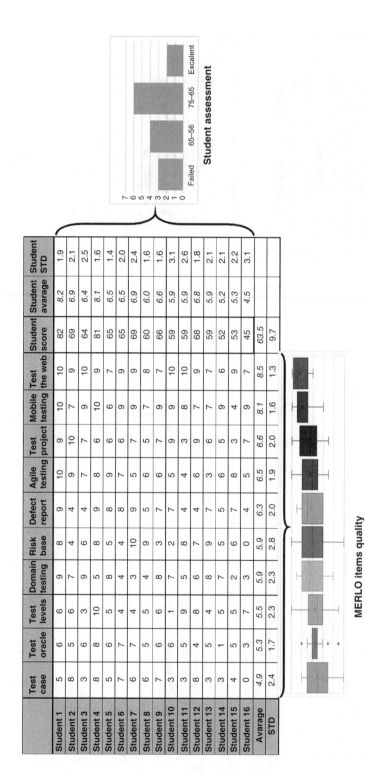

| | Test case | Test oracle | Test levels | Domain testing | Risk base | Defect report | Agile testing | Test project | Mobile testing | Test the web | Student score | Student avarage | Student STD |
|---|---|---|---|---|---|---|---|---|---|---|---|---|---|
| Student 1 | 5 | 6 | 6 | 9 | 8 | 9 | 10 | 9 | 10 | 10 | 82 | 8.2 | 1.9 |
| Student 2 | 8 | 5 | 6 | 7 | 4 | 4 | 9 | 10 | 7 | 9 | 69 | 6.9 | 2.1 |
| Student 3 | 3 | 6 | 3 | 9 | 6 | 4 | 7 | 7 | 9 | 10 | 64 | 6.4 | 2.5 |
| Student 4 | 8 | 8 | 10 | 5 | 8 | 9 | 8 | 6 | 10 | 9 | 81 | 8.1 | 1.6 |
| Student 5 | 5 | 6 | 5 | 8 | 5 | 8 | 9 | 6 | 6 | 7 | 65 | 6.5 | 1.4 |
| Student 6 | 7 | 7 | 4 | 4 | 4 | 8 | 7 | 6 | 9 | 9 | 65 | 6.5 | 2.0 |
| Student 7 | 6 | 7 | 4 | 3 | 10 | 9 | 5 | 7 | 9 | 9 | 69 | 6.9 | 2.4 |
| Student 8 | 6 | 5 | 5 | 4 | 9 | 5 | 6 | 5 | 7 | 8 | 60 | 6.0 | 1.6 |
| Student 9 | 7 | 6 | 6 | 8 | 3 | 7 | 6 | 7 | 9 | 7 | 66 | 6.6 | 1.6 |
| Student 10 | 3 | 6 | 1 | 7 | 2 | 7 | 5 | 9 | 9 | 10 | 59 | 5.9 | 3.1 |
| Student 11 | 3 | 5 | 9 | 5 | 8 | 4 | 4 | 3 | 8 | 10 | 59 | 5.9 | 2.6 |
| Student 12 | 8 | 4 | 8 | 6 | 7 | 4 | 6 | 9 | 7 | 9 | 68 | 6.8 | 1.8 |
| Student 13 | 3 | 5 | 4 | 8 | 9 | 7 | 3 | 6 | 7 | 7 | 59 | 5.9 | 2.1 |
| Student 14 | 3 | 1 | 5 | 7 | 5 | 5 | 6 | 5 | 9 | 6 | 52 | 5.2 | 2.1 |
| Student 15 | 4 | 5 | 5 | 2 | 6 | 7 | 8 | 3 | 4 | 9 | 53 | 5.3 | 2.2 |
| Student 16 | 0 | 3 | 7 | 3 | 0 | 4 | 5 | 7 | 9 | 7 | 45 | 4.5 | 3.1 |
| Avarage | 4.9 | 5.3 | 5.5 | 5.9 | 5.9 | 6.3 | 6.5 | 6.6 | 8.1 | 8.5 | 63.5 | | |
| STD | 2.4 | 1.7 | 2.3 | 2.3 | 2.8 | 2.0 | 1.9 | 2.0 | 1.6 | 1.3 | 9.7 | | |

MERLO items quality

Student assessment

Figure 6.15 MERLO results.

---

**Example 6.7 Risk-Based Testing MERLO Items**

These two MERLO items address risk-based testing. It is recommended to get students involved in their preparation as a study activity.

Scoring a single MERLO item was discussed in Section 6.3; the following is the instruction proposing the answering routine for a MERLO item:

1) Identify a topic and subtopic.
2) Phrase target statements in local language.
3) Identify alternative representations with equivalent meaning and surface similarity.
4) Fill out the MERLO item.

---

We analyze MERLO scores as a teaching aid improvement mechanism and an opportunity for the teacher to improve their teaching skills. Figure 6.15 lists results from a MERLO examination. Presenting the scores by topics, in addition to actual student scores, enables the evaluation of the quality of the MERLO items prepared – or alternatively evaluation of the class knowledge of each subtopic.

MERLO items are a very powerful tool for assessing student learning. It is standard practice to combine MERLO item examinations with an open practical implementation assignment – there are students who express their knowledge and skills better in a problem-solving environment. A final examination may have two parts: a MERLO item quiz and a real-world practical assignment. Example 6.8 is an example of an open examination question.

---

**Example 6.8 Open Examination Question**

**General:** Answer the following questions precisely (only what you are being asked). Each subquestion will be evaluated separately. The use of tables and charts is recommended. If you are not sure on a specific detail, make an assumption (and report it in tour answer).

**Open question #1: Defect management during Agile development**

**Background:** As preparation for the next product development in an Agile/Scrum SDLC it is your assignment to formulate the defect management mechanism and standards.

1) List the variables, parameters and codes (static and dynamic) you recommend to store, manage, and maintain within the defect management system (consider and express the nature of the project).
2) Present the control points and workstation that must be addressed and maintained for the suggested system. Assign an official and stakeholders for each of them, and suggest a defect management form to accompany each defect.
3) Using all the previous suggestions, draw a defect flow chart representing the defect management system.
4) Specify what would have been different in the defect management scheme if the SDLC was a waterfall-style project.

---

*(Continued)*

> **Example 6.8 (Continued)**
>
> **Instructors note:**
>
> Topic of course: Quality and testing projects
> Subtopics: Software testing and SDLC
>     Item: Agile/Scrum development
>     Item: Waterfall development
> Subtopic: Defect management

## 6.8 Summary

In summary, we highlight the main points presented in this chapter.

- Mutual dependence of academia and industry: Industry/academia interaction works in two directions. One needs to synchronize efforts for mutual benefits.
- Dynamically update content, scope, and techniques: The question of what to teach in SE is discussed in this chapter. It is like chasing a rapidly moving target. It seems vital, but not obvious, to collect, organize, and store the teaching materials. Another consideration is to consider the engineering and practical nature of software testing. This affects the practical versus conceptual educational trade-offs.
- Soft skills are a key consideration: This chapter presents the soft skills demanded of software test engineers such as presentation and discussion skills, teamwork communication, etc. This aspect of engineering education is typically dealt with outside of the teaching curriculum. Here we present techniques and practices that consider soft skills by adapting several tactics such as project orientation, gaming, and MERLO pedagogy.
- The MERLO approach contributes to upgraded teaching of testing: Adopting a deeper understanding of the scope content provides the foundation for a new implementation of software testing teaching. Considering the practical demand from software testing professionals is supported by combining MERLO item assessment together with an open practical implementation of exercises and assignments.
- Flexible, innovative, open: All of these qualities must be considered and addressed when building a software testing course. It is all tied up with the ability to expose and influence the rapid changes in the industry.

This chapter has covered various aspects of teaching and knowledge acquisition. It provides concrete examples and aggregated experience that can be considered as door-openers to further advances in the teaching of software engineering in general, and testing in particular.

## References

Altbach, P.G. (2014) MOOCs as neocolonialism: Who controls knowledge? *International Higher Education*, 75, 5–7.

Ardis, M., Budgen, D., Hislop, G.W., Offutt, J., Sebern, M., and Visser, W. (2015). SE 2014: Curriculum guidelines for undergraduate degree programs in software engineering. *Computer*, 48(11), 106–109.

Arzarello, F., Kenett, R.S., Robutti, O., Shafrir, U., Prodromou, T., and Carante, P. (2015a) Teaching and assessing with new methodological tools (MERLO): A new pedagogy? In M.A. Hersh and M. Kotecha (eds) *Proceedings of the IMA International Conference on Barriers and Enablers to Learning Maths: Enhancing Learning and Teaching for All Learners, Glasgow, UK*.

Arzarello, F., Carante, P., Kenett, R.S., Robutti, O., and Trinchero, G. (2015b) MERLO project: A new tool for education, in *Proceedings of IES 2015: Statistical Methods for Service Assessment, Bari, Italy*.

Barata, G., Gama, S., Jorge, J., and Gonçalves, D. (2013) Improving participation and learning with gamification, in *Proceedings of the First International Conference on Gameful Design, Research, and Applications*, ACM, pp. 10–17.

Black, R. (2007) *Pragmatic Software Testing: Becoming an Effective and Efficient Test Professional*, Chichester:Wiley.

Bloom, B.S., Krathwohl, D.R., and Masia, B.B. (1956) *Taxonomy of Educational Objectives: The Classification of Educational Goals*. New York: D. McKay.

P. Bourque and R.E. Fairley (eds) (2014) *Guide to the Software Engineering Body of Knowledge, Version 3.0*, New York: IEEE Computer Society.

Cheng-lin, H. and Jian-wei, C. (2016) SWOT analysis on the development of MOOC in China's higher education. *American Journal of Educational Research*, 4(6), 488–490.

Day, R. and Mulligan, B. (2015) Making free online learning sustainable through reduction of production costs, in *Proceedings of Higher Education in Transformation Conference*, Dublin, pp. 273–280.

Dhiraj K. (2013) *What is GBL (game-based learning)?* http://edtechreview.in/dictionary/298-what-is-game-based-learning [accessed 30 January, 2018].

Downes, S. (2006) Learning networks and connective knowledge, in *Collective Intelligence and E-Learning 2.0: Implications of Web-Based Communities and Networking* (eds H.H. Yang and S.C.-Y. Yuen), Hershey, PA: IGI Global, pp. 1–26.

Etkind M., Kenett R.S. and Shafrir U. (2010) The evidence-based management of learning: Diagnosis and development of conceptual thinking with meaning equivalence reusable learning objects (MERLO), in *Proceedings of the 8th International Conference on Teaching Statistics (ICOTS)*, Ljubljana, Slovenia.

Felderer, M. and Beer, A. (2013) Using defect taxonomies to improve the maturity of the system test process: Results from an industrial case study, in *Software Quality: Increasing Value in Software and Systems Development*, New York: Springer, pp. 125–146.

Felderer, M., and Schieferdecker, I. (2014) A taxonomy of risk-based testing. *International Journal on Software Tools for Technology Transfer*, 16(5), 559–568.

Fosnot, C. T. (2013) *Constructivism: Theory, Perspectives, and Practice*. New York: Teachers College Press.

Hazzan, O. and Har-Shai, G. (2013) Teaching computer science soft skills as soft concepts, in *Proceedings of the 44th ACM Technical Symposium on Computer Science Education*, ACM.

Hazzan, O., Lapidot, T., and Ragonis, N. (2015) *Guide to Teaching Computer Science: An Activity-Based Approach*. New York: Springer.

Hoffman, D. (1999) Heuristic test oracles. *Software Testing & Quality Engineering*, (March/April), 29–32.

Hoffman, D. (2003) Using test oracles in automation, in *Spring 2003 Software Test Automation Conference*, Software Quality Methods, LLC.

IEEE (2010) Systems and Software Engineering: Vocabulary, *ISO/IEC/IEEE* 24765:2010.

IEEE (2015) *Software Engineering Curriculum Guidelines for Undergraduate Degree Programs in Software Engineering*, Joint Task Force on Computing Curricula IEEE Computer Society Association for Computing Machinery. https://www.acm.org/binaries/content/assets/education/se2014.pdf [accessed 30 January, 2018].

Jones, K.M., Stephens, M., Branch-Mueller, J., and de Groot, J. (2016) Community of practice or affinity space: A case study of a professional development MOOC. *Education for Information*, 32(1), 101–119.

Kaner, C., Padmanabhan, S., and Hoffman, D. (2013) *The Domain Testing Workbook*. Context Driven Press.

Kenett, R.S. (2013) Implementing SCRUM using business process management and pattern analysis methodologies, *Dynamic Relationships Management Journal*, 2(2), 29–48.

Kenett, R.S. and Baker, E. (2010) *Process Improvement and CMMI for Systems and Software*, Boca Raton, FL: Auerbach CRC Publications.

Kenett, R.S. and Zacks, S. (2014) *Modern Industrial Statistics: With Applications using R, MINITAB and JMP*, 2nd edition, Chichester: Wiley.

Lederman, D., Affirmative action, innovation and the financial future: A survey of presidents, *Inside Higher Ed*, March 1, 2013. http://www.insidehighered.com/news/survey/affirmative-action-innovationandfinancial-future-survey-presidents [accessed 30 January, 2018].

Leon Urrutia, M., White, S., Dickens, K., and White, S. (2015) *Mentoring at scale: MOOC mentor interventions towards a connected learning community*. European MOOC Stakeholders Summit, Belgium.

McLoone, S., Lawlor, B., and Meehan, A. (2013). On project-oriented problem-based learning (POPBL) for a first-year engineering circuits project, in *Proceedings of Irish Signals and Systems Conference 2014 and 2014 China–Ireland International Conference on Information and Communications Technologies (ISSC 2014/CIICT 2014)*, pp. 386–391.

Matturro, G. (2013) Soft skills in software engineering: A study of its demand by software companies in Uruguay, in *Proceedings of 6th International Workshop on Cooperative and Human Aspects of Software Engineering (CHASE)*, pp. 133–136.

Mulligan, B., (2013) Opening education through competency-based assessment, in *Call for Vision Papers on Open Education 2030 Workshop*, Seville, April 29–30. http://goo.gl/mcSYSy [accessed 30 January, 2018].

Porat, A. (2009) *On the human and personal demand from V&V engineer. Guest lecture at Ben Gurion University*, Beer Sheva, Israel.

Robles, G., Plaza, H., and González-Barahona, J.M. (2014) Free/open source software projects as early MOOCs, in *Proceedings of the 2014 IEEE Global Engineering Education Conference (EDUCON)*, pp. 878–883.

Schawbel, D. (2012) How recruiters use social networks to make hiring decisions now, *Time Magazine*.

Shafrir, U. and Etkind, M. (2006) eLearning for depth in the semantic web. *British Journal for Educational Technology*, 37(3), 425–444.

Shafrir, U., and Etkind, M. (2010) *Concept Science: Content and Structure of Labeled Patterns in Human Experience*. Presentation, version 31.0.

Shafrir, U. and Kenett, R.S. (2010) Conceptual thinking and metrology concepts, *Accreditation and Quality Assurance*, 15(10), 585–590.

Shafrir U. and Kenett R.S. (2015) Concept science evidence-based MERLO learning analytics, in *Handbook of Applied Learning Theory and Design in Modern Education*, Hershey, PA: IGI Global.

Sobel, A. (2013) *Software Engineering Education Knowledge (SEEK)*, The Educational Activities Board of the IEEE Computer Society and the ACM Education Board. http:// sites.computer.org/ccse/know/FinalDraft.pdf [accessed 30 January, 2018].

Sommerville, I., Cliff, D., Callinescu, R., Keen, J., Kelly, T., Kwiatkowska, M., Mcdermid, J., and Paige, R. (2012) Large-scale complex IT systems. *Communications of the ACM*, 55(7), 71–77.

Terenzini, P.T., Cabrera, A.F., Colbeck, C.L., Parente, J.M., and Bjorklund, S.A. (2001) Collaborative learning vs. lecture/discussion: Students' reported learning gains. *Journal of Engineering Education*, 90(1), 123–130.

Tillmann, N., De Halleux, J., Xie, T., Gulwani, S., and Bishop, J. (2013) Teaching and learning programming and software engineering via interactive gaming, in *Proceedings of the 35th International Conference on Software Engineering (ICSE)*, pp. 1117–1126.

Torrente, J., Moreno-Ger, P., Martínez-Ortiz, I., and Fernandez-Manjon, B. (2009) Integration and deployment of educational games in e-learning environments: The learning object model meets educational gaming, *Educational Technology & Society*, 12(4), 359–371.

Vizoso, C.M. (2013) Los MOOCs un estilo de educación 3.0, in *SCOPEO INFORME N° 2. MOOC: Estado de la situación actual, posibilidades, retos y futuro*, pp. 239–261.

Xie, T., Tillmann, N., and De Halleux, J. (2013). Educational software engineering: Where software engineering, education, and gaming meet, in *Proceedings of the 3rd International Workshop on Games and Software Engineering: Engineering Computer Games to Enable Positive, Progressive Change*, pp. 36–39.

# Part II

# Statistical Models

# 7

# Non-homogeneous Poisson Process Models for Software Reliability

*Steven E. Rigdon*

## Synopsis

The non-homogeneous Poisson process (NHPP) has long been used as a model for the reliability of repairable hardware systems. Such models allow for the improvement or deterioration of the system. By its nature, software acts differently than hardware. We introduce the NHPP and discuss its applicability to software reliability. One class of models that is closely related to the NHPP is the class of piecewise exponential models. We discuss this class and its hybrids, and how they are related to the NHPP.

## 7.1 Introduction

Models for software reliability often resemble models for the reliability of repairable systems. Both models must describe the occurrence of events in time, and both should accurately describe how the reliability changes as a function of time. Classes of models for software reliability (and for repairable systems as well) are usually based on how the reliability changes. Does the reliability change continuously? Or, does the reliability change only at the failure times? The former model leads to the non-homogeneous Poisson model (NHPP); the latter leads to the class of piecewise exponential (PEXP) models, or some generalization of this class. Section 7.2 covers the NHPP and some of its properties, and Section 7.3 covers the PEXP models. Generalizations, or hybrids, of these models are described in Section 7.4.

All models will involve unknown parameters, which must be estimated given data that were collected. There are two basic paradigms for collecting data from software testing. One involves the recording of the exact time of each failure, or equivalently, the times between failures. The other is to record, for several time intervals, the time tested along with the number of failures in that interval. The failure times could be measured in operating time, or in calendar time. Operating time is preferable, because it is more closely related to the stress placed on a system; the models we describe here are applicable to both, however. We discuss in Section 7.5 statistical inference for both approaches.

Two examples are discussed in Section 7.6. One involves the recording of the times between failure, and the other involves counting how many failures fell in each interval. Section 7.7 describes some other models that have been applied to software reliability.

*Analytic Methods in Systems and Software Testing*, First Edition.
Edited by Ron S. Kenett, Fabrizio Ruggeri, and Frederick W. Faltin.
© 2018 John Wiley & Sons Ltd. Published 2018 by John Wiley & Sons Ltd.

## 7.2 The NHPP

Software reliability models describe the failure processes when software is being tested. The NHPP has been widely used to model the failure occurrences of hardware systems (Crow, 1974; Rigdon and Basu, 2000). It has also been applied in the context of software reliability, beginning with the work by Goel and Okumoto (1979).

The non-homogeneous Poisson process (NHPP) is a counting process that describes the occurrence of events in time. The random function $N(t)$ is defined to be the number of failures in the interval $[0, t]$. A variation of this notation is $N(a, b)$, which equals the number of failures in the interval $(a, b)$; closed intervals, as well as half-open intervals, can be used as well. The *mean function* of a counting process is the expected number of failures through time $t$; that is,

$$\Lambda(t) = E(N(t)). \tag{7.1}$$

The assumptions behind the NHPP are the following.

- $N(0) = 0$; that is, we begin with zero failures.
- If $t_1 < t_2 \leq t_3 < t_4$, then $N(t_1, t_2)$ and $N(t_3, t_4)$ are independent random variables. This is called the *independent increments* property.
- There is a function $\lambda(t)$ such that

$$\lambda(t) = \lim_{\Delta t \to 0} \frac{P(N(t, t + \Delta t) = 1)}{\Delta t}.$$

The function $\lambda(t)$ is called the *intensity function*.

- $\lim_{\Delta t \to 0} P(N(t, t + \Delta t) \geq 2)/\Delta t = 0$. This assumption precludes the possibility of simultaneous failures.

These assumptions lead to the result that the number of failures in an interval, say $(a, b)$, has a Poisson distribution, with mean $\mu = \int_a^b \lambda(t) \, dt$. Thus,

$$P(N(a, b) = x) = \frac{\exp\left(-\int_a^b \lambda(t) \, dt\right) \left(\int_a^b \lambda(t) \, dt\right)^x}{x!}, \qquad x = 0, 1, 2, \ldots$$

This result gives the Poisson process its name.

Implicit in these assumptions is that the failure intensity is the same just after a failure as it was just before the failure, so long as the intensity function is continuous. For hardware repairable systems, this concept is called *minimal repair*, since the repair puts the item in the same condition as just before the failure. Thus, under the NHPP model the system reliability is unchanged by a failure and subsequent repair; this is true for both software and hardware systems. The reliability changes continuously, irrespective of the pattern of failure. This assumption may be questionable in the software reliability context, where during failure-free periods the system remains in an identical condition (since no changes have been made to it). It could be argued, however, that the stress placed on the system varies with time, and therefore the failure process should reflect this time dependence. At the time of a failure, the software is presumably changed to correct the error or "bug." This would suggest that models where the intensity "jumps" at each failure may be more appropriate.

The NHPP is really a class of models. The choice of a functional form of the intensity creates a specific model, although even this involves unknown parameters. One of the models most commonly applied to hardware reliability is the power-law process (PLP) described by Crow (1974) and Rigdon and Basu (1989), which has intensity function

$$\lambda(t) = \frac{\beta}{\theta}\left(\frac{t}{\theta}\right)^{\beta-1}, \qquad t > 0. \tag{7.2}$$

A value of $\beta > 1$ leads to an increasing intensity function, indicating deterioration. On the other hand, a value of $\beta$ satisfying $0 < \beta < 1$ leads to a decreasing intensity function, indicating reliability improvement. If $\beta = 1$, then we obtain the homogeneous Poisson process which has a constant intensity. The mean function for the PLP is

$$\Lambda(t) = \left(\frac{t}{\theta}\right)^{\beta}. \tag{7.3}$$

The PLP is closely related to one of the first models for the reliability of repairable systems, the Duane model (Duane, 1964), although the two models are not identical (see Rigdon, 2002).

The other commonly applied model is the exponential intensity model described by Cox and Lewis (1966). This model has intensity function

$$\lambda(t) = \exp(\alpha + \beta t), \qquad t > 0. \tag{7.4}$$

Goel and Okumoto (1979) were the first to apply an NHPP model to the reliability of software. They began with the assumption that the mean function satisfies

$$\Lambda(t + \Delta t) - \Lambda(t) = b[a - \Lambda(t)]\Delta t + o(\Delta t). \tag{7.5}$$

In other words, the expected number of failures in the interval $(t, t + \Delta t)$ is proportional (with constant of proportionality $b$) to the number of failures that remain undetected, that is, $a - \Lambda(t)$. Dividing by $\Delta t$ and taking the limit leads to the differential equation

$$\Lambda'(t) = b[a - \Lambda(t)],$$

with boundary conditions $\Lambda(0) = 0$ and $\Lambda(\infty) = a$, where $a$ is the number of faults or errors in the system. The solution of Equation (7.5) is

$$\Lambda(t) = a(1 - e^{-bt}). \tag{7.6}$$

For an NHPP, the intensity function is the derivative of the mean function, so for the Goel and Okumoto model, the intensity is

$$\lambda(t) = \Lambda'(t) = abe^{-bt} = \exp(\log ab - bt). \tag{7.7}$$

Thus, we can see that the Goel and Okumoto model is equivalent to the Cox and Lewis (1966) model with $\alpha = \log ab$ and $\beta = -b$.

A very general NHPP was developed by Pham, Nordmann, and Zhang (1999); we call this the PNZ model. They began by allowing $a$ and $b$ in the Goel–Okumoto model to be time dependent. This led to the differential equation

$$m'(t) = b(t)[a(t) - \Lambda(t)]. \tag{7.8}$$

The solution to Equation (7.8) that satisfies the initial condition $\Lambda(0) = 0$ is

$$\Lambda(t) = \exp(B(t)) \int_0^t a(u)b(u) \exp[B(u)] \, du, \tag{7.9}$$

where

$$B(t) = \int_0^t b(u)\, \mathrm{d}u.$$

The possibility of introducing additional software errors while debugging the system can be introduced into the model by allowing $b$ to be time dependent; that is, a function $b(t)$. Depending on the parametric choices made for $a(t)$ and $b(t)$, the PNZ model reduces to a number of other models that had been previously studied. For example, if $a(t) = a$ and $b(t) = b$ are constant, then the PNZ model becomes the Goel–Okumoto model. Other choices led to the models introduced by Yamada and Osaki (1985) and Yamada, Tokuno, and Osaki (1992). For example, the choice of

$$a(t) = a^*(1 + \alpha t) \quad \text{and} \quad b(t) = \frac{b^*}{1 + \beta \exp(-b^* t)}$$

leads to an NHPP model with mean function

$$\Lambda(t) = \frac{a^*}{1 + \beta(1 - \exp(-b^* t))}[(1 - \exp(-b^* t))(1 - \alpha/b^*) + \alpha t].$$

This is often called the PNZ software reliability growth model.

## 7.3 Piecewise Exponential Models

Another class of models assumes that the times between failure have independent exponential distributions, but the means of the exponential distributions are allowed to vary. Thus, the times between failure are independent but not identically distributed.

The PEXP model of Sen and Bhattacharyya (Sen, 1998; Sen and Bhattacharyya, 1993) is one such model. In this model the times between failure $X_1, X_2, \ldots$ are independent exponential random variables satisfying

$$E(X_i) = \frac{\delta}{\mu} i^{\delta - 1}, \qquad i = 1, 2, \ldots$$

The expected times between failure are proportional to $i^{\delta - 1}$, so if $\delta > 1$ then the expected times between failure are getting larger (reliability improvement), and if $\delta < 1$, then the times between failure are getting smaller (reliability deterioration). Of course, if $\delta = 1$, then the model reduces to the homogeneous Poisson process.

Jelinski and Moranda (1972) suggested a model based on the following assumptions. The software initially contains a fixed but unknown number $N$ of faults or "bugs." As the software is tested, each fault has a constant rate of occurrence; call this rate $\phi$. Thus, initially, with $N$ faults each discoverable with rate $\phi$ the fault discovery rate is $N\phi$. Once a fault is discovered, there remain $N - 1$ faults, again with each having discovery rate $\phi$; the overall discovery rate is now $(N - 1)\phi$. This continues, until there is a single fault remaining, which has discovery rate $\phi$. Thus, the $i$th fault has discovery rate $(N - i + 1)\phi$, so that the time between the $(i - 1)$st and $i$th faults has an exponential distribution with mean

$$E(X_i) = \frac{1}{(N - i + 1)\phi}.$$

This is also a piecewise exponential model, although it is different from the model in Sen (1998) and Sen and Bhattacharyya (1993). Thompson (1988) pointed out how this model is related to the order-statistics process, another point process model.

## 7.4   Hybrid Models

The main difference between the piecewise exponential models and the NHPP models is the domain over which reliability growth occurs. In the NHPP models, reliability growth (or deterioration in the case of some hardware systems) occurs continuously, and is completely unaffected by a failure and subsequent repair. By contrast, the piecewise exponential model has the property that no reliability growth occurs during failure-free operating time. At the time of a failure, however, the software reliability improves if the discovered fault is removed and no new bugs are introduced. If more bugs are introduced than removed, the software reliability might actually be worse than before.

Figure 7.1 illustrates the differences. The three plots on the left side are graphs of the *complete intensity function*, that is, the intensity function conditioned on the past history of the process. The three graphs in the right-hand column are those of the conditional mean functions. The first row shows the intensity function and the mean function for the Goel–Okumoto NHPP model. Note how the intensity and the mean function change continuously, irrespective of the pattern of failures. The middle row shows the intensity and mean functions for a piecewise exponential model. For this model, note that the intensity function is flat over failure-free intervals, indicating no reliability growth (or deterioration) between failures, but at each failure there is a downward jump in the intensity. This jump indicates that the reliability has changed for the better at each failure. The mean function shown on the right is continuous, but is piecewise linear. A hybrid model is shown in the bottom pair of graphs. Here, the reliability gets worse at each failure, which would indicate that more faults were introduced than removed. Reliability does improve continuously during failure-free periods.

## 7.5   Statistical Inference

All software models will involve unknown parameters that must be estimated from observed data. Most estimation methods involve either maximum likelihood or least squares.

We will focus on maximum likelihood for the case where we observe the exact failure times. For the NHPP, where testing stops at the $n$th failure, the likelihood given the $n$ failure times $0 < T_1 < T_2 < \cdots < T_n$ is

$$L(\theta|\mathbf{y}) = \left( \prod_{i=1}^{n} \lambda(t_i) \right) \exp\left( -\int_0^{t_n} \lambda(u)\, du \right). \tag{7.10}$$

Here, $\theta$ indicates the parameter of the parametric form of the intensity. Note that this likelihood was derived under the assumption that testing stopped at the time of the $n$th failure; this is called *failure truncation*. If testing is stopped at a predetermined time $t$, then we have *time truncation*. We will not discuss time truncation further, but details

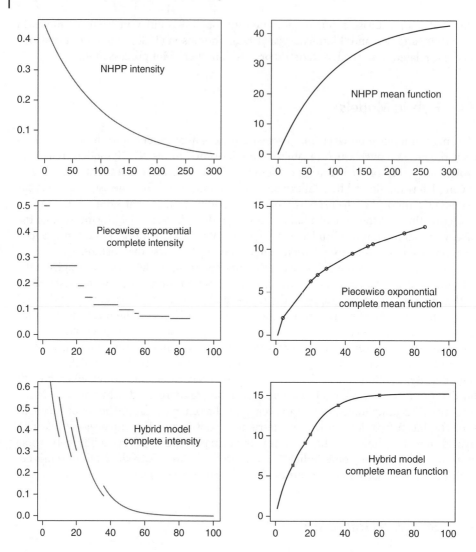

**Figure 7.1** Complete intensity functions, and complete mean functions for three models: (top) NHPP, (middle) piecewise exponential, and (bottom) a hybrid model. The complete intensity function is the failure intensity conditioned on the past history of the failure process.

on the difference can be found in Rigdon and Basu (2000). With the Goel–Okumoto model, which has intensity function $\lambda(t) = abe^{-bt}$, the log-likelihood becomes

$$\ell(a, b|\mathbf{y}) = n\log a + n\log b - b\sum_{i=1}^{n} t_i - a + a\exp(-bt_n).$$

To obtain the maximum likelihood estimations (MLEs), we would take the partial derivatives $\partial\ell/\partial a$ and $\partial\ell/\partial b$, yielding

$$\frac{\partial\ell}{\partial a} = \frac{n}{a} - 1 - \exp(-bt_n)$$

and

$$\frac{\partial \ell}{\partial b} = \frac{n}{b} - \sum_{i=1}^{n} t_i - a t_n \exp(-b t_n),$$

and set these results equal to 0. The first equation can be solved directly for $a$ in terms of $b$ to obtain

$$a = \frac{n}{1 + \exp(-b t_n)}, \tag{7.11}$$

which can be substituted into the second equation. This second equation must be solved numerically for $b$; once found, the MLE for $b$ can be substituted into Equation (7.11).

The likelihood function for the Jelinski–Moranda model is

$$L(\phi, N) = \prod_{i=1}^{n} [(N - i + 1)\phi] \exp[-(N - i + 1)\phi x_i],$$

yielding the log-likelihood function

$$\ell(\phi, N) = \sum_{i=1}^{n} \log(N - i + 1) + n \log \phi - \phi \sum_{i=1}^{n} (N - i + 1) x_i.$$

The partial derivatives are

$$\frac{\partial \ell}{\partial \phi} = \frac{n}{\phi} \sum_{i=1}^{n} (N - i + 1) x_i$$

and

$$\frac{\partial \ell}{\partial N} = \sum_{i=1}^{n} \frac{1}{N - i + 1} - \phi \sum_{i=1}^{n} x_i.$$

When we set these results equal to zero, once again, we see that the first equation can be solved for one of the variables, $\phi$ in this case, to give

$$\phi = \frac{n}{\sum_{i=1}^{n} (N - i + 1) x_i}.$$

When this is substituted into the second equation, we obtain the equation

$$\sum_{i=1}^{n} \frac{1}{N - i + 1} - \frac{n \sum_{i=1}^{n} x_i}{\sum_{i=1}^{n} (N - i + 1) x_i} = 0$$

in just the variable $N$. This equation can be numerically solved for $N$, which can then be substituted into the former to obtain the MLE for $\phi$. This assumes that $N$ is a continuous variable, so for example we could have an estimate of $\hat{N} = 35.4$ total bugs in a system. If we wish to restrict $N$ to being an integer, then we could evaluate the profile log-likelihood function

$$\ell(N) = \max_{\phi} \ell(\phi, N) \tag{7.12}$$

for each $N$ over a reasonable range. The value $\hat{N}$ of $N$ that maximizes the profile likelihood function is then the MLE of $N$, subject to the constraint that $N$ is an integer, and

$$\hat{\phi} = \arg\max \ell(\phi, \hat{N})$$

is the MLE of $\phi$. Often, the MLE from the continuous parameter case can be used to get a starting point for evaluating the profile likelihood function.

Suppose now that we observe the numbers of failures in several time intervals. If the cutoff points for the $m$ intervals are

$$0 = s_0 < s_1 < s_2 < \cdots < s_m$$

then the software system was run for $s_i - s_{i-1}$ time units during the $i$th interval. Let $n_i, i = 1, 2, \ldots, m$, denote the number of observed failures in interval $i$, that is from time $s_{i-1}$ to $s_i$. If the failure process is governed by an NHPP with intensity function $\lambda(t)$ and mean function $\Lambda(t)$, then

$$n_i \sim \text{Poisson}(\Lambda(s_i) - \Lambda(s_{i-1})).$$

The likelihood function is then

$$L(\theta | n_1, n_2, \ldots, n_m) = \prod_{i=1}^{m} \frac{1}{n_i!} [\Lambda(s_i) - \Lambda(s_{i-1})]^{n_i} \exp[-\Lambda(s_i) + \Lambda(s_{i-1})],$$

and the log-likelihood function is

$$\ell(\theta | n_1, n_2, \ldots, n_m) = -\sum_{i=1}^{m} \log n_i! + \sum_{i=1}^{m} n_i \log[\Lambda(s_i) - \Lambda(s_{i-1})] - \Lambda(s_m).$$

Here we assume that we begin with exactly zero failures, so that $\Lambda(0) = 0$. Also, $\theta$ indicates the parameter of whatever particular form is assumed for the intensity function. For example, for the Goel–Okumoto model with mean function $\Lambda(t) = a(1 - \exp(-bt))$, the log-likelihood function is

$$\ell(a, b | n_1, n_2, \ldots, n_m) = -\sum_{i=1}^{m} \log n_i! + \sum_{i=1}^{m} n_i \log[-a \exp(-bs_i) + a \exp(-bs_{i-1})]$$
$$-a(1 - \exp(-bs_m)). \tag{7.13}$$

Numerical methods are required to approximate the maximum likelihood estimates.

## 7.6 Two Examples

We describe two examples of real software testing. One involves the recording of the exact failure times; the other involves counting the use (in terms of CPU usage) and the number of failure in unequally spaced intervals. Recording the exact failure times will, of course, provide more information. There is naturally some information lost when data are accumulated into blocks of time.

The first involves the failure times of a software system for the Naval Tactical Data System (NTDS). This classic data set was originally given by Jelinski and Moranda (1972). There were 26 failures during the production phase, and eight more failures in the post-production phase. The raw data are shown in Table 7.1. We will focus on just

**Table 7.1** Failure data for the Naval Tactical Data System.

| Phase | Error number | Time between errors (days) | Cumulative time between errors |
|---|---|---|---|
| Production | 1 | 9 | 9 |
| Production | 2 | 12 | 21 |
| Production | 3 | 11 | 32 |
| Production | 4 | 4 | 36 |
| Production | 5 | 7 | 43 |
| Production | 6 | 2 | 45 |
| Production | 7 | 5 | 50 |
| Production | 8 | 8 | 58 |
| Production | 9 | 5 | 63 |
| Production | 10 | 7 | 70 |
| Production | 11 | 1 | 71 |
| Production | 12 | 6 | 77 |
| Production | 13 | 1 | 78 |
| Production | 14 | 9 | 87 |
| Production | 15 | 4 | 91 |
| Production | 16 | 1 | 92 |
| Production | 17 | 3 | 95 |
| Production | 18 | 3 | 98 |
| Production | 19 | 6 | 104 |
| Production | 20 | 1 | 105 |
| Production | 21 | 11 | 116 |
| Production | 22 | 33 | 149 |
| Production | 23 | 7 | 156 |
| Production | 24 | 91 | 247 |
| Production | 25 | 2 | 249 |
| Production | 26 | 1 | 250 |
| Test | 27 | 87 | 337 |
| Test | 28 | 47 | 384 |
| Test | 29 | 12 | 396 |
| Test | 30 | 9 | 405 |
| Test | 31 | 135 | 540 |
| User | 32 | 258 | 798 |
| Test | 33 | 16 | 814 |
| Test | 34 | 35 | 849 |

the production failures; that is, the first 26 failures. Figure 7.2 shows the raw data on the top, and the fitted mean functions (using the Goel–Okumoto and PLP models) on the bottom. The MLEs of the parameters of the Goel–Okumoto model are

$$\hat{a} = 33.999 \quad \text{and} \quad \hat{b} = 0.005791.$$

The MLEs for the PLP are

$$\hat{\beta} = 0.2178 \quad \text{and} \quad \hat{\theta} = 0.0000795.$$

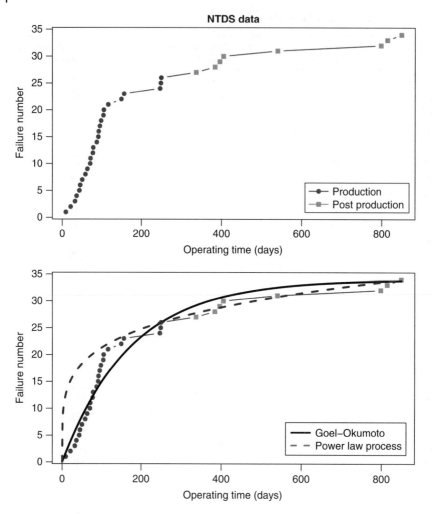

**Figure 7.2** (Top) Plot of cumulative operating time vs. failure number for the NTDS data. (Bottom) The same data with the Goel–Okumoto and PLP models fit to the mean function.

Clearly, the Goel–Okumoto model fits better over the production period, which is the set of failures that went into building the model. Interestingly, both models give a nearly identical fit around 800 hours. There is a fundamental difference between the two models, however, regarding the ultimate number of failures that the system will experience. The Goel–Okumoto model would predict that about 34 failures would ultimately be discovered, while the PLP predicts that there will be infinitely many.

If we fit the Jelinski–Moranda model to this data set, we find that the MLEs are

$$\hat{N} = 31.222 \quad \text{and} \quad \hat{\phi} = 0.00684.$$

The plot of the estimated complete intensity function is shown in the top of Figure 7.3. This is, of course, a step function, because the complete intensity function is assumed to be constant between failures. The estimated mean function, is the accumulated or

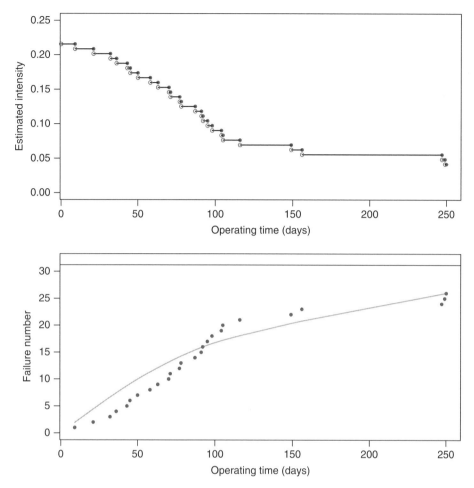

**Figure 7.3** (Top) Plot of estimated complete intensity function (top) and estimated mean function (Bottom) for the JM model applied to the NTDS data set.

integrated intensity function, and is shown in the bottom of Figure 7.3. This function is piecewise linear, although with 26 data points, the curve looks rather smooth.

If we restrict the unknown number of bugs $N$ to be an integer, then we can take the profile likelihood approach described in Section 7.5. The profile likelihood is shown in Figure 7.4 for $26 \leq N \leq 50$ (note that $N$ must be greater than or equal to the observed number of failures $n$). The maximum of the profile likelihood occurs when $N = 31$, and in this case $\hat{\phi} = 0.00695$. In this case, the MLEs differ very little from the "continuous $N$" case.

The second example comes from Wood (1996), who studied the failure times for software developed at Tandem Computers. This data set, shown in Table 7.2, gives the testing time (cumulative CPU hours) and the number of failures in that interval. The exact failure times are not known. The data are displayed in Figure 7.5. The decrease in the failure rate (failures per week) is clear from the bottom graph.

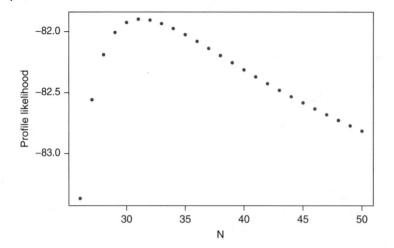

**Figure 7.4** Profile log-likelihood function for NTDS data set using the JM model.

**Table 7.2** Failure times for software developed at Tandem Computers.

| Week | CPU hours | Cumulative errors found | Week | CPU hours | Cumulative errors found |
|---|---|---|---|---|---|
| 1 | 519 | 16 | 11 | 6539 | 81 |
| 2 | 968 | 24 | 12 | 7083 | 86 |
| 3 | 1430 | 27 | 13 | 7487 | 90 |
| 4 | 1893 | 33 | 14 | 7846 | 93 |
| 5 | 2490 | 41 | 15 | 8205 | 96 |
| 6 | 3058 | 49 | 16 | 8564 | 98 |
| 7 | 3625 | 54 | 17 | 8923 | 99 |
| 8 | 4422 | 58 | 18 | 9282 | 100 |
| 9 | 5218 | 69 | 19 | 9641 | 100 |
| 10 | 5823 | 73 | 20 | 10 000 | 100 |

Consider now the Goel–Okumoto model applied to this data set. Since the data are grouped, we must maximize the log-likelihood function given in Equation (7.13). We find that the MLEs of $a$ and $b$ are

$$\hat{a} = 122.02 \quad \text{and} \quad \hat{b} = 0.0001712.$$

Testing for this system stopped after 100 errors were discovered, and it was estimated that there were about 122 errors, so the estimate of the number of remaining errors was 22. Testing could continue until more failures were uncovered, but Figure 7.5 suggests that quite a bit more testing might be required, since the weekly failure rates were nearly 0 after week 17.

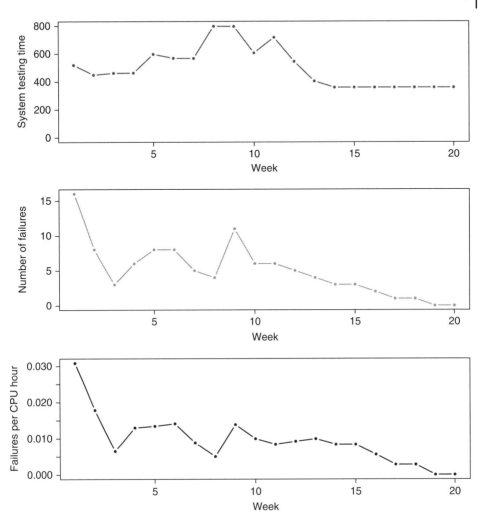

**Figure 7.5** Tandem Computers software failure data. (Top) Testing time measured in CPU hours. (Middle) Number of failures per week. (Bottom) Failure rates (failures per CPU hour) for each week.

## 7.7 Summary

Two commonly used classes of models for software reliability are the non-homogeneous Poisson process, with various parametric forms for the intensity function, and the piecewise exponential model. Two of the most well-known models, the Goel–Okumoto (Goel and Okumoto, 1979) model and the Jelinski–Moranda (Jelinski and Moranda, 1972) are examples of NHPP and piecewise exponential models, respectively. There have been generalizations of these models, for example to allow the intensity to assume an "S" shape (which is not possible in the original Goel–Okumoto model). Another model that allows for imperfect debugging using a hidden Markov model was given by Pievatolo, Ruggeri, and Soyer (2012). A general model which subsumes both the Goel–Okumoto and Jelinski–Moranda models was described by Langberg and Singpurwalla (1985). Singpurwalla and Soyer (1996) discussed a number of general models for software reliability.

Comparisons of efficiencies, mostly among NHPP models, have been done by Sharma et al. (2010) and Anjum, Haque, and Ahmad (2013). Sharma et al. (2010) use a distance-based approach, while Anjum, Haque, and Ahmad (2013) use a weighted average of a number of criteria, such as mean squared error (MSE), mean error of prediction (MEOP), etc.

In Table 1 of Sharma et al. (2010) they list 16 software reliability models, and they give the functional forms for the intensity function $\lambda(t)$ and the mean function $\Lambda(t)$. The number of unknown parameters ranges from two, for simple models like the Goel–Okumoto model, to six, for the Zhang–Teng–Pham (Zhang et al., 2003) model. They find that the Goel–Okumoto model is one of the best-fitting models, despite its simplicity with just two parameters. Many of the other models that involve more parameters have greater flexibility, but the additional parameters introduce considerable variation in the estimates of all parameters, causing them to have a worse fit than the simpler models. The Goel–Okumoto model is still one of the best-fitting models for software reliability.

# References

Anjum, M., Haque, M.A., and Ahmad, N. (2013) Analysis and ranking of software reliability models based on weighted criteria value. *International Journal of Information Technology and Computer Science (IJITCS)*, 5(2), 1.

Cox, D. and Lewis, P. (1966) *The Statistical Analysis of Series of Events*, Chichester: Wiley.

Crow, L.H. (1974) Reliability analysis for complex repairable systems, in F. Proschan and R. Serfling (eds), *Reliability and Biometry*, Philadelphia, PA: SIAM, pp. 379–410.

Duane, J. (1964) Learning curve approach to reliability monitoring. *IEEE Transactions on Aerospace*, 2(2), 563–566.

Goel, A.L. and Okumoto, K. (1979) Time-dependent error-detection rate model for software reliability and other performance measures. *IEEE Transactions on Reliability*, 3, 206–211.

Jelinski, Z. and Moranda, P.B. (1972) Software reliability research, in W. Freiberger (ed.), *Statistical Computer Performance Evaluation*, New York: Academic Press, pp. 465–484.

Langberg, N. and Singpurwalla, N.D. (1985) A unification of some software reliability models. *SIAM Journal on Scientific and Statistical Computing*, 6(3), 781–790.

Pham, H., Nordmann, L., and Zhang, Z. (1999) A general imperfect-software-debugging model with S-shaped fault-detection rate. *IEEE Transactions on Reliability*, 48(2), 169–175.

Pievatolo, A., Ruggeri, F., and Soyer, R. (2012) A Bayesian hidden Markov model for imperfect debugging. *Reliability Engineering & System Safety*, 103, 11–21.

Rigdon, S.E. (2002) Properties of the Duane plot for repairable systems. *Quality and Reliability Engineering International*, 18(1), 1–4.

Rigdon, S.E. and Basu, A.P. (1989) The power law process: A model for the reliability of repairable systems. *Journal of Quality Technology*, 21(4), 251–260.

Rigdon, S.E. and Basu, A.P. (2000) *Statistical Methods for the Reliability of Repairable Systems*, New York: Wiley.

Sen, A. (1998) Estimation of current reliability in a Duane-based reliability growth model. *Technometrics*, 40(4), 334–344.

Sen, A. and Bhattacharyya, G. (1993) A piecewise exponential model for reliability growth and associated inferences, in A.P. Basu (ed.), *Advances in Reliability*, Amsterdam: Elsevier, pp. 331–355.

Sharma, K., Garg, R., Nagpal, C., and Garg, R. (2010) Selection of optimal software reliability growth models using a distance based approach. *IEEE Transactions on Reliability*, 59(2), 266–276.

Singpurwalla, N.D. and Soyer, R. (1996) Assessing the reliability of software: An overview, in S. Ozekici (ed.) *Reliability and Maintenance of Complex Systems*, New York: Springer, pp. 345–367.

Thompson, W. (1988) *Point Process Models with Applications to Safety and Reliability*, New York: Springer.

Wood, A. (1996) Predicting software reliability. *Computer*, 29(11), 69–77.

Yamada, S. and Osaki, S. (1985) Software reliability growth modeling: Models and applications. *IEEE Transactions on Software Engineering*, 11(12), 1431.

Yamada, S., Tokuno, K., and Osaki, S. (1992) Imperfect debugging models with fault introduction rate for software reliability assessment. *International Journal of Systems Science*, 23(12), 2241–2252.

Zhang, X., Teng, X., and Pham, H. (2003) Considering fault removal efficiency in software reliability assessment. *IEEE Transactions on Systems, Man, and Cybernetics-Part A: Systems and Humans*, 33(1), 114–120.

# 8

# Bayesian Graphical Models for High-Complexity Testing: Aspects of Implementation

*David Wooff, Michael Goldstein, and Frank Coolen*

## Synopsis

The Bayesian graphical models (BGM) approach to software testing was developed in close collaboration with industrial software testers. It is a method for the logical structuring of the software testing problem, where focus was on high reliability final-stage integration testing. The core methodology has been published and is briefly introduced here, followed by discussion of a range of topics for practical implementation. Modeling for test–retest scenarios is considered, as required after a failure has been encountered and attempts at fault removal have been made. The expected duration of the retest cycle is also considered; this is important for planning of the full testing activity. As failures often have different levels of severity, we consider how multiple failure modes can be incorporated in the BGM approach, and we also consider diagnostic methods. Implementing the BGM approach can be a time-consuming activity, but there is a benefit of re-using parts of the models for future tests on the same system. This will require model maintenance and evolution, to reflect changes over time to the software and the testers' experiences and knowledge. We discuss this important aspect, which includes consideration of novel system functionality. End-to-end testing of complex systems is a major challenge which we also address, and we end by presenting methods to assess the viability of the BGM approach for individual applications. These are all important aspects of high-complexity testing which software testers have to deal with in practice, and for which Bayesian statistical methods can provide useful tools. We present the basic approaches to these problems – most of these topics will need to be carefully extended for specific applications.

## 8.1 The Bayesian Graphical Models Approach to Testing: A Brief Overview

Bayesian graphical models (BGM), also known as Bayesian networks and a variety of further terms, are graphical representations of conditional independence structures for random quantities, which have proven to be powerful tools for probabilistic modeling and related statistical inference (Jensen, 2001). The BGM approach for supporting software testing was developed by the authors in close collaboration with an industrial

*Analytic Methods in Systems and Software Testing*, First Edition.
Edited by Ron S. Kenett, Fabrizio Ruggeri, and Frederick W. Faltin.

partner; the core methodology was presented in 2002). Some additional aspects of this approach were briefly discussed by 2007). The BGM approach presents formal mechanisms for the logical structuring of software testing problems, the probabilistic and statistical treatment of the uncertainties to be addressed, the test design and analysis process, and the incorporation and implication of test results. Once constructed, the models are dynamic representations of the software testing problem. They may be used to answer what-if questions, to provide decision support to testers and managers, and to drive test design. The models capture the knowledge of the testers for further use. The main ingredients of the BGM approach are briefly summarized in this section and illustrated via part of a substantial case study (Wooff et al., 2002).

Suppose that the function of a piece of software is to process an input number, e.g. a credit card number, in order to perform an action, and that this action might be carried out correctly or incorrectly. The tests that might be run correspond to choosing various numbers and checking that the *software action* is performed correctly for each number. Usually it will not be possible to check all inputs, but instead one checks a subset from which one, hopefully, may conclude that the software is performing reliably. This involves a subjective judgment about the functionality of the software over the collection of inputs not tested, and the corresponding uncertainties. One must choose whether to explicitly quantify the uncertainties concerning further failures given test results, or whether one is content to make an informal qualitative judgment for such uncertainties. In many areas of risk and decision analysis, uncertainties are routinely quantified as subjective probabilities representing the best assessments of uncertainty of the expert carrying out the analysis (Bedford and Cooke, 2001).

In the BGM approach, the testers' uncertainties for software failure are quantified and analyzed using subjective probabilities, which involves investment of effort in thinking carefully about prior knowledge and modeling at a level of detail appropriate for the project. The rewards of such efforts include probabilistic statements on the reliability of the software taking test results into account, guidance on optimal design of test suites, and the opportunity to support management decisions quantitatively at several stages of the project, including deciding on time and cost budgets at early planning stages.

The simplest case occurs when the tester judges all possible test results to be exchangeable, which implies that all such test results are judged to have the same probability of failing, and that observing each test result gives the same information about all other tests. Qualitatively, exchangeability is a simple judgment to make: either there are features of the set of possible inputs which cause the tester to treat some subsets of test outcomes differently from other subsets or, for him, the collection of outcomes is exchangeable. Some practical aspects, including elicitation of such judgments, are discussed later. Of course, testers may not be "correct" in their judgments, but attempts to model the software in far greater detail, relying on many different sources, tend to be very time consuming and are not part of common practice in many software testing situations. The approach discussed here starts with the judgments of the testers, and helps in selecting good test suites in line with those judgments. An important further aspect is the use of diagnostic methods and tests, aimed at discovering discrepancies between those judgments and the actual software performance. Later, aspects of such diagnostic testing are also discussed. In the example, suppose that the software is intended to cope with credit cards with both short (S) and long (L) numbers, and that the tester judges test success to be exchangeable for all short numbers and exchangeable

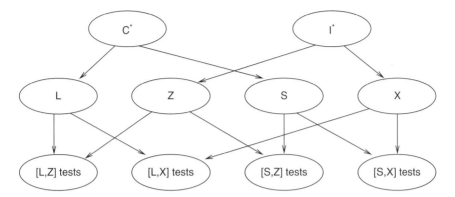

**Figure 8.1** Bayesian graphical model (reduced form) for a software problem with two failure modes.

for all long numbers. For example, it might be the case that dealing with long numbers is newly added functionality. In addition, suppose that the tester judges test success to be exchangeable for numbers starting with zero (Z), and exchangeable for numbers starting with a non-zero digit (X).

Figure 8.1 is a BGM, in reduced form, reflecting such judgments. Each node represents a random quantity, and the arcs represent conditional independence relations between the random quantities (Cowell et al., 1999; Jensen, 2001; Wooff et al., 2002). There are four subgroups of tests, resulting from the combinations of number length and starting digit given the exchangeability judgments. For example, node "[L,Z] tests," with attached probability, represents the probability that inputs of the type [L,Z] lead to an output error when tested. Such an error could be caused by three different problems, namely general problems for long numbers (node L), general problems for numbers starting zero (node Z), or problems specific for this combination. This last cause is not explicitly represented in this reduced BGM, as it would be a single parent node feeding into the node "[L,Z] tests," but for the sake of simplicity all nodes with a single child and no parents have been deleted. Similarly, the node $C^*$ represents general problems related to the length of the numbers, so problems in L can be either caused by problems in $C^*$, or in a second parent node of L (not shown), for problems specifically arising for long numbers. Upon testing, the random quantities represented by these nodes are typically not directly observable. Their interpretation is as binary random quantities, indicating whether or not there are faults in the software, particularly leading to failures when a related input is tested. Finally, the actual test inputs are also not represented by nodes in this reduced BGM; they would be included in the form of a single node corresponding to each input tested, feeding off the corresponding node at the lowest level presented in Figure 8.1.

To build such a model, exchangeability assumptions are required over all relevant aspects, and many probabilities must be assigned, both on all nodes without parents, and for all further nodes in the form of conditional probability tables, for all combinations of the values of the parent nodes. For larger applications, which typically included several thousands of nodes, methods to reduce the burden of specification to manageable levels are described by 2002). For example, the tester was asked to rank nodes according to "faultiness," and then to judge overall reliability levels to assign probabilities consistent with these judgments – see Wooff, Goldstein, and Coolen (2002) for more details, where in particular further theoretical, modeling, and practical aspects are discussed. After the test results have been observed, all these probabilities are updated by the rules of BGMs (Cowell et al., 1999; Jensen, 2001), which provides an explicit quantification of the effect of the test results on the overall reliability of the software system, within the subjectively quantified reliability model reflecting the judgments of the testers. Therefore one can assess the value of a test suite from this perspective, because either it will indicate faults, which must be fixed and then retested, or one observes no failures and then has the decision as to whether to release the software. Thus, one may value the test suite according to the probability of no faults remaining, if one enters successful test results for all inputs of the test suite into the model.

The main case study presented in Wooff, Goldstein, and Coolen (2002), where this method was applied in a project similar to the management of credit card databases, involved a major update of existing software, providing new functionality and with some earlier faults fixed. It could be called a "medium-scale" case study; there were 16 separate major testable areas, and initial structuring by the tester led to identification of 168 different domain nodes (in Figure 8.1 there are four domain nodes, namely those at the lowest level presented), contained in 54 independent BGMs. A single test typically consisted of a variety of inputs spread over several of these BGMs. For example, one test could be "change of credit card," which would consist of a combination of delete an existing card from the database and add a new card to the database with appropriate credit limits. The practical problem for the testers is the optimal choice of the test suite: assessment of the information that may be gained from a given test suite is very difficult unless a formal method is used. In this case study, the tester had originally identified a test suite consisting of 233 tests. Application of the BGM approach revealed that the best 11 of these tests reduced the prior probability of at least one fault remaining by about 95%, and that 66 tests could not add any further information, according to the tester's judgments, assuming that the other tests had not revealed any failures. The BGM approach was also used to design extra tests to cover gaps in the coverage provided by the existing test suites. One extra test was designed which had the effect of reducing the residual probability of failure by about 57%, mostly because one area had been left completely untested in the originally proposed test suite. The value of this test was agreed by the senior tester. Subsequently, the authors have developed a fully automated approach for test design, providing test suites without using a tester's test suite as input. This leads to even better designs, and allows the inputs to be tested in sequences according to a variety of possible optimization criteria, where for example aspects of test–fix–test can be taken into account, e.g. minimizing total expected test time including time to fix faults shown during testing and the required retesting, as is discussed later.

A second case study in Wooff, Goldstein, and Coolen (2002) involved software for renumbering all the records in a database, required because the present number of digits was insufficient to meet an expansion in customer demand. Similar actions had been carried out in the past, but this software was newly created and to be used only once. The sole tester had no prior expertise in the reliability of this software and only vague notions about the software operations. This was a "small-scale" case, with three separate major testable areas represented by three independent BGMs, with a total of 40 domain nodes. It was assumed that, although the tester must choose an input from a possibly large range of allowable inputs, each such test was expected to fully test the domain node. The tester's test suite consisted of 20 tests, constructed (but not run) before the

BGMs were elicited. The BGM analysis showed that this original test suite, if no failure were revealed, would reduce the probability of at least one fault remaining from 0.8011 to 0.2091, so reducing the original probability by about 74%. Of these 20 original tests, at least 11 were shown to be completely redundant, under the elicited assumptions in the BGM approach, in the sense that their test coverage was fully covered by (some combinations of) the remaining nine tests. The BGM approach was used to automatically design an efficient test suite, given only the tester's basic specification of the operations to be tested and the BGM so constructed. This test suite contained only six tests, but would fully test the software according to the judgments provided by the tester. Consequently, the tester agreed that the automatically designed test suite was more efficient at testing the software, and that it tested an area he had missed. The tester had not originally considered that this area needed testing, but agreed that it was sensible to test it, which came to light only through the initial BGM structuring process. It should be emphasized that, although several of the 20 originally suggested tests were redundant, they were of value from the perspective of diagnostics for the elicited model and probabilities, as is discussed later.

The following sections focus on aspects of implementation of the BGM approach, and corresponding management issues, in addition to those discussed by Wooff, Goldstein, and Coolen (2002). These general discussions are based on the industrial collaboration, and motivation is sometimes given in terms of database-related software functions, in line with the case studies mentioned above.

## 8.2    Modeling for the Test–Retest Process

One of the most challenging aspects of the testing process is modeling and design in the sequence where software is tested, faults are (supposedly) fixed, and the system is partially retested, to protect against new errors that might have been introduced in the fixing process.

Full modeling of the complete retest cycle for all possible failures that we might possibly observe may be very complex, due to the many different types of fault that we may encounter. Further, much of this detailed modeling will usually be wasted, as the great majority of tests will not fail. However, choices as to the amount of retesting that we must do will be key determinants of overall testing costs and thus of our ability to complete testing within budgetary constraints, usually against strict deadlines. Therefore, we cannot avoid some form of retest modeling, to guide the design and sequencing of the test suite through estimation of the cost and time to completion of the full testing process. We suggest a pragmatic simplification for retest modeling which should roughly correspond to the judgments of the tester, and give sensible guidelines both for choosing the test suite and assessing total budget implications, with the understanding that for the, hopefully small, number of faults which are actually found, we may choose to reconsider the implications of the individual failures.

### 8.2.1    Retest Modeling

Suppose that we have already carried out test suite $T_{[r-1]} = T_1, \ldots, T_{r-1}$, all tests proving successful. We now carry out test $T_r$, and an error occurs on a test attached to domain node $J$. To simplify what follows, let us suppose that the first test we run on retest is the test $T_r$ which previously failed, and that this test is successful (as otherwise we would return the software for further fixing and then repeat the procedure). It seems reasonable to select $T_r$ as our first test, and this avoids unnecessary complications in the following account. (Of course, in practice, we might need to run some of the other tests in order to carry out the test $T_r$, but this does not affect the general principle of our approach.) What we need to decide is whether further errors have been introduced into the model during the software fix, and to update our probability model accordingly, thus revealing which tests need to be rerun. A simple way to model new errors is to divide such errors into two types:

- *similar* errors to the observed error (for example, errors in nodes which are linked to the node $J$ in the model);
- *side-effect* errors (i.e. errors that could occur anywhere in the model). These side-effect errors are errors specifically introduced by the fixing of another bug, and which in principle we can find by further appropriate testing.

A simple way of handling such errors is to return the probability description to the state that it was in before testing began.

- If there are *similar* errors then we cannot trust the results of the subset of previous tests with outcomes linked to the node with the error, and so these results must be ignored.
- If there are *side-effect* errors then we cannot trust any of the previous test results.

We might combine this approach with a more sophisticated analysis of the root probabilities. For example, if we find many errors, then we might feel that we should have raised the prior probability of errors in all root nodes by some fixed percentage. Doing so does not change the general idea of the remodeling.

### 8.2.2 Consequences of a Test Failure

We denote by $T_S$ the subset of the previous test suite $T_{[r-1]}$ which tests nodes which are *similar* to the node $J$ which failed on test $T_r$. We let $T_G$ be the corresponding subset of remaining tests in $T_{[r-1]}$ which tests those nodes which are dissimilar to node $J$. We must take into account three possibilities.

$R_1$ The error detected by $T_r$ is fixed and no new errors are introduced.
$R_2$ The error detected by $T_r$ is fixed, but, possibly, *similar* errors have been introduced.
$R_3$ The error detected by $T_r$ is fixed, but, possibly, *side-effect* errors have been introduced, possibly together with further *similar* errors.

Consideration of tests that result in failures in multiple unrelated nodes is left as a topic for future research. A simple way of proceeding is to postulate that these three scenarios have the following consequences.

$R_1$ Our beliefs about the model are just as though we had carried out on the original model test suite $T_{[r]} = T_1, \ldots, T_r$ and all tests had been successful.
$R_2$ Our beliefs about the model are just as though we had carried out test suite $T_G$ and test $T_r$ successfully, but we have not carried out any of the tests in $T_S$.
$R_3$ Our beliefs about the model are just as though we had carried out on the original model the single test $T_r$ successfully, but we have not carried out any of the other tests.

These consequences are simplifications in the following sense. For scenario $R_2$, the subset of tests $T_S$ will be testing many nodes which are *dissimilar* to the failure node, in addition to testing the failure node and nodes similar to it. For these *dissimilar* nodes, the beliefs which pertain to them at the time following the completion of test suite $T_{[r]}$ may be more appropriate than the corresponding prior beliefs. For scenario $R_3$, we might believe it appropriate to handle *similar* and *dissimilar* nodes distinctively; dealing with these refinements, which offer some benefits, is left as a topic for future research.

### 8.2.3 Further Specification

In order to handle this problem, we need to understand the relative probability of each of these three possibilities. Therefore, to complete the specification we introduce the three numbers $p_1, p_2, p_3$ to represent the probabilities that we assign to the three possibilities $R_1$, $R_2$, and $R_3$, respectively, given that test $T_r$ succeeded. We suppose that we have already tested $T_r$ successfully, as otherwise we would need three more possibilities corresponding to $T_r$ failing, which would unnecessarily complicate the model, as we would not carry out further tests given failure of $T_r$, but would instead return the software for further correction.

$$p_i = \mathrm{P}(R_i), \quad i = 1, 2, 3; \quad p_1 + p_2 + p_3 = 1.$$

### 8.2.4 Prior Specification

It may be appropriate that these three values, and the division into *similar* and *side-effect* errors, can be assigned in advance for each possible node to be tested. For example, such a specification might be guided by knowledge of the abilities of the software engineer responsible for fixing the software for which the error occurred. If there are $n$ nodes in the model being tested, this would require at most a specification of $3n$ probabilities and $n$ divisions into error types, although there will usually be many simplifications to be exploited.

Otherwise, it may be appropriate simply to specify in advance three basic probabilities $p_1, p_2, p_3$ which we use for each node to be tested, along with a simple rule of thumb for distinguishing the test subsets $T_S$ and $T_G$. For example, a simple rule is to assign those nodes which are connected to the observed failure node in the BGM as *similar*, and the remainder as *dissimilar*. Both the full specification and the simple specification allow an analysis of expected completion time for the test suite. Such prior specification is mostly useful to provide for order-of-magnitude appropriate calculations, and for simulation purposes.

### 8.2.5 Wait-and-See Specification

Instead of specifying the probabilities and divisions into *similar* and *side-effect* a priori, it is better to wait until we see a particular test failure and specify the probabilities $p_1, p_2, p_3$ and the separation into subsets $T_s$ and $T_G$ which are relevant for that particular test. This might be appropriate, for example, if the decision as to which software engineer will be responsible for fixing a bug is only taken when a bug is found. However, a wait-and-see specification will not allow an analysis of expected completion time for the test suite.

### 8.2.6 Informative Failure

In practice, when the software is returned after an error is found, there will usually be some auxiliary information which will be helpful in determining which of the three possibilities $R_1, R_2, R_3$ has occurred. Such information is usually difficult to model in advance. Therefore, we shall concentrate in what follows on evaluating upper and lower bounds on such information. The upper bound corresponds to the situation where we obtain no additional information on failure, which we call *uninformative failure*. The lower bound comes from supposing that, having found an error and sent the software for retesting, then we will obtain sufficient additional information to allow us to determine precisely which of the three possibilities $R_1, R_2, R_3$ will apply in the retesting situation. We term this *informative failure*. In practice, we will usually obtain some, but not perfect, additional information about the retest status of the model on failure, so that the analysis of informative and uninformative versions of the retest model will provide bounds on the expected time for the retest cycle, and identify the importance of such additional information for the given testing problem.

### 8.2.7 Modifying the Graphical Model

We now describe how to incorporate the above probabilistic specification into a modified graphical model. We begin with the original graphical model. We add each test $T_i$ that we carried out previously as a node that is linked to the corresponding domain nodes which determine the probability that the test is successful. We now introduce a further *retest* root node, $R$. The node $R$ is linked to each test node $T_i$ except node $T_r$. We now need to describe the probability distribution of each $T_i$ given all parents, i.e. the original parents plus node $R$. Let $T_i^*$ be the event that test $T_i$ succeeds. Suppose that the original parents of $T_i$ have possible combined values $A_1, \ldots, A_m$. In the original model, we therefore determined $P(T_i^*|A_j), j = 1, \ldots, m$, and used these probabilities to propagate the observation of the test successes across the model. In the new model, we must describe the probability for the event that $T_i$ succeeds given each combination of $A_j$ and $R_k$. We do this by assigning the probability for $T_i$ to succeed as given by the original parents if either $R_1$ occurs or the test in question is not similar to $T_r$, but assigning the probability for $T_i$ success to be the prior probability $P(T_i^*)$ of success in all other cases, as in such cases the observed probability is not affected by the current states of the parent nodes. Formally, we have

$$
\begin{aligned}
P(T_i^*|A_j, R_1) &= P(T_i^*|A_j), \\
P(T_i^*|A_j, R_2) &= P(T_i^*|A_j) & T_i \in T_G, \\
P(T_i^*|A_j, R_2) &= P(T_i^*) & T_i \in T_S, \\
P(T_i^*|A_j, R_3) &= P(T_i^*).
\end{aligned}
\tag{8.1}
$$

Given the new model, we can now feed in the observations of success for each of the original tests $T_{[r-1]}$ and the new test $T_r$, and this will give our updated probabilities for the new model. Each additional test that we carry out is added to the model in the usual way as a fresh node; for example, a new test which repeats an earlier test is added as an additional node to the model whose success updates beliefs across the model in the usual way. Observe that while no further test nodes are attached to retest node $R$, each retest will update beliefs over $R$ and thus update beliefs about the information contained in the original tests.

### 8.2.8   Expected Duration of the Retest Cycle

In principle, we should choose a test suite in order to minimize the time to completion of the full retest sequence. Further, after each fault is found, we should choose a new test suite optimized against the revised model. As there will usually be a substantial period while the software is being fixed, it will often be practical to rerun the test selection algorithm in that period to select a new test suite. However, to assess the expected time for completion of the cycle in this way would be very computer intensive for large test suites. Therefore, we find instead a simple upper bound on the test length which corresponds well to much of current practice, by considering the expected length of time to complete successfully any given fixed test suite. Our test procedure is as follows. We carry out each test in sequence. Either test $j$ passes, in which case we go on to the test $j+1$, or test $j$ fails, in which case the software is fixed and retested, where we repeat the tests on that subset of the first $j$ tests which are required under the model in Equation (8.1). Either all these tests are successful, in which case we can go on to test $j+1$, or one of the intermediate tests fails, in which case it must be fixed and retested, and so forth.

To assess the effect of retesting, we must consider, at the minimum, the amount of time that it is likely to take to fix a particular fault, and the amount of retesting that is required when a fault is found and supposedly fixed. Therefore, let us suppose that with every test $T_j$, we associate an amount of time $c_j$ to carry out the test and a random quantity $X(T_j)$, which is the length of time required to fix a fault uncovered by that test. We could similarly attach costs for carrying out the tests and fixing any faults. We elicit the prior expectation and variance for each $X(T_j)$. We may, if appropriate, consider the time for fixing more carefully, by treating $X(T_j)$ to be a vector of times taken for error fixing, depending on the nature and severity of the error, or we may instead simplify the model by choosing a single random quantity, $X$ representing the time to fix given any fault. Different levels of detail in the description will affect the effort in eliciting the inputs into the model, but will not affect the overall analysis.

Suppose that we have chosen and sequenced a test suite $S = (T_1, \dots, T_r)$. Suppose that tests $T_1, \dots, T_{j-1}$ are successful, and then test $T_j$ fails. Suppose then that the software is fixed. Suppose now that test $T_j$ is repeated and is successful, so that the current model is as described by Equation (8.1). Now, some of the tests $T_1, \dots, T_{j-1}$ will need to be repeated, while some may not be necessary. We can check whether any of the first $j-1$ tests may be eliminated by running a step-wise deletion algorithm on the subset of tests $T_1, \dots T_{j-1}$ by finding at each step the minimum value of the posterior probability for the criterion function on which the original design was constructed when we attempt to delete one of the first $j-1$ tests, where all other tests in $S$ are successful. We denote by $R(j)$ the *retest set* given failure of $T_j$, namely the subcollection of $T_1, \dots T_{j-1}$ which must be retested when we have deleted all of the tests that we can by stepwise deletion under the retest model. For the uninformative failure model the retest sets required for Equation (8.3) are extracted by direct calculation on the model in Equation (8.1). For the informative failure model, if $T_j$ fails, then with probabilities $p_{1j}, p_{2j}, p_{3j}$, there will occur the possibilities $(R_1, R_2, R_3)$ of Section 8.2.1, and after the software has been fixed we will obtain sufficient additional information to allow us to determine precisely which of the three possibilities $(R_1, R_2, R_3)$ will apply in the retesting situation. Under $(R_1)$, no retests are required; under $(R_2)$, the set $T_{Sj}$ of similar tests to $T_j$ are repeated; and under $(R_3)$ all tests are repeated.

We denote by $E_+(T_j)$ the expected length of time between starting to test $T_j$ for the first time and starting to test $T_{j+1}$ for the first time, so that the expected time to complete testing for the suite $S = (T_1, \ldots, T_r)$ is

$$E_+(S) = \sum_{j=1}^{r} E_+(T_j). \tag{8.2}$$

To evaluate each $E_+(T_j)$ fully is an extremely computationally intensive task. An approximate calculation which will give roughly the right order of magnitude for test time is as follows. When we first run test $T_j$, there are two possibilities. With probability $(1 - q_j)$, $T_j$ is immediately successful, and, with probability $q_j$, $T_j$ fails, in which case the software must be fixed and the subset $R(j)$ of tests must be repeated, along with $T_j$, where $q_j$ is the probability that test $j$ fails given that tests $1, \ldots, j-1$ were successful. The full analysis of the length of retest time under the retest model is usually too complicated to be used directly as a design criterion. A simple approximation, on the assumption that the retest will definitely fix the error that was found, that will usually give roughly the same result, is to define recursively

$$E_+(T_j) \approx c_j + q_j(c_j + E(X(T_j)) + \left( \sum_{i:\, T_i \in R(j)} E_+(T_i) \right). \tag{8.3}$$

If each retest set is obtained from the uninformative failure model, then Equation (8.3) gives an upper bound for the expected length of the test cycle. Under the informative failure model, we have approximately that

$$E_+(T_j) \approx c_j + q_j(c_j + E(X(T_j)) + p_{2j}\left( \sum_{i:\, T_i \in T_{Sj}} E_+(T_i) \right) + p_{3j}\left( \sum_{i:\, i<j} E_+(T_i) \right), \tag{8.4}$$

which we use to establish our lower bound for the expected time to complete the test sequence.

Now suppose that we have already selected and sequenced a test suite $S = (T_1, \ldots, T_r)$ using the BGM approach described by Wooff, Goldstein, and Coolen (2002). We would prefer to run as early as possible in the sequence those tests which, if they fail, will require a large number of retests. A practical approximation to an optimal sequencing for the full retest cycle for the selected test suite is to choose a stepwise search algorithm that proceeds by switching adjacent pairs of tests in $S$ in ways that we expect to shorten the overall testing time. In particular, if $T_{i-1} \in R(i)$, then, to a good approximation, the only effect of switching the order of $T_{i-1}$ and $T_i$ is to reduce testing time if, having switched the order of the two tests, we would not need to include $T_i$ in the retest set for $T_{i-1}$. We may choose to do this under either the informative or the uninformative form of the model. Therefore, we continue to make pairwise transpositions until we can no longer remove an item from any retest suite by such transpositions.

Finally, when we have constructed our test suite, we may choose to carry out a probabilistic simulation to assess the full distribution of the time to completion of the test procedure. We may choose to incorporate a more realistic description of the test schedule into the simulation, for example allowing the category of errors which we would choose to fix to depend on the length of time remaining before the proposed release date of the software. Simulation under informative failure is straightforward. Simulation under uninformative failure is more complicated as the retest model increases in complexity with multiple failures, but in principle the simulation is again straightforward.

### 8.2.9 Remarks

There are many further complexities to take into account; for example, the role of provocative testing and the way in which batching of tests is handled. Additionally, the elicitation of the specifications $X(T_j)$ needs further consideration. For example, the sensitivity to simplifying choices (such as choosing $X(T_j)$ to be the same for all $j$) needs careful examination. One possibility here, for example, might be to choose lower and upper choices $X_L$ and $X_U$ to correspond to the lower and upper bound arguments.

## 8.3 Multiple Failure Modes

Treating all faults as equally important is appropriate for many applications. However, for some applications, there may be many kinds of possible faults at the domain node level, ranging from errors which are sufficiently severe that they would certainly need to be fixed before the software could be released to errors that certainly would not be worth the cost of fixing, where the only purpose of finding the errors is to be able to alert users in the documentation. Many errors are of intermediate severity, such that they would be fixed if found sufficiently early in the test cycle, but would not be fixed if uncovered too near the projected release date. Thus, we now consider how to allow for a range of consequences of error, how to allow for different costs for different types of tests, and the implications for test design.

### 8.3.1 Nodes for Multiple Failures

To handle the range of different errors, we need to describe in more detail the consequences of failure. Wooff, Goldstein, and Coolen (2002) partly addressed these features, in the sense that we have allowed different domain nodes to have different consequences of failure. While there are many ways to classify the different types of fault, one approach is to rate each fault on a four-point scale, as follows:

4 (show-stopper): corresponds to a fault which must be fixed before release;
3 (major): a fault we would very much want to fix if allowed by time constraints;
2 (minor): a less serious fault which we might fix given time;
1 (trivial): an essentially cosmetic fault that we would usually not bother to fix.

Generally, the level of granularity in the scale adopted should be chosen to reflect the amount of detail that is available and considered relevant by the expert testers.

There are various ways to incorporate different levels of faults into the graphical model. At present, each domain node within a BGM is linked to a single failure node which carries, inter alia, the probability that the entire BGM contains at least one bug. One way to handle multiple failures is to replace the single failure node by one node for each failure level. Thus, if the testers have specified four failure levels (show-stopper, major, minor, cosmetic) then we replace the BGM's single failure node by four failure nodes, $F_1, F_2, F_3, F_4$, and add a directed arc from each domain node to each failure level node for which it is possible that a test in the domain might uncover an error at the corresponding level. In summary, this is a mechanism for decomposing each BGM according to the different failure modes contained therein.

### 8.3.2 Specifications and Test Implications

If each test of a domain node can only reveal errors at one particular level, then we do not need to carry out any further modeling. If it is possible that a test in a domain node might uncover errors of different degrees of severity, then we must model not just the probability of error in a given node, but the probability of each different level of error in that node. We model the different fault levels in the explanatory nodes, and propagate the different errors to the domain and test nodes. While it may require more effort to construct the corresponding conditional probability tables, introducing the different types of error that we can observe may actually simplify the resulting elicitation process as the knowledge of the expert tester may be based on previous experiences of finding the different kinds of error.

In order to modify the BGM to take account of multiple failure modes, we need to construct each failure node $F_i$ and specify $P(F_i|N_j)$ for every domain node $N_j$ in the BGM. $P(F_i|N_j)$ is the probability, given that node $N_j$ fails a test, that the failure is of degree $i$. Note that $\sum_i P(F_i|N_j) = 1$ for each $N_j$. Given these specifications, we can calculate the prior probability of each kind of failure. We write $S_0$ for the null test set, and write $P(F_i|S_0)$ for the prior probability of failure of each kind before testing.

Each successful test modifies the probabilities at the four failure nodes. We denote by $P(F|S+)$ the vector of probabilities $(P(F_1|S+), P(F_2|S+), P(F_3|S+), P(F_4|S+))$, where $P(F_i|S+)$ is the posterior probability that there is at least one failure in the failure node at level $i$ given that all members of the test suite $S$ are successful.

### 8.3.3 Considerations for Design

There are two complementary approaches to testing that we may take. First, we may take a decision-theoretic view, in which each error that is not found incurs a cost, on a utility scale, to the company. Therefore, we extract from the graphical model at any stage the expected utility for releasing the software given the current state of knowledge about potential errors.

For many problems, however, there will not be an agreed utility structure, partly because there will be many stakeholders each of whom have a different view of the costs, and partly because the test team may not have access to the detailed judgments required in order to assign precise utility values. In such cases, we may instead identify the different risks involved in the software release, and extract from the graphical model the probability associated with each particular risk. In this view, the model provides the precise quantitative information required in order to implement the risk management procedures related to the software release.

### 8.3.4 Test Design Criteria for Multiple Failure Modes

Our intention, in high-reliability testing, is to reduce the posterior probability of even one serious fault to an acceptably small level. Suppose that for each level of fault, $i$, we specify the utility cost for releasing the software with at least one fault of this type to be $U_i$. A simple way to design for the extended fault classification is to construct the *release cost*:

$$C(S+) = \sum_i U_i P(F_i|S+),$$

which is the expected utility cost of release given that all tests in suite $S$ are successful. If we accept this test criterion, then it is straightforward to design a corresponding test suite, using identical methods to the approach given in Wooff, Goldstein, and Coolen (2002), but replacing the criterion of minimizing $P(F|S+)$ with the criterion of minimizing $C(S+)$.

If we do not want to place costs on the various fault nodes, then we may instead place targets on the individual elements of $P(F|S+)$. We may choose designs to meet these targets using similar step-wise search algorithms. Although we are designing on several criteria, we have classified the errors in order of importance, so that a simple approach is first to design the test suite to control the posterior probability for level 4 errors, then to add tests to control the posterior probability for level 3 errors, and so forth. Regular step-wise deletion stages are used to weed out superfluous tests. Alternately, we may seek to optimize some similar criterion to the decision-theoretic criterion described above, and monitor the individual failure probabilities to ensure that the test design controls the individual error probabilities to the required levels.

### 8.3.5 Implications of Resource Limitations

If, due to constraints of money and time, it becomes apparent that we cannot sufficiently reduce the overall error probabilities, then we could further subdivide the collection of possible faults by introducing a fuller collection of fault nodes so that we can track errors in different core functions of the software. This will give more information on which to assess the suitability of any test suite to identify the various areas in which substantial levels of risk remain, so that a reasonable compromise suite of tests can be made. Incorporating the corresponding utility costs, we design for the modified version of $C(S+)$ as above.

The implications of resource limitations may become apparent prior to testing, through using the analyses we have described here and elsewhere. It is also frequently the case that resource constraints become apparent during the testing process. For example, the testing process may reveal bugs which take unexpectedly long to fix and so delay further testing. When this is the case, it may similarly be advisable to further subdivide the collection of possible faults and so enable any further testing to concentrate on whatever are, by that stage, the most crucial areas in the software still to be tested.

### 8.3.6 Remarks

In the above description, we have treated each test as though it has the same cost. If test costs differ, then the algorithm that we follow is as above, but at each step-wise addition or deletion, we measure the benefit that can be gained for a fixed level of cost.

All of the test procedures that we have described above are reliant on the underlying graphical model. In certain circumstances, we may reserve a portion of our budget for diagnostic testing of the judgments in the model, which essentially corresponds to placing rather more tests on very important areas of the software functionality than appear to be necessary given the form of the model. This is addressed later.

Wooff, Goldstein, and Coolen (2002) described how we carry out sensitivity analyses to check the sensitivity of the conclusions we reach to variations in the prior specification. It is straightforward here to carry out similar sensitivity analyses to test the effect of varying the prior specification on the four probabilities for each kind of failure. This is useful particularly when testers have a good feel for the number of the most serious kinds of bugs thought to be in the software, prior to release. Further, historic information from earlier testing of similar software, or of software sourced from the same supplier, can be used to help calibrate the models for the software being tested.

We have addressed test design for multiple failure modes in the sense of targeting design at the most probable locations of serious failures. This leaves open the question of how to handle the test–retest cycle where a test fails and where we allow the possibility of determining the level of failure at that time.

A related issue is that, even if the test process is uncovering no faults, the importance of certain kinds of error can be upgraded or downgraded as time proceeds. For example, it might be decided late on in the testing process that certain functionality will not, after all, be released. In this case, what might have been very serious possible errors are substantially downgraded. One possibility is simply to reoptimize what remains of the testing process at this stage. An alternative is to have a mechanism that takes account of the downgrading or upgrading of certain kinds of error without going to full reoptimization of the test process.

Finally, we have not considered how we may take advantage of observing the kind of error that occurs, and propagating this information over the network. For example, if we see a show-stopper, one possibility would be to propagate this information over the models and thereby raise the probability of there being a show-stopper in untested nodes. This would require some extra modeling, in particular because the number of states in each domain node, at present essentially the two states [good, bad], would have to be decomposed further, for example into the states [good, show-stopper, major fault, minor fault, cosmetic fault]. Beyond such more complicated modeling, there are few extra difficulties associated with this extension.

## 8.4   Diagnostics for the BGM Approach

The main purpose of diagnostic assessments is to compare actual performance (either through testing or by use of software in the field) to the performance predicted by the models. In the BGM approach, the BGMs are representations of the tester's knowledge, so that the value of good diagnostic procedures is (a) to provide feedback on the quality of the representation for an individual tester for a specific software testing problem, and more generally (b) to help calibrate the general BGM approach through continued feedback on the success of the modeling and test design processes.

With regard to the quality of representation for an individual tester (or team of testers) for a specific problem, our concerns fall into two areas: whether or not the structuring process gives rise to adequately structured models; and whether or not the probabilistic specification appears to be in line with actual testing and field performance. These two issues are affected by the diagnostics that we attach to (1) test performance, and (2) field performance, which we describe separately.

### 8.4.1   Diagnostics during Testing

#### 8.4.1.1   Test Performance

Typically, there are three scenarios we might consider. For these, we suppose that the testing process does not include purely diagnostic tests (which we describe below), and that a test suite has been designed to achieve desired levels of reliability. The testing process should be represented by one of the following scenarios.

- Performance of the software is about as expected. That is, testing reveals roughly the number of faults which the models predicted, and in expected places. The implication is that the structuring process delivered about the right structures, and the specification process delivered about the right specification.
- Performance of the software is better than expected. That is, testing reveals fewer faults than the models predicted. The implication is that the software was actually more reliable than specified, and thus possibly that resources were expended unnecessarily on too much testing or structuring to too fine a level of detail.
- Performance of the software is worse than expected. That is, testing reveals more faults than the models predicted, and may also have found faults in unexpected places. The implication is that the specification process over-stated the reliability of the software, or that the structuring process provided models which were inadequate.

#### 8.4.1.2  Test Diagnostics

Two kinds of diagnostic are available from observing tests from a test suite. First, there is a global measure available at the end of testing which compares the number of faults found with the number of faults expected at the start of testing. Such information is mostly useful for calibrating the general BGM approach (for example, testers will obtain an improved "feel" for the specification process) and for updating or evolving the BGM for further use.

Secondly, diagnostics can be calculated as each successive test in a test suite passes or fails. Each test has a certain additional chance of revealing a fault. Assuming the tests are scheduled to test the software areas with highest probability of failure first, we compare the results of successive tests with the marginal reduction in probability of failure (or expected number of faults remaining) that each such test provides. For each test (and in particular each sequence of tests) we expect a roughly matching number of faults found and reduction in expected faults. Too high a proportion of faults found provides an early indication that the prior probability of failure is too low, and vice versa. Later in the testing cycle, it may be that a test with very low probability of revealing an error fails. This may be an indication of deficiency in the structure of the model.

#### 8.4.1.3  Sensitivity Analyses and Diagnostic Tests

Information from tests is propagated over the various BGMs. Depending on the testers' judgments, the results of tests will have direct implications for some nodes (that is, the nodes directly observed), and indirect implications for some other nodes (that is, the nodes linked to those directly observed). For many nodes, depending on the strengths of the links between observed nodes and the efficiency of the test design, the model will suggest (according to the testers' judgments) that the nodes will have been fully tested. It will be useful to measure how important it is to have observed such nodes directly. This measure will provide us with a diagnostic as to which further diagnostic tests we might carry out if we do not entirely trust the links between nodes. One possibility is to construct a measure within each node which partitions the evidence in a node into direct and indirect evidence. This would probably be straightforward in a Bayes linear paradigm (Goldstein and Wooff, 2007), but would require new research in this setting.

In addition to diagnostic tests that arise from identifying areas of the model which have been only indirectly tested, there is a straightforward way to identify further diagnostic tests. At some point the main testing process ends, either because the resources available for testing have been used up, or because testing indicates that the software has achieved the desired levels of reliability. In the former case, there are presumably no resources available for further diagnostic tests, so we need only consider the situation where the software is ready for release according to specified levels of reliability, and where there may be some resources available for further diagnostic tests. We now run as a first diagnostic test the test which we expect to resolve most uncertainty in the model (in the same way that we identify useful tests for the test design process). This might require no extra effort because the test design process will typically result in an ordered sequence of tests that we wish to run. Testing will have stopped part way through this sequence, and so the first diagnostic test is simply the first test not to have been run in the original ordered sequence. (Alternatively, we might simply design a new suite of diagnostic tests, or design a single diagnostic test at every iteration.) We now check whether there are sufficient resources to run an additional diagnostic test, and proceed in this way until the resources have been spent, or the model has been fully tested.

The advantage of this procedure is that the diagnostic tests continue (in some optimal sense) to test the software to higher and higher levels of reliability, while also helping to confirm the validity of the model's structure through testing areas which should have no, or a very low, probability of failure.

Whether or not diagnostic tests can be run depends on whether or not there is resource available, after normal testing has been completed. Such decisions need to be addressed taking into account the viability issues discussed later. For example, if there are resources available within the budget for testing a particular piece of software after the normal testing stage has been completed, it may be that these resources are better diverted to other needier projects.

### 8.4.2 Diagnostics after Software Release

#### 8.4.2.1 Field Performance

We suppose that the software is released for field use following testing to an adequate level of reliability, and in particular that the models suggest that there are no (or there is only a small probability of) serious faults remaining in the software. The possible scenarios arising from field use are as follows.

- Performance of the software is about as expected. That is, no serious extra faults were revealed. The implication is that the structuring process delivered about the right structures, and the specification process delivered about the right specification, that the chosen test design was effective, and that the test-bed provided an adequate representation of field conditions.
- Performance of the software is better than expected. That is, no serious extra faults were revealed and there were fewer other faults than expected (assuming the software to have been released with some probability of remaining faults). The implication is that the structuring process delivered about the right structures, and the specification process may be understating the reliability of the software, that the chosen test design was effective, and that the test-bed provided an adequate representation of field conditions.

- Performance of the software is worse than expected. That is, serious faults were revealed in field use, or more than expected less serious faults occurred. The implication is that the specification process overstated the reliability of the software, or that the structuring process delivered structures which were not sufficiently finely detailed, or that the chosen test design was ineffective, or that the test-bed did not provide an adequate representation of field conditions.

### 8.4.2.2 Diagnostic Feedback from Field Use

Diagnostic feedback from field use can be used in a number of different ways. Field faults can be classified as follows.

- Faults which could have been found by the testing process, implying that the testing process was not sufficiently effective.
- Faults which could not have been found by the testing process, implying that the test conditions were inadequate to represent field conditions.

We concentrate on the former kind of faults. We will assume that the nature of the fault can be precisely identified, and that the mechanism by which that fault could have been found by testing can be precisely identified. For example, we assume that we can identify a specific test or test type which could have found the fault. Such faults can be further classified as follows.

- A fault is category A if the fault occurs in a location which the model suggests has been fully tested.
- A fault is category B if the software was released with a relatively low probability of such a fault occurring, and in particular that the test that could have tested for this fault was not carried out.
- A fault is category C if the software was released with a relatively high probability of at least one fault remaining, and the test which could have identified the fault is just one of a number of tests which might have been carried out, but were not.

Category A faults are the most serious as they imply that the tester has omitted to consider parts of the software to be tested, or that the BGMs are not to a sufficiently fine level of detail. An example would be that the tester judges that one test of software intended to process 8- to 16-digit numbers is appropriate to fully test the software, and that a single test (say, using an eight-digit number) passes. Later, we discover that there is a fault in processing nine-digit numbers. The conclusion is (a) that the tester was wrong to consider that one test tests all, and (b) that the tester might consider partitioning the input domain further, say into 8-, 9–14-, 15-, and 16-digit numbers. Category A faults thus have an implication for the underlying BGMs.

Category B faults have particular implications for the probabilistic specification, as an error has occurred that the model suggests as having been unlikely. This provides information to update the current model, perhaps by down-scaling the probabilities that the software is fault free, or perhaps by directing the tester to reconsider the judgments used to quantify the model.

Category C faults are, in some sense, expected. Thus, they do not have special implications for the model or the approach.

In general, in addition to the implications already considered, the number of field failures might also be used for general calibration of the BGM approach. This includes the following: choices of tuning parameters for those areas of the BGMs concerned with sporadic error propagation, as considered in Section 8.6; and choices of probabilities for the probability of introducing side-effect errors during the test–retest cycle, as considered in Section 8.2.

### 8.4.3 Fault Severity

We have addressed the implications of faults from test-bed and field use for diagnostics. It is also useful to take into account the severity of such faults. At present we have simply described these as serious or non-serious, but it is clearly desirable to formalize them, and to link the severity of a fault with our recommendations on multiple failure modes, as described in Section 8.3.

## 8.5 Model Maintenance and Evolution

The BGM encodes the knowledge of the tester (a) for further use of the tester, and (b) for the use of other testers. To enable changes to the BGM in the light of new information or changed functionality, it is important to distinguish between the actual uncertainty judgments made by the tester and the reasons for making these judgments. Particularly in circumstances where testers involved in the creation of the BGM are not involved in any future revision (for example restructuring) of the BGM, it is important that the information on which previous judgments were based is still available.

When a BGM is developed to assist software testers, the structuring of the software actions according to the tester's view of its functionality is not essentially a Bayesian activity. Instead, such structuring of software functionality extracts from testers a core framework for the testing process which is of considerable practical use whatever the testing strategy adopted. One advantage is that the corpus of structures elicited provides an explicit representation of the tester's reasoning concerning the functionality of the software to be tested. A second advantage, of course, is that we can construct from them BGMs to provide the full panoply of the BGM approach. These are the twin themes of what we must take into account with regard to model maintenance:

- how the structures, which carry the tester's judgments as to how the software functions, can be modified or evolved to take account of new or modified functionality;
- how the probabilistic information specified over the corresponding BGMs can be modified or evolved.

For the remainder of this section, we suppose that we have constructed a Bayesian model for the previous release of some software, and wish to update to a new release of the software. We describe the sources of information about software reliability that may help to construct the new model, and the actions that should be taken to use such information.

### 8.5.1 Documenting the Construction

If BGMs for a software system are to be maintained in the sense outlined above, it is essential first to add an extra layer to the structuring process. This consists of (a) careful documentation of the grounds for structuring the software functionality in the first place, and (b) careful documentation of the reasoning underlying the probability specification process. The former is naturally recommended whether or not the testing strategy employs the BGM approach.

Such careful documentation might be of use at several stages. First, when creating the initial BGM, it is useful to have a clear idea of the information available, to ensure that all relevant information is taken into account correctly. Secondly, if at a later stage the need occurs to re-examine the BGM, then well-recorded information will be of great value. It should be emphasized that such information may originate from several sources, ranging from actual reliability data to tester's judgments based mostly on experience. The benefits of recording such information include the following.

- The availability of rigorous documentation should simplify and improve the process of making prior probability judgments.
- Experiences arising throughout the testing process can be referred back to the underlying groundwork, thus providing for sanity checks and assessment of the testing methodology (whether or not via the BGM approach).
- Proper documentation as to why certain judgments were made is available for auditing purposes. When software turns out to have been badly tested, this presumably leads to large costs of some kind. Therefore it is useful to have audit procedures for identifying the reasons why various decisions were made. It is likewise possible to identify when software has been very competently tested, and it is useful then to be able to use any documentation to identify good practice.
- When the models need to evolve, there is a careful record as to how the models were constructed initially, so that no extra uncertainty as to original software functionality need be introduced at the later stage of model evolution, and any further evolution can (if desired) take place along the lines chosen for the initial structuring.

### 8.5.2 Evolving the Structures

#### 8.5.2.1 New Functionality

The evolution of structures is necessary when the functionality of the software is extended or new functionality added. Note that careful documentation of the earlier structuring for testing is vital in deciding what is new and what is extended, and by how much it is extended. We define new functionality to be extensions to the software for actual new functionality, so that testing such functionality is not informative to remaining functionality. We handle such new functionality just as we would handle any other software to be tested for the first time, as described in Wooff, Goldstein, and Coolen (2002); for example, we can judge which functions are intrinsically difficult to program and so identify areas with high failure probabilities. Note, in addition, that we also have the advantage of being able to take into account the general level of reliability of the remaining pre-existing software being tested, which will help the specification process.

#### 8.5.2.2 Extended Functionality

Our approach differs according to whether the extended functionality is minor or major. By minor extended functionality we mean that (according to the tester's judgment or to explicit knowledge of any required software rewriting) the software is modified in a minor way to accommodate minor changes in functionality. A possible example is that software dealing with eight-digit numbers might be revised so that numbers ending with a zero are treated slightly differently. Such extended functionality should be able to be handled at the stage where we partition a software action for its inputs. This will involve adding to the current partition structure, which is straightforward, and adding new nodes to existing BGMs. Within a tool, this should be simple to accomplish. Finally, we will need to make a probability specification over the new and old parts of the structure.

The probability specification over the old parts of the structure can be handled as described below where we address evolving the probability specification. In particular, it will normally be advisable to take into account that the software extension may have introduced new faults into the pre-existing (and previously tested) areas. The probability specification over the new parts of the structure is guided by the specification made for the old. Typically, the tester would be guided to probabilities of the same order of magnitude as for the old parts of the structure, but generally larger to reflect the novel nature of the new functionality. Sensitivity analyses can then be run to assess the implications for test design and so forth, and calibration analyses can be run to help ensure that the BGMs are in line with the tester's judgments.

For major extended functionality, we have the choice of dealing with the software as though it is entirely novel or of repartitioning and evolving existing BGMs, as for a minor extension of functionality. An example is provided by extending old software capable of dealing with eight-digit numbers to software dealing with up to 16-digit numbers. If we choose to regard this as extended functionality, new nodes are added to the existing BGMs, with corresponding scaling up of root probabilities to express the possibility of new errors being introduced in old functionality. Alternatively, we can choose to regard the software changes as so wide-ranging as to merit ab initio modeling, but where we can take advantage (a) of the existing BGMs for the software being extended, to give order-of-magnitude guides as to the likely reliability outputs from the new model; (b) of the careful documentation process outlined above, which details how the existing BGMs were formed, and whether or not there are diagnostic features from live testing which need taking into account.

### 8.5.2.3 Knowledge at Code Level

Depending on how much information is available, we may have direct access to code information or be able to request such information from the code producers. The types of information that we might find informative are knowledge of the formal call structure or the observed call structure, or information as to which aspects of the code have been substantially modified since the previous release. If this is the case, any information about the code that is newly available is now used to check whether the modeling has overlooked any important features of the new release. Information about the code for earlier releases of the software should already be reflected in the previous generation of graphical model.

### 8.5.3 Evolving the Probability Specification

In this section we assume that there is no substantial new functionality, and that we are dealing with a revision of existing software with the same basic functionality as before. (It is straightforward to allow for minor extensions of functionality.) The probability evolution needs to take into account information from previous testing of previous versions of the software, information from field usage of previous versions of the software, and the scale of the software revision.

### 8.5.3.1 Exploiting the Test History

Test records for previous releases of the software contain information as to which tests passed and which failed, both initially and on retest. These records are directly relevant for criticizing current modeling strategies and for constructing the new models. Therefore, diagnostic comparisons between the previous model and the corresponding test results are used to identify where the tester was over-confident or under-confident of the software reliability in the previous release, and by how much. The diagram for the new release of the software might start by recreating the diagram for the preceding release, as a simple scaling (at the level of the root probabilities) of the original prior probabilities using the above heuristic to express confidence in the original model.

### 8.5.3.2 Exploiting Previous Field Experience

Commonly, the previous release of the software will have been tested or used live. From such use, new faults may have been identified. Failure reports from the field identify which tests were overlooked and which *provocative* tests should have been carried out; also, by implication, areas with no reported failures increase confidence in the testing process and the true software reliability for the previous release. Therefore, one approach is to make an overall assessment of software reliability for the previous software release, taking into account the evidence from its field usage. This can take the form of a confidence assessment in the previous test model and also a level of confidence in the performance of the software immediately prior to the latest rewrite. This can be used to rescale the root probabilities derived above (after considering the previous test history) and possibly suggests further nodes representing provocative tests.

### 8.5.3.3 Knowledge about the Nature of the Software Rewrite

If the new release is a minor upgrade to the old release, then the most important consideration is whether the field errors have been fixed, while if there has been a major rewrite, what matters is largely the new errors that may have been introduced. The outputs from the two previous sections, taking into account both the testing history and field experience, will have produced revised models for the initial state of uncertainty and the posterior state of uncertainty for the original model. These are now combined in light of knowledge about the nature of the software rewrite. For example, we might choose a linear weighting of the prior and posterior root uncertainties at each node (near prior for major rewrite, near posterior for minor rewrite), possibly choosing different weightings for different regions of the model.

## 8.6  End-to-End Testing

By end-to-end testing, we mean testing of a complete software system, including all its subsystems and the links between them. The end-to-end testing problem is broadly similar to the testing problem we have already covered, and much of the methodology we have provided already can be applied straightforwardly to the end-to-end testing problem. It is particularly characterized by the requirement to test many subsystems, and the links between them, simultaneously, and typically requires more testing in the way of testing the flow of software processing; this has implications for the test design process, as there are far more possibilities to consider.

### 8.6.1 Modeling for Integration of Subsystems

We view a software system as a collection of subsystems linked together. The full testing problem can be separated into two stages. First, we apply the BGM approach, as far as practicable, to each of the subsystems, which we treat as independent modules. We consider dependent subsystems below. Therefore, whatever the present level of testing, we consider that all the subsystems have been structured for testing according to the BGM approach. We deal with the case where this is not so, together with viability issues, later. It should be noted that if we have applied the BGM approach to all the subsystems in the system, then each such subsystem provides a full probabilistic summary of the current state of testing within that subsystem, whether or not any testing has taken place within that subsystem. In principle, there is no difficulty in designing either end-to-end tests or within-subsystem tests which will have an impact on the current probabilistic description of reliability within a subsystem. However, it may be more natural (and easier to manage) to undertake testing within subsystems as modules, especially if teams are given special responsibility for particular subsystems. Note that end-to-end tests will continue to provide fresh information about the state of reliability within each subsystem.

Consequently, we focus here on testing the integration of subsystems. The structuring required for the integration problem is similar to the structuring involved for the subsystem testing problem. The team responsible for end-to-end testing identifies the software actions which link the subsystems (and also identifies any reliability dependencies between subsystems – see below), and provide judgments as to their reliability. This team must also specify the inputs which form the domain of each software action and partition them in the usual way. Further, the team must complete the mapping of tests to observables, and must specify, for example, any constraints on the ranges of inputs provided for tests. The testing of the integration of subsystems needs to address two issues.

- The links between subsystems need testing. This is relatively straightforward to achieve, as all the links should be representable as simple software actions dealing with transmission and reception of data. The case study addressed in Wooff, Goldstein, and Coolen (2002) contains several examples of such links.
- The integration of subsystems. The kinds of errors we want to detect here are the kinds of errors which would not ordinarily occur if the testers are correct in their judgments about how the software is structured. Such errors may occur intermittently or be difficult to replicate. Other such errors may occur because there were unforeseen (and therefore unmodeled) relationships between subsystems.

Of these two issues, the second is much harder to address. We propose to deal with it by allowing a general model for sporadic errors of this kind. We begin by specifying a high probability for sporadic error across the system. Each test pass reduces this probability. However, the observable for each such test is that each database must be examined for side effects before and after each test, and examined to ensure that the only differences are those required by the instructions implied by the test. Such checks may, or may not, be an intolerable burden, but do genuinely test the integration. Handling of such sporadic errors is considered briefly in Section 8.2, in which we also consider side-effect errors introduced by the test–fix–retest process.

If the burden of checking databases for side effects is too costly, simpler approaches may be employed. However, these would provide less information. One such approach is to consider a hierarchy of sporadic errors for databases within a BGM framework. For example, we might represent each database used in the system by a node in a BGM, each connected to a common parent node which carries the general level of sporadic error across the system. This general level of sporadic error will need to be specified at the outset to reflect the tester's judgments, but would be a tunable parameter which could be subjected to sensitivity and calibration analyses. Similarly, each of the nodes for the single databases carries the general level of sporadic error for that database. The nodes on the BGM are connected by arcs to represent possible flow of information, to whatever level of sophistication is deemed appropriate to meet the tester's judgments. For example, a pair of databases might be judged as having a higher pair-wise degree of unforeseen side effect, and this can be straightforwardly modeled via a BGM.

### 8.6.2 Input Arrival Testing

Suppose that testing has been carried out to some satisfactory level. Typically, the tests we design for end-to-end testing will pass through a series of subsystems. For each of these subsystems, the inputs will arrive as intended and be of the correct type. That is, if a subsystem expects an eight-digit number to process, part of the testing process focuses on delivering that eight-digit number correctly. Furthermore, each subsystem is typically tested for the range of inputs it is intended to process.

End-to-end testing should also test what happens if one subsystem passes on inputs which are correct for that subsystem, but inappropriate for a subsequent subsystem. This might happen for two different reasons. First, it may be that the systems design has overlooked an incompatibility between one subsystem and a subsequent subsystem. This will be handled by at the stage at which tests are mapped to the observables, as the input constraints for the various subsystems are considered at that point. Secondly, it may be that the inputs arriving for a subsystem are inappropriate because of an error occurring in the previous subsystem, or because of a transmission error. However, both of these are modeled via software actions in the usual way, and so require no separate treatment.

### 8.6.3 Clusters of Subsystems

It may be necessary to handle situations where the software in one subsystem is related, in terms of reliability, to software in another subsystem. This might occur, for example, because the software engineers copy code from one application and reuse it for another. An example is where encryption and decryption algorithms exist as identical duplicates in different subsystems. This then implies that there may be situations where nodes within BGMs for one subsystem may be connected to BGMs within another subsystem. In principle this can be handled in the way we describe in earlier documents, which is to connect such BGMs via a clustering factor. However, this has the drawback of being more difficult to manage, particularly if the subsystems are being tested by different teams.

An alternative is to continue to test independently within subsystems. However, when we come to use any reliability summaries from such modularized testing, the end-to-end model may understate the degree of reliability in each cluster of subsystems. In this respect, modularized testing offers a conservative picture. If the test budget allocated to such testing is sufficiently high, we may choose to undertake more detailed modeling in order to detail dependencies more carefully. Such a choice can be guided by viability considerations as discussed below.

We should, of course, take advantage of any such software relationships. Therefore, one of our recommendations is that, if practicable, the tester responsible for the full system should coordinate the BGM approach to structuring the software actions in the two subsystems to minimize the effort of duplication of structuring and documentation; and that test failures directly attributable to those software actions in one subsystem be notified to the testers responsible for testing the related subsystem.

### 8.6.4 Viability

As we describe in Section 8.7, it may not be worthwhile to construct BGMs for every subsystem in the system being tested. The guidelines contained therein can be used to decide whether or not to fully model each given subsystem. For example, if one of the subsystems is known with high confidence to be highly reliable, and if it would also be very expensive to construct the BGMs for it and then test them, we might decide not to construct them. We would, of course, continue to test links between this and the other subsystems.

In order to obtain numerical summaries over the full system, it will be necessary to specify directly appropriate summaries for subsystems which we do not test by full BGM modeling. If we can observe whether or not the outputs from such subsystems are correct, we might further introduce some crude mechanism for updating such numerical summaries. For example, we might introduce a beta prior distribution to represent the probability that the full subsystem works correctly, and update for each successful test observed. Otherwise, it might be best to treat such a probability of failure for such subsystems as irreducible without further detailed attention. Clearly, as elsewhere, the cost of paying more attention to modeling is offset by payoff in knowledge gained.

### 8.6.5 Implications for Design

End-to-end testing can subsume within it the testing within individual subsystems (for example every end-to-end test will help to update the BGMs for separate subsystems), but is probably best tailored according to these principles.

- Testing proceeds first within subsystems, to agreed levels of reliability.
- Testing proceeds until the level of sporadic error is deemed sufficiently low. This can be guided by whatever benefit and resource constraints are involved, as described in other documents for the general BGM approach.
- Testing proceeds until the links between subsystems have been adequately tested. There are BGMs representing the software actions involved in the links and interfaces between subsystems, so the methodology to do this has already been described in other documents.
- Testing proceeds until the various combinations of flow of control between subsystems have been adequately tested. This extra remark is made because there are typically many more such combinations than for processing within subsystems, and although each such control has typically a very high level of prior reliability, there may be a considerable cost for failure.

It is worth noting that it may be more efficient to regard every problem as an end-to-end testing problem, with no modular testing of subsystems, so that the problem of test design is how to test end-to-end tests that test the whole system and all the subsystems satisfactorily. Alternatively, subsystems might be tested modularly to some specified level of reliability, but not to a releasable state of reliability, and then further end-to-end tests can be carried out with the partial aim of completing the subsystem testing *en passant*. Whilst this would be efficient (and possibly practicable for small systems), it might be much more difficult to tackle the testing problem in this way.

## 8.7 Viability of the BGM Approach for Individual Applications

We now discuss how the viability assessment process can be achieved for individual testing situations within a company or institution; this was also presented by Coolen, Goldstein, and Wooff (2003). The first part of the process is to link the BGM approach with the company's strategy for making decisions and assessing risks. This depends strongly on context and so it is more meaningful to explain how to match the test effort required in using the BGM approach to the requirements of the problem, so that, if all goes well, managers will always have a procedure that suits its needs. To do so, we begin with a qualitative assessment by the company of the costs and benefits of their existing approaches, and then contrast these with the costs and benefits of the BGM approach.

### 8.7.1 Assessment of an Existing Testing Approach

We focus on an imaginary forthcoming project at a company, and attempt to see what the assessment must cover. There are two aspects to consider:

**General:** features which will be relevant to the assessment whichever testing approach is applied;
**Specific:** features which depend on the approach that is used, one such approach being the BGM approach.

Each of these two aspects can be cross classified according to:

**Cost:** the cost of overlooking errors for the project (general) and of carrying out testing (specific);
**Error level:** the level of error in the software to be tested before testing (a *general* feature) and after testing (a feature *specific* to the approach used).

### 8.7.2 General Assessments

#### 8.7.2.1 Project Cost
There are many relevant considerations for the cost of overlooking errors in the software before release. This is intimately linked to the general importance of the project for the company, and underpins all the effort that it is worth taking to do a good job on the testing. Informally, we may classify the importance that the project performs appropriately as follows.

**High Importance.** A particularly important project at the heart of company profitability for which it would be extremely costly (in time, effort, prestige) if the project were to develop major failures in the field, and where it is very important that the software is good.

**Medium Importance.** A fairly typical project, for which poor performance will be "typically" costly, and where it is important that the software is good.

**Low Importance.** The kind of project where it does not really matter whether the software is good. For example, the project might still be in a trial phase, or concern areas with little consequence, or the software may be a simple stop-gap, until a better solution is found; code quality is not a major concern.

The effect of this classification is to determine one, or several, multipliers on the consequences of different types of failure after release. Of course, it is really a crude discretization of (several) continuous quantities. If we were intending to carry out a full decision analysis, then such a quantification would doubtless be important. However, we are really trying to illustrate the qualitative structure of the assessment. Further, as soon as we try to elicit precise quantitative effects, we will tend to face many practical, psychological, and political types of issue, whereas it is often quite quick and easy to get agreement to a broad categorization as above.

Therefore, for each factor, here and below, we suggest three levels of importance: high, medium, and low. By *high importance* we mean that improved testing is always a major issue. By *low importance* we mean that testing is not a major concern, according to the particular aspect under consideration. If we eventually implement the initial recommendations presented in this document, then we may reword or reclassify or add new levels to fit better into the company's particular milieu. Although it is not difficult to make more detailed distinctions, the relevance of the different levels should be fairly clear.

If we do wish to refine the judgment, then we might introduce a cost vector, representing, say, the expected cost of each minor, major, and show-stopper error found after release.

### 8.7.2.2 Initial Error Level of Software

The importance of testing will obviously depend on how many errors the software contains initially. Clearly this is unknown, but an assessment can be made into the following categories.

**High error level.** The software is likely to contain important faults. Testing is very important.

**Medium error level.** Software quality is uncertain. Some testing is required.

**Low error level.** Testing is mainly a formality.

The effect of this classification is to scale the probability of faults existing a priori in the software, and so likely to be remaining in the software after testing. We could further break down our information about the software according to the following considerations.

- Is the software substantially new, or largely a routine rewrite of existing, good quality software?
- Is the area of application of the software relatively straightforward (and so more likely to be error free), or novel and difficult to program reliably (and so more likely to contain errors)?
- Is the software long and elaborate, or short and simple?
- Is the software produced by a reliable team, or not?

We could build up a simple assessment procedure based on this type of consideration to give probabilities for each level of software deficiency. Again, we might introduce a probability vector for the different types of error for each category.

### 8.7.3 Specific Assessments

#### 8.7.3.1 Test Cost
We suppose that the company assigns a testing budget, $B$, using their current methods of assessment. $B$ has two components: money, $M$, and time, $T$. Each of these is a prior expectation, which may be modified, but they are both set fairly early in the process, to some order of magnitude. For simplicity, we suppose that each element $M$ and $T$ may be classified as follows.

**High expense.** Sufficiently expensive that reducing test cost is a major concern.
**Medium expense.** There may be some scope for cost reduction, but this is not a driving consideration.
**Low expense.** Cost reduction is not an issue.

Of course, these classifications are relative to the constraints of the problem. For example, a testing budget is expensive in time if the expected testing time gets very close to the promised release date of the software. Time enters the costings indirectly as a constraint. For example, if time cost is high and overall test quality is poor then we will seek to use time more efficiently to improve testing.

#### 8.7.3.2 Error Level of Testing
Within their current approach, suppose that the company now assesses how effective they believe their testing regime is likely to be. Presumably most testing is not planned to be poor, though time constraints and so forth will mean that full testing (whatever that might be) is not always possible. Again, to keep things simple for the purposes of illustration, suppose that the company considers three categories of test deficiency. Note that we are concerned here with errors that could be caught by the testing process. Any further errors that can only be detected after release are beyond the scope of this analysis.

**High deficiency.** There is large uncertainty as to whether testing will find all high-priority faults. For example, some important areas of functionality might only be superficially tested. Improved testing is a priority.
**Medium deficiency.** There is reasonable confidence that all high-priority faults will be found, but less confidence that all major errors will be found. There is scope for improved testing.
**Low deficiency.** There is confidence that all highest-priority and other major faults will be found, in sufficient time to fix them before release. Improving test effectiveness is not considered important.

This classification is made by comparison with similar testing problems addressed by the company. In practice, this is essentially how budgets are assigned, in that expert testers make assessments of the resources required for careful testing, by analogy with previous testing experiences, and then by negotiation settle on a time frame and resource budget with which either they are happy (presumably as they feel this will lead to low test deficiency) or which they accept as a compromise involving some degree of test deficiency. As with each of the other inputs, we could refine the categories by discussion with test managers if we wanted to implement this approach.

Note that test error level refers to lack of confidence in the *test procedures*, not in the quality of the software after testing. For example, if we did no testing at all, then we would have no confidence in the testing (obviously), but if we had high confidence that the software was of good quality then we would still be happy to release the software. Effectively, we are dividing the belief specification problem into two parts. First, the chance of errors in the software (initial error level), and secondly, the probability of finding such errors given that they are in the software (testing error level). This decomposition is made to make it easier for the company to specify their judgments, and because software deficiency will be a common feature to the problem irrespective of the testing approach, but test deficiency strictly applies to what the company can achieve without the BGM approach. In the same way, in financial considerations, project importance is common to all approaches but current test budget only relates to the company's choice without the BGM approach. Again, we may specify a probability vector for the ability to find each type of error.

### 8.7.4 Use of Assessments

We now consider the potential for the BGM approach to be worth considering for a new project. The financial considerations are summarized by the cost assessments, while the confidence in releasing good-quality software is summarized in the initial and test error levels. The advantage of starting with simple classifications is that for many cases the conclusion will be fairly obvious without detailed calculation; for example, if all costs and error likelihoods are assessed as high then we would expect the BGM approach to be very helpful, while if all are assessed as low then presumably we would expect much less benefit from more sophisticated approaches. In practice, we suspect that for many problems this cross-classification will quickly reveal whether the BGM approach should be explored, as the potential for improvement can be gauged from the number and nature of high values in the specification. If we do seek a quantification, the overall expected cost to the company is of the form

$$[\text{Project cost}] * [\text{Initial code error}] * [\text{Test error}] + [\text{Budget cost}],$$

suitably scaled, and possibly based on vectors of costs and probabilities. However, it is more fruitful to use the above assessments to direct the amount of effort we put into the BGM approach analysis, as follows.

### 8.7.5 Assessment of the BGM Approach

Set against the various advantages of the BGM approach (some of these advantages, such as the availability to managers of detailed justifiable assessments of software reliability before and after testing, are not easily expressible in monetary terms) are the costs of implementing the approach. Essentially, the extra cost that is incurred is the additional effort required to construct the Bayesian models for the testing problem. Of course, how easy or difficult it is to construct the model depends strongly on the effectiveness of the tools that are put in place by the company to implement the approach. Further, much of the work required to construct the BGM approach model is required of any approach to construct the test suite, and, given good tools, the cost, in complicated testing problems, may actually be less than for the alternative approaches, as the BGM approach may make it straightforward to generate efficient test designs. Finally, the BGM approach may be carried out either in full, or in cost-saving approximations, depending on the amount of time and thought that goes into the structuring and specification.

The BGM approach exploits the relationships between different test outcomes to transfer information from the outcome of one test to beliefs about potential failures elsewhere in the software. The more complex the diagram, the more the potential to exploit such transfers of information, but also the higher the cost in producing the diagram. Rather than making a somewhat arbitrary guess at the potential savings that a general BGM approach might yield, it seems more sensible to suggest how to match the amount of effort that should go into the BGM approach modeling with the requirements of the problem at hand, so that the BGM approach should always give good value.

In any case, we start with a description of the possible tests that can be run. This is essentially cost neutral, as we must make such an assessment whichever approach we take. We would imagine, however, that the general features of the problem (project cost, software quality) will influence the level of detail at which it is thought worth describing the observation space.

Now we describe various choices for the level of detail for the BGMs that we shall construct. We identify three aspects to this, and for each aspect we describe three levels: *full*, *partial*, and *simple*.

**Outcome space description.** The level of detail at which we *describe* the test outcome.
   **Full:** the finest level of detail that we might consider;
   **Partial:** an intermediate level of detail for the outcome space, aggregating outcomes which are almost equivalent;
   **Simple:** makes as few distinctions as possible.
   As a general principle, the more important it is not to overlook errors (that is, the higher the project cost), then the more detailed we will want the outcome space description to be, as it will be important not to overlook error types.
**Model linkage.** The level of detail of the *links* between different parts of the model.
   **Full:** makes all of the links between outcome nodes, through higher-order structuring, that we can identify;
   **Partial:** keeps the main links, but drops link nodes which appear to have little influence (as they are very unlikely, or are likely to have small arc values).
   **Simple:** only adds unavoidable links to the diagram, minimizes the number of parent nodes.
   As a general principle, the more important it is to reduce the testing cost, i.e. the higher the test budget, then the more detailed we will want the linkage to be, as this allows us to reduce probabilities in parts of the diagram based on tests in other parts of the diagram, and so reduce the number of tests.
**Probability elicitation.** The care with which we *elicit* probabilities for the model.
   **Full:** belief specification is made carefully for each node;
   **Partial:** some use is made of of exchangeability (similarity) to simplify specifications which are almost exchangeable;
   **Simple:** belief specification is based on a minimal number of exchangeable judgments and simple scaling arguments across parent nodes.

As a general principle, the more we suspect that we will not be able to carry out full testing (that is, the higher the test error level), then the more careful we will need to be in our prior elicitation, as we will not be able to drive all error probabilities to near zero, so some of our assessments may be strongly influenced by prior values.

The higher the initial error level of the software, the higher the level we prefer for each of the above three criteria.

### 8.7.6  Choosing an Appropriate Bayesian Approach

We now describe how to put the various ingredients together. There are three essentially different ways that the BGM approach may be valuable.

- For the same test budget as for the existing company approach, it may provide a more effective test suite and thus reduce errors in software release.
- It may achieve the same effectiveness as an existing company test suite constructed according to current practice, but using less test effort and so reducing test budget.
- The company may judge that there is an intrinsic "good practice" value in following an approach which offers a more rigorous and structured approach to testing, which constructs models for the test process that can be maintained over a series of software releases, which provides probabilistic assessments of software reliability that can be fed directly into any risk assessment that must be made about software release, and which can be used to monitor and forecast time until completion of the testing process and so give advance warning of problems in meeting test deadlines.

Our main objective is to reduce error when the project cost is high and to reduce budget when the budget cost is high. We suggest the following choices for setting the levels for the three aspects of the Bayesian specification.

1) Start with medium levels for outcome space description, model linkage, and elicitation.
2) If the initial software error level is high (low), then raise (lower) each of the three levels.
3) If the project cost is high (low), then raise (lower) the level of outcome space description.
4) If budget is high (low), then raise (lower) the level of model linkage.
5) If test error level is high (low), then raise (lower) the level of detail of prior elicitation.
6) Consider raising the various levels to realize more fully the "good practice" value of the testing.

Obviously, we may suggest further rules, and add further criteria. However, this is intended to be illustrative of our intention to produce an approach that is appropriate to each problem, rather than a use or do not use rule with some arbitrary cutoff.

## 8.8  Summary

This chapter has presented a brief review of the Bayesian graphical models (BGM) approach to software testing, which the authors developed in close collaboration with industrial software testers. It is a method for the logical structuring of the software testing problem, where the focus was on high-reliability final-stage integration testing. The chapter presents discussion of a range of topics for practical implementation of the BGM approach, including modeling for test–retest scenarios, the expected duration of the retest cycle, incorporation of multiple failure modes, and diagnostic methods. Furthermore, model maintenance and evolution is addressed, including consideration of novel system functionality. Finally, end-to-end testing of complex systems is discussed and methods are presented to assess the viability of the BGM approach for individual applications. These are all important aspects of high-complexity testing which software testers have to deal with in practice, and for which Bayesian statistical methods can provide useful tools. This chapter presents the basic approaches to these important issues. These should enable software testers, with support from statisticians, to develop implementations for specific test scenarios.

## Acknowledgment

The authors gratefully acknowledge the provision of funding for much of this work from British Telecommunications plc and input on its industrial direction.

## References

Bedford, T. and Cooke, R. (2001) *Probabilistic Risk Analysis: Foundations and Methods.* Cambridge: Cambridge University Press.

Coolen, F.P.A., Goldstein, M., and Wooff, D.A. (2003) Project viability assessment for support of software testing via Bayesian graphical modelling, in *Proceedings ESREL'03, Maastricht*, pp. 417–422.

Coolen, F.P.A., Goldstein, M., and Wooff, D.A. (2007) Using Bayesian statistics to support testing of software systems. *Journal of Risk and Reliability* 221, 85–93.

Cowell, R.G., Dawid, A.P., Lauritzen, S.L., and Spiegelhalter, D.J. (1999) *Probabilistic Networks and Expert Systems*. New York: Springer.

Goldstein, M. and Wooff, D.A. (2007) *Bayes Linear Statistics: Theory and Methods.* Chichester: Wiley.

Jensen, F.V. (2001) *Bayesian Networks in Decision Graphs*. New York: Springer.

Wooff, D.A., Goldstein, M., and Coolen, F.P.A. (2002) Bayesian graphical models for software testing. *IEEE Transactions on Software Engineering*, 28, 510–525.

# 9

# Models of Software Reliability

*Shelemyahu Zacks*

## Synopsis

An item of software is a complex system of programs, routines, and symbolic language that controls the functioning of the hardware of a computer. A given piece of software might have faults among its units. Not all faults lead to immediate failure. The reliability of a system is defined as a function of time, $R(t)$, which gives the probability that the given system will function for at least $t$ time units without failure. Software reliability is similarly defined. However, if a software fault is in some rarely used function, with very small probability of demand, the reliability of the software might still be high, provided other types of functions have no faults or a very small number of them. There are two types of software reliability model: time domain models and data domain models. Time domain models provide survival functions or hazard functions of time, which depend on parameters that have to be estimated. Data domain models are not time-dependent models, but sampling models from finite populations that provide estimates of the number of faults or other type of units in the software. Examples of these models and their applications are presented in this chapter. This includes a relatively simple time domain model, indicates current trends with non-homogeneous models, and discusses two different data domain models. The chapter provides the formulation and properties of estimates in these models.

## 9.1 Introduction

The reliability of a system is defined as a function of time, $R(t)$, which gives the probability that the given system will function for at least $t$ time units without failure. Software reliability could be defined similarly. Wikipedia defines software reliability as "the probability of failure-free software operation for a specified time in a specified environment." An item of software is a complex system of programs, routines, and symbolic language that controls the functioning of the hardware of a computer. A given piece of software might have *faults* among its units. For a definition of faults, see Kenett (2007). Not all faults lead to immediate *failure* or a wrong answer. For example, suppose that the software is designed to yield values of mathematical functions. If the fault is in the routine that computes the logarithmic function then there is a high probability that it will

*Analytic Methods in Systems and Software Testing*, First Edition.
Edited by Ron S. Kenett, Fabrizio Ruggeri, and Frederick W. Faltin.
© 2018 John Wiley & Sons Ltd. Published 2018 by John Wiley & Sons Ltd.

be demanded by a customer early, and a wrong answer will be given. In this case the reliability of the software is low. On the other hand, if the fault is in some rarely used function, with a very small probability of demand, the reliability of the software might still be high, provided other types of functions have no faults or a very small number of them. There are several books on software reliability; we mention Xie (1991), Lyu (1996), Pham (2000), and Xie, Dai, and Poh (2004), among others.

There are two types of software reliability model: *time domain* models and *data domain* models. Time domain models provide survival functions or hazard functions of time, which depend on parameters that have to be estimated. Data domain models are not time-dependent models, but sampling models from finite populations that provide estimates of the number of faults or other type of units in the software. Examples of these models and their applications will be discussed later in this chapter.

## 9.2 Time Domain Models

As noted by Xie and Xie and Yang (2007), over 100 software reliability models have been proposed since 1970. One early model is that of Jelinski and Moranda (1972). In this model it is assumed that, given initially $N$ faults among the units (modules) of the software, the time it takes to detect any one of these faults is a random variable, $X$, having an exponential distribution with mean $\beta$. Moreover, it is assumed that these $N$ random variables $\{X_i, i = 1, \ldots, N\}$ are identically and independently distributed. Furthermore, a detected fault is removed (corrected) so that the number of faults is reduced by one. This is a Markovian "death" model, or a "reliability growth" model. Indeed, $\min\{X_1, \ldots, X_N\}$ is distributed exponentially, with mean $\beta/N$. If there are, indeed, no more than $N$ faults, and the other assumptions are valid, the reliability of the system after $i$ faults have been detected and removed is

$$R_i(t; \beta, N) = \exp\{-(N - i)t/\beta\}, \quad i = 1, \ldots, N. \tag{9.1}$$

Under this model, the expected length of (operational) time, $T$, to remove all faults is

$$E\{T; N\} = \beta \sum_{i=0}^{N-1} \frac{1}{N - i}. \tag{9.2}$$

The functions in Equations (9.1) and (9.2) cannot be evaluated if the parameters $(N, \beta)$ are unknown. This Markovian model is simple and nice, but might not be valid. The times until faults are detected might not be independent, their distributions might not be exponential, etc. Later we discuss some data domain models for estimating the size $N$ of a (sub)population. If the model is valid and data is available, maximum likelihood estimators (MLE) might be used to estimate the parameters. If prior information is available on the distributions of the parameters, Bayesian estimation might be used. We illustrate this in the following subsection.

### 9.2.1 Estimation of $(N, \beta)$ in the Jelinski–Moranda Model

Maximum likelihood or Bayesian estimation depend on the available data at time $t$ (sufficient statistics), on the likelihood function of the sufficient statistics, on the prior distribution of the unknown parameters, and on a loss function for erroneous estimation. The parameter $N$ is bounded by the number of units, $N^*$, in the system. We show first that MLEs of $(N, \beta)$ do not exist. We then derive possible Bayesian estimators.

### 9.2.1.1   Maximum Likelihood Estimation

Let $\{T_n, n = 1, \ldots, N\}$ denote independent random variables, which are the times between consecutive arrivals of faults. As mentioned before, $T_n$ is distributed exponentially with mean $\beta/(N - n)$. Let $S_n = \sum_{j=0}^{n} T_j$, where $T_0 = 0$. The random number of arrivals in the interval $(0, t)$ is

$$K(t) = \max\{n \geq 0 : S_n \leq t\}.$$

The data at time $t$ is $\Delta(t) = \{K(t), T_0, \ldots, T_{K(t)}\}$. Notice that if $K(t) = 0$ there is no data for estimation – one has to wait until $K(t) \geq 1$. According to this model, if $N = 1$ then a fault will be detected sooner or later, following the exponential distribution with mean $\beta$. On the other hand, if $N = 0$ a rule is required to stop the search. The problem is that the values of $N$ and $\beta$ are unknown. A stopping time is suggested as follows.

Let $\hat{\beta}(t)$ be an estimator of $\beta$ at time $t$. Let $0 < \epsilon < 1$ be an error probability (risk level) for a wrong decision. Let $T_{1-\epsilon}$ denote the $(1 - \epsilon)$-quantile of the exponential distribution with mean $\beta$, i.e. $T_{1-\epsilon} = -\beta \log(\epsilon)$. (Note, all logarithms are natural). Thus, we adopt the stopping rule

$$T_s = \inf\{t > 0 : t \geq -\hat{\beta}(t) \log(\epsilon)\},$$

where $\hat{\beta}(t)$ is the estimator of $\beta$, based on $\Delta(t)$.

Since the inter-arrival times are independent, the conditional likelihood function, given by $\Delta(t)$, when $k \geq 1$ is

$$L(\beta, N | K(t) = k) = I\{N \geq k\} \left( \prod_{i=1}^{k} \frac{N - i + 1}{\beta} \right) \exp\left\{ -\sum_{i=1}^{k} \frac{N - i + 1}{\beta} T_i \right\}.$$

The MLE is a point $(\hat{N}, \hat{\beta})$ in the parameter space that maximizes the likelihood function. The log-likelihood function is

$$l(\beta, N | K(t) = k) = I\{N \geq k\} \left[ \sum_{i=1}^{k} \log\left( \frac{N - i + 1}{\beta} \right) - \sum_{i=1}^{k} \frac{N - i + 1}{\beta} T_i \right]. \tag{9.3}$$

For a given $\{N \geq k\}$, the MLE of $\beta$ is

$$\hat{\beta}_k(N) = \frac{1}{k} \sum_{i=1}^{k} (N - i + 1) T_i. \tag{9.4}$$

Notice that $\hat{\beta}_k(N)$ is also the best unbiased estimator of $\beta$, having a variance $V\{\hat{\beta}_k(N)\} = \beta^2/k$.

If $N$ is unknown we could try to obtain an MLE of $N$ by substituting Equation (9.4) into Equation (9.3). We obtain the function

$$l_k^*(N) = I\{N \geq k\} \left[ \sum_{i=1}^{k} \log\left( \frac{N - i + 1}{\hat{\beta}_k(N)} \right) - k \right].$$

The MLE of $N$ is the value of $\hat{N} \geq k$ that maximizes $l_k^*(N)$. Notice that $\hat{\beta}_k(N)$ is distributed around $\beta$ for any $N \geq k$. On the other hand, $\sum_{i=1}^{k} \log(N - i + 1) \to \infty$ as $N \to \infty$. Thus, the MLE of $N$ is equal to the upper bound $N^*$, and hence the MLE of $(N, \beta)$ is $(N^*, \hat{\beta}(N^*))$, which is of no value. This is an unacceptable solution.

### 9.2.1.2 Bayesian Estimation

Another approach is to derive Bayesian estimators of $(N, \lambda)$, where $\lambda = 1/\beta$. We derive these Bayesian estimators for the special case of prior joint density

$$\pi(\lambda, N) = I\{N \geq k\}\Lambda e^{-\lambda\Lambda}p(N, \mu)/(1 - P(k - 1, \mu)),$$

where

$$p(n, \mu) = e^{-\mu}\mu^n/n!$$

is the probability mass function of the Poisson distribution, and $P(k - 1, \mu) = \sum_{j=0}^{k-1} p(j, \mu)$ is the corresponding cumulative distribution function.

The posterior joint density of $(\lambda, N)$, given $\Delta(t)$, is

$$\pi(\lambda, N|\Delta(t)) = I\{N \geq k\}\pi(\lambda, N)L(\lambda, N)/ \sum_{N=k}^{\infty} p(N, \mu)$$

$$\times \prod_{i=1}^{k}(N - i + 1) \int_0^{\infty} \Lambda e^{-\lambda\Lambda}\lambda^k \exp\left\{-\lambda \sum_{i=1}^{k}(N - i + 1)T_i\right\} d\lambda.$$

Furthermore,

$$\int_0^{\infty} \Lambda e^{-\lambda\Lambda}\lambda^k \exp\left\{-\lambda \sum_{i=1}^{k}(N - i + 1)T_i\right\} d\lambda = \frac{\Lambda k!}{\left(\Lambda + \sum_{i=1}^{k}(N - i + 1)T_i\right)^{k+1}}. \quad (9.5)$$

Thus, the denominator of Equation (9.5) is

$$\Lambda k!e^{-\mu} \sum_{N=k}^{\infty} \frac{\mu^N}{(N - k)!}\left(\Lambda + \sum_{i=1}^{k}(N - i + 1)T_i\right)^{-(k+1)}$$

$$= \Lambda k!\mu^k \sum_{j=0}^{\infty} p(j, \mu)\left(\Lambda + \sum_{i=1}^{k}(k + j - i + 1)T_i\right)^{-(k+1)}.$$

This is obviously a finite positive quantity. Hence, the posterior density of $(N, \lambda)$ is

$$\pi(\lambda, N|\Delta(t)) = I\{N \geq k\}e^{-\lambda\Lambda}\lambda^k \exp\left\{-\lambda \sum_{i=1}^{k}(N - i + 1)T_i\right\}$$

$$\times p(N - k, \mu)/\left[k! \sum_{j=0}^{\infty} p(j, \mu)\left(\Lambda + \sum_{i=1}^{k}(k + j - i + 1)T_i\right)^{-(k+1)}\right].$$

The Bayesian estimator of $\beta$ for the squared error loss is

$$\hat{\beta}(t) = \sum_{N=k}^{\infty} \int_0^{\infty} \lambda^{-1} \pi(\lambda, N | \Delta(t)) d\lambda$$

$$= \frac{\frac{1}{k} \sum_{j=0}^{\infty} p(j, \mu) \left( \Lambda + \sum_{i=1}^{k} (k + j - i + 1) T_i \right)^{-k}}{\sum_{j=0}^{\infty} p(j, \mu) \left( \Lambda + \sum_{i=1}^{k} (k + j - i + 1) T_i \right)^{-(k+1)}}.$$

The Bayesian estimator of $N$ for the squared error loss is

$$\hat{N}(t) = \sum_{N=k}^{\infty} N \int_0^{\infty} \pi(\lambda, N | \Delta(t)) d\lambda$$

$$= \frac{\sum_{j=0}^{\infty} (j + k) p(j, \mu) \left( \Lambda + \sum_{i=1}^{k} (k + j - i + 1) T_i \right)^{-(k+1)}}{\sum_{j=0}^{\infty} p(j, \mu) \left( \Lambda + \sum_{i=1}^{k} (k + j - i + 1) T_i \right)^{-(k+1)}}.$$

This is an example of a finite Bayesian estimator of $N$.

### 9.2.1.3 Additional Time Domain Models

Several models were proposed to improve somewhat on the above Markovian model of Jelinski and Moranda. One could mention Schick and Wolverton (1978), Shanthikumar (1981), Xie (1987), Whittaker, Rekab, and Thomason (2000), and Boland and Singh (2003). A class of time domain models which is still in fashion (see Yoo (2015)) is based on non-homogeneous Poisson processes (NHPPs). The regular, homogeneous, Poisson process is based on failure times that are exponentially distributed with the same mean. The hazard function of such a time to failure is fixed at $\lambda = 1/\beta$ for all $t$. In the non-homogeneous Poisson process the intensity of the process is a function $\lambda(t)$. The expected number of failures, $N(t)$, in the interval $(0, t)$ is $m(t) = \int_0^t \lambda(s) ds$. The motivation to apply non-homogeneous Poisson processes is the danger that while removing a fault other mistakes might be inserted. The danger of this might increase in time, since faults that are detected late might have a complicated structure, and therefore be more difficult to remove. The models are now called "infinite NHPP failure models." It is implicitly assumed that faults might always appear. In other words, there are infinitely many faults (a very large number). Yoo (2015) lists several models with monotonic intensity functions, and shows how to derive and compute their maximum likelihood estimators. The first model is the "power-law function"

$$\lambda_{\text{PL}}(t) = \alpha \beta t^{\beta - 1}.$$

The second model is a "logarithmic function" (Musa and Okumoto (1984)),

$$\lambda_{MO}(t) = \alpha \log(1 + \beta t).$$

The third model is the "Gompertz intensity function," which is

$$\lambda_{GO}(t) = \alpha \beta e^{\beta t}.$$

The parameters $(\alpha, \beta)$ in these three models are positive. It is much easier to compute maximum likelihood estimators of $\alpha$ and $\beta$ for these three models, in comparison to those of the Jelinski–Moranda model. For each one of these models we have distinct $m(t)$ functions. The probability density function of the corresponding non-homogeneous Poisson process is

$$p(k, m(t)) = e^{-m(t)} \frac{m(t)^k}{k!}, \quad k = 0, 1, 2, \ldots$$

After each arrival of a fault, a new reliability function is computed. Thus, given the data $\Delta(t)$, with $K(t) = k, k \geq 0$, the reliability function is

$$R_k(t; S_k) = \exp\{-[m(S_k + t) - m(S_k)]\}.$$

## 9.3 Data Domain Models

Data domain models are sequential sampling procedures from finite populations, designed to find the defective units (with faults). As noted by Runeson (2007), "the focus of sampling and statistical quality control is on the development process." The inspection process starts with sampling of documents for pre-inspection. This pre-inspection provides estimates of fault content. The documents are then ranked according to fault content, and the remaining inspection effort is spent on the documents with the highest ranks.

We describe here two sequential procedures, published by Chernoff and Ray (1965) and by Zacks, Pereira, and Leite (1990). The Chernoff–Ray procedure is to select units from the population, according to their size and cost index, until a stopping time is reached. Zacks, Pereira, and Leite studied the capture–recapture method for estimating the size of a finite population. The capture–recapture technique was first applied to software inspection by Eick et al. (1992).

### 9.3.1 The Chernoff–Ray Procedure

Let $\Omega$ be a finite collection of units (distinct elements of the software to be tested for faults). Let $N$ be the size of $\Omega$. Let $\theta, 0 < \theta < 1$, be the proportion of defective units. All defective units found are replaced by good ones (rectified). The cost of releasing (not finding) a defective unit and rectifying it in the field is $\$K$. On the other hand, the cost of inspecting a unit is $\$c$. The risk after inspecting $n$ units is

$$R(\theta, n) = (N - n)\theta K + cn.$$

If $\theta K > c$ then it is optimal to inspect every unit. The problem is that $\theta$ is unknown. Often we have covariates for each unit of $\Omega$, say $\underline{x} = (x_1, \ldots, x_p)$, that can assist in predicting whether a unit is defective or not. For each unit, let $Y = 1$ if the unit is defective, and $Y = 0$ otherwise. Given $\underline{x}$, let $\theta(\underline{x}) = P\{Y = 1 | \underline{x}\}$. We apply the response model $\theta(\underline{x}) = F(\underline{x}^T \underline{\gamma})$, where $F$ is a standard distribution. If the regression coefficients $\underline{\gamma}$ are known, the risk of not inspecting $\Omega$ is

$$R_0 = \sum_{i=1}^{N} K_i F(\underline{x}_i^T \underline{\gamma}).$$

If we inspect a sample of $n$ units $s = \{u_{i_1}, \ldots, u_{i_n}\}$, the risk is

$$R_1(s) = R_0 - \sum_{i \in s} (K_i F(\underline{x}_i^T \underline{\gamma}) - c_i).$$

Accordingly, only elements for which $F(\underline{x}_i^T \underline{\gamma}) > \frac{c_i}{K_i}$ should be inspected. The problem is to devise an adaptive selection procedure and a stopping rule when $\underline{\gamma}$ is unknown. For example, suppose that each unit has one covariate $x$, which is its size (number of lines of code) or complexity. We consider the model

$$\theta(x; \gamma) = \exp\{-\gamma/x\}, \quad \gamma > 0, \quad x > 0.$$

Thus, the probability of a fault is an increasing function of $x$. Furthermore,

$$c_i = cx_i, \, i = 1, \ldots, N,$$
$$K_i = Kx_i, \, I = 1, \ldots, N.$$

Hence, the optimal procedure of selection is to select the $n$ units having the largest $x$ values. The MLE of $\gamma$ is the root of the equation

$$\sum_{i=1}^{n} e^{-\gamma/x_i} [x_i (1 - e^{-\gamma/x_i})]^{-1} = \sum_{i=1}^{n} Y_i [x_i (1 - e^{-\gamma/x_i})]^{-1}.$$

The MLE does not exist if $\sum_{i=1}^{n} Y_i = 0$ or $\sum_{i=1}^{n} Y_i = n$. If $0 < \sum_{i=1}^{n} Y_i < n$ then there is a unique solution, which can be obtained numerically.

The Fisher information function is

$$I_n(\gamma) = \sum_{i=1}^{n} e^{-\gamma/x_i} [x_i^2 (1 - e^{-\gamma/x_i})]^{-1}.$$

Let $\hat{\gamma}_n$ be the MLE of $\gamma$. The lower confidence limit of $\gamma$, for large $n$, is

$$\hat{\gamma}_{n,\alpha}^{(L)} = \hat{\gamma}_n - Z_{1-\alpha} \left( \frac{1}{I_n(\hat{\gamma})} \right)^{1/2}.$$

One could use the stopping rule:

$$M = \min\{n \geq m : K \exp\{-\hat{\gamma}_{n,\alpha}^{(L)} / x_{(n+1)}\} < c\},$$

where $x_{(1)} \geq x_{(2)} \geq \cdots \geq x_{(n)} \geq x_{(n+1)}$.

## 9.3.2 The Capture–Recapture Method

The problem of estimating the size of a finite population is of interest in ecology, wildlife management, fisheries, and more. There are different approaches for studying this problem. We discuss here the capture–recapture approach. There are several papers in the literature on this topic – see Freeman (1972), Samuel (1968, 1969), and Pollock and Otto (1983). The material in this section is based on Zacks, Pereira, and Leite (1990).

Suppose there is a finite population of size $N$. There are sampling procedures based on sequential samples of size one (for big wild animals, for example), or samples of large (random) size (in fisheries, for example). Let $M_k$ denote the size of the $k$th sample. The population is closed during the study. Units which have been previously observed are tagged, and their number, $U_k$, is recorded.

Let $T_k = \sum_{i=1}^{k} U_i$. Notice that $U_1 = M_1$ and $U_i \leq M_i$. Also, $\max\{M_i, i = 1, \dots, k\} \leq N$. We assume that different samples are conditionally independent, given $\{M_i, i = 1, \dots, k\}$. The data after $k$ samples are $\Delta_k = \{(U_i, M_i), i = 1, \dots, k\}$. The likelihood of $N$ given $\Delta_k$ is

$$L(N; \Delta_k) = \prod_{i=1}^{k} \binom{N-T_{i-1}}{U_i} \binom{T_{i-1}}{M_i-U_i} / \binom{N}{M_i}$$

$$= \frac{\prod_{i=2}^{k} \binom{T_{i-1}}{M_i-U_i}}{\prod_{i=1}^{k} U_i!} K(N; \Delta_k),$$

where the kernel of the likelihood is

$$K(N; \Delta_k) = I\{T_k \leq N\} \frac{N!}{(N - T_k)! \prod_{i=1}^{k} \binom{N}{M_i}}.$$

Let $\pi(n)$ be the prior probability density function of $N$. The posterior probability distribution of $N$ given $\Delta_k$ is

$$\pi(n|\Delta_k) = I\{N \geq T_k\} = \pi(n)K(n; \Delta_k) / \sum_{j=T_k}^{\infty} \pi(j)K(j; \Delta_k).$$

It is proved in Zacks, Pereira, and Leite (1990) that $\sum_{j=T_k}^{\infty} \pi(j) K(j; \Delta_k) < \infty$. Hence, $B(k, t)$ is the Bayesian estimator of $N$, for the squared error loss, when $t = T_k$. If $\pi(n) = e^{-\lambda} \lambda^n / n!$, then

$$B(k, t) = t + \lambda \left[ \sum_{n=0}^{\infty} \lambda^n \frac{1}{n!} \prod_{j=1}^{M} \left( \frac{t+2-j}{n+t+1-j} \right)^k \right] \left[ \sum_{n=0}^{\infty} \lambda^n \frac{1}{n!} \prod_{j=1}^{M} \left( \frac{t+1-j}{n+t+1-j} \right)^k \right]^{-1}.$$

The Bayesian risk is

$$R(k, t) = [B(k, t) - t][B(k, t + 1) - B(k, t)].$$

In Table 9.1 we present some simulation results in computing the Bayesian estimator $B(k, t)$ and the Bayesian risk $R(k, t)$. In these simulations we assumed that $N = 50$ and $M = 1$. The Poisson prior distribution has mean $\lambda = 20$. This is a situation of wrong prior assumption. We see that if each sample is of size 1, one needs about 150 samples in order to obtain an estimate close to the true value of $N$.

**Table 9.1** Simulation results for Bayesian estimator and Bayesian risk.

| $k$ | $T_k$ | $B(k,T_k)$ | $R(k,T_k)$ |
|-----|-------|------------|------------|
| 70  | 33    | 35.926     | 3.458      |
| 100 | 42    | 44.128     | 2.383      |
| 150 | 48    | 48.969     | 1.025      |

## 9.4 Summary

The present chapter does not provide a comprehensive review of all possible attempts at modeling software reliability. An item of software might be a very complicated system of components, each with specific roles. The reliability of each component might be defined differently, and modeled by a different approach. This chapter illustrates a relatively simple time domain model, indicates current trends with non-homogeneous models, and discusses two different data domain models.

## References

Boland, P.J. and Singh, H. (2003) A birth-process approach to Moranda's geometric software-reliability model. *IEEE Transactions on Reliability*, 52, 168–174.

Chernoff, H. and Ray, S.N. (1965) A Bayes sequential sampling inspection plan. *Annals of Mathematical Statistics*, 36, 1387–1407.

Eick, S.G., Loader, C.R., Long, M.D., Votta, L.G., and Wiel, S. (1992) Estimating software fault content before coding: Software engineering, in *Proceedings International Conference on Software Engineering*, Melbourne, Australia.

Freeman, P.R. (1972) Sequential estimation of the size of a population. *Biometrika*, 59, 9–17.

Jelinski, A. and Moranda, P.B. (1972) Software reliability research, in W. Freiberger (ed.) *Statistical Computer Performance Evaluation*, New York: Academic Press, pp. 465–497.

Kenett, R.S. (2007), Software failure data analysis, in F. Ruggeri, R.S. Kenett, and F.W. Faltin (eds) *Encyclopedia of Statistics in Quality and Reliability*, Chichester: Wiley.

Lyu, M. (ed.) (1996) *Handbook of Software Reliability Engineering*, New York: McGraw-Hill.

Musa, J.D. and Okumoto, K. (1984) A logarithmic Poisson execution time model for software reliability measurement, in *Proceedings of the 7th International Conference on Software Engineering*, Orlando, FL, pp. 230–238.

Pham, H. (2000) *Software Reliability*, Singapore: Springer-Verlag.

Pollock, K.H. and Otto, M. (1983) Robust estimation of population size in closed animal population from capture–recapture experiments. *Biometrics*, 39, 1035–1049.

Runeson, P. (2007) Sampling in software development, in F. Ruggeri, R.S. Kenett, and F.W. Faltin (eds) *Encyclopedia of Statistics in Quality and Reliability*, Chichester: Wiley.

Samuel, E. (1968) Sequential maximum likelihood estimation of a population. *Annals of Mathematical Statistics*, 39, 1057–1068.

Samuel, E. (1969) Comparison of sequential rules for estimation of the size of a population. *Biometrics*, 25, 517–527.

Shanthikumar, J.G. (1981) A general software reliability model for performance prediction. *Microelectronics and Reliability*, 21, 671–682.

Schick, G.J. and Wolverton, R.W. (1978) An analysis of competing software reliability models. *IEEE Transactions on Software Engineering*, SE-4, 104–120.

Whittaker, J., Rekab, K., and Thomason, M.G. (2000) A Markov chain model for predicting reliability of multi-built software, *Information and Software Technology*, 42, 889–894.

Xie, M. (1987) A shock model for software failures. *Microelectronics and Reliability*, 27, 717–724.

Xie, M. (1991) *Software Reliability Modeling*. Singapore: World Scientific Publishers.

Xie, M. and Yang, B. (2007) Software reliability modeling and analysis, F. Ruggeri, R.S. Kenett, and F.W. Faltin (eds) *Encyclopedia of Statistics in Quality and Reliability*, Chichester: Wiley.

Xie, M., Dai, Y.S., and Poh, K.L. (2004) *Computing Systems Reliability*, Boston, MA: Kluwer Academic.

Yoo, T.-H. (2015) The infinite NHPP software reliability model based on monotonic intensity function. *Indian Journal of Science and Technology*, 8(14).

Zacks, S., Pereira, C.A., and Leite, J.G. (1990) Bayes sequential estimation of the size of a finite population. *Journal of Statistical Planning and Inference*, 25, 363–380.

## 10

# Improved Estimation of System Reliability with Application in Software Development

*Beidi Qiang and Edsel A. Peña*

## Synopsis

A complex piece of software, for example an operating system, could be viewed as a coherent system composed of several components (subprograms) configured according to some structure function. When the software is under operation, of interest is the time to failure – for instance, the time to encountering a bug in a certain operating system. It is therefore of importance to learn about the distribution of this system's time to failure. This distribution may be estimated by simply using observations of the system lifetimes. However, if component lifetime data is available, possibly some of them right-censored by the system life, a better estimator of the system failure time may be obtained. In this paper, we demonstrate that shrinkage ideas may further be implemented to obtain additional improvement in the estimation of the system or software reliability function.

## 10.1  Introduction

Arguably, computer software could be considered in our era as the engine that powers society's progress. Its reliability, which is the probability of it being able to successfully complete a given task, is therefore of paramount importance. Think, for instance, of the importance of computer software in global financial and security systems, its indispensability in automated systems for industrial production processes, in flying airplanes and spaceships, in aviation traffic control systems, and very soon in running autonomous driverless cars. It is critical for safety purposes and for the proper functioning of many of society's systems that we have up-to-date knowledge of the reliability of computer software in order to assess its effectiveness and to manage and monitor safety and risk issues.

This chapter is concerned with the problem of estimating the reliability of a software system either during the software development stage or during real-time operation. Being able to estimate its reliability will be instrumental in making a decision on whether the software could be deployed in the field or further debugging is required. A complex piece of software, for example an operating system, could be viewed as a coherent system composed of several components configured according to some structure function. The components could be units, sections of code, or subprograms, which are "connected" according to some structure. Software testing involves the execution of a software component or the whole system to evaluate the quality of the

*Analytic Methods in Systems and Software Testing*, First Edition.
Edited by Ron S. Kenett, Fabrizio Ruggeri, and Frederick W. Faltin.
© 2018 John Wiley & Sons Ltd. Published 2018 by John Wiley & Sons Ltd.

product. The goal of such testing is to allow the assessment of the risk of failure of the software when deployed in the field.

There are numerous approaches available in software testing. Traditionally, the approaches can be divided into two major categories: white-box and black-box testing (Patton, 2005; Ammann and Offutt, 2008). White-box testing (also known as clear-box testing) refers to testing procedures that utilize information on internal performance and the source code. This is analogous to testing a system, when component-level data is observed. Black-box testing treats the software as a "black box." It tests the performance of the whole system without knowledge of its internal structure, or the source code. Thus, only system-level data is available in black-box testing.

There are also different types of tests that can be performed on software, such as compatibility testing, destructive testing, and others. One important type of test that will be discussed in this chapter is performance testing, which is used to determine whether a system can continuously function beyond a certain period of time or above a particular workload (Kenett and Pollak, 1986; Patton, 2005; Ammann and Offutt, 2008). This type of testing is essentially a test on system reliability over time (or workload), where the system of interest is the software and the question is to find the probability of surviving (or, equivalently, the risk of failure) of the system.

In this chapter, we look at the situation where white-box testing is used in software performance assessment, which means component-level failure times (or workloads) are observed. Of course, the simplest way, though clearly less efficient, is to estimate the system reliability using only the observed system lifetimes. Depending on the situation, this could be done via parametric estimation methods, though this may not be feasible for complicated systems because of non-identifiable parameters when only system-level data are available, or via a non-parametric approach, which is always implementable. When component-level data are available, the traditional way of estimating system reliability is to find component-level maximum likelihood estimators (MLEs) and then utilize these MLEs according to the system structure in order to obtain an estimator of the system reliability (Boland and Samaniego, 2001; Samaniego, 2007; Coolen and Coolen-Maturi, 2012). For instance, this approach was utilized by 1989) using non-parametric component-level MLEs. In this chapter, we propose improved estimators of system reliability by applying shrinkage estimators for each of the component reliabilities, and then utilize the structure function to combine these estimators to obtain the system reliability estimator. Some numerical examples with different system structures and component lifetimes are provided. Simulation results are also presented to examine and compare the performance of the different estimators.

## 10.2 A Review of Basic Concepts and Methods

### 10.2.1 Reliability Concepts

Detailed discussions of the concepts and ideas presented below are available in Barlow and Proschan (1975) and Kenett and Zacks (2014).

Consider a component with a lifetime, denoted by $T$, measured in some unit of "time." Usually, the lifetime will be measured in literal time, but this need not always be the case. It could, for instance, represent the load for which the component fails. For example, in software testing, $T$ could be the load to fail of a particular subprogram or module that is under a load test. Such a $T$ is a non-negative random variable. The reliability function of this component is defined via

$$R(t) = 1 - F(t) = \Pr\{T > t\},$$

where $F(\cdot)$ is the corresponding distribution function. We assume that lifetime variables are continuous. For $T$ with distribution function $F$, its probability density function (pdf) is

$$f(t) = \frac{dF(t)}{dt} = -\frac{dR(t)}{dt}.$$

Its hazard rate function $\lambda(t)$ is defined as

$$\lambda(t) \equiv \lim_{dt \downarrow 0} \frac{1}{dt} \Pr\{t \leq T < t + dt \mid T \geq t\} = \frac{f(t)}{R(t)},$$

which can be interpreted as the rate of failure of the component at time $t$, given that the component is still working just before time $t$. Given the hazard rate function, the cumulative hazard function is

$$\Lambda(t) = \int_0^t \lambda(v) dv.$$

For a continuous lifetime $T$, we have the following relationships:

$$f(t) = \lambda(t) \exp\{-\Lambda(t)\} \quad \text{and} \quad R(t) = \exp\{-\Lambda(t)\}.$$

For a component with lifetime $T$, its associated state process is $\{X(t) : t \geq 0\}$, where $X(t) = I\{T > t\}$ is a binary variable taking values of 1 or 0 depending on whether the component is still working (1) or failed (0) at time $t$. The function $I(\cdot)$ denotes an indicator function.

Consider a system composed of $K$ components, where this system is either in a working (1) or failed (0) state. The system could be a software system with $K$ subprograms or modules; it could be a computer with $K$ items of software; or, more generally, it could be a network of $K$ computers that work together as a system. The functionality of a system is characterized by its structure function. This is the function

$$\phi : \{0, 1\}^K \rightarrow \{0, 1\},$$

from the components' state space to the system's state space, with $\phi(x_1, x_2, \ldots, x_K)$ denoting the state of the system when the states of the components are $\mathbf{x} = (x_1, x_2, \ldots, x_K) \in \{0, 1\}^K$. The vector $\mathbf{x}$ is called the component state vector. Such a system is said to be *coherent* if each component is relevant and the structure function $\phi$ is non-decreasing in each argument. The $i$th component is *relevant* if there exists a state vector $\mathbf{x} \in \{0, 1\}^K$ such that $\phi(\mathbf{x}, 0_i) = 0 < 1 = \phi(\mathbf{x}, 1_i)$, with the notation that $(\mathbf{x}, a_i) = (x_1, \ldots, x_{i-1}, a_i, x_{i+1}, \ldots, x_n)$. We will only consider coherent systems in this chapter. Four simple examples of coherent systems are the series, parallel, three-component series–parallel, and five-component bridge systems.

A series system is a system in which all the components are "serially connected," and the system fails when at least one component fails. For example, consider a simple computer system that only consists of the operating system software and an internet browser (a simple version of the Chromebook, for instance). If at least one of these two items of software fails, the computer system stops functioning since users can no longer browse web pages.

On the other extreme, a parallel system only fails when all of its components fail. In this system, the components are in essence "backed up" by each other. Think of a security system with several layers – for example, facial, fingerprint, and PIN recognition modules. The security system can only be cracked when all three modules are cracked.

As a combination of the previous two types of system, a series–parallel system can be viewed as a series system with some of the components redundantly replicated. For instance, consider a simple computer system with an operating system and several internet browsers (e.g., Chrome, Firefox, and Internet Explorer). The computer would fail its task of being able to browse online if the operating system fails, but would still be functional if the operating system is still working and at least one of the three browsers is still functional. This is usually the case for a computer system consisting of system software and application software. If the system software fails, the computer stops functioning immediately. But the applications are usually backed up by each other, so if one of the application fails, the whole system can still function by using an alternative application.

A more complicated type of system is the bridge system. Consider a system consisting of two local area networks (LANs), each having a sender and a receiver, and these are structured according to a bridge setup. To complete a signal transmission, the signal can go through the sender and receiver in the same LAN, or it can go from the sender of one LAN, pass though the bridge, and then reach the receiver in the other LAN. The system will fail when there is no sender or receiver available, or when the bridge is down and the available sender and receiver are in different LANs.

Schematic block diagrams for these four system types are given in Figure 10.1, and their respective structure functions are given below.

$$\phi_{\text{ser}}(x_1, \ldots, x_K) = \prod_{i=1}^{K} x_i, \tag{10.1}$$

$$\phi_{\text{par}}(x_1, \ldots, x_K) = \coprod_{i=1}^{K} x_i \equiv 1 - \prod_{i=1}^{K}(1 - x_i), \tag{10.2}$$

$$\phi_{\text{serpar}}(x_1, x_2, x_3) = x_1(x_2 \vee x_3), \tag{10.3}$$

$$\phi_{\text{br}}(x_1, x_2, x_3, x_4, x_5) = (x_1 x_3 x_5) \vee (x_2 x_3 x_4) \vee (x_1 x_4) \vee (x_2 x_5). \tag{10.4}$$

The binary operator $\vee$ means taking the maximum, i.e. $a_1 \vee a_2 = \max(a_1, a_2) = 1 - (1 - a_1)(1 - a_2)$ for $a_i \in \{0, 1\}, i = 1, 2$.

Let $X_i, i = 1, \ldots, K$, be the state (at a given point in time) random variables for the $K$ components, and assume that they are independent. This is a realistic assumption under modular programming, which separates a software program into independent and interchangeable modules. Denote by $p_i = \Pr\{X_i = 1\}$, $i = 1, \ldots, K$, and let $\mathbf{p} = (p_1, p_2, \ldots, p_K) \in [0, 1]^K$ be the components' reliability vector (at a given point in time). Associated with the coherent structure function $\phi$ is the reliability function defined via

$$h_\phi(\mathbf{p}) = E[\phi(X)] = \Pr\{\phi(X) = 1\}.$$

This reliability function provides the probability that the system is functioning, at the given point in time, when the component reliabilities at this time are the $p_i$. For the first three concrete systems given above, these reliability functions are, respectively:

**Figure 10.1** Schematic block diagrams of series, parallel, series–parallel, and bridge systems.

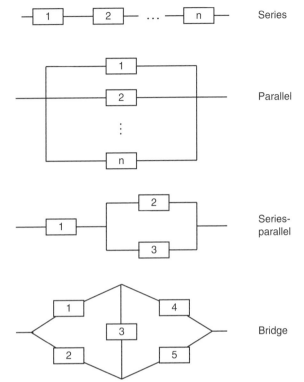

Series

Parallel

Series-parallel

Bridge

$$h_{\mathrm{ser}}(p_1, \dots, p_K) = \prod_{i=1}^{K} p_i, \tag{10.5}$$

$$h_{\mathrm{par}}(p_1, \dots, p_K) = \coprod_{i=1}^{K} p_i \equiv 1 - \prod_{i=1}^{K} (1 - p_i), \tag{10.6}$$

$$h_{\mathrm{serpar}}(p_1, p_2, p_3) = p_1[1 - (1 - p_2)(1 - p_3)]. \tag{10.7}$$

For the bridge structure, its reliability function at a given point in time, obtained first by simplifying the structure function, is given by

$$h_{\mathrm{br}}(p_1, p_2, p_3, p_4, p_5) = (p_1 p_4 + p_2 p_5 + p_2 p_3 p_4 + p_1 p_3 p_5 + 2 p_1 p_2 p_3 p_4 p_5)$$
$$- (p_1 p_2 p_3 p_4 + p_2 p_3 p_4 p_5 + p_1 p_3 p_4 p_5 + p_1 p_2 p_3 p_5 + p_1 p_2 p_4 p_5). \tag{10.8}$$

Of more interest, however, is viewing the system reliability function as a function of time $t$. Denoting by $S$ the lifetime of the system, we are interested in the function

$$R_S(t) = \Pr\{S > t\},$$

which is the probability that the system survives or does not fail in time period $[0, t]$. Let $\mathbf{T} = (T_1, \dots, T_K)$ be the vector of lifetimes of the $K$ components. The vector of component state processes is $\{\mathbf{X}(t) = (X_1(t), \dots, X_K(t)) : t \geq 0\}$. The system lifetime is then

$$S = \sup\{t \geq 0 : \phi[X_1(t), \dots, X_K(t)] = 1\}.$$

The component reliability functions are $R_i(t) = E[X_i(t)] = \Pr\{T_i > t\}, i = 1, \ldots, K$. If the component lifetimes are independent, then the system reliability function becomes

$$R_S(t) = E[\phi(X_1(t), \ldots, X_K(t))] = h_\phi(R_1(t), \ldots, R_K(t)). \tag{10.9}$$

That is, under independent component lifetimes, to obtain the system reliability function we simply replace the $p_i$ in the reliability function $h_\phi(p_1, \ldots, p_K)$ by $R_i(t)$. For the concrete examples of coherent systems given in Equations (10.1)–(10.4), we therefore obtain:

$$R_{\mathrm{ser}}(t) = \prod_{i=1}^{K} R_i(t), \tag{10.10}$$

$$R_{\mathrm{par}}(t) = 1 - \prod_{i=1}^{K}(1 - R_i(t)), \tag{10.11}$$

$$R_{\mathrm{serpar}}(t) = R_1(t)[1 - (1 - R_2(t))(1 - R_3(t))]. \tag{10.12}$$

For the bridge structure, in Equation (10.8) we replace each $p_i$ by $R_i(t)$ to obtain its system reliability function. As an illustration of these system reliability functions for the four examples of coherent systems, with $R_i(t) = \exp(-\lambda t), i = 1, \ldots, K$, the system reliability functions are

$$R_{\mathrm{ser}}(t; \lambda, K) = \exp(-K\lambda t),$$
$$R_{\mathrm{par}}(t; \lambda, K) = 1 - [1 - \exp(-\lambda t)]^K,$$
$$R_{\mathrm{serpar}}(t; \lambda) = \exp(-\lambda t)[1 - (1 - \exp(-\lambda t))^2],$$
$$R_{\mathrm{br}}(t; \lambda) = 2\exp\{-2\lambda t\} + 2\exp\{-3\lambda t\} + 2\exp\{-5\lambda t\} - 5\exp\{-4\lambda t\}.$$

These system reliability functions are plotted in Figure 10.2 for $\lambda = 1$ and $K = 5$.

An important concept in reliability is measuring the relative importance of each of the components in the system. There are several possible measures of component importance (Barlow and Proschan, 1975). We focus on the so-called reliability importance measure, as this will play an important role in the improved estimation of the system reliability. The reliability importance of component $j$ in a $K$-component system with reliability function $h_\phi(\cdot)$ is

$$I_\phi(j; \mathbf{p}) = \frac{\partial h_\phi(p_1, \ldots, p_{j-1}, p_j, p_{j+1}, \ldots, p_K)}{\partial p_j} = h_\phi(\mathbf{p}, 1_j) - h_\phi(\mathbf{p}, 0_j). \tag{10.13}$$

This measures how much system reliability changes when the reliability of component $j$ changes, with the reliabilities of the other components remaining the same. For a coherent system, the reliability importance of a component is positive. As examples, the reliability importance of the $j$th component in a series system is

$$I_{\mathrm{ser}}(j; \mathbf{p}) = \frac{h_{\mathrm{ser}}(\mathbf{p})}{p_j}, \quad j = 1, \ldots, K,$$

showing that in a series system the weakest (least reliable) component is the most important ("the system is as good as its weakest link"). For a parallel system, the reliability importance of the $j$th component is

$$I_{\mathrm{par}}(j; \mathbf{p}) = \frac{1 - h_{\mathrm{par}}(\mathbf{p})}{1 - p_j}, \quad j = 1, \ldots, K,$$

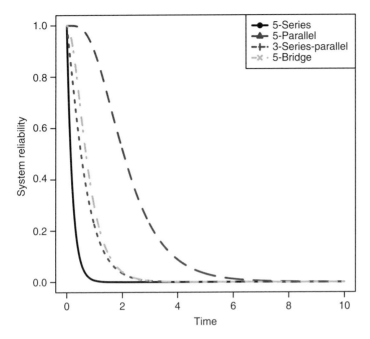

**Figure 10.2** System reliability functions for the series, parallel, series–parallel, and bridge systems when the components have common unit exponential lifetimes and there are five components in the series and parallel systems.

indicating that the most reliable component is the most important component in a parallel system. For the three-component series–parallel system, the reliability importance of the three components are

$$I_{\text{serpar}}(1; \mathbf{p}) = 1 - (1 - p_2)(1 - p_3),$$
$$I_{\text{serpar}}(2; \mathbf{p}) = p_1(1 - p_3),$$
$$I_{\text{serpar}}(3; \mathbf{p}) = p_1(1 - p_2).$$

Evaluated at $\mathbf{p} = (p, p, p)$, they become $I_{\text{serpar}}(1; p) = p(2 - p)$ and $I_{\text{serpar}}(2; p) = I_{\text{serpar}}(3; p) = p(1 - p)$, which confirms the intuitive result that when the components are equally reliable, the component in series (component 1) is the most important component. For example, in a computer system, its operating system software is more important compared to its application programs, when they are all equally reliable, though of course in reality the operating system software will usually be more reliable than application software. In general, however, in a series–parallel system, component 1 (the component in series) need not always be the most important. For instance, if components 2 and 3 are equally reliable with reliability $p_2$, then the reliability importance of components 1, 2, and 3 become

$$I_{\text{serpar}}(1; (p_1, p_2, p_2)) = p_2(2 - p_2),$$
$$I_{\text{serpar}}(2; (p_1, p_2, p_2)) = I_{\text{serpar}}(3; (p_1, p_2, p_2)) = p_1(1 - p_2).$$

In this case, component 2 (and 3) is more important than component 1 whenever

$$p_1(1 - p_2) > p_2(2 - p_2),$$

or, equivalently, when

$$p_1 > \frac{p_2(2 - p_2)}{1 - p_2}.$$

For example, if $p_2 = 0.2$, then if $p_1 > 0.45$, component 2 (and 3) is more important than component 1.

### 10.2.2 The Decision-Theoretic Approach

A decision problem has the following elements:

$$(\Theta, \mathcal{A}, \mathcal{X}, \mathcal{F} = \{F(\cdot|\theta), \theta \in \Theta\}, L, D).$$

Here, $\Theta$ is the parameter space containing the possible values of some parameter $\theta = (\theta_1, \theta_2, \ldots)$, which could be finite dimensional or infinite dimensional; $\mathcal{A}$ is the action space consisting of all possible actions that the decision maker could take; $L : \Theta \times \mathcal{A} \to \mathcal{R}$ is the loss function, with $L(\theta, \mathbf{a})$ denoting the loss incurred by choosing action $\mathbf{a}$ when the parameter is $\theta$. The observable data $X$ takes values in the sample space $\mathcal{X}$, with $X$ given $\theta$ having distribution $F(\cdot|\theta)$, which belongs to the family of distribution functions $\mathcal{F}$. Non-randomized decision functions are (measurable) mappings $\delta : \mathcal{X} \to \mathcal{A}$, and the totality of such decision functions is the decision function space $D$. To assess the quality of a decision function $\delta \in D$, we utilize the risk function given by

$$R(\theta, \delta) = E[L(\theta, \delta(X)|\theta],$$

which is the expected loss incurred by using decision function $\delta$ when the parameters are $\theta$. Good decision functions are those with small risks whatever the value of $\theta$. In particular, a decision function $\delta_1$ is said to dominate a decision function $\delta_2$ if, for all $\theta \in \Theta$, $R(\theta, \delta_1) \leq R(\theta, \delta_2)$, with strict inequality for some $\theta \in \Theta$. In such a case the decision function $\delta_2$ is inadmissible and should not be used at all. The statistical inference problem of point estimation of functionals of a parameter falls into this decision-theoretic framework. In this case the decision functions are called *estimators*.

### 10.2.3 Simultaneous and Shrinkage Estimators

This decision-theoretic framework carries over to simultaneous decision making; in particular, simultaneous estimation. Consider, for instance, the situation where $\theta = (\mu_1, \mu_2, \ldots, \mu_K) \in \Theta = \mathcal{R}_K$, the action space is $\mathcal{A} = \mathcal{R}_K$, and the data observable is $\mathbf{X} = (X_1, X_2, \ldots, X_K) \in \mathcal{X} = \mathcal{R}_K$, where the $X_j$ $(j \neq i)$ are independent of $X_i$, given $\mu_i$, having a normal distribution with mean $\mu_i$ and variance $\sigma^2$, assumed known. For the simultaneous estimation problem, we could use the loss function given by

$$L(\theta, \mathbf{a}) = ||\theta - \mathbf{a}||^2 = \sum_{i=1}^{K} (\theta_i - a_i)^2, \quad (\theta, \mathbf{a}) \in \Theta \times \mathcal{A}.$$

This loss function is referred to as a quadratic loss function. The maximum likelihood (ML) estimator of $\theta$ is $\delta_{\mathrm{ML}}(\mathbf{X}) = \mathbf{X}$, whose risk function is given by

$$R(\theta, \delta_{\mathrm{ML}}) = E[||\theta - \mathbf{X}||^2|\theta] = K\sigma^2.$$

Incidentally, this ML estimator is also the least-squares (LS) estimator of $\theta$. When $K = 1$ or $K = 2$, $\delta_{\mathrm{ML}}$ is the best (risk-wise) estimator of $\theta$. However, when $K \geq 3$, James and Stein (1961) demonstrated that there is a better estimator of $\theta$ than $\delta_{\mathrm{ML}}(\mathbf{X}) = \mathbf{X}$.

An estimator that dominates the ML estimator is their so-called shrinkage estimator of $\theta$, given by

$$\delta_{JS}(\mathbf{X}) = \hat{\theta}_{JS} = \left[1 - \frac{(K-2)\sigma^2}{\|\mathbf{X}\|^2}\right]\mathbf{X}.$$

More generally, if $\mathbf{X} = (\mathbf{X}_1, \dots, \mathbf{X}_n)$ are independent and identically distributed $1 \times p$ multivariate normal vectors with mean vector $\theta$ and common covariance matrix $\sigma^2 I_K$, then the James–Stein estimator of $\theta$ is given by

$$\delta_{JS}(\mathbf{X}) = \hat{\theta}_{JS} = \left[1 - \frac{(K-2)\sigma^2}{n\|\overline{\mathbf{X}}\|^2}\right]\overline{\mathbf{X}},$$

where $\overline{\mathbf{X}} = \frac{1}{n}\sum_{j=1}^{n}\mathbf{X}_j$ denotes the vector of sample means.

The James–Stein shrinkage estimator is one which utilizes the *combined* data for estimating each component parameter, even though the component variables are independent, to improve the simultaneous estimation of the components of $\theta$. It shows that optimizing (i.e., minimizing) a global loss is not the same as optimizing individually the loss of each component estimator. In essence, there is an advantage in the borrowing of information from each of the component data items, demonstrating that when dealing with a combined or global loss function, it may be beneficial to borrow information in order to improve the estimation process. Observe that the James–Stein type of shrinkage estimator is of the form

$$\delta_c(\mathbf{X}) = \hat{\theta}_c = c\overline{\mathbf{X}},$$

for some $c > 0$, which in this case is data dependent. We will employ this shrinkage idea later to improve the estimation of the system reliability function.

The application of shrinkage estimation in survival analysis has been partly addressed in the literature. Siu-Keung and Geoffrey (1996) examined three versions of shrinkage estimators under exponential lifetimes. Later, Prakash and Singh (2008) presented shrinkage estimators under the LINEX loss function in the situation with exponential lifetime censored data. The shrinkage estimation of the reliability function for other lifetime distributions was also studied by Chiou (1988) and Pandey (2014). Extension to the estimation of system reliability has also been discussed under certain system structures and lifetime distributions; see, for instance, Pandey and Upadhyay (1985, 1987).

## 10.3  Statistical Model and Data Structure

We now consider the problem of estimating the system reliability function on the basis of observed data from the system or its components. We suppose that the software system has $K$ components and has been in the white-box test situation. As pointed out earlier, time here need not be literal time but could be the amount of workload for the system or components. In this setting, the observed data will be the $K$ components' time to failure, but this could be right-censored by the system lifetime or an upper bound to the monitoring period. We let $T_j$ denote the time to failure of component $j$, and we assume that the $T_j$ are independent of each other. We denote by $R_j(\cdot; \theta_j)$ the reliability function of $T_j$, where $\theta_j \in \Theta_j$ and the parameter space $\Theta_j$ is an open subset of $\mathfrak{R}_{m_j}$. Furthermore, note that it is possible that there could be common parameters among the $(\theta_j, j = 1, 2, \dots, K)$.

Since, in the white-box setting, we have information on the internal structure of the software, we assume that the system structure function $\phi(\cdot)$ is known, hence we also know the reliability function $h_\phi(\mathbf{p})$. As discussed in the review portion of this chapter, the system reliability function $R_S(\cdot)$ can therefore be expressed via

$$R_S(t) = R_S(t; \theta_1, \theta_2, \ldots, \theta_K)$$
$$= h_\phi[R_1(t; \theta_1), R_2(t; \theta_2), \ldots, R_K(t; \theta_K)]. \qquad (10.14)$$

Suppose the performance testing is conducted $n$ times, so that the observable component lifetimes are

$$\{T_{ij} : i = 1, 2, \ldots, n; \ j = 1, 2, \ldots, K\}.$$

We shall assume that the $T_{ij}$ are independent, and that for each $j$, $(T_{ij}, i = 1, 2, \ldots, n)$ are identically distributed with reliability function $R_j(\cdot; \theta_j)$. However, in practice, the exact values of the $T_{ij}$ are not all observable. Rather, they could be right-censored by either the system life or by the monitoring period. The observable right-censored data is

$$\mathbf{D} = \{(Z_{ij}, \delta_{ij}) : i = 1, 2, \ldots, n; \ j = 1, 2, \ldots, K\}, \qquad (10.15)$$

where $\delta_{ij} = 1$ means that $T_{ij} = Z_{ij}$, whereas $\delta_{ij} = 0$ means that $T_{ij} > Z_{ij}$. We shall suppose that for each $i \in \{1, 2, \ldots, n\}$, we have a random variable $C_i$ (e.g., the upper bound of monitoring time for the $i$th system) and also the system life $S_i$, and

$$Z_{ij} = \min\{T_{ij}, \min(C_i, S_i)\} \quad \text{and} \quad \delta_{ij} = I\{T_{ij} \leq \min(C_i, S_i)\}.$$

On the basis of this observable data $\mathbf{D}$, it is of interest to estimate the system reliability function given in Equation (10.14).

## 10.4 Estimation of a System Reliability Function

### 10.4.1 Estimation Based Only on System Lifetimes

A simple estimator of the system reliability function is to utilize only the observed system lifetimes, the $S_i$. However, these system lives may be right-censored by the $C_i$, so that we may only be able to observe $Z_i = \min(S_i, C_i)$ and $\delta_i = I\{S_i > C_i\}$ for $i = 1, 2, \ldots, n$. This setting is usually realistic in black-box testing, where the information on the internal structure is unknown and component lifetimes are unobserved. If the component lifetime distributions are governed by just one parameter vector, then a parametric approach to estimating this parameter may be possible, for example when all the component lifetime distributions are exponential with rate parameter $\lambda$; see the expressions of the system reliability functions for the four concrete systems presented earlier.

But if the the component lifetime distributions are governed by different parameters, then using only the system lifetimes for inferring these parameters, and hence the system lifetime distribution, may not be possible due to the non-identifiability of the parameters when only the right-censored system lifetimes are available. A way to circumvent this problem is to use a non-parametric estimator of the system life distribution based on the right-censored system lifetimes. In this approach we simply assume that the system reliability function belongs to the space of all system reliability functions and is not necessarily parametrized by some parameter vector. This problem has a well-known solution using the Kaplan–Meier (KM; Kaplan and Meier, 1958) or the product-limit

(PL) estimator of the system reliability function (Kalbfleisch and Prentice, 1980; Fleming and Harrington, 1991; Andersen et al., 1993).

Denote by

$$t_{(1)} < t_{(2)} < \cdots < t_{(m)}$$

the ordered $m$ distinct observations among the $Z_i$ with $\delta_i$ equal to 1, i.e., the uncensored observations. Let

$$d_l = \sum_{i=1}^{n} I\{Z_i = t_{(l)}, \delta_i = 1\} \quad \text{and} \quad n_l = \sum_{i=1}^{n} I\{Z_i \geq t_{(l)}\}.$$

Note that $d_l$ is the number of the $n$ systems that have failed at $t_{(l)}$, whereas $n_l$ is the number at risk at time $t_{(l)}$. Then, the PL estimator of $R_S(\cdot)$ is defined via

$$\hat{R}_S(t) = \prod_{\{l:t_{(l)} \leq t\}} \left[1 - \frac{d_l}{n_l}\right], \quad t \geq 0. \tag{10.16}$$

This estimator is the non-parametric ML estimator of $R_S(\cdot)$ based on the right-censored system lifetimes. It is well known (Kalbfleisch and Prentice, 1980; Fleming and Harrington, 1991; Andersen et al., 1993) that, under some regularity conditions and when $n \to \infty$, $\hat{R}_S(t)$ is asymptotically normal with asymptotic mean $R_S(t)$ and an estimate of its asymptotic variance, called Greenwood's formula, given by

$$\widehat{\text{Avar}[\sqrt{n}\hat{R}_S(t)]} = [\hat{R}_S(t)]^2 \sum_{\{l:t_{(l)} \leq t\}} \frac{d_l}{n_l(n_l - d_l)}.$$

### 10.4.2 Estimators Using Component Data

When component-level data is available, we could improve the estimation of system reliability by utilizing the information on the internal structure. Following the model setting in the previous section, the lifetimes of the components are independent with reliability function

$$R_j(t; \theta_j) = 1 - F_j(t; \theta_j), \quad j = 1, \ldots, K.$$

Classically, the estimator of the system reliability function, $R_S(t)$, is given by

$$\hat{R}_S(t) = h_\phi[\hat{R}_1(t), \ldots, \hat{R}_K(t)],$$

where $\hat{R}_j(\cdot)$ is maximum likelihood estimator of component reliability based on $(\mathbf{Z}_j, \delta_j) = \{(Z_{ij}, \delta_{ij}) : i = 1, 2, \ldots, n\}$. In the parametric setting, $\hat{R}_j(t) = R_j(t, \hat{\theta}_j)$, where $\hat{\theta}_j$ is the MLE of $\theta_j$.

For component $j$, to find $\hat{\theta}_j$, denote the density associated with component $j$ by $f_j(t; \theta_j)$, so the likelihood function based on the completely observed lifetimes of component $j$ is given by

$$L(\theta_j | t_{1j}, \ldots, t_{nj}) = \prod_{i=1}^{n} f_j(t_{ij} | \theta_j).$$

In the presence of right-censoring, the likelihood function based on the observed censored data for component $j$ becomes

$$L_j(\theta_j; (\mathbf{Z}_j, \delta_j)) = \prod_{i=1}^{n} f(z_{ij} | \theta_j)^{\delta_{ij}} R(z_{ij} | \theta_j)^{1-\delta_{ij}}.$$

This likelihood could be maximized with respect to $\theta_j$ to obtain the ML estimate $\hat{\theta}_j$, which will be a function of the $(z_{ij}, \delta_{ij}), i = 1, 2, \ldots, n$. The resulting system reliability function estimator is

$$\tilde{R}_S(t) = h_\phi[R_1(t; \hat{\theta}_1), R_2(t; \hat{\theta}_2), \ldots, R_K(t; \hat{\theta}_K)]. \tag{10.17}$$

From the theory of ML estimators for right-censored data (Andersen et al., 1993) under parametric models, as $n \to \infty$, we have that

$$(\hat{\theta}_1, \ldots, \hat{\theta}_K) \sim \text{AN}\left((\theta_1, \ldots, \theta_K), \frac{1}{n}\text{BD}[\mathfrak{I}_1^{-1}, \ldots, \mathfrak{I}_K^{-1}]\right), \tag{10.18}$$

where $\mathfrak{I}_j^{-1}$ is the inverse of the Fisher information matrix for the $j$th component, and BD means "block diagonal." Assuming no common parameters among the $K$ components, by using the Delta-Method, we find that, as $n \to \infty$,

$$\tilde{R}_S(t) \sim \text{AN}\left[R_S(t), \frac{1}{n}\sum_{j=1}^{K} I_\phi(j; t)\overset{\bullet}{R}_j{}^{\text{T}}(t)\mathfrak{I}_j^{-1}\overset{\bullet}{R}_j(t)I_\phi(j; t)\right],$$

with

$$\overset{\bullet}{R}_j(t) = \frac{\partial}{\partial\theta_j}R(t; \theta_j).$$

In principle, this asymptotic variance could be estimated, though the difficulty may depend on the structure function and/or distributional form of the $R_j$. Later we will instead perform comparisons through numerical simulations.

## 10.5 Improved Estimation Through the Shrinkage Idea

### 10.5.1 Idea and Motivation of the Estimator

Suppose that the component-level data is available. The estimation of the parameters $(\theta_j, j = 1, 2, \ldots, K)$ becomes a problem of simultaneous estimation. In this context, the problem is to estimate simultaneously the reliability of all modules in the software system. Thus, we propose an improved estimator following the idea of James and Stein. Consider estimators of component reliabilities of the form

$$\tilde{R}_j(t) \equiv \tilde{R}_j(t; c) = [\hat{R}_j(t)]^c, \quad j = 1, \ldots, K, \quad c \in \mathbb{R},$$

where $\hat{R}_j(t) = R_j(t, \hat{\theta}_j)$ is the ML estimator of $R_j(t)$ based on $(\mathbf{Z}_j, \delta_j)$ under the assumed parametric model. The system reliability estimator then becomes

$$\hat{R}_S(t; c) = h_\phi[\tilde{R}_1(t; c), \ldots, \tilde{R}_K(t; c)]$$
$$= h_\phi\{[R_1(t; \hat{\theta}_1)]^c, \ldots, [R_K(t; \hat{\theta}_K)]^c\}.$$

Notice that when $c = 1$, we obtain the standard ML estimator discussed in the preceding subsection. If $\hat{\Lambda}_j(t)$ denotes the estimator of the cumulative hazard function for component $j$, then we have

$$[\hat{R}_j(t)]^c = \exp[-c\hat{\Lambda}_j(t)].$$

Thus, we are essentially putting a shrinkage coefficient on the ML estimators of the cumulative hazard functions. We remark at this point that it is not always the case that the optimal $c^*$ is less than 1, so that in some cases, instead of shrinking, we are expanding the estimators!

The goal is to find the optimal $c$ in terms of a global risk function for the system reliability estimator. If the optimal shrinkage coefficient is $c^*$, the improved estimator of system reliability becomes

$$\hat{R}_S(t; c^*) = h_\phi\{[R_1(t, \hat{\theta}_1)]^{c^*}, \dots, [\hat{R}_K(t, \hat{\theta}_K)]^{c^*}\}.$$

However, this optimal coefficient $c^*$ may depend on the unknown parameters $(\theta_1, \dots, \theta_K)$. Therefore, we also need to find an estimator of $c^*$, denoted by $\hat{c}^*$. An intuitive and simple way to obtain an estimator of $c^*$ is to simply replace the unknown $\theta_j$ in the $c^*$ expression with their corresponding MLEs. The final estimator becomes

$$\check{R}_S(t) = h_\phi\{[R_1(t, \hat{\theta}_1)]^{\hat{c}^*}, \dots, [\hat{R}_K(t, \hat{\theta}_K)]^{\hat{c}^*}\}. \tag{10.19}$$

### 10.5.2 Evaluating the Estimators

We implement a decision-theoretic viewpoint in the determination of the optimal multiplier $c$. The performance of an estimator $a(\cdot)$ of $R(\cdot) = R_S(\cdot)$ will be evaluated under the following class of loss functions:

$$L(R, a) = -\int |R(t) - a(t)|^k h(R(t)) dR(t),$$

where $h(\cdot)$ is a positive function. Note that the negative sign is because the differential element $dR(t)$ is negative since $R(\cdot)$ is a non-increasing function. We will interchangeably use the notation $-dR(t) = dF(t)$, where $F$ is the associated distribution function. The decision problem of estimating $R_S$, when using this class of loss function, becomes invariant with respect to the group of monotonic increasing transformations on the lifetimes. A special member of this class of loss functions is the weighted Cramer-von Mises loss function given by

$$L(R, a) = -\int \frac{[R(t) - a(t)]^2}{R(t)(1 - R(t))} dR(t).$$

This loss function is a global loss function and can be viewed as the weighted squared loss functions aggregated (integrated) over time.

The expected loss or risk function is then given by

$$\text{Risk}(a) = \text{E}(L(R, a)) = \text{E}\left[-\int \frac{[R(t) - a(t)]^2}{R(t)(1 - R(t))} dR(t)\right],$$

with the expectation taken with respect to the random elements in the estimator $a(t)$. To find the optimal coefficient $c$, we plug $\hat{R}_S(t; c)$ for $a(t)$ into the risk function and then minimize this risk with respect to $c$. After this substitution, we simply denote the risk function by $\text{Risk}(c)$ to simplify the notation.

First, notice that as $c \to \infty$,

$$\text{Risk}(c) = E\left[\int \frac{[h_\phi\{[R_1(t;\hat{\theta}_1)]^c, \ldots, [R_K(t;\hat{\theta}_K)]^c\} - R(t)]^2}{R(t)(1 - R(t))} dF(t)\right]$$

$$\to E\left[\int \frac{[0 - R(t)]^2}{R(t)(1 - R(t))} dF(t)\right]$$

$$= \int \frac{R(t)}{1 - R(t)} dF(t)$$

$$= \infty.$$

Similarly, when $c \to 0$,

$$\text{Risk}(c) \to \int \frac{1 - R(t)}{R(t)} dF(t) = \infty.$$

Therefore, since for every finite $c > 0$, Risk(c) is finite, and since Risk(c) is a continuous function of $c$, there exists a value of $c$, denoted by $c^* \in [0, \infty)$, that minimizes the risk function Risk(c). Thus the existence of an optimal $c$ is demonstrated.

Using properties of maximum likelihood estimators, as $n \to \infty$, $\hat{\theta}_j$ converges in probability to $\theta_j$, $j = 1, \ldots, K$. Thus, when $n \to \infty$, the optimal $c^*$ that minimizes the expected loss converges in probability to $c^* = 1$. Therefore, for large $n$, the improved estimator becomes probabilistically close to the ML estimator.

### 10.5.3 Estimating the Optimal $c^*$

To find an approximation of the optimal shrinkage coefficient $c$, we first use a linear approximation for the expected loss based on the asymptotic properties of maximum likelihood estimators and a first-order Taylor expansion. Using a first-order Taylor expansion on the estimated system reliability function

$$\hat{R}_S(t; c) = h_\phi[\tilde{R}_1(t), \ldots, \tilde{R}_K(t)]$$

at $(\tilde{R}_1(t) = R_1(t), \ldots, \tilde{R}_K(t) = R_K(t))$ we obtain that

$$\hat{R}_S(t; c) \approx h_\phi[R_1(t), \ldots, R_k(t)] + \sum_{j=1}^{K} I_\phi[j, R_1(t), \ldots, R_k(t)][\tilde{R}_j(t) - R_j(t)],$$

where $I_\phi(j, R_1(t), \ldots, R_k(t)) \equiv I_\phi(j, t)$ denotes the reliability importance of component $j$ at time $t$ under system structure $\phi$. See Equation (10.13).

The loss function can now be written as

$$L(R, \tilde{R}) = \int_0^\infty \frac{[R(t) - \tilde{R}(t)]^2}{R(t)(1 - R(t))} dF(t)$$

$$\approx \int_0^\infty \frac{\{\sum_{j=1}^{K} I_\phi(j, t)[\tilde{R}_j(t) - R_j(t)]\}^2}{R(t)(1 - R(t))} dF(t)$$

$$= \int_0^\infty \sum_{j=1}^{K} \frac{I_\phi^2(j, t)[\tilde{R}_j(t) - R_j(t)]^2}{R(t)(1 - R(t))} dF(t)$$

$$+ \int_0^\infty \sum_{j \neq k} \frac{I_\phi(j, t)I_\phi(k, t)[\tilde{R}_j(t) - R_j(t)][\tilde{R}_k(t) - R_k(t)]}{R(t)(1 - R(t))} dF(t).$$

To evaluate the expected loss, we take expectation with respect to the $\tilde{R}_j(t)$. First we examine the following expectations:

$$E[\tilde{R}_j(t) - R_j(t)]^2 \approx \text{Var}[\tilde{R}_j(t)] + [R_j(t)^c - R_j(t)]^2$$
$$= \text{Var}[\hat{R}_j(t)^c] + [R_j(t)^c - R_j(t)]^2.$$

Employing the Delta-Method and using the asymptotic distribution of MLEs, $\text{Var}[\hat{R}_j(t)^c]$ can be approximated by $c^2 R_j^{2c-2}(t) \text{Var}[\hat{R}_j(t)]$. Thus,

$$E[\tilde{R}_j(t) - R_j(t)]^2 \approx c^2 R_j^{2c-2}(t) V_j + [R_j(t)^c - R_j(t)]^2,$$

where $V_j = \text{Var}[\hat{R}_j(t)]$. Since the components are independent, we have

$$E[\tilde{R}_j(t) - R_j(t)][\tilde{R}_k(t) - R_k(t)]$$
$$\approx \text{Cov}[\hat{R}_j(t)^c, \hat{R}_k(t)^c] + [R_j(t)^c - R_j(t)][R_k(t)^c - R_k(t)]$$
$$= [R_j(t)^c - R_j(t)][R_k(t)^c - R_k(t)].$$

Rearranging terms, the expected loss can be expressed as

$$\text{Risk}(c) \approx \int_0^\infty \sum_{j=1}^K \frac{I_\phi^2(j,t)\{c^2 R_j^{2c-2}(t) V_j + [R_j(t)^c - R_j(t)]^2\}}{R(t)(1 - R(t))} dF(t)$$
$$+ \int_0^\infty \sum_{j \neq k} \frac{I_\phi(j,t) I_\phi(k,t)[R_j(t)^c - R_j(t)][R_k(t)^c - R_k(t)]}{R(t)(1 - R(t))} dF(t).$$

To minimize this approximate risk function with respect to $c$, note the approximations

$$R_j^c(t) \approx R_j(t) + (c-1) R_j(t) \log R_j(t),$$

$$R_j^{2c}(t) = [R_j^2]^c(t) \approx [R_j(t)]^2 + 2(c-1)[R_j(t)]^2 \log R_j(t).$$

Then the approximate risk function becomes a polynomial in $c$, given by

$$\text{Risk}(c) \approx A \cdot c^2 + B \cdot c^2(c-1) + D \cdot (c-1)^2, \tag{10.20}$$

where

$$A = \int_0^\infty \sum_{j=1}^K \frac{I_\phi^2(j,t) V_j}{R(t)(1 - R(t))} dF(t),$$

$$B = \int_0^\infty \sum_{j=1}^K \frac{2 I_\phi^2(j,t) V_j \log R_j(t)}{R(t)(1 - R(t))} dF(t),$$

$$D = \int_0^\infty \sum_{k=1}^K \sum_{j=1}^K \frac{I_\phi(j,t) I_\phi(k,t) R_j(t) R_k(t) \log R_j(t) \log R_k(t)}{R(t)(1 - R(t))} dF(t).$$

The approximate optimal shrinkage coefficient $c^*$ based on the preceding approximation is the minimizer of the polynomial in Equation (10.20), which is of the form

$$c^*[R_1(t), \ldots, R_K(t)] = \frac{2B - 2A - 2D + \sqrt{(2B - 2A - 2D)^2 + 24BD}}{6B}. \tag{10.21}$$

As mentioned before, in real-life studies, the true component reliability functions, the $R_j(t)$, are unknown because of the unknown parameters $\theta_j$. Thus, when estimating $c^*$ empirically, we replace the occurrences of $R_j(t)$ in the expressions in Equation (10.21) by $R_j(t, \hat{\theta}_j)$ for $j = 1, \dots, K$.

Notice that as $n \to \infty$, $V_j = \text{Var}[\hat{R}_j(t)]$ converges to 0 according to the asymptotic property of MLEs. Thus, when $n \to \infty$, both $A$ and $B$ converge to 0. As a consequence, for large $n$,

$$\text{Risk}(c) \approx D \cdot (c - 1)^2.$$

This again demonstrates that as $n \to \infty$, the optimal $c$ converges 1.

## 10.6 Examples and Numerical Illustration

To illustrate the performance of the different estimators, we provide some numerical examples under different system structures. We generated system and component lifetime data based on series and parallel structures with ten components. A sample of ten systems was created, and component lifetimes generated according to exponential distributions with different means. The component and system lifetimes were randomly right censored. Figures 10.3 and 10.4 present the estimates for one replication, along with the true system reliability function for the series and the parallel structures, respectively.

Since this is only based on one replication, we obviously cannot make definitive comparisons of the performance of the different estimators. However, from these plots, one could see that for the series system the estimates do track the true system reliability function well. However, for the parallel system, there appears to be a big discrepancy between the true system reliability function and the three estimates. In the next section

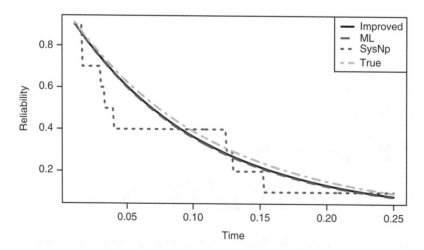

**Figure 10.3** Estimated and true system reliability functions over time for a series system with ten components. The estimators are the PLE, ML based on component data, and the improved estimator based on a system sample of size 10.

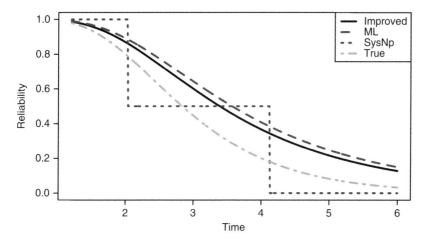

**Figure 10.4** As Figure 10.3, but for a parallel system.

**Table 10.1** Average losses of the estimators in the series system under different lifetime distributions. These are based on 1000 simulation replications. The systems had ten components, and the sample size for each replication was 10. Also indicated are the average value of the estimated shrinkage coefficient *c*.

|  | Non-censored | | Censored | |
| --- | --- | --- | --- | --- |
|  | Exp | Weibull | Exp | Weibull |
| System level | 0.0801 | 0.0858 | 0.0801 | 0.1142 |
| ML | 0.0076 | 0.0098 | 0.0103 | 0.0182 |
| Improved | 0.0069 | 0.0090 | 0.0096 | 0.0168 |
| *c* | 0.9906 | 0.9905 | 0.9856 | 0.9847 |

we present the results of simulation studies to compare the performance of the different estimators of the system reliability function.

## 10.7 Simulated Comparisons of the Estimators

In this section, we demonstrate the advantage of the improved estimator over the standard MLE and PLE using simulated data. Tables 10.1 and 10.2 present comparisons of the performance of the system-data-based PL estimator, the ML estimator, and the improved estimator, in terms of the average loss based on 1000 replications. The simulated data were for $K = 10$ components each with $n = 10$ complete or randomly right-censored observations. Different lifetime models (exponential and Weibull) and system structures were considered in these simulations. The means of the estimated shrinkage coefficients $c$ are also presented.

When component-level data is available, the component-data-based estimators perform much better then the system-data-based estimator in terms of average loss. The improved estimator dominates the ML estimator in the context of the estimated global

**Table 10.2** As Table 10.1, but for the parallel system.

| | Non-censored | | Censored | |
|---|---|---|---|---|
| | Exp | Weibull | Exp | Weibull |
| System level | 0.0871 | 0.1220 | 0.1413 | 0.1385 |
| ML | 0.0249 | 0.0300 | 0.0594 | 0.0430 |
| Shrinkage | 0.0228 | 0.0278 | 0.0524 | 0.0386 |
| c | 1.0219 | 1.0234 | 1.0289 | 1.0263 |

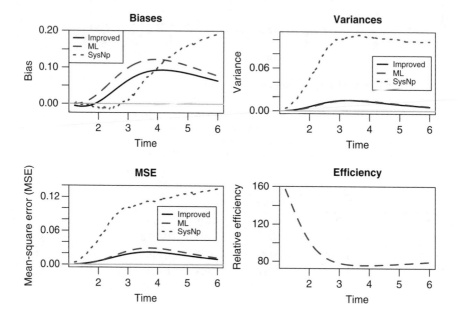

**Figure 10.5** Comparing the performance of system-data- and component-data-based estimators at different values of $t$ for a ten-component parallel system. One thousand replications were used, with each replication having a sample size of 10.

risk. The improvement is more significant when the sample size $n$ is small, in the parallel system, and when there is censoring. Observe that for the series system, the average of the $c$ values is less than 1, indicating that on average there is shrinkage going on. However, for the parallel system, the average value of $c$ is greater than 1. This is a rather surprising phenomenon to us since it indicates that we are in a sense *expanding*, in contrast to *shrinking*, the component hazard estimators.

Figure 10.5 presents the bias, variance, and mean-squared error of the estimators of the system reliability at each time point. The estimators are based on a parallel system with ten components that follow exponential lifetimes. Although there was some under-performance in the tail region, the improved estimator dominated the ML estimator in general. The relative efficiency at time $t$ is calculated via

$$\text{RelEff}(t) = [\text{MSE}(t; \text{Shrink})/\text{MSE}(t; \text{ML})] \times 100.$$

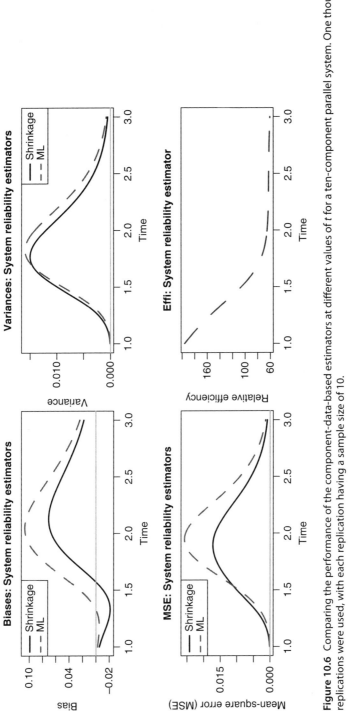

**Figure 10.6** Comparing the performance of the component-data-based estimators at different values of *t* for a ten-component parallel system. One thousand replications were used, with each replication having a sample size of 10.

Since the non-parametric method is used in the system-data-based estimation, it is unfair to compare this estimator with those using the component data. Thus, in Figure 10.6 we simply focus on the comparison between the component-data-based ML estimator and the improved estimator. Again, the estimators are based on a parallel system of ten components, with the component reliability functions being of exponential form.

## 10.8 Summary

In this chapter, we addressed the question of determining the reliability of a system with emphasis on software testing. Such reliability performance of software systems for a given time period or a workload threshold is a highly important one since with unreliable software systems, undesirable consequences such as loss of life or a negative economic impact could ensue. This chapter on system reliability views a piece of software as a coherent system and discusses ways to improve the estimation of software reliability in the setting of white-box testing. Based on the results of the simulation studies, the estimators that exploit the internal structure and component-level data dominate the estimator that only involves system-level data. Moreover, if a global assessment is used to determine the performance of estimators, it is beneficial to incorporate the idea of simultaneous estimation instead of treating each component individually. Although some approximations, based on asymptotic properties, were used in developing the improved simultaneous estimators of the component reliabilities, the resulting system reliability estimator indeed dominated the usual ML estimator, at least for the series and parallel structures, and for the component lifetime distributions (exponential and Weibull distributions) considered in the simulation studies.

Notice that in our formulation of the problem, parametric models of component lifetimes are actually *not* required, since the "shrinking" coefficient $c$ was not specified with respect to the model parameters $\theta$, but instead was just a multiplier of the cumulative hazard function. Thus, in future work, we will extend the shrinkage idea utilized here into the non-parametric framework, where the parametric ML estimators of component reliabilities will be replaced by their corresponding non-parametric PL estimators.

## Acknowledgments

We are grateful to the reviewers for their careful reading of the manuscript and for their suggestions which led to improving the manuscript. We acknowledge NIH grants P30GM103336-01A1 and NIH R01CA154731 which partially supported this research. We also thank Taeho Kim, Piaomu Liu, James Lynch, Shiwen Shen, Jeff Thompson, and Lu Wang for their comments.

## References

Ammann, P. and Offutt, J. (2008) *Introduction to Software Testing*, Cambridge: Cambridge University Press.

Andersen, P.K., Borgan, Ø., Gill, R.D., and Keiding, N. (1993) *Statistical Models Based on Counting Processes*, Springer Series in Statistics, New York: Springer-Verlag.

Barlow, R.E. and Proschan, F. (1975) *Statistical Theory of Reliability and Life Testing*, New York: Holt, Rinehart and Winston, Inc.

Boland, P.J. and Samaniego, F.J. (2001) *The Signature of a Coherent System. Mathematical Reliability: An Expository Perspective*, Boston, MA: Kluwer Academic Publishers.

Chiou, P. (1988) Shrinkage estimation of scale parameter of the extreme-value distribution in reliability. *IEEE Transactions*, 37(4), 370–374.

Coolen, F.P.A. and Coolen-Maturi, T. (2012) *Generalizing the Signature to Systems with Multiple Types of Components*, Berlin: Springer, pp. 115–130.

Doss, H., Freitag, S., and Proschan, F. (1989) Estimating jointly system and component reliabilities using a mutual censorship approach. *Annals of Statistics*, 17(2), 764–782.

Fleming, T.R. and Harrington, D.P. (1991) *Counting Processes and Survival Analysis*, Wiley Series in Probability and Mathematical Statistics: Applied Probability and Statistics, Chichester: Wiley.

James, W. and Stein, C. (1961) Estimation with quadratic loss, in *Proceedings of the Fourth Berkeley Symposium on Mathematical Statistics and Probability*, pp. 361–379.

Kalbfleisch, J.D. and Prentice, R.L. (1980) *The Statistical Analysis of Failure Time Data*, Chichester: Wiley.

Kaplan, E.L. and Meier, P. (1958) Nonparametric estimation from incomplete observations. *Journal of the American Statistical Association*, 53, 457–481.

Kenett, R. and Pollak, M. (1986) A semi-parametric approach to testing for reliability growth with an application to software systems. *IEEE Transactions on Reliability*, 35(3), 304–311.

Kenett, R. and Zacks, S. (2014) *Modern Industrial Statistics: With applications in R, MINITAB and JMP*, Chichester: Wiley, 2nd edn.

Pandey, M. and Upadhyay, S. (1985) Bayesian shrinkage estimation of reliability in parallel system with exponential failure of the components. *Microelectronics Reliability*, 25(5), 899–903.

Pandey, M. and Upadhyay, S. (1987) Bayesian shrinkage estimation of system reliability with Weibull distribution of components. *Microelectronics Reliability*, 27(4), 625–628.

Pandey, R. (2014) Shrinkage estimation of reliability function for some lifetime distributions. American *Journal of Computational and Applied Mathematics*, 4(3), 92–96.

Patton, R. (2005) *Software Testing*, Indianapolis, IA: Sams Publishing, 2nd edn.

Prakash, G. and Singh, D. (2008) Shrinkage estimation in exponential type II censored data under the LINEX loss function. *Journal of Korean Statistical Society*, 37(1), 53–61.

Samaniego, F.J. (2007) *System Signatures and their Applications in Engineering Reliability*, New York: Springer.

Siu-Keung, T. and Geoffrey, T. (1996) Shrinkage estimation of reliability for exponentially distributed lifetimes. *Communications in Statistics – Simulation and Computation*, 25(2), 415–430.

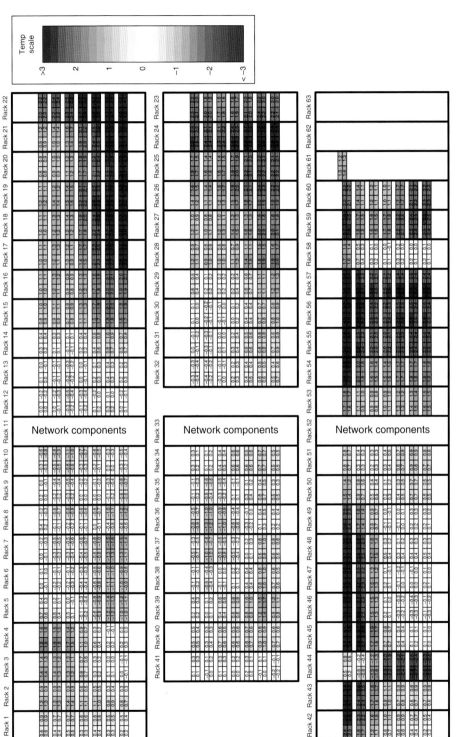

**Figure 13.2** Estimated posterior mean effect on Mustang nodes of the deployment of Wolf and other changes to the DC. This figure from Storlie et al. (2017), published by Taylor and Francis Ltd. in the *Journal of the American Statistical Association* (http://www.tandfonline.com/toc/uasa20/current), reprinted by permission of Taylor and Francis Ltd. and the American Statistical Association.

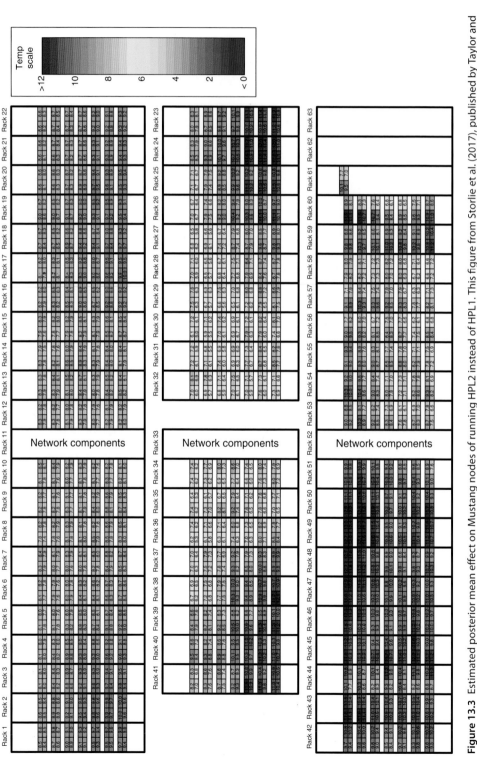

**Figure 13.3** Estimated posterior mean effect on Mustang nodes of running HPL2 instead of HPL1. This figure from Storlie et al. (2017), published by Taylor and Francis Ltd. in the *Journal of the American Statistical Association* (http://www.tandfonline.com/toc/uasa20/current), reprinted by permission of Taylor and Francis Ltd. and the American Statistical Association.

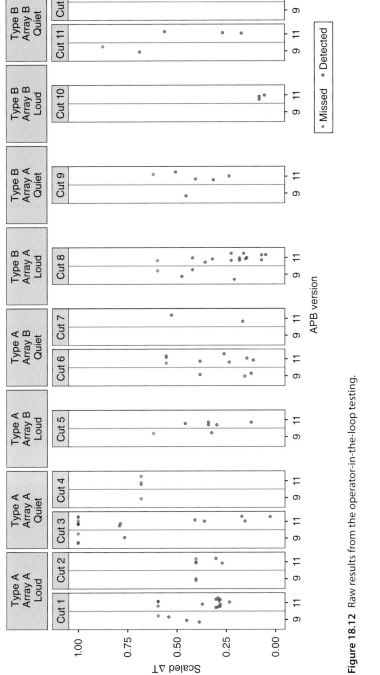

**Figure 18.12** Raw results from the operator-in-the-loop testing.

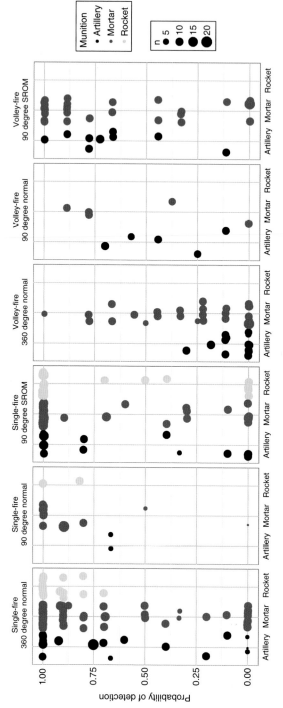

**Figure 18.17** Detection probabilities for 323 fire missions conducted during the Q-53 IOT&E.

# 11

# Decision Models for Software Testing

*Fabrizio Ruggeri and Refik Soyer*

## Synopsis

In this chapter we consider decision problems that arise during software testing/debugging processes. We present game- and decision-theoretic frameworks to develop optimal software testing strategies. We discuss the optimal choice of release time for software under different scenarios: minimization of an objective function based on testing and failure costs as well as software reliability at release time, minimization of costs subject to the achievement of an acceptable reliability level, and constrained simultaneous optimization of costs and reliability. We also present a game-theoretic approach where a software implementer and a tester are involved in an adversarial setting with both interested in producing quality software as well as maximizing their respective rewards.

## 11.1 Introduction

As described in Chapters 7, 9, and 10, software reliability models are often used for inference and prediction. Singpurwalla (1989), among others, stated that software reliability models can also be used for decision making. Chapter 3 of this volume is an example: the authors are interested in determining the optimal time between software (Mozilla Firefox, in their case) releases based on bug detection data. The development of optimal testing strategies is an important component of software reliability modeling. In particular, it is important to find those strategies determining how long software should be tested and which test cases should be used, as pointed out by Singpurwalla and Wilson (1999). In this chapter we consider different approaches, starting from the most common one based on decision theory. In Section 11.2 we first present a decision model based on the optimization of an objective function which combines testing and repair costs (that is, costs before and after release), as well as software reliability at its release time. This is followed by a sequential decision model where corrections and modifications are made to the software at the end of each test stage with the hope of increasing its reliability. In Section 11.3 we present a game theoretic approach where two players, namely an implementer and a tester, are involved in adversarial testing. The two parties are adversaries, where maximization of individual payoffs do not necessarily

*Analytic Methods in Systems and Software Testing*, First Edition.
Edited by Ron S. Kenett, Fabrizio Ruggeri, and Frederick W. Faltin.
© 2018 John Wiley & Sons Ltd. Published 2018 by John Wiley & Sons Ltd.

lead to best-quality software. In Section 11.4 a multi-objective optimization approach is presented, where the interest of the software producer is not represented by a unique objective function as in Section 11.2 but the simultaneous optimization of costs and reliability. Finally, a few concluding remarks are presented in Section 11.5, mostly based on a new paradigm, Adversarial Risk Analysis, which has been an area of increasing interest for researchers in recent years. Interested readers can find thorough illustrations of decision and game theory in many books, such as French and Rios Insua (2000) and Barron (2013).

## 11.2 Decision Models in Software Testing

The choice of an optimal release time to stop testing of software has been addressed in many papers. We will provide a review of some of the contributions in the literature, including their merits and shortcomings. The interested reader is encouraged to refer to those works for a thorough mathematical development and illustration.

The first works addressing the issue of stopping the testing process and releasing the software were not based on a formal decision-theoretic approach. Such works include Forman and Singpurwalla (1977, 1979), Okumoto and Goel (1980), Yamada, Narihisa, and Osaki (1984), and Ross (1985). Dalal and Mallows (1986) were among the first to consider a decision-theoretic approach to the problem, providing an exact but complicated solution as well as an asymptotic one. Kenett and Pollak (1986) considered parametric distributions (exponential for illustrative purposes) for the times between failures, and set an a priori threshold value of the parameter corresponding to acceptable reliability: testing is performed until the parameter reaches the threshold. Singpurwalla (1989, 1991) also followed a Bayesian decision-theoretic approach, considering a two-stage problem where the preposterior analysis requires complex computations. Later work by Ozekici and Catkan (1993) provided characterizations of the optimal release policy, whereas McDaid and Wilson (2001) considered, from a Bayesian viewpoint, the case of single-stage testing using a nonhomogeneous Poisson process model. The latter authors also considered a sequential testing problem, but the solution was not analytically tractable. Ozekici and Soyer (2001) considered optimal testing strategies for software with an operational profile as discussed in Musa (1993). It was assumed that the software was tested sequentially for a given time duration under each of the operations, and optimal testing time for each operational profile was obtained by the authors using a Bayesian approach. Boland and Singh (2002) considered a geometric Poisson model for the release time, whereas Morali and Soyer (2003) considered the testing process as a sequential decision problem in a Bayesian framework, providing characterizations of the optimal release policy. The work of the latter will be presented in Section 11.2.2.

### 11.2.1 Minimization of Expected Cost

The most common decision model in software testing considers the minimization of an objective function, i.e., an expected cost, which combines (sometimes conflicting) losses due to testing and repair costs as well as software reliability at release time. A particular, but detailed, representation of such an expected cost function at time $t$ is provided by Li, Xie, and Ng (2012):

$$E[C(t)] = c_0 + c_1 t^\kappa + c_2 \mu_y m(t) + c_3 \mu_w [m(t + t_w) - m(t)] + c_4[1 - R(x|t)], \quad (11.1)$$

where

- $c_0$ is the set-up cost for software testing
- $c_1$ is the cost of testing per unit testing time
- $\kappa, 0 < \kappa \leq 1$, is the discount rate of testing cost over time
- $c_2$ is the cost of removing a fault per unit time during the testing phase
- $c_3$ is the cost of removing a fault per unit time during the warranty phase
- $\mu_y$ is the expected time to remove a fault during the testing phase
- $\mu_w$ is the expected time to remove a fault during the warranty phase
- $t_w$ is the warranty period
- $c_4$ is the cost due to software failure
- $m(t)$ is the mean value function of the nonhomogeneous Poisson process (NHPP) describing the failure process
- $R(x|t) = e^{-[m(t+x)-m(t)]}$ is the software reliability at time $t$.

Since fixing a fault during the warranty period is more expensive than during testing, $c_3 > c_2$. Furthermore, the parameter $\kappa$ models the learning process of the testing team. Many NHPPs can be chosen as illustrated in Chapters 7 and 9.

The actual function to be minimized is obtained following a frequentist approach, by replacing parameters with their maximum likelihood estimators, whereas integration with respect to the posterior distribution of the parameters is needed when following the Bayesian paradigm.

Minimization of equations like Equation (11.1), or its expectation in a Bayesian framework, leads to determination of the optimal release time. Although very practical, such an approach is a cause for concern for some authors since uncertainty could arise from both statistical estimation errors and misspecification of costs. The problem has been addressed, e.g., by Li, Xie, and Ng (2010), who investigated the sensitivity of the software release time through various methods, like a one-factor-at-a-time approach, design of experiments, and global sensitivity analysis. As an alternative, construction of credible (confidence, in a frequentist setup) intervals about optimal release times could be pursued as in Okamura, Dohi, and Osaki (2011).

A constrained optimization problem arises when considering minimization of costs subject to the achievement of a minimal reliability level $R_0$. An example of such a problem is given by the minimization of

$$E[C(t)] = c_0 + c_1 t^\kappa + c_2 \mu_y m(t) + c_3 \mu_w [m(t + t_w) - m(t)], \quad (11.2)$$

subject to $R(x|t) \geq R_0$.

The constrained optimization could also arise as a consequence of converting the problem to maximizing the reliability $R(x|t)$ subject to a constraint $E[C(t)] \leq C^*$, where $C^*$ is the maximum allowable cost level.

Simultaneous optimization of multiple objectives will be discussed in Section 11.4.

### 11.2.2 A Sequential Decision Model

We now consider the protocol where testing is done sequentially in stages until fault detection or a prespecified testing time, whichever occurs first. Software is then corrected and modified at the end of each test stage, aiming to increase its reliability. Let $T_i, i = 1, 2, \ldots$ denote the (possibly censored) life-length of the software during the $i$th testing stage, i.e. after the $(i-1)$st modification has been made to it. Each $T_i$ follows the same distribution but, in general, with different parameters. In particular, we consider, for illustrative purposes, an exponential model with failure rate $\lambda_i$ which changes from one stage to another as a result of the modifications made to it after each stage.

At the end of each stage, after modifying the software, a decision is taken about termination of the debugging process, based on the accumulated information $T^{(i)} = (T_i, T^{(i-1)})$, where $T^{(0)}$ is the available information before testing. Morali and Soyer (2003) assumed that the evolution of the $\lambda_i$ can be described by a Markovian model and considered, after each stage $i$, the loss function

$$\mathcal{L}_i(T^{(i)}, \lambda_{i+1}) = \sum_{j=1}^{i} \mathcal{L}_T(T_j) + \mathcal{L}_S(\lambda_{i+1}), \tag{11.3}$$

where $\mathcal{L}_T(\cdot)$ is the loss related to the life-length for each individual stage, and $\mathcal{L}_S(\cdot)$ is the loss associated with stopping and releasing the software after the stage. The latter loss depends on the current value of the parameter (a proxy for the current software reliability), and it is an increasing function of this since the parameter is proportional to the number of bugs present in the software. It should be observed that the loss due to releasing the software before any testing, i.e., $\mathcal{L}_0$, is just a function of $\lambda_1$.

Morali and Soyer (2003) presented the stopping problem as a sequential decision problem described by the $m$-stage decision tree given in Figure 11.1. The solution of the decision problem requires dynamic programming, taking expectation at random nodes and minimizing the expected loss at the decision nodes. At each decision node $i$, the additional expected loss associated with the STOP and the TEST decisions are given by $E[\mathcal{L}_S(\lambda_{i+1})|T^{(i)}]$ and $E[\mathcal{L}_T(T_{i+1})|T^{(i)}] + L_{i+1}^*$, respectively, where

$$L_i^* = \min\{E[\mathcal{L}_S(\lambda_{i+1})\,|\,T^{(i)}]\,,\ E[\mathcal{L}_T(T_{i+1})\,|\,T^{(i)}] + L_{i+1}^*\} \tag{11.4}$$

for $i = 0, 1, \ldots$ It can be shown that the optimal decision at decision node $i$ is the one associated with $L_i^*$.

In Figure 11.1, the maximum number of testing stages, $m$, can be considered infinite with $L_{m+1}^* = \infty$. It is worth mentioning that even for the case of finite $m$ the calculation of $L_i^*$ in Equation (11.4) is not trivial as it involves implicit computation of expectations and

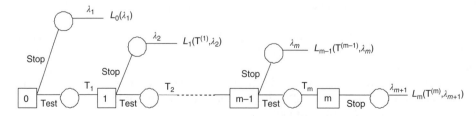

**Figure 11.1** The $m$-stage decision tree for the optimal release problem.

minimizations at each stage. Morali and Soyer (2003) studied the possibility of developing one-stage-ahead optimal stopping rules by using results from van Dorp, Mazzuchi, and Soyer (1997), and illustrated implementation of their approach using simulated as well as actual software failure data.

## 11.3   Games in Software Testing

Previously we discussed approaches based either on optimal stopping rules minimizing the total cost in testing, possibly combined with specific reliability constraints, or on optimal allocation of testing efforts. In literature, minor emphasis has been placed on the aspect of competition between rival producers of software. The adversarial nature of the problem has been considered by Zeephongsekul and Chiera (1995) and Dohi, Teraoka, and Osaki (2000), who used a game-theoretic approach. The former authors considered (for simplicity) the case of only two competitors, labeled $i$, $i = 1, 2$, producing software performing the same set of tasks and with life cycle length not exceeding $T$.

Player $i$, $i = 1, 2$, can decide to release the software at any time $t$ in $[0, T]$, and succeeds in selling the product with probability $A_i(t)$. The functions $A_i(t)$, $i = 1, 2$, are supposed to be continuously differentiable, concave, and such that $A_i(0) = A_i(T) = 0$ with a unique maximum at time $\eta_i$. The choice of such functions is made not only for mathematical convenience but is also justified by the actual behavior. More specifically, the success probability is expected to be close to 0 both at the beginning and the end of the life cycle $[0, T]$, because of initial poor reliability and final obsolescence, respectively.

Zeephongsekul and Chiera (1995) made an assumption typical of game theory papers, i.e. that both players know the functions $A_i(t)$, $i = 1, 2$. Such an assumption is sometimes unrealistic, and it could be addressed within an Adversarial Risk Analysis (ARA) framework where each player has just guesses about the other's probabilities. In this case the problem would be seen from the viewpoint of one player, say 1, and she would have her opinion on the probability $A_1(t)$, either as a unique function or a distribution on it, whereas she should elicit a distribution on the space of the possible functions $A_2(t)$. More details on the ARA approach can be found in Banks, Rios, and Rios Insua (2015).

Zeephongsekul and Chiera (1995) considered a cost function quite similar to the one proposed by Okumoto and Goel (1980), without assuming an infinite cycle length. They considered the expected cost $c_i(t)$ incurred by player $i$ in releasing the software at time $t$ as

$$c_i(t) = c_{1i}t + c_{2i}m(t) + c_{3i}(m(T) - m(t)),\qquad(11.5)$$

where $c_{1i}$ is the cost of testing per unit time, $c_{2i}$ the cost of removing a fault during testing, $c_{3i}$ the cost of removing a fault during operation, and $m(t)$ the expected number of faults detected up to time $t$. Since fixing an error is more expensive after release than before it, $c_{3i} > c_{2i}$ is assumed. Such an assumption, combined with the choice of an increasing, concave, and differentiable $m(t)$, with $m(0) = 0$, implies that the function $c_i(t)$ is convex with a minimum at $\gamma_i$ such that

$$m_i'(\gamma_i) = \frac{c_{i1}}{(c_{3i} - c_{2i})}.$$

The authors assumed that $T$ is sufficiently large such that $\gamma_i < T$.

The novelty of the work by Zeephongsekul and Chiera was that they introduced the notion of competition where a player has to consider not only her cost but also the action of a competitor, and his costs. If player 1 releases the software at time $x$ and player 2 at time $y$, then $M_i(x, y)$ is the expected unit profit to player $i$. Such profit is the consequence of the difference between the unit price $p_i > 0$ of the software produced by player $i$ and the cost incurred $c_i$ given by Equation (11.5). In general, it holds that $M_1(x, y) \neq M_2(x, y)$, where

$$M_1(x, y) = \begin{cases} p_1 A_1(x) - c_1(x) & 0 \leq x < y \leq T, \\ p_1(1 - A_2(y))A_1(x) - c_1(x) & 0 \leq y < x \leq T, \end{cases}$$

and $M_2(x, y)$ can be described similarly. It is obvious that this is a non-zero sum game.

The simplifying assumption behind this model is that the success of a player in selling her product implies the impossibility for the other player to sell his own. The model could be acceptable when there is just one customer, interested in a unique purchase. More complex models could be possible in other scenarios, e.g. when the existence of many customers can provide opportunities for both players. A possibility could be offered by lowering the price of the last marketed software. Therefore, $p_i$ should in this case be a function of $A_{3-i}(x)$, $i = 1, 2$.

The expected utility for each player depends on four factors:

- testing cost
- fault removal cost (with a higher one for faults after release)
- reliability of the released product
- release of software by the other player.

In general, late releases imply higher reliability and higher costs combined with the risk of an earlier successful release by the competitor.

The final goal of Zeephongsekul and Chiera (1995) consists of finding the optimal release policies among Nash equilibrium points. Dohi, Teraoka, and Osaki (2000) found their solution restricted to just a particular case and computationally quite intractable, and proposed a different approach addressing those issues.

Alternative approaches have been proposed in the literature. It is worth mentioning the work by Feijs (2001), who considered a game where the two players have very specific, distinct roles. One is an implementer ($I$) who is rewarded if she delivers an (almost) error-free piece of software, while the other is a tester ($T$) who is rewarded only if he performs a thorough testing job. Feijs considers an Idealized Testing Game (ITG), which is a two-player strategic game where each player has two choices about performing their jobs: *bad* (B) or *good* (G) quality. Intermediate quality levels are also possible, as discussed in Feijs (2001). A pair $(x, y)$ of payoffs is associated with each combination of job quality, where $x$ is the payoff for $I$ and $y$ for $T$. The payoff matrix is given by

|   | B | G |
|---|---|---|
| B | (2, 2) | (0, 3) |
| G | (3, 0) | (1, 1) |

The payoffs $(0, 3)$ and $(3, 0)$ have a clear interpretation: this is the case where one player ($I$ in the first case) is unable to detect faults whereas the other succeeds. This explains why one player gets the lowest payoff in the matrix and the other gets the highest. The payoffs $(2, 2)$ and $(1, 1)$ have a less evident explanation. The former pair corresponds to the case of a poor-quality job by both players: they are of course penalized because they are unable to discover many faults, but they are rewarded since they did not make significant effort. The latter pair corresponds to the opposite situation: the players are rewarded because they were able to discover many causes of faults but that occurred because of a large effort (and related cost).

The idea behind this model is that implementer $I$ and tester $T$ choose their performance levels simultaneously. Once the problem has been structured and the software specification is available, $I$ starts implementing the software whereas $T$ starts looking for test cases and describing them thoroughly. In this way, the overall project duration is reduced as much as possible, leaving just the actual testing phase after the implementation of the software.

An alternative model, briefly described in Feijs (2001), corresponds to the case in which $I$ chooses first the quality of her job and then $T$ decides what to do after observing what $I$ has done. This is another common situation where the tester can decide to make extra efforts if he believes the implementer did a poor job (or vice versa). Considering the previous payoff matrix, then $I$ has two possible choices: B or G. In the former case $T$ is left with just the first row of the matrix, and he can choose between payoffs $(2, 2)$ and $(0, 3)$. Of course he would choose the second one, leaving $I$ a payoff of 0. Should $I$ choose good quality, then the choice of $T$ would be between the payoffs $(3, 0)$ and $(1, 1)$, with an obvious preference for the second one. From the viewpoint of the implementer, she gets a payoff of 0 if she performs a poor-quality job and 1 if the quality is good. Therefore, rational behavior leads to a payoff pair $(1, 1)$ corresponding to good quality jobs by both players.

Those games, resembling the famous Prisoner's Dilemma, could be rethought in an ARA framework, as described earlier.

As pointed out by Feijs (2001), the actual software development is more complex and requires balancing of different aspects like development time, code size, reusability, etc. Another critical aspect is the large number of possible test cases and the search for a restricted number of them that could lead to a significant improvement of the software quality and, at the same time, can be performed in a reasonable amount of time: good quality and quick delivery are often clashing goals!

Kukreja, Halfond, and Tambe (2013) considered software testing as a security game where a Defender uses her limited resources for protecting public infrastructure (e.g. airports) from an Attacker (e.g. a terrorist). In particular, they define a testing game in which the tester $T$ is playing the role of Defender and the implementer $I$ is the Attacker.

The tester $T$ is willing to ensure high software quality and, therefore, is interested in developing a testing strategy that will execute the most efficient test cases, given constraints on resources and/or time. Therefore, $T$ is the Defender of the software quality, whereas $I$ is treated as an Attacker who might produce software full of bugs that could have a negative impact on its quality. The tester should detect such bugs before software release. Implementers are Attackers not because they get a reward from poor-quality software but because they might get credit for the quick development of the software rather than for a delayed one, even if due to careful testing. Kukreja, Halfond, and Tambe (2013) associated utilities, for both players, to the test cases and computed a distribution that maximizes the tester payoff. As in security games, the defender $T$ may employ nondeterministic strategies, i.e., selecting a particular action with some probability. This use of a distribution decreases the predictability of $T$, making the task of the Attacker $I$ harder.

## 11.4 Multi-Objective Optimization in Software Testing

In Section 11.2.1 we formulated the optimization problem to determine the optimal release time under three different scenarios: an unconstrained problem where the objective function depended on both (testing and failure) costs and software reliability, and two constrained problems where either costs were minimized subject to the achievement of a minimal reliability level or reliability was maximized subject to a fixed cost level. Although simple to formulate, these approaches can hardly describe management's attitude, especially the unconstrained problem where there is the issue of how much reliability should be weighted with respect to cost. A more natural approach, although computationally more complex, consists of minimizing costs and maximizing reliability simultaneously. Therefore, the problem becomes a multi-objective optimization one, where the goal is to find the optimal release time $t^*$ solution of

$$\max_{t>0} R(x|t) \,\&\, \min_{t>0} E[C(t)], \tag{11.6}$$

where $x$ is the useful life or warranty time of the software once released.

Different approaches have been presented in the literature, as discussed by Li, Xie, and Ng (2012). A first approach is based on trade-off analysis, which has the goal of identifying nondominated actions which are solutions to Equation (11.6). In the current context, we say that an action $a$ (in this case a release time) is nondominated if there is no other action $b$ such that $R_b(x|t) \geq R_a(x|t)$ and $E[C_b(t)] \leq E[C_a(t)]$, with strict inequality for at least one of them. In this case the subscripts describe which action we are referring to. The nondominated solutions, also called Pareto optimal solutions, are not inferior to any other solution. The approach simplifies the management decision process since it reduces the search from all feasible solutions to a subset where a rational compromise can be made among the different solutions.

Multi-attribute utility theory (MAUT) addresses the problem of having different objectives in different scales and units; it solves the problem by considering weights and a single utility function. For each attribute $d_i$, $i = 1, \ldots, n$ (reliability and costs in our context), a utility function $u(d_i)$ is specified and then the multiattribute function

$$U(d_1, \ldots, d_n) = \sum_{i=1}^{n} w_i u(d_i)$$

is considered, where the $w_i$, $i = 1, \ldots, n$, are the importance weights assigned to each utility function. In this way each attribute is converted to a value in $[0, 1]$ through its utility function, where the weights, adding up to 1, determine the relative importance of each attribute. Methods for the elicitation of each utility function and the corresponding weight have been proposed, although it should be remarked that they are subject to the same sensitivity concerns discussed earlier. In practice, the utility is often chosen as a linear function for each attribute, and management just has to provide upper and lower values on the corresponding attribute. On the other hand, the choice of the weights could be obtained by comparing a certain scenario and a lottery. A thorough illustration of the approach can be found in Keeney and Raiffa (1976).

## 11.5 Summary

In this chapter we have presented several game- and decision-theoretic formulations of problems in software testing. We have not taken a firm position about the choice between a frequentist or a Bayesian approach, since we recognize merits in both. Nonetheless, we believe that the Bayesian approach can make better use of available information and preferences, also providing a more coherent theoretical approach. As mentioned in the chapter, Adversarial Risk Analysis is an emerging field, thoroughly described in Banks, Rios, and Rios Insua (2015), which could be successfully applied in this context.

## References

Banks, D., Rios J., and Rios Insua D. (2015) *Adversarial Risk Analysis*. Boca Raton, FL: Chapman and Hall/CRC.

Barron, E.N. (2013) *Game Theory: An Introduction*, 2nd edn. Chichester: Wiley.

Boland, P.J. and Singh, H. (2002) Determining the optimal release time for software in the geometric Poisson reliability model. *International Journal of Reliability Quality and Safety Engineering*, 9, 201–213.

Dalal, S.R. and Mallows, C.L. (1986) When should one stop testing software? *Journal of the American Statistical Association*, 83, 872–879.

Dohi, T., Teraoka, Y., and Osaki, S. (2000) *Journal of Optimization Theory and Applications*, 105, 325–346.

Feijs, L. (2001) Prisoner's dilemma in software testing, in *Proceedings of 7e Nederlandse Testdag*, TU/e, Eindhoven, The Netherlands, pp. 65–80.

Forman, E.H. and Singpurwalla N.D. (1977) An empirical stopping rule for debugging and testing computer software. *Journal of the American Statistical Association*, 72, 750–757.

Forman, E.H. and Singpurwalla N.D. (1979) Optimal time intervals for testing hypotheses on computer software errors. *IEEE Transactions on Reliability*, R-28, 250–253.

French, S. and Rios Insua, D. (2000) *Statistical Decision Theory: Kendall's Library of Statistics 9*. London: Arnold.

Keeney, R.L. and Raiffa, H. (1976) *Decisions with Multiple Objectives*. New York: Wiley.

Kenett, R. and Pollak, M. (1986) A semi-parametric approach to testing for reliability growth, with application to software systems. *IEEE Transactions on Reliability*, R-35, 304–311.

Kukreja, N., Halfond, W.G.J., and Tambe, M. (2013). Randomizing regression tests using game theory, in *Proceedings of the 28th IEEE/ACM International Conference on Automated Software Engineering (ASE)*.

Li, X., Xie, M., and Ng, S.H. (2010) Sensitivity analysis of release time of software reliability models incorporating testing effort with multiple change-points. *Applied Mathematical Modelling*, 34, 3560–3570.

Li, X., Xie, M., and Ng, S.H. (2012) Multi-objective optimization approaches to software release time determination. *Asia-Pacific Journal of Operational Research*, 29, 1240019-1–1240019-19.

McDaid, K. and Wilson, S.P. (2001) how long to test software. *Statistician*, 50, 117–134.

Morali, N. and Soyer, R. (2003) Optimal stopping in software testing. *Naval Research Logistics*, 50, 88–104.

Musa, J.D. (1993) Operational profiles in software reliability engineering. *IEEE Software*, 10, 14–32.

Okamura, H., Dohi, T., and Osaki, S. (2011) Bayesian inference for credible intervals of optimal software release time, in Kim, T. et al. (eds.), *Software Engineering, Business Continuity, and Education*. Berlin: Springer.

Okumoto, K. and Goel, A.L. (1980) Optimum release time for software systems, based on reliability and cost criteria. *Journal of Systems and Software*, 1, 315–318.

Ozekici, S. and Catkan N.A. (1993) A dynamic software release model. *Computational Economics*, 6, 77–94.

Ozekici, S. and Soyer, R. (2001) testing strategies for software with an operational profile. *Naval Research Logistics*, 48, 747–763.

Ross, S.M. (1985) Software reliability: The stopping rule problem. *IEEE Transactions on Software Engineering*, SE-11, 1472–1476.

Singpurwalla, N.D. (1989) Preposterior analysis in software testing, in Dodge, Y. (ed.), *Statistical Data Analysis and Inference*. New York: Elsevier.

Singpurwalla, N.D. (1991) Determining an optimal time interval for testing and debugging software. *IEEE Transactions on Software Engineering*, SE-17, 313–319.

Singpurwalla N.D. and Wilson S.P. (1999) *Statistical Methods in Software Engineering*. New York: Springer-Verlag.

van Dorp, J.R, Mazzuchi, T.A., and Soyer, R. (1997) Sequential inference and decision making during product development. *Journal of Statistical Planning and Inference*, 62, 207–218.

Yamada, S., Narihisa, H., and Osaki, S. (1984) Optimum release policies for a software system with a scheduled software delivery time. *International Journal of Systems Science*, 15, 905–914.

Zeephongsekul, P. and Chiera, C. (1995) Optimal software release policy based on a two-person game of timing. *Journal of Applied Probability*, 32, 470–481.

# 12

# Modeling and Simulations in Control Software Design

*Jiri Koziorek, Stepan Ozana, Vilem Srovnal, and Tomas Docekal*

## Synopsis

This chapter focuses on using various modeling and simulation approaches in a process of control system design. During control system design, important steps could be realized by modeling instead of using a real system. Modeling and simulation can increase the effectiveness of the design and can help to avoid faults in control software. It can also enable verification and testing of control software when testing on the real system is not possible. There are a lot of approaches to building and using models; the most typical approaches for use in control system design are described. There are also examples of modeling tools that are used in industrial area. The chapter also presents two case studies focused on processor-in-the-loop and software-in-the-loop modeling.

## 12.1  Control System Design

Control system design is a complex process that has a major impact on the quality of control applications, system reliability, sustainability, and further extension of the system. Control systems can be understood as a subset of the general concept of the system, which is defined in the IEEE (1990) standard glossary as "a collection of components organized to accomplish a specific function or set of functions." Control systems have certain specifics that distinguish them from other systems. These typically include the following features:

- Interaction with the environment, where the environment can be represented by other systems, hardware and software entities, people, the physical world, etc. The interaction with the environment is often done by sensors and actuators, which transmit information from the physical environment to the control system and vice versa. The interaction with the environment or other systems is usually implemented using communication systems.
- Controllers typically have to react with the surrounding environment in real time. A real-time system is understood as "a system or mode of operation in which computation is performed during the actual time that an external process occurs, in order that the computation results can be used to control, monitor, or respond in a timely manner to the external process" (IEEE, 1990). Real-time systems can be divided into two

*Analytic Methods in Systems and Software Testing*, First Edition.
Edited by Ron S. Kenett, Fabrizio Ruggeri, and Frederick W. Faltin.
© 2018 John Wiley & Sons Ltd. Published 2018 by John Wiley & Sons Ltd.

main groups: soft real-time systems, where the usefulness of a result degrades after its deadline, thereby degrading the system's quality of service, and hard real-time, when missing a deadline is a total system failure (Schlager, 2011). The fundamental property of a real-time system is its response time, on which there are high demands, especially for hard real-time systems.

The design of control systems can be based on general methods and approaches of engineering system design, while respecting the specific characteristics of control systems. An example of such techniques is an approach called the V-model. This is a graphical representation of the process of system design showing the group of models used from facilitating the understanding of a complex system to a detailed description of the various stages of the system lifecycle. It has a characteristic V shape in which the left side represents the decomposition and the specification of the system, while the right side represents system integration and validation.

While using the model for the control system design, or respectively control software design, the model describes the various steps from the definition of the system requirements to the implementation of the software. But it also describes the testing phase of each step. The V-model can be understood as an extension of the so-called Waterfall model, which is also used in control software design (Pressman, 2010). An example of the V-model structure for the design of control software is shown in Figure 12.1.

During the verification/development phase it is evaluated whether the control system or the control software fulfills the requirements, specifications, and criteria. The validation/testing phase deals with evaluating whether the system meets the requirements of the customer or the client.

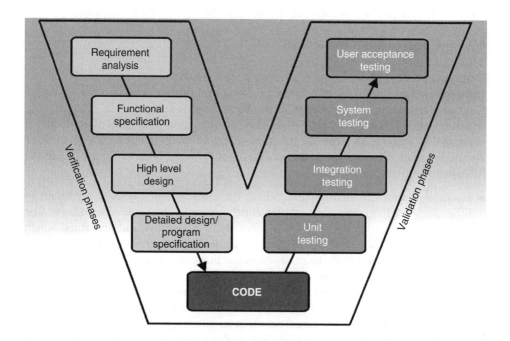

**Figure 12.1** Control system design by the V-model (systems development lifecycle).

Testing begins at the lowest level when the individual modules of the designed system are tested. These blocks are the results of low-level design, and could be program blocks of some control software. Then, integration tests follow. Groups of several blocks are tested simultaneously, including their cooperation and communication. These structures of blocks are the results of high-level design. The next step is the testing of whole designed system to see if it meets the functional specifications. The final step of the validation phase is the acceptance test, which tests whether the system meets the requirements of the users or clients. This phase is followed by installation, commissioning, operation, and maintenance.

Various simulation techniques and simulation tools can be used in the process of validation/testing. Some of these techniques are described in following text.

## 12.2 Modeling and Simulation of Systems

Analysis of the properties of large and complex systems is virtually impossible using analytical computation; it is often necessary to examine the behavior of different devices in extreme situations which must not occur at the actual devices as they could lead to serious damage, or it is necessary to analyze the properties of large machines before their production. In such cases it is very effective to work with a model instead of the actual device (Zeigler, 2000). Modeling reality is one of the major goals of human knowledge. Such models can be divided into physical and mathematical; a physical model is a representation by a real system, e.g. a small car in an aerodynamic tunnel, whereas a mathematical model consists of an abstract system of mathematical relations that describe the properties of the examined device, e.g. a system of differential equations. The implementation of mathematical models and experimenting with them is the subject of simulation, which has nowadays developed into a separate science discipline that uses knowledge from system theory, cybernetics, control theory, and mathematics.

Simulation represents a tool of a human knowledge, as it allows analysis of the properties and behavior of objects of human knowledge – objects of simulation. The subjects of system simulation are systems that are defined with the objects of knowledge. The basic principle of simulation is replacement of the original system by another system, the so-called simulation model, and retrospective application of findings gained from the simulation model to the original system. Application of general principles of simulation systems for solving specific tasks requires a systematic approach.

The development of computer technology has meant a significant extension of the solvability of mathematical models; it has made it possible to automate the calculation of mathematical model relations. The technical implementation of a mathematical model of an object by means of a computer is referred to as a computer model. Previously, the application of a mathematical model of an object used to be limited by the existence of analytic solutions. Computer realization of a mathematical model of a system makes it possible to automate the calculation of solutions of model equations, whereby the user only takes care of providing inputs to the mathematical model and processing model outputs generated by the computer (post-processing). With sufficient software support at our disposal we can make experiments with the mathematical model of a system in the same manner as with a real object. We can observe the behavior of the system under various conditions, in emergency situations, and in states that must not occur (especially important in the case of critical technology units such as nuclear reactors). Working with a mathematical model of a system and its computer model then has the same character as experimenting with the actual device. A characteristic feature of this activity is imitation of real phenomena by their computer model and experimenting with them, and therefore it is referred to as a simulation.

The use of computer models and simulation methods cannot be reduced just to the implementation of a mathematical model, i.e. the solution of the equations of a mathematical description of the problem. The term *simulation* means all phases of a discovery process, which results in equivalence between a computer model and the investigated object in terms of the properties and manifestations considered as significant, with a precision sufficient for its intended purpose. The main phases of the simulation process are as follows:

- definition of a system as the analyzed object
- construction of a computer model, its implementation
- verification of the correspondence between the computer model and the object
- customized experiments with the computer model
- application of the results of the simulation experiments to the analyzed object.

Genuine experimentation with a computer model as a substitute for a real sample of the object can only be performed if the calculation itself is a reliable, reasonably fast, and easy-to-use automated procedure. The better the equipment of a simulation workplace, the easier and more reliable it is for a worker to cope with a computer model as a substituted sample object and gain valuable information.

### 12.2.1 The Concept of Modeling

Modeling is one of the commonest ways to represent the outside world, undertaken to explore existing objective laws within it. Modeling is an experimental information process in which the explored system (original, object, part) is unambiguously represented by another system, physical or abstract, called the model, according to certain criteria. The modeling of dynamic systems with direct or indirect feedback effects on the examined object is called simulation.

The introduction of the concept of simulation helped to narrow down the set of models having importance in the development of scientific knowledge from the very large set of different models. This includes models of dynamic systems and processes, including models of the limit steady states in which the studied system may be operated. It does not matter whether the model is implemented on a computer or by other technical means. By definition, the term simulation can always be replaced by the more general term modeling, but not vice versa. The most significant issue that applies for simulation is a transfer of knowledge acquired by a simulation back to the analyzed object. Note that the terms simulation and computation are often mistakenly confused. Computation is only a routine sequence of computing operations without expressing a connection to the solved process. Improper use of the word computation instead of simulation thus reduces the practical significance of the simulation based on a model's relation to the explored object.

Various types of modeling are associated with various forms of similarity and analogy; similarities can be seen as an unambiguous mutual assignment between various systems in terms of their structure, properties, and behavior. Physical resemblance expresses the similarity between systems and processes of the same physical nature, and apart from geometric similarity it also includes similarity in parameters and system state variables. Mathematical resemblance expresses the similarity between systems and processes having the same mathematical description. In the case of physically different systems and processes it is called analogy. Cyber (functional) resemblance expresses mathematical similarity in terms of the outward behavior of systems. According to these three kinds of similarities it is possible to divide modeling into physical, mathematical, and cybernetic modeling. Cybernetic modeling uses black-box models. This concept, introduced by N. Wiener, means that the systems do not provide information about their internal structure, but only their outward behavior. The opposite of a black box is a white box, or rather transparent, giving information about the internal structure of the system and the ongoing process. Similarly, you can implement, e.g., a gray box, providing partial information about the internal structure of the system.

A superior general term for modeling is experiment. This is a purposeful practical activity leading to the development of knowledge. An experiment can be physical or mental. It may be scientific, industrial, educational, or something else, each characterized by a different level of abstraction. In terms of society, particular importance is given to scientific experiments, based on measurements and modeling. Measurement is understood as the information process having its essence in the sensing, transfer, and transmission of information from the object to the user. An important property of an experiment is its repeatability under the same conditions. A model experiment is used to detect information about the investigated object. The experimental technique involves measuring, computing, and control technologies. Integrating the discipline in relation to the analysis and control of complex systems is technical cybernetics. Experimental pieces of information (data) are symbolically expressed by the results of the experiment. The plan of an experiment includes an algorithm for the experiment, work procedures, and the nature and magnitude of the changes of variables. The algorithm of an experiment is understood as a set of unambiguous rules whose fulfillment leads to achievement of the goal of the experiment. A computer experiment has three basic roles: processing large amounts of experimental data, serving as a model agent, and controlling the process of experimentation.

Reliable determination of the properties of the investigated system has a fundamental importance for modeling. This is closely related to the credibility of a mathematical simulation model, and therefore of the modeling results. Therefore, identification represents an important phase in the modeling and simulation of systems, on which the effectiveness of the modeling largely depends. Identification is an experimental method of determining the essential characteristics of a process, allowing the construction of a mathematical model. The characteristics of a system are particularly understood as the different physical properties, structure, and parameters of the system and the ongoing process within. Diagnosis means identifying the system status, and in particular deviations from the expected state.

Obtaining a general mathematical model requires a detailed analysis of the physical nature of the problem and of the model knowledge gained. Only in this way is it possible to switch to simulation after transformation, to gather reliable information about system behavior.

### 12.2.2 Motivation – Modeling as a Discipline

The emergence of the new discipline of modeling systems on computers was induced by the practical needs of applied cybernetics. For many years, in addition to the ordinary requirements for computations for different tasks, assignments of the following types appeared: the client wants to study a system for which they have a mathematical description. They want to know whether the mathematical description is sufficiently accurate compared to the original, how to modify or extend it, and what changes should be made to the system in order to achieve the required behavior. Other common issues are, for example, questions of how much you can trust the results that we get after the solution of the mathematical representation of a system, whether the mathematical description is sufficient, how far the behavior of a system is from optimal behavior, etc. These qualitatively new requirements change the role of a programmer and a computer. The programmer becomes a co-investigator of the problem, and the computer is no longer a mere computing device itself, but it becomes a model of the original built on the basis of a mathematical description of the system. The work involved in creating a mathematical description and its implementation on a computer has been named modeling, cybernetic modeling, simulation systems, and the like (Woods, 1997).

An essential feature of modeling is the connection of computational operations with the study of the system. The programmer becomes a co-creator of the mathematical description of the system, monitoring and participating in all analyses. For smaller jobs it is optimal if the client is a programmer as well. For complex modeling tasks it is optimal to have a team of workers (programmers, engineers, mathematicians).

Over the years, a number of findings have been accumulated regarding the application of simulation, and their gradual generalization has formed a basis for the creation of modeling as an independent discipline within technical cybernetics. The theoretical bases of modeling are the theory of computers and their programming, the theory of automatic control, probability and statistics, numerical analysis, the theory of the stability of differential equations, game theory, queuing theory, and other theoretical disciplines.

### 12.2.3 Development of the Digital Computer as an Optimal Means of Computation

The digital computer is universal, and the most common means for creating digital models. An operational unit has a fixed structure, and apart from various logical operations and the transfer of numbers it only allows implementation of basic arithmetic operations with binary numbers. All information about investigated systems, about examined processes, and about the strategy and progress of a simulation is included in the program management of the operational units. The preparation of a simulation model thus consists of the transformation of a mathematical model into a form that can be described by means of formulation in the chosen programming language.

Conversion from that programming language into machine language is entrusted to a computer. A program in machine language is a model of certain classes of problems. Along with the specific input data, which can be stored in computer memory or prepared in an input device, the program then creates a digital simulation model of the examined process.

The speed of the simulation when compared to dedicated digital and analog models is limited both by the serial arrangement of the simulation model and the serial performing of logic operations upon execution of particular operations with arithmetic numbers. The serial arrangement of a model is given by the concept of digital computers, whereas the method of implementation of the various operations is given by the chosen technical solution. However, it is regularly serial with respect to costs, size, power consumption, and reliability. In terms of the needs of modeling, all these properties are disadvantageous. On the other hand, the digital computer assures high accuracy of the results and allows extensive automation of the preparation of a simulation model using programming languages of different levels. When implementing the model directly in machine language or assembler, the user must directly transform the mathematical model into a form suitable for implementation. This procedure is necessary when using certain types of microprocessors. When using some of the higher-level programming languages a user implements a mathematical model partly transformed into an operational model. The routine tasks involved in the transfer of a model implemented in a programming language into a simulation model are performed by a computer, or an assembler, respectively. Higher-level programming languages allow symbolic notation of a large range of different functions, and thus they partly automate a transformation of a mathematical model into an operational model. They allow the processing of complex mathematical objects such as multidimensional arrays, complex numbers, logical variables, or matrices as necessary. Creation of the program itself can also be automated – it is possible to implement quite complex algorithms from a flow diagram. Various languages contain extensive built-in libraries with subroutines. Inclusion of well-tested subroutines enables the use of partial transformations of a mathematical model.

The use of a higher-level simulation language significantly helps to reduce the effort needed for the transformation of a model. Simulation languages allow the automation of the transformation of the mathematical model of a certain type of task. They make it easier to cope with time management, generation of random variables, statistical processing of results, etc. The aim is to automate all transformations of a mathematical model. However, as a single language cannot cover all types of mathematical models, hundreds of simulation languages have been gradually created, each oriented to a specific, sometimes very specialized, type of simulation task.

Languages for continuous simulation are typically designed to simulate problems described by systems of ordinary differential equations. One of the drivers for their creation was the testing and verification of analog models on a digital computer. Such languages allow the numerical simulation of the activity of the essential elements of an analog model. However, they may implement other elements, without the inaccuracies of the analog solution. For some languages there is no need to convert the equation into the integral form since they allow a direct notation of derivatives, with certain limitations. Advanced languages support variable-step solvers. Regarding the simulation of physical fields described by partial differential equations there is no universal suitable method. Some let the user choose from several possible methods, while others transform partial differential equations into a set of ordinary differential equations.

Languages for discrete simulation involve several very different groups. The first group includes languages for the discrete simulation of discrete systems. These are used to simulate the behavior of finite automata, of control loops, of circuits for the implementation of operations of digital computers, and for modeling of computers themselves. The second group consists of languages for the simulation of systems described by difference equations, thus also for discretized continuous systems.

Languages for combined simulation merge the advantages of both the abovementioned groups. They are used for the simulation of systems containing continuous and discrete variables, such as in biology, medical research, and other disciplines. A universal programming language makes it possible to automate the process of creating a simulation model based on a mathematical model for a defined class of problems; it allows the description of the selected algorithm in different ways. The user defines the objects and operations, and uses them later on.

Nowadays, modeling and simulation almost exclusively uses digital (personal) computers in conjunction with modern higher-level programming languages such as C++, C#, Java, Python, and many others. More often, however, universal comprehensive integrated solutions providing comfortable user environments and allowing the definition of problems in many different ways (textually or graphically) have become used widely (Sokolowski, 2009).

The essence of simulation languages is different; for example, there are equation-oriented languages (programmable by writing the equations of the model), block-oriented languages (programmable by the interconnection of functional blocks that implement basic operations such as integration and various mathematical functions), module-oriented languages (containing modules representing ready-made models of technical components or processes which are then connected by signals representing the flow of materials and energy) and the like. They usually include various built-in numerical methods to solve the model equations, provide excellent graphical presentation of simulation results, and have the capability of working with input and output data files. They are often industry oriented.

Simulation programming languages facilitate the creation of models and experiments. The following areas may use advanced simulation programming languages effectively:

- work with abstract systems (knowledge base, etc.)
- programming of simulation models (simulation systems, languages, libraries)
- experimenting with simulation models (simulators)
- visualization and evaluation of results.

### 12.2.4 State of the Art

Since the beginning of the 21st century modeling using digital computers has become universal and widely applicable in all fields of human activity, especially in the field of technical cybernetics. In addition to simulations performed with a digital computer itself there are lots of methods used based on the connection between software and specific hardware, yet this is not a hybrid system in the classical sense, but a modern highly sophisticated approach especially for the purpose of accurate simulation of complex systems in industrial practice. As examples, real-time processor-in-the-loop (PIL) or hardware-in-the-loop (HIL) simulations are widely and effectively used in these areas.

In recent decades a lot of higher-level simulation languages have been developed, usually based on a graphical definition of the tasks in the form of a block diagram, as their problems with insufficiently powerful computers, particularly for real-time simulations, have been satisfactorily resolved a long time ago. These allow the solution of very intricate tasks in the area of engineering practice: nonlinear and multi-physics problems described by systems of ordinary and partial differential equations in combination with algebraic relations.

On a global scale, Matlab and Simulink are considered the top-ranked leaders in the category of general-purpose computing, modeling, and simulation tools, having a philosophy that inspired a lot of third parties to create a series of similar free products (i.e. "Matlab-like," "Simulink-like") tools, such as Scilab, Octave, or a simulation tool for complex control systems, REX (www.rexcontrols.com). Furthermore, there are many highly specialized environments focused on categories such as computer-aided engineering (CAE), finite element analysis (FEA), and finite element methods (FEM); for example, COMSOL Multiphysics (Zimmerman, 2006; Pryor, 2009), ANSYS (ANSYS Fluent, ANSYS Workbench), AMESim, or SolidWorks. Finally, there are products in the computer-aided design (CAD) category available on the market, primarily designed for the creation of project documentation with integrated simulation tools (typically the analysis of material stress, modeling of thermal processes, etc.), usually also based on the finite element method, such as Autodesk Inventor.

### 12.2.5 Prospective Future Developments in the Field of Simulation

Generally speaking, methods, algorithms, and computational resources in modeling and simulation are still under constant development, moving forward quickly, and so it will be in the near future. On the other hand, the mathematical methods that have laid the foundations in the past dozens or even hundreds of years are still valid and applicable without substantial changes, therefore progress can focus on further development and optimization of computational resources.

One of the clues that indicate where one of the points of interest may reside is, for example, the emergence of some modules (toolboxes) in existing software environments; the Parallel Computing Toolbox for Matlab is a typical example. Applications that use so-called parallel computational clusters are undoubtedly one modern development in modeling and simulation, as it systematically addresses the optimal distribution of computational complexity and efficient utilization of computing capacity. A similar vision is shared by the concepts of "cloud computing" and "grid computing," sometimes broadly referred to as "distributed computing."

An up-to-date and relatively young discipline in this area is the effort toward the development, implementation, and future large-scale spread of so-called quantum computers, for which there is a reasonable assumption of significant speed-up in solutions for some categories of problems (apart from modeling and simulation, this applies to cryptography, for example). Since 1999 there have been intensive attempts at many leading scientific institutions, particularly QUIC (Quantum Information Center), generously supported by DARPA, as well as LANL (Los Alamos National Laboratory), MIT (Massachusetts Institute of Technology), and Caltech (California Institute of Technology). Developers will certainly have to overcome initial problems related to the hardware architecture of a quantum computer, problems with interactions between qubits and the environment that may lead to decoherence and errors, as well as other future problems with the construction of a quantum computer that are expected to arise. At present, we have been witnessing the birth of quantum hardware. They are still very primitive, but it is clear that useful quantum computers are not far away.

One of the first truly functional quantum computers was built in 2000 by IBM. It was a five-qubit computer built to test the technology. In 2011 the D-Wave One was introduced to the world as the first commercial quantum computer, carrying a 128-qubit processor. In April 2013, *Science* magazine published an article with the news that scientists at Stanford University had launched the first biological computer – they announced that their new type of transistor is made entirely from genetic material and works inside bacteria.

The need to increase computer performance remains an important issue connected to the challenges that will appear in the near future during the further development of software in the area of modeling and simulation.

## 12.3    Testing of Control Systems

The testing of a control system is a key activity that, in conjunction with a well-executed development phase, minimizes risks and errors in the designed software, assuring the quality of a design, effective design, and other benefits.

While designing control systems or control software, an engineer often faces the problem that it is not possible to test the designed system with real technology. The reasons may be various, but most often it could be that the controlled technology is not available at the time of the control system design. Another reason could be that the control system is designed in another location, to be united with the controlled system when complete. Even if the controlled technology exists and is available in the location of the designed control system, direct testing can be disadvantageous for safety, operational, financial, and other reasons. Even when testing is carried out using the target system, it is often not possible to test all possible failures and conditions of the control system, which arise only during standard operation.

Therefore, testing a designed control system using modeling and simulation tools seems to be very useful, sometimes more useful than testing with the real system. Process simulation in industrial practice greatly simplifies and makes more effective the design, implementation, testing and commissioning of control systems.

To test the control software, tools and programs for modeling and simulation of manufacturing process technology or other controlled systems can be used. They are used for the validation and optimization of control algorithms, testing sequential logic, user interface testing, and examining the response to disturbances and failures in the controlled system. Testing using simulation tools requires some additional steps compared with testing on real technology, such as the creation of models, the development of applications in simulation tools, etc. The costs required will, however, lead to a better quality of designed control system, a shorter commissioning period, and easier maintenance and expansion of the system.

Testing control software may from a practical point of view be divided into two basic steps:

- Testing during development / testing of results of development. When compared with the phases of testing defined in the V-model, this step comprises a unit test, an integration test, and a system test. Different types of simulation tools, described in Section 12.5, could be used for these tests. The basic function of process simulation is to provide system feedback to the control system that is close to the response of the real system, to test all aspects of the designed control application. The basic step is testing of program blocks, which are elementary parts of the control software. These particularly include verifying the acquisition of process values, the control algorithms and sequences for actuators, interlocking according to technological conditions, the automatic control sequences, the control algorithms, the feedback from the various devices, etc. The ability to maintain defined process values usually directly affects the output quality of the production process, or the quantity of resources needed, and therefore the financial demands of the production process. Therefore, the emphasis is on optimizing the control algorithms and their parameters. All the failure and process conditions are tested. After the testing of individual blocks, their interactions are tested by integration tests. The next step is the testing of the whole control system to evaluate its overall functionality.

- Testing with the user of the system, which includes tests with the client's experts and with the end user. Within the V-model, this step is mentioned as "user acceptance testing." For these tests, it is also advantageous to use simulation tools as a substitute for the real technology to allow demonstration of the full functionality of the control system. Process technology experts can see the simulation of the behavior of the manufacturing process technology. The operation of the control system is further validated using functional description of technology. Process technology experts can then gain an idea of the overall operation of the system and can verify all states of the controlled technology. Use of these tests can be also detect and solve a possible incomplete functional description of the technological unit before commissioning the system.

From the above text, it is clear that for the two basic steps of testing it is very advantageous to use a model of the controlled system and simulation tools, instead of testing a real technology. Once the model is created within these steps and applied in some simulation tool, it can also be used for other activities such as installation, commissioning, operation, maintenance, and, particularly, operator training. The operator can try to control the technology before its implementation and to learn appropriate reactions to technology failures that cannot easily be tested in real operation. Experienced operators may also be able to give advance warning of other potential issues for the technological process prior to commissioning. Another advantage is using the simulations to support the workers who do the commissioning of the system at the installation site, allowing the tuning of the control loops and their optimization in a simulated mode. Required modifications to the control software, as well as the optimization of control structures, can be tested safely in simulation mode by using parameters detected in the real system on site.

## 12.4   Testing of Control Systems during the Development Process

Testing a control system during development is crucial for quick and easy use in practice. For these purposes engineers often use techniques from model-based design (MBD), mathematical graphically oriented methods for solving specific problems associated with designing complex control systems, including signal processing and communication systems. Model-based design brings a flexible solution approach that is used in many sectors and improves product quality and reduces development time. This methodology is applied in the design of many embedded software systems, and provides an efficient approach to establishing a common framework for communication throughout the design process while supporting the development cycle. The development can be divided into six stages (Papp et al., 2003):

- analyzing and designing a model of the plant
- verification and tuning the plant
- analyzing and synthesizing a controller for the plant
- verification and tuning of the controller
- simulation of a model with controller and plant
- implementing and testing the controller for specific hardware.

The first way of modeling with a model-based design is called model in the loop (MIL), which means that we propose models of the plant and controller in the loop. After model design, it is possible to run a simulation process to check if the system behavior is comparable to the real system. The MIL method for testing is shown in Figure 12.2. This testing method runs only on a host system where neither the controller nor the plant operates in real time. This way of modeling is used toward the beginning of the process, and allows development engineers to check the modeled performance of the system and design the control algorithms in a simulation environment (Shokry and Hinchey, 2009). Model in the loop helps in the design of control strategies using a representative system model where is possible to co-simulate models of the control device and system for function and block specification. For good system accuracy it is possible to evaluate several levels of models (Bringmann and Krämer, 2008).

The second way of modeling with a model-based design is software in the loop (SIL). The actual control blocks in the simulation are replaced by a software code representation, which means the inclusion of compiled production software code into a simulation model. It is possible to use automatic code generation for functions from specific control blocks. The blocks from the simulation can be coded in any programming language (C, C++, VHDL, Verilog, etc.) and then compiled using a standard compiler on the host system (Visual C/C++, gcc/g++, etc.). The final code is packaged in a shared library that is reused back in the model using custom functions (for example, S-functions in Matlab/Simulink). This is done to permit the inclusion of software functionality for which no model exists. This design technique should enable faster simulation runs on the host operating system. The main purpose is enable the inclusion of control algorithm functionality for which no model exists, and to increase the simulation speed by including

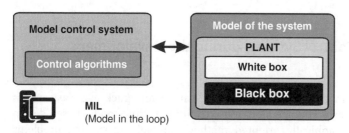

**Figure 12.2** Model in the loop – simulation process.

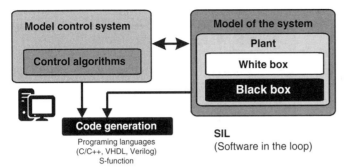

**Executing compiled blocks on host system in the model (Non-RT)**

**Figure 12.3** Software in the loop.

compiled code in place of interpreted models. A further reason for SIL is the need to verify that code generated from a model will function identically to the model, which will be a guarantee that an algorithm in the modeling environment will function identically to that same algorithm executing in a production controller. Neither the controller nor the plant operates in real time. In SIL model design, as shown in Figure 12.3, just as for MIL, neither the controller nor the plant operates in real time (Burns and Rodriguez, 2002).

The third way of modeling using model-based design is processor in the loop (PIL). This development and testing phase of model-based design is shown in Figure 12.4. Processor-in-the-loop simulation code is generated for either the top model or part of the model. The proposed code is cross-compiled and downloaded as a control algorithm (control blocks in the model) into an embedded target processor and communicates directly with the plant model via standard communications such as Ethernet. Through a communication channel it is possible to send signals to the code on the host or target processor for each sample period of the simulation. In this stage of development we do not use the I/O devices or bus interfaces that are used for communication with the real plant. A PIL simulation process involves cross-compiling and running production object code on a target processor or on an equivalent instruction set simulator (for example, products from OPAL-RT technology or National Instruments).

The embedded software component is tested and linked with the plant model and then is reused in a test suite across the simulation process. The same production code is then compiled for the target hardware device. This approach avoids the time-consuming process of development in a software environment and verifying production code on a separate test infrastructure. Through SIL and PIL modeling it is possible achieve early verification and defect fixing in the embedded code being developed (Astuti et al., 2008).

Hardware in the loop (HIL) is the fourth way of modeling in model-based design, and uses a technique for combining a mathematical simulation model of a system with actual physical hardware. The HIL technique is used in the development and testing of complex real-time embedded systems, where the hardware performs as though it were integrated into the real system. An HIL simulation should include electrical emulation of sensors and actuators, if they are used in the real system. These electrical emulations act as the interface between the plant simulation and the embedded system under test. The HIL design and testing technique is shown in Figure 12.5 (Li et al., 2004).

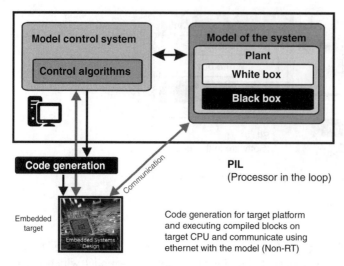

**Figure 12.4** Processor in the loop.

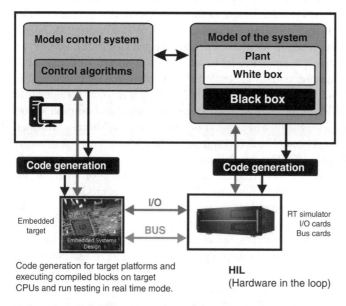

**Figure 12.5** Hardware in the loop.

Testing and development based on HIL techniques use embedded electronic controllers, the hardware controller, and associated software that are connected to a mathematical simulation model of the system plant, which is executed on a real-time simulator. To connect the real-time model to the hardware controller, the real-time simulator receives electrical signals from the embedded controller as actuator commands to drive the plant, and converts these signals into the physical variables connected to the plant model. The plant model calculates the physical variables that represent the outputs of the plant, which are converted into electrical signals that represent the voltages produced by the sensors that feed the controller (Yan et al., 2005).

Rapid control prototyping (RCP) is a process that lets engineers quickly design, test, verify, and iterate their control strategies on a real-time computer or real-time simulators with real input/output devices or communication cards. The RCP process is shown in Figure 12.6; it differs from HIL in that the control strategy is simulated in real time and the controlled system is the real system. Rapid control prototyping is a typical method for finding errors at the start of your project to save time and cost, and to increase overall quality; it provides easily built real-time execution of the control design using efficient tools for tuning and validation of the entire process, such as power electronics, electric drive, and power system control strategies. The other simple reason for using RCP is that the control software, which might be, for example, in engine and transmission control units, is difficult and time-consuming to modify easily. The RCP technique has been adopted in many branches of industry such as automotive, rail, aerospace, energy, industrial automation, etc. For research and development, RCP is a powerful tool for technology demonstrations, which are made possible even at an early stage of a project, without coding or complex implementation work. New or modified functionality is simply added to the production code in the controller embedded target processor to verify the additions or changes in the next phase of the project. Once all the functions have been developed and tested on a real-time simulator or PC, the production code is finally implemented (Schlager et al., 2006).

It is also possible to use a real-time simulator or computer for identification in the loop (IIL) as the real plant model to compare it with the proposed model. This identification method is shown in Figure 12.7. Production code is implemented for both parts of the simulation, control system, and model of the plant in the real-time simulator. The plant model parameters are tuned and verified by Ethernet communication.

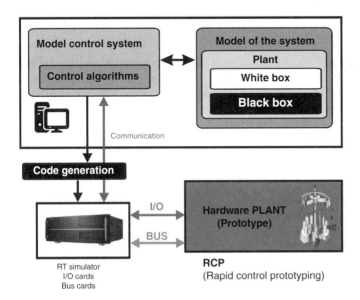

**Figure 12.6** Rapid control prototyping.

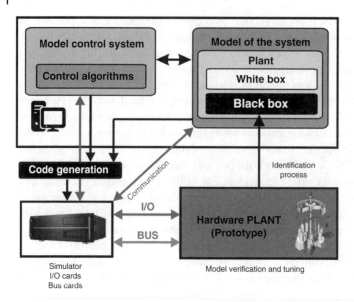

**Figure 12.7** Rapid control prototyping and identification in the loop.

Another way to extend model-based design is to use RCP combined with HIL. This combination is called component in the loop, where an entire system is connected to a source emulating the rest of the real system (vehicle, craft, aircraft, spaceship, etc.). Component-in-the-loop evaluation, shown in Figure 12.8, allows researchers and developers to study component technologies and their system-level impact without building and disassembling an entire system each time a component is changed. Development engineers can perform prototype component evaluations within a real system context (focused on component experiments) and real system-level evaluations that require real components (focused on real system experiments) that permit the study and implementation of physical components by using a "virtual system / computer model" to simulate controls and drive cycles. The further evaluation of model components, such as motors, turbines, engines, batteries, and so on, is possible as if they were operating in a real system. Thanks to some of these methods we are able to shorten research and development time, leading to more rapid commercialization of new technologies (Kain et al., 2011).

## 12.5 Tools for Modeling and Simulations

There are innumerable modeling and simulation tools available in the market at this time. Many are commercial in nature, but we can also find quality tools under free software licenses. Certainly, one of the decisive factors for the choice of tools will be the target application area. Choosing the right modeling and simulation tools may be crucial to the success of the project, whether at the time of preparation of the project or throughout its implementation. Before choosing a suitable instrument for the project (or often a combination of several instruments), it is necessary to analyze all the areas in which the project will need modeling and simulation.

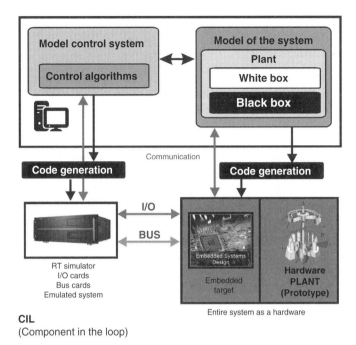

**Figure 12.8** Component in the loop.

The tools discussed for modeling and simulation concern a wide range of applications. In this chapter the main emphasis is on the area of industrial automation. If, based on an analysis of the project, the conclusion is that a robust system that is able to cover a wide number of fields is needed, it is probably advisable to choose a commercial solution that offers a much more comprehensive and coherent solution. In contrast, free software tools can often meet the requirements of narrow fields of specialization (National Instruments Corporation, 2003).

Nowadays, most manufacturers of modeling and simulation tools allow the customer to find and put together just the software they need. This means choosing core simulation software, and after that selecting expansion modules for the area of use. Another important criterion for the selection of tools for modeling and simulation is support for running simulations in real time. "Real time" for a variety of industries and applications can vary greatly, so it is important to take this factor thoroughly into account, and before buying discuss this with the software vendor. In this case it is also important to select the correct hardware on which a given simulation model will run (Schlager et al., 2007).

If the first option is chosen in the form of complex commercial software, a number of solutions are offered:

- Matlab/Simulink from MathWorks
- COMSOL Multiphysics from Comsol
- ANSYS from ANSYS
- LabVIEW from National Instruments
- RT-LAB from OPAL-RT Technologies
- 20-sim and 20-sim 4C from ControlLab

- LMS Amesim from Siemens
- SIMIT from Siemens
- WinMOD from Mewes & Partner GmbH.

When choosing the right tool, we also need to check that the appropriate design methods are supported. For example, Matlab/Simulink enables testing of MIL, SIL, and PIL using an expansion module (toolbox Embedded Coder). Software-in-the-loop simulation in this tool involves compiling and running the source code on the development workstation and verifying model functionality. For PIL simulation Matlab/Simulink enables code transfer to the target device and then runs the simulation on the particular test hardware. An important factor is also the way the simulation runs. If we return to our example with Matlab/Simulink, then the SIL and PIL simulation methods run in normal mode simulation and are exclusively designed for process simulation and testing, and do not allow reduction of the time required for the simulation. For these purposes in Matlab a block can be switched to Rapid Accelerator mode. Testing takes place via Matlab, which sends to the target device or simulator stimulating signals that are generated with a given sampling period, and the SIL/PIL algorithm is executed in a step of the sampling period, returning a signal with the computed results. Another important factor may be whether the software is certified for the design of algorithms with a given functional safety, and whether it can be tested during the design (Youn et al., 2006).

## 12.6   Case Studies

### 12.6.1   SIL and PIL Models of a Rotary Inverted Pendulum using REX Control System

#### 12.6.1.1   Concept of REX Control System

The REX Control System is a family of software products for automation projects (Balda et al., 2005). It is an industrial control system that can be used in all fields of automation, robotics, measurement, and feedback control (www.rexcontrols.com). The main features of the REX Control System are as follows:

- graphical programming without hand-coding
- programming control units on a standard PC or laptop
- user interface for desktop, tablet, and smartphone
- wide family of supported devices and input–output units
- industry-proven control algorithms
- easy integration in business IT infrastructure (ERP/BMS).

REX Control System supports both computer-controlled and remote-controlled experiments, including MIL, SIL, PIL, and HIL. A typical case study concerning a virtual lab and the PIL technique is demonstrated below.

The basic idea of PIL is to provide the possibility of testing the functionality of the designed control system with a plant model instead of the real plant. The control system with its algorithms is running on the so-called embedded target. Under real conditions it controls the real plant via an input–output (I/O) interface. The PIL technique makes it

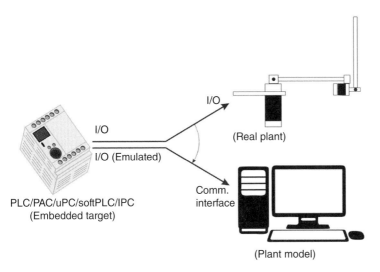

I/O

I/O (Real plant)

I/O (Emulated)

PLC/PAC/uPC/softPLC/IPC
(Embedded target)

Comm.
interface

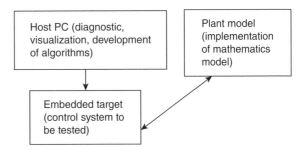

(Plant model)

**Figure 12.9** Idea of PIL simulation for rotary inverted pendulum.

**Figure 12.10** Block scheme of PIL simulation components for rotary inverted pendulum.

Host PC (diagnostic, visualization, development of algorithms)

Plant model (implementation of mathematics model)

Embedded target (control system to be tested)

possible to substitute a plant model instead, providing that all the I/O signals are emulated on a communication line, as shown in Figure 12.9 for the rotary inverted pendulum example. The embedded target is usually connected to a so-called host computer, so the entire block scheme can be represented by Figure 12.10.

The definition of PIL claims that the simulation of the plant model is synchronized with the control algorithms running on the embedded target at full rate (as fast as possible). This is what PIL is designed for, namely the choice of an appropriate sampling period and verification of whether the required performance of the embedded target is achieved. However, the real-time synchronization on both sides (embedded target and simulation of the plant model) does not clash with the definition. It is secured by Rex-Core – a runtime core running on the target device (Linux IPC, WinPAC, Raspberry Pi, etc.). It handles the timing and execution of all algorithms, and provides various services. The individual tasks are executed using preemptive multitasking.

From a REX point of view, the following configurations cover these typical situations (see Figure 12.11):

- Two targets (A, B) are linked via a communication line, both with the REX runtime, with A considered as the embedded target hosting the control algorithms (C), and B considered as the plant model (S), with a monitor or touchscreen attached to simulate the feeling of observing the real plant; this concept is suitable for situations where REX

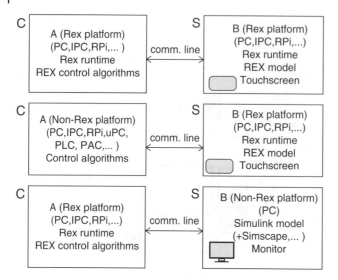

**Figure 12.11** Typical REX PIL configurations.

Control System is used both for control and for simulating a mathematical model of the real plant.

- Two targets (A, B) are linked via a communication line, with A considered as the embedded target hosting the control algorithms (C) with no need of a REX runtime of any type (PLC, IPC, etc.), and B considered as the plant model (S), with a monitor or touchscreen attached to simulate the feeling of observing a real plant; this concept is suitable for situations where the control algorithms (C) are implemented on a non-REX platform and should be tested with a mathematical model (S) of the real plant created in REX.
- Two targets (A, B) are linked via a communication line, with A considered as the embedded target with a REX runtime hosting the control algorithms (C), and B considered as a plant model (S) implemented on a PC and Simulink, with the monitor attached to simulate the feeling of observing a real plant; this concept is suitable for situations where the control algorithms (C) are implemented in some REX platform and should be tested with a mathematical model (S) of the real plant created in Simulink (plus additional Simscape-family blocks for physical modeling) or any other simulation software environment.

Additionally, the following features of REX Control System determine the possibility of using it for the concept of virtual laboratories:

- It is capable of generating and performing web-based and customized visualizations.
- It contains its own embedded web server, therefore the only thing one needs to perform remote experiment is a web browser.
- The web visualization can be password protected and secured.
- From a technical point of view there is no difference between a computer-controlled and remote-controlled experiment (Ozana and Docekal, 2017).
- It is multi-platform and allows many possibilities for attaching various I/O modules.

Apart from web access, REX also allows remote access by the Windows-based RexView diagnostic tool, providing hierarchical information about the running control algorithm.

With this proposed approach, both control algorithms and the model of the plant can run on a PC, miniPC (such as a Raspberry Pi), PLC, or IPC, moreover under different operating systems. The model of the plant should be created in REX Control System, but the control algorithms can be implemented in various software environments running on various hardware platforms, while communication between these two parts is established via standard protocols such as RS, I2C, Modbus, Profinet, and others. As for visualization, the only tool needed to display visualization is a standard web browser running on a PC, smartphone or tablet. Moreover, there are more types of visualization available. The basic visualization can be autogenerated via a single click based on the block scheme created in the REX environment (RexDraw). It is displayed within a web browser, showing the same block structure as in the source scheme, displaying live values of inputs and outputs of the blocks and allowing the setting of the blocks' parameters on the fly. In the case of more advanced visualization screens, the so-called RexHMI Designer is available. It makes it possible to use both embedded and custom graphical elements, and to let them be easily bound to particular REX variables. Composition of the entire visualization is carried out in the same environment (RexDraw), as both algorithms and visualization are integrated into one common project.

The usage methodology is as follows:

- Set up the mathematical model of the plant and control algorithms.
- REX implementation of the mathematical model of the plant using REX blocks.
- Implementation of the control algorithms on a chosen platform (it is also a REX-based platform; the SIL technique is very useful to try before the final PIL).
- Implementation of the PIL technique (emulating all physical signals via a communication protocol).
- Implementation of web-based visualization, binding of individual visualization variables to REX signals.
- Testing the PIL model, changing control algorithms, changing plant's parameters on the fly.
- Transfer of the solution to control of real plant.

### 12.6.1.2 Basic Introduction to a Rotary Inverted Pendulum

Control of a simple rotary inverted pendulum has been chosen as the case study in this section for several reasons. First of all, it is a challenging and interesting task, quite similar to classical simple inverted pendulum control, but with some important differences. It is also attractive to students and researchers, and is a very good example of a feedback control system, combining classical algorithms (PID) and more advanced algorithms based on so-called modern control theory. The topic of inverted pendulum control (most often the linear case) has been treated so many times, much more than any other physical educational model. This section does not describe the deep details of the control approach, but yet it shows some novelty. It very clearly shows the concept of the mathematical model and control algorithm through inputs from two encoder signals (two link angles) and a PWM output signal representing a manipulated value brought to a power electronic unit and a DC motor. The physical setup used for this case study is shown as a three-dimensional model in Figure 12.12.

**Figure 12.12** Three-dimensional model of a rotary inverted pendulum.

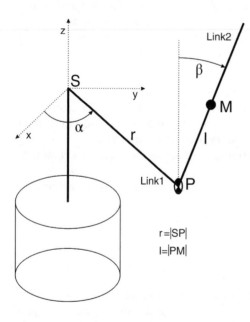

**Figure 12.13** Notation for model of rotary inverted pendulum.

### 12.6.1.3 Modeling a Rotary Inverted Pendulum

There are several possible approaches and equations describing the rotary pendulum model. The physical setup shown in Figure 12.12 can be adapted to the form shown in Figure 12.13 to show the notation used.

It is possible to derive Equation 12.1 for the pendulum dynamics:

$$l \cdot \ddot{\beta} - g \cdot \sin \beta + b \cdot \dot{\beta} + u \cdot r \cdot \cos \beta = 0, \tag{12.1}$$

where

| $r$ | (m) | pendulum link 1 length, $|SP|$ |
| $l$ | (m) | pendulum link 2 length to the mass center, $|PM|$ |
| $\alpha$ | (rad) | pendulum link 1 angle |
| $\beta$ | (rad) | pendulum link 2 angle |
| $g$ | $(m \cdot s^{-2})$ | gravity acceleration |
| $b$ | $(m \cdot s^{-1})$ | damping coefficient |
| $u$ | $(rad \cdot s^{-2})$ | link 1 angular acceleration. |

In order to get the full state-space description of the system, Equation 12.1 can be extended with two state variables representing the angular position and speed of link 1, assuming $u$ as the controlled input. Using the usual formal notation, the entire state-space description of the inverted pendulum can be described by Equations 12.2–12.5, which relates to the IP block in Figure 12.16.

$$\dot{x}_1 = x_2, \tag{12.2}$$

$$\dot{x}_2 = \frac{g}{l} \cdot \sin x_1 - \frac{r}{l} \cdot u \cdot \cos x_1 - \frac{b}{l} \cdot x_2, \tag{12.3}$$

$$\dot{x}_3 = x_4, \tag{12.4}$$

$$\dot{x}_4 = u, \tag{12.5}$$

where

| $x_1$ | (rad) | pendulum link 2 angle ($\beta$ in Figure 12.13) |
| $x_2$ | $(rad \cdot s^{-1})$ | pendulum link 2 angular speed |
| $x_3$ | (rad) | pendulum link 1 angle ($\alpha$ in Figure 12.13) |
| $x_4$ | $(rad \cdot s^{-1})$ | pendulum link 1 angular speed |
| $u$ | $(rad \cdot s^{-2})$ | link 1 angular acceleration. |

### 12.6.1.4 Closed-Loop Speed Control

Though the mathematical model gives information about the angular speed of both links, only two variables are available under real conditions – the angles of both links, $x_1$ and $x_3$ (shown in Figure 12.14). The remaining unmeasured states $x_2$ and $x_4$ representing the links' angular speeds are approximated by numeric derivatives.

Figure 12.14 also shows the speed closed loop $G_{Rv}$ with the angular speed setpoint $v_{sp}$. It is used with the real model for achieving the speed required by the control system, and therefore application of the control signal in the form of acceleration.

The actuator (a DC motor and power electronic unit for speed control) can be considered as the part of the entire system S with its dynamic approximated by a fast first-order system – see the $G_{Rv}$ block in Figure 12.16. The time constant $\tau$ has been identified experimentally. It represents the dynamical behavior of the inner speed loop $G_{Rv}$ implemented as a fast PI controller $R_v$ within the real setup. First the actuator transfer function was identified, and then the PI controller was designed by the modulus optimum method;

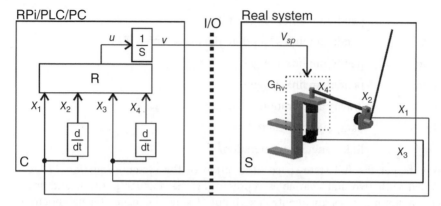

**Figure 12.14** Control of the real system of the rotary inverted pendulum.

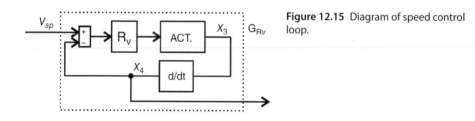

**Figure 12.15** Diagram of speed control loop.

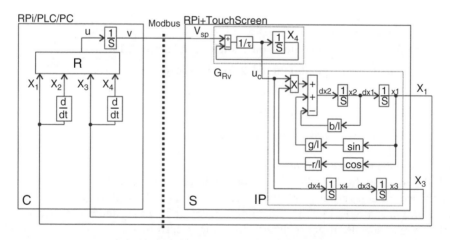

**Figure 12.16** PIL modeling diagram for the rotary inverted pendulum.

a diagram of this loop is shown in Figure 12.15. The final step response settling time of the closed-loop system was used for evaluation of the time constant $\tau$.

If this closed-loop system meets the requirement of a zero steady-state control error and the response is fast enough, the angular speed of link 1 is almost the same as the speed setpoint. Also note that if the speed controller is considered as ideal ($G_{Rv} = 1$), then $u \equiv u_c$ in Figure 12.16.

Under real conditions, the controller R generates a manipulated value $u(t)$ which is integrated and used as the speed setpoint, which results in the generation of the needed acceleration $u_c(t)$ while $u \approx u_c$. The entire model of the system considered for the PIL technique is then shown in Figure 12.16.

Note that the control system C hosting the control algorithms is identical both in the control of the real system (Figure 12.14) and in its PIL representation (Figure 12.16), which complies with the main conceptual idea of the PIL technique.

### 12.6.1.5 The Control System Based on a Two Degrees of Freedom LQR Structure

The detailed structure of the entire control system C depicted in Figures 12.14 and 12.16 is shown in Figure 12.17. The inner part consists of the controller R. Its structure corresponds to the two degrees of freedom scheme as described in Graichen, Hagenmeyer, and Zeitz (2005) and Milam, Mushambi, and Murray (2000). The two degrees of freedom design consists of a trajectory generator and a linear feedback controller. Based on a desired objective, the trajectory generator (TG) provides a feasible feed-forward control $u^*(t)$ and reference trajectory $x^*(t)$ in the presence of system and actuation constraints. Given the inherent modeling uncertainty, a feedback controller $K(t)$ (time-variant LQR) provides stability around the reference trajectory. This approach has the advantage that the system is tracking a feasible trajectory along which the system can be stabilized.

### 12.6.1.6 Generation of Reference Control and Reference Trajectories

When handling a nonlinear system, there is another problem separated from the two degrees of freedom controller design – the trajectory generation problem. This means finding appropriate state trajectories and feed-forward controls for the desired movement. The entire state space may contain various areas or points (singularities) that must be avoided.

There are lots of methods for reference trajectory generation; it is also possible to use some tools for these purposes that are available on the market. One of the very effective available tools is PyTrajectory (Schnabel et al., 2016), a package of libraries created in Python. The main goal of this tool is to generate trajectories for nonlinear system by solving the boundary value problem via collocation methods. According to this, PyTrajectory is able to find trajectories between two equilibriums (or universally any points in state space) defined by state vectors without applying any extra specific cost function. However, possible constraints for the control signal and states can easily be applied.

The reason for the generation of reference trajectories is to find the transition between the given states. In the case of this rotary inverted pendulum, it is possible to use this

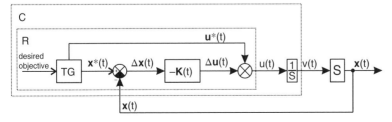

**Figure 12.17** Two degrees of freedom control structure.

approach to swing the pendulum up or down, in other words to find the appropriate reference control signal and reference state trajectories between the lower stable and upper unstable equilibriums.

The first step when using this library is to enter the dynamics of the nonlinear system into the new Python script, which includes the appropriate libraries.

```
def f(x,u):
    x1, x2, x3, x4 = x       # system variables
    u1, = u                  # input variable
    r = 0.3                  # length of the link1
    l = 0.15                 # length of the link2 to mass center
    g = 9.81                 # gravitational acceleration
    tau = 0.1                # time constant of speed loop
    b = 0.0225               # dumping coefficient
    ff = [ x2,
           g/l*sin(x1)-r/l*u1*cos(x1)-b/l*x2,
           x4,
           u1]
    return ff
```

The next step specifies the boundary conditions, which means the start and end times (*a* and *b*) and the system state boundary values at these times. It is possible to add some constraints for the state variables too. After these settings the problem can be solved. Here is the routine to be run in PyTrajectory:

```
a = 0.0 #start time
# system state boundary values for a = 0.0 [s] and b = 2.5 [s]
xa = [pi, 0.0, 0.0, 0.0] # x1(0)=pi; x2(0)=0; x3(0)=0; x4(0)=0;
b = 2.5 #final time
xb = [0.0, 0.0, 0.0, 0.0] # x1(T)=0; x2(T)=0; x3(T)=0; x4(T)=0;
# boundary values for the control signal u(t)
ua = [0.0] #u(0)
ub = [0.0] #u(T)
# next, this is the dictionary containing the constraints
con = { 3 : [-2, 2]} # constraint values for x4 (speed of the link 1)
# task definition with some settings according PyTrajectory documentation
Sol = ControlSystem(f, a, b, xa, xb, ua, ub, constraints=con, kx=2,
use_chains=False, eps=4e-2, maxlt=7, su=20, sx=20)
# launch task solution stored into Sol variable
Sol.solve()
```

If the solution *Sol* is found, the reference state trajectories and feed-forward control signal are available, as shown in Figure 12.18. These signals were used for further control design of the time-variant LQR controller on a finite time horizon.

### 12.6.1.7 SIL Simulation

The generated reference trajectories and designed control algorithm should be verified before applying to the real setup to avoid any possible failures. The first test was done in Simulink, which complies with the MIL simulation technique. It is often used for basic verification to check if there are some fundamental mistakes in the designed control strategy.

The SIL simulation shows the behavior of the simulated model and control system in real time. REX Control System was used for the implementation of the model of the pendulum and control system together, separated into individual blocks for better clarity

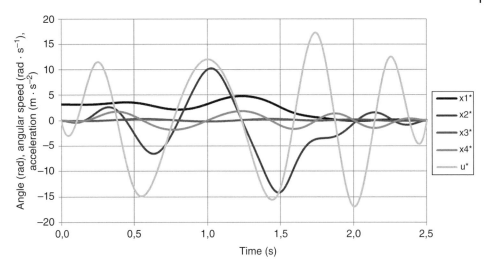

**Figure 12.18** Swing-up reference state trajectories $x^*(t)$ and feed-forward control signal $u^*(t)$.

**Figure 12.19** Block diagram of SIL simulation designed in REX Control System.

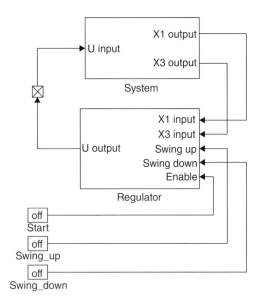

(see Figure 12.19). It is possible to use the same control system with a real plant in later phases of the design.

The control system was based on the structure shown in Figure 12.17. It uses special block with customized code for loading previously stored reference trajectories and controller values from CSV files. These sequences are applied to the SIL model once the swing-up (or swing-down) command is triggered. Due to the higher accuracy of the SIL model and the real-time operation, there are visible deviations of the current state trajectories from the reference ones. The controller should be robust enough to deal with them. Otherwise, the swing-up action would not be functional and the pendulum

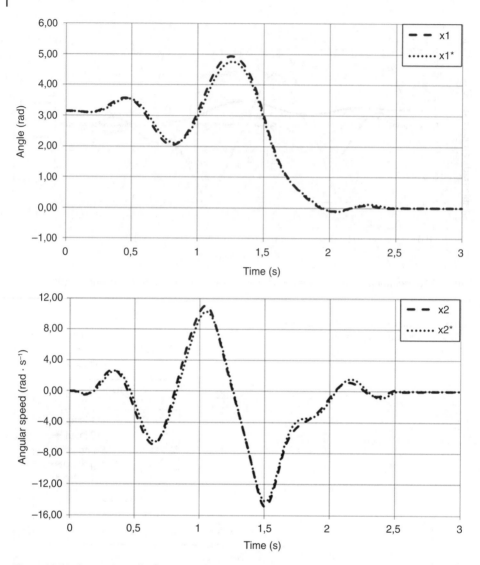

**Figure 12.20** Comparison of reference and current trajectories $x_1, x_2$ for SIL simulation.

would not get into the upright position. Figure 12.20 shows the most important simulation result: the verification experiment of the swing-up action. Link 2 successfully moves to the upright position where $x_1(T) = x_1(2.5) = 0$, and it is stabilized in terms of the position and the speed for $t \geq 2.5$ s. Similarly, it would be possible to analyze the other states, $x_3(t)$ and $x_4(t)$.

### 12.6.1.8 PIL Simulation
When using PIL simulation, it is necessary to separate the control system and the simulated mathematical model of the rotary inverted pendulum to two devices connected by some communication interface used to emulate the physical signals. In order to comply with the PIL concept, the control system should be deployed on the same target as the

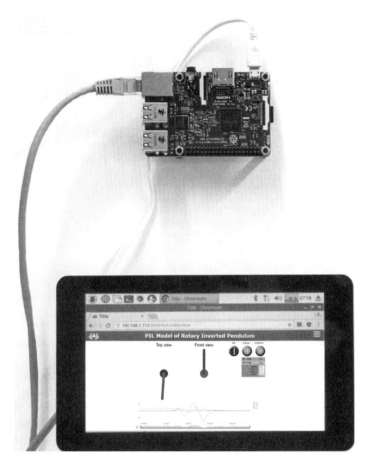

**Figure 12.21** Photo of testing stand for the PIL model of the control of the rotary inverted pendulum.

one planned to control the real setup. Two Raspberry Pi 3 single-board computers are used for these purposes using Modbus over Ethernet, as shown in Figure 12.21 (upper part: Rpi1; lower part: Rpi2 and touchscreen). Because the SIL model was prepared so that the simulated system and control system were separated into individual blocks, it was easy to divide it into individual tasks connected by Modbus communication.

The difference between the PIL and SIL simulations is in the division and using two devices, where the control system is the same as for the real system. There are some extra delays in the control loop that correspond more to real situations. The communication between the two hardware platforms causes more significant deviations than in SIL. Once the control algorithms are verified by PIL simulation, they can be deployed to the real system, or the HIL technique can possibly be applied as a very last phase before the deployment.

### 12.6.1.9 Web Visualization

The web visualization contains both text and visual information about the angular position of link 1 and link 2, using dynamic graph components and text boxes. Visualization of the movement of both links' angles are displayed in real time. It is composed of two

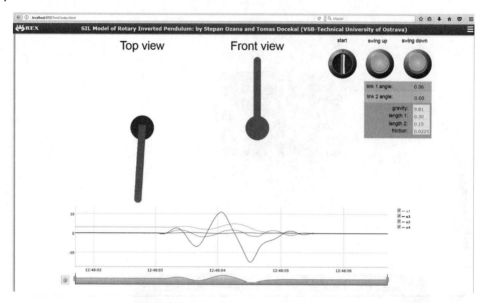

**Figure 12.22** Web visualization screenshot.

objects representing both links in individual views. The rotation properties of the links are bound to the $x_1$ and $x_3$ REX variables, respectively.

It is also important to have the possibility of data export and archiving. The RexHMI designer makes it possible to use a TRND web component capable of selecting the signals to be displayed, viewing the current trends, or pausing the simulation in order to analyze the time waveforms. The export of raw values can be added to the HTML source code, or easily via the RexView remote client. The experiment is performed via the switches responsible for swing-up, regulation in the upright position, and swing-down. A detailed screenshot of the website in a web browser corresponding to Figure 12.21 showing regulation in the upright position is given in Figure 12.22. The visualization can be used with both mentioned simulation types – SIL and PIL. It is also possible to connect it with the real model and its control system to adapt this concept to a so-called remote laboratory.

### 12.6.2 Heating Furnace

This case study demonstrates the use of modeling and simulation to control industrial processes that are implemented in Matlab/Simulink, and the subsequent code written and linked with the industrial programmable logic controllers. The example, as will be discussed, is focused on modeling a real heating furnace (located in Ostrava, Czech Republic, in the company ArcelorMittal Ostrava a.s.), which is divided into several heating zones – see Figure 12.23. The furnace model is implemented using a neural network that simulates the behavior of a real furnace, where the furnace behavior was designed from real furnace data. This furnace is controlled using several fuzzy controllers which regulate its temperature in the individual zones.

Because it was not possible to test the control of the furnace in real operation, a fully featured model using existing parameters and measured data was designed. To build the

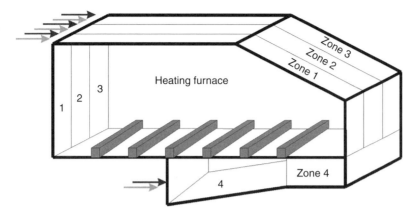

**Figure 12.23** Heating furnace with four zones.

model of the furnace, which was divided into four parts (zone 1, zone 2, zone 3, zone 4), we used four neural networks for each zone that learned data from the real furnace. Each neural network contains three hidden layers of neurons, containing 10, 20, and 30 neurons, respectively, and then one output layer with three neurons (Hagen et al., 2014) – see Figure 12.24.

Part of the proposed solution technique used the PIL method, where the behavior of the control object was replaced by artificial intelligence in the form of a neural network. As a control actuator, a Takagi–Sugeno-type fuzzy controller was used. For each part of the furnace a pair of fuzzy controllers was used with automatic correction of the combustion ratio. An assembly of these fuzzy controllers then set up a control actuator for the furnace. The input value to each fuzzy controller was the actual temperature of the furnace zone, and the output values were the flow of gas and air.

As target testing hardware, two powerful industrial programmable logic controllers (PLCs) S71516-3 PN/DP from Siemens were chosen. One PLC featured the simulated heating furnace system represented by the neural network, and the second PLC held the automatic control actuator system in the form of fuzzy controllers. The PLC machines were connected to each other using an industrial Profinet network that provides real-time communication. The connection between the PLCs and Matlab/Simulink was ensured using Ethernet and the S7 communication protocol, as shown in Figure 12.25.

Drafts of the master and slave models came from various stages of development. The first method of component-based modeling focused on setting up the basic objects in an MIL model of the furnace controller. After implementing a model of the furnace using a neural network, this model was verified and subsequently tested on a set of real data from the furnace, without previously being trained. After several iterations and adjustments, the tests were passed and we could go on to the development of a control model with a system of fuzzy controllers. The furnace and fuzzy controller algorithms were written in the Matlab script language and subsequently were compiled and implemented in MEX (DLL) binary file form. After the implementation and verification of the control model, this model was linked with the simulated furnace system and passed into the testing phase using SIL techniques. In this way, the weight constants of the fuzzy controllers were tuned. After tuning of the resulting simulations, the final stage of testing was undertaken using PIL techniques. This stage of testing was divided into two parts:

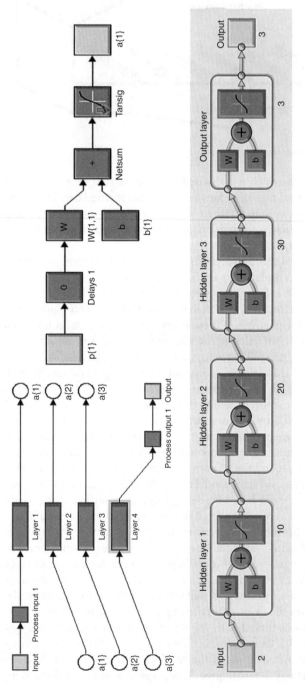

**Figure 12.24** MIL model of neural network for heating zone 4.

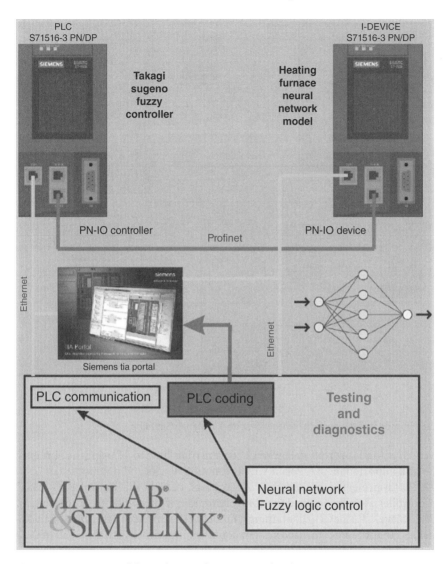

**Figure 12.25** Diagram of the PIL heating furnace control and testing system.

the first was independent testing of the furnace control system, and the second was subsequent testing of the actuator in the form of a system of fuzzy controllers together with a model of the heating furnace (Moreira et al., 2011).

The first step in the PIL modeling was to translate the furnace model and the controller into a programming language that the PLC would understand. This programming language was chosen to be structured text (ST) conforming to the IEC 61131-3 standard. To load this code into both PLCs it was necessary to use the TIA Portal V13 development environment and programming tool from Siemens. There were four neural networks for each heating zone implemented using the ST programming language and then 12 fuzzy controllers, where each furnace zone used three controllers in this project. The model

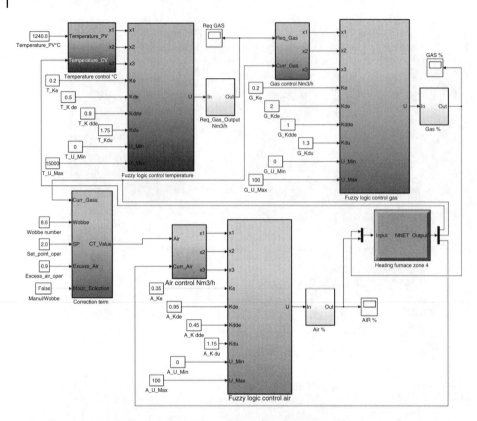

**Figure 12.26** SIL model of the heating furnace for zone 4 in Matlab/Simulink.

blocks (written in Matlab script) in Figure 12.26 were translated to ST using the Simulink PLC coder and imported in TIA Portal V13 as function blocks and functions.

The TIA Portal project was divided into two parts, namely the control PLC (Master – IO Controller), which included the control program with a set of fuzzy controllers, and the PLC (Slave – IO Device) containing a model of the furnace in the form of a neural network. So that the PLC furnace system could communicate with the control PLC (fuzzy controllers) it was necessary that the second PLC operate in the so-called I-Device mode where the PLC can act as an IO Device member. The communication interface between the PLCs was chosen to be real-time industrial Ethernet – Profinet. Fuzzy controllers are implemented in the TIA Portal environment as they were designed in Matlab and as functional blocks. The individual functional blocks of the fuzzy controllers are embedded in the organizational unit, which performs the appropriate sampling period, as it did during the testing of the model in Matlab/Simulink. The main difference is that the target hardware is running the control system composed of the group of fuzzy controllers in real time. Testing was performed using a 100 ms sampling period.

Communication with Matlab/Simulink was provided via Ethernet, namely the industrial communication protocol S7. Individual blocks in Simulink as "Fuzzy controller for controlling the temperature, the amount of gas and air, and then the correction term and neural network" were replaced by input/output blocks consisting of S-functions, as shown in Figure 12.27, which were connected with the S7 protocol driver via shared DLL

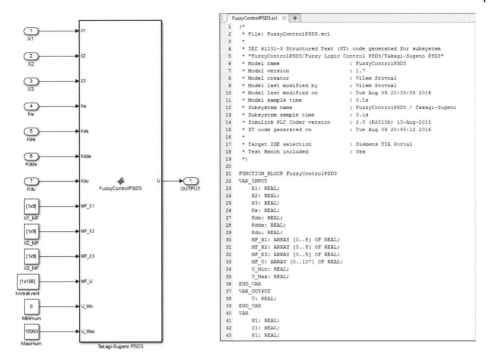

**Figure 12.27** Fuzzy control function block in Matlab/Simulink, and sample of the PLC source code in ST format.

libraries for the Windows OS and shared.so libraries for the Linux OS. The input and output blocks in the S-functions were connected to the data blocks DB in each PLC via parameters that could be changed and thus change the behavior of the PIL model in real time. The Matlab/Simulink simulation starts to run both PLCs, which are synchronized with each other in real time, where the synchronization jitter is shorter than 1 µs.

The control PLC IO Controller sends a setting for the position of the gas and air valves in [% Nm$^3$/h] in the zone corresponding to the neural network, which is part of the second PLC IO Device (I-Device) and returns the readings for the current temperature of the furnace and the gas and air to the control PLC. Individual parameters for the fuzzy controller weighting constants can then be selected as constants, as shown in Figure 12.26, or choose as variables and debugged and tested using simulation in Matlab/Simulink.

The waveforms of the three controlled variables, namely the temperature (indicated by dotted line), the air (indicated by normal line), and the gas (indicated by dashed line), are shown in Figures 12.28 and 12.29. Fig 12.28 shows the results for the heating furnace control using the PIL design model on a set of real test data. Figure 12.29 shows the controller tuning design for the real furnace using an RCP model with algorithms implemented in the PLC as a controller of the furnace. As can be seen in Figure 12.30, comparison of the two parts of the control design is not entirely tangible, but in the context of the end-to-end tuning of the real controller, it saves much time. It should be added that technology upgrade time is severely limited by technology shutdown, often no longer than one week.

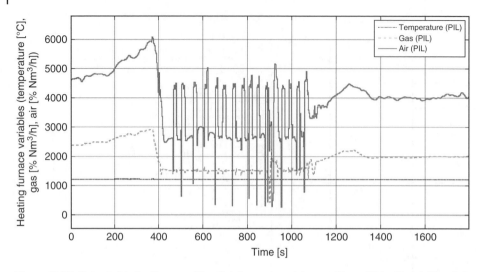

**Figure 12.28** Output data for the zone 4 heating furnace model control loop (PIL) – Matlab/Simulink and Siemens PLC (dotted line: temperature [°C]; normal line: air [% Nm³/h]; dashed line: gas [% Nm³/h]).

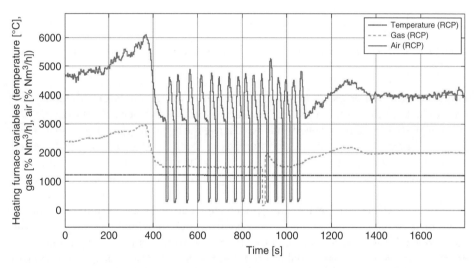

**Figure 12.29** Output data for zone 4 of the real heating furnace in the control loop with the Siemens PLC (RCP) (dotted line: temperature [°C]; normal line: air [% Nm³/h]; dashed line: gas [% Nm³/h]).

An important part of the testing is also the processor time that the fuzzy controllers spend on a PLC sampling cycle. Measurement of the load of both PLCs when running the entire simulation showed that the hardware was fully adequate and the tasks consumed only about 5% of the CPU time in the clock-running PLC.

Finally, it must be said that the changes that will be made in the simulation model during PIL testing can be considered as real-time changes for sampling periods greater than 20 ms. Decreasing the sampling period means changes in the simulation may not occur in the same sampling cycle of the PLC, but in the following sampling cycle, due to

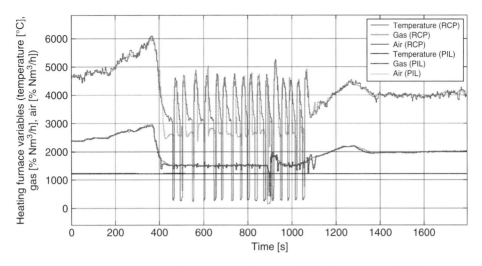

**Figure 12.30** Comparison of results from the PIL and RCP models (red: temperature [°C]; blue: air [% Nm³/h]; orange: gas [% Nm³/h]). See color section.

the overhead of communication between the PLC and the PC, and also depending on the selected operating system.

## 12.7 Summary

MIL, SIL, PIL, and HIL are all effective techniques for the design of control systems used during the validation phases described by the V-model in Figure 12.1.

The main relevance of the documented approach in the rotary inverted pendulum case study is the thorough design and test of the swing-up algorithms, as it is quite difficult to compute a functional control strategy concerning constraints given by physical reality, mainly due to the limits of the movement and a possible saturation limit for a speed controller. Unsuccessful attempts during control design applied in the real system without PIL simulation may result in damage to the real setup. For a double or a triple pendulum, design and verification of the control algorithms using SIL, PIL, or HIL become a necessity.

A very important additional benefit of the proposed solution of PIL simulation of the inverted pendulum is the possibility of introducing customizable interactivity in the web visualization components. For example, the graphical objects can be draggable, or assigned to other actions allowable by JavaScript events. As applied for this case study, it is possible to simulate a disturbance in the control circuit by dragging the links, corresponding to a manual touch of the pendulum in the upright position under real conditions. Moreover, important parameters playing a crucial role in the system dynamics (such as gravity, link lengths, friction coefficient) can be entered via web interface edit boxes.

On the other hand, the second case study of a heating furnace is a relatively slower regulation than the rotary inverted pendulum, and the individual parts of the furnace are divided into individual linearized zones. The first furnace design was performed using a mathematical model with the individual parameters of the furnace, which was replaced by the neural network model for the furnace zones trained directly on technological data from actual operation, due to the different operating and technical nonlinearities with respect to the age of the furnace. The main reason for using modeling techniques of control-based model design in this case study was the impossibility of designing the control directly on the real heating furnace due to its continuous operation.

Having applied the PIL simulation technique, successful verification of the control strategy for the rotary inverted pendulum was carried out. In the final step of the control design, the same control algorithms were used for control of a real rotary inverted pendulum. Due to the complexity of the system and its strong nonlinear character, it is easy to assess the effectiveness, usability, and accuracy of the PIL simulation. Particular computed trajectories and reference control signals are only valid and functional if the parameters of the inverted pendulum are determined accurately enough. Simulation experiments have been performed in terms of changing the chosen rotary inverted pendulum parameters, and they show a good correspondence with real behavior in practice.

The PIL model of the rotary inverted pendulum was demonstrated using REX Control System. Besides industrial applications, this development environment is especially useful for the education process due to favorable licensing and technical capabilities. It is widely applicable in the analysis and synthesis of control systems. Moreover, reconfiguration of SIL towards PIL is easy and fast.

The PIL design technique used to design the heating furnace control and the rotary inverted pendulum controller had to be transferred to the industrial automation environment and to meet the requirements of the environment. As a target control circuit, an industrial Siemens PLC had to be used and the proposed control algorithm had to be in accordance with the IEC 61131-3 standard.

Another requirement was for the controller systems located in different PLCs to be able to communicate in real time using the Profinet industrial network. Due to the not very precise control of the mixed gas and air supply valves in the heating furnace technology, due to the age of the technology, it has been decided to replace the existing PSD controllers with more robust fuzzy controllers. The weight constants of the cascade fuzzy regulators would have been difficult to match without a PIL design model, which also used optimal design techniques using genetic algorithms.

The degree of idealization in case of PIL and SIL is not sustainable for more complex cases. These models do not respect many other influences such as nonlinearities, computational power in the control hardware used, transmission rate of communication lines, etc. When simulating PIL in real time it is possible to verify the function of the control system outside the software simulation tool, with respect to factors that will be significant in practice. It is possible to verify that the performance of the control system hardware is sufficient to control the process, and to determine the shortest possible period of the control cycle. It is also possible to evaluate the quality of the regulation. Hardware-in-the-loop simulation is typically used in complex control tasks, characterized by large amounts of financial resources and serious consequences in the case of failure in the control process; HIL simulation is closest to the real world as the control system and its connection to the sensors as well as action interventions in the controlled system are carried out as under real conditions. Thanks to its independence from the real-world controlled system, it is possible to simulate failures that would be hard to afford.

The range of use of SIL, PIL, and HIL simulation methods is very wide. They are used in aviation, army, and automotive contexts, and basically wherever complex control algorithms are involved in the control of complex systems. Their great benefit is the ability to detect many hidden problems before the control system is actually implemented and used; this will be positively reflected in many inconsequential aspects. Creating a background for these simulation methods requires a major initial effort and investments, but it will pay off in the future.

## References

Astuti, G., Longo, D., Melita, C.D., Muscato, G., and Orlando, A. (2008) HIL tuning of UAV for exploration of risky environments. *International Journal of Advanced Robotic Systems*, 5(4), 419–424.

Balda, P., Schlegel, M., and Štětina, M. (2005) Advanced control algorithms + Simulink compatibility + Real-time OS = REX. *IFAC Proceedings Volumes*, 38(1), 121–126.

Bringmann, E. and Krämer, A. (2008) Model-based testing of automotive systems, in *International Conference on Software Testing, Verification, and Validation*, pp. 485–493.

Burns, D.J. and Rodriguez, A.A. (2002) *Hardware-in-the-Loop Control System Development Using MATLAB and xPC*, Department of Electrical Engineering, Center for System Science and Engineering, Arizona State University.

Graichen, K., Hagenmeyer, V., and Zeitz, M. (2005) A new approach to inversion-based feedforward control design for nonlinear systems. *Automatica*, 41(12), 2033–2041.

Hagan, M., Demuth, H., Beale, M., and De Jesus, O. (2014) *Neural Network Design*. Martin Hagan.

IEEE (1990) *IEEE Standard Glossary of Software Engineering Terminology, 610.12-1990*. New York: IEEE.

Kain, S., Schiller, F., and Dominka, S. (2011) Methodology for reusing real-time HiL simulation models in the commissioning and operation phase of industrial production plants, in G. Naik (ed.) *Intelligent Mechatronics*. Rijeka: InTech.

Li, Z., Kyte, M., and Johnson, B. (2004) Hardware-in-the-loop real-time simulation interface software design, in *Proceedings of the IEEE Intelligent Transportation Systems Conference*, Washington, D.C., USA, pp. 1012–1017.

Milam, M.B., Mushambi, K., and Murray, R.M. (2000) A new computational approach to real-time trajectory generation for constrained mechanical systems, in *Proceedings of the IEEE Conference on Decision and Control*, pp. 845–851.

Moreira, E., Pantoni, R., and Brandão, D. (2011) Equipment based on the hardware in the loop (HIL) concept to test automation equipment using plant simulation, in F. Silviu (ed.) *LabVIEW: Practical Applications and Solutions, Chapter 7*. Rijeka: InTech.

National Instruments Corporation (2003) *LabVIEW FPGA in Hardware-in-the-Loop Simulation Applications*. Austin, TX: National Instruments Corporation.

Ozana, S. and Docekal, T. (2017) The concept of virtual laboratory and PIL modeling with REX Control System, in *Proceedings of the 2017 21st International Conference on Process Control, PC 2017*, pp. 98–103.

Papp, Z., Dorrepaal, M., and Verburg, D.J. (2003) Distributed hardware-in-the-loop simulator for autonomous continuous dynamical systems with spatially constrained interactions., in *Proceedings of the IEEE International Parallel and Distributed Processing Symposium*, Nice, France.

Pressman R.S., (2010) *Software Engineering: A Practitioner's Approach*. New York: McGraw-Hill.

Pryor, R.W. (2009) *Multiphysics Modeling Using COMSOL®: A First Principles Approach*. Burlington, MA: Jones & Bartlett Learning.

Schlager M., (2011) *Hardware-in-the-Loop Simulations: A Scalable, Component-Based, Time-Triggered Hardware-in-the-Loop Simulation Framework*. Saarbrücken: VDM Verlag Dr. Muller GmbH & Co. KG.

Schlager, M., Elmenreich, W., and Wenzel, I. (2006) Interface design for hardware-in-the-loop simulation, in *Proceedings of the IEEE International Symposium on Industrial Electronics (ISIE'06)*, Montréal, Canada, pp. 1554–1559.

Schlager, M., Obermaisser, R., and Elmenreich, W. (2007) A framework for hardware-in-the-loop testing of an integrated architecture, SEUS 2007. *Lecture Notes in Computer Science*, 4761, 159–170.

Schnabel, O., Kunze, A., and Knoll, C. (2016) *PyTrajectory User's Guide*. https://pytrajectory.readthedocs.io/en/master/ [accessed 30 January, 2018].

Shokry, H. and Hinchey, M. (2009) Model-based verification of embedded software. *Computer*, 2(4), 53–59.

Sokolowski, J.A. and Banks, C.M. (2009) *Principles of Modeling and Simulation: A Multidisciplinary Approach*. Chichester: Wiley.

Woods, R.L. and Lawrence, K.L. (1997) *Modeling and Simulation of Dynamic Systems*. Upper Saddle River, NJ: Prentice Hall.

Yan, Q., Oueslati, F., Bielenda, J., and Hirshey J. (2005) *Hardware in the Loop for a Dynamic Driving System Controller Testing and Validation, SAE technical paper 2005-01-1667*, Warrendale, PA: Society of Automotive Engineers.

Youn, J., Ma, J., Sunwoo, M., and Lee, W. (2006) *Software-in-the-Loop Simulation Environment Realization using Matlab/Simulink, SAE Technical Paper 2006-01-1470*, Warrendale, PA: Society of Automotive Engineers.

Zeigler, B.P. (2000) *Theory of Modeling and Simulation*, 2nd edn. Amsterdam: Elsevier.

Zimmerman, W.B.J. (2006) *Multiphysics Modeling with Finite Element Methods, Series on Stability, Vibration and Control of Systems*. Singapore: World Scientific Publishing Co. Pte. Ltd.

Part III

Testing Infrastructures

# 13

## A Temperature Monitoring Infrastructure and Process for Improving Data Center Energy Efficiency with Results for a High Performance Computing Data Center[*]

*Sarah E. Michalak, Amanda M. Bonnie, Andrew J. Montoya, Curtis B. Storlie, William N. Rust, Lawrence O. Ticknor, Laura A. Davey, Thomas E. Moxley III, and Brian J. Reich*

## Synopsis

To increase the energy efficiency of its high performance computing (HPC), i.e. supercomputer, data centers (DCs), Los Alamos National Laboratory has developed an approach to energy efficiency improvement that includes a temperature monitoring infrastructure and a process that enables changes to DC cooling systems, infrastructure, and compute equipment for increased energy efficiency. This work describes the temperature monitoring infrastructure and process, as well as results from the first DC in which energy efficiency improvement efforts were undertaken. The approach enables changes to improve energy efficiency even when a DC is experiencing frequent changes that could affect its cooling requirements, a common occurence. Despite ongoing changes to the DC's compute equipment and physical layout, during the first six months of work four computer room air conditioning (CRAC) units were turned off, CRAC unit control settings were adjusted four times, and the average CRAC unit return air temperature increased by just under 2 °C. The approach was further applied to field experiments to quantify the effects of changes to HPC platform infrastructure on node or server temperatures and for root cause identification when high temperatures were suddenly reported from an HPC platform. Our approach to improving energy efficiency is applicable to DCs that house a range of compute technologies and infrastructures.

[*]This work has been authored by an employee or employees of Los Alamos National Security, LLC under contract with the US Department of Energy. Accordingly, the US Government retains an irrevocable, nonexclusive, royalty-free license to publish, translate, reproduce, use, or dispose of the published form of the work and to authorize others to do the same for US Government purposes. Los Alamos National Laboratory strongly supports academic freedom and a researchers right to publish; as an institution, however, the Laboratory does not endorse the viewpoint of a publication or guarantee its technical correctness. This work is published under LA-UR-18-22956. Figures 13.1–13.3 originally appeared in Storlie, C., Reich, B., Rust, W., Ticknor, L., Bonnie, M., Montoya, M., and Michalak, S. (2017) Spatiotemporal modeling of node temperatures in supercomputers, *Journal of the American Statistical Association*, 112(517), 92–108, published by Taylor and Francis Ltd., www.tandfonline.com, and are reprinted by permission of the publisher and the American Statistical Association, www.amstat.org.

*Analytic Methods in Systems and Software Testing*, First Edition.
Edited by Ron S. Kenett, Fabrizio Ruggeri, and Frederick W. Faltin.
© 2018 John Wiley & Sons Ltd. Published 2018 by John Wiley & Sons Ltd.

## 13.1   Introduction

A significant fraction of the power required to operate a high performance computing (HPC) platform, i.e., a supercomputer, as described in Section 13.3, may be used for cooling. Given the power required to cool HPC platforms of increasing sizes, the efficient cooling of such systems has become crucial. Similar concerns exist for DCs housing other compute infrastructure such as those used for cloud computing and by companies such as Facebook (Manese, 2014).

Traditionally, HPC DCs have been kept quite cool as this was thought to be conservative in terms of avoiding heat-related compute equipment damage. However, newer architectures may be more robust and able to run at higher temperatures without adverse effects. This presents an opportunity to increase energy efficiency.

Recently, a team at Los Alamos National Laboratory (LANL) developed an approach to increasing a DC's energy efficiency that can be applied to DCs of a variety of sizes and with a variety of types of compute equipment and cooling infrastructures. It can further be applied to a range of changes that increase energy efficiency including changes made to the physical layout of compute equipment in a DC and to its cooling system and infrastructure. This approach to energy efficiency is discussed in the context of an HPC DC, but is more broadly applicable.

The approach is grounded in instrumented temperature data collection that enables the development of statistical models for compute equipment operating temperatures and other temperature measurements. The results of the statistical models can be used to baseline equipment temperatures before changes are made and to investigate the effects of changes to the DC's cooling configuration, its infrastructure, and the compute resources it houses. Graphical communication of results, e.g. predicted compute equipment temperatures, is a crucial aspect because easily interpretable and actionable information is needed, and the team undertaking the energy efficiency improvements may include individuals with varying knowledge of statistics.

The approach was developed in the context of improving energy efficiency in an almost 1162 m$^2$ LANL DC that uses computer room air conditioning (CRAC) units for cooling and contains multiple air-cooled HPC platforms, storage systems, and support servers. Specifically, the approach permitted the supply air temperature (SAT) of CRAC units to be increased and certain CRAC units to be turned off. This enabled increased cooling efficiency of air-cooled HPC platforms because the efficiency of CRAC units tends to increase linearly with warmer return air temperatures (RAT), up to a certain maximum RAT, and because fewer CRAC units were used to cool the DC. Reducing the number of CRAC units used to cool a DC, i.e. turning off CRAC units previously used for cooling which are not needed for adequate cooling, leads to decreased energy consumption and, moreover, decreased heat load in the DC that must be cooled since each CRAC unit has a motor that generates heat. At the same time, enough CRAC units must remain operational to maintain adequate static air pressure through the perforated tiles that provide cool air for air cooling of HPC platforms. Changes to the compute systems in the DC for increased energy efficiency, i.e. the removal of unnecessary drive trays discussed in Section 13.4.10, were also implemented.

This work discusses our approach to improving DC energy efficiency and outcomes for the DC mentioned above, including results based on controlled experiments performed in a production setting and root cause identification for sudden extreme temperatures. It stresses the use of statistical modeling and graphical communication of results in the context of teams that include individuals with different backgrounds in statistics. The focus is system testing, specifically the testing of HPC platforms, for DC energy efficiency improvement and management, the temperature monitoring infrastructure that supports such testing, and the energy efficiency improvements that such testing enables. A companion paper (Storlie et al., 2013) details the statistical models used as part of the temperature monitoring infrastructure. Throughout, all temperatures are in degrees Celsius (°C).

## 13.2   Related Work

Several recent US government initiatives have emphasized the need for increased DC energy efficiency. In the summer of 2014 the US Department of Energy (DOE), as part of the its Better Buildings Challenge (US Department of Energy, 2014, 2017), launched the Data Center Energy Challenge, with a goal of improving the efficiency of federal enterprise data centers by 20% by 2020. Relatedly, a Center of Expertise for Energy Efficiency in Data Centers was instituted at Lawrence Berkeley National Laboratory (Lawrence Berkeley National Laboratory, 2012). Various US government organizations have undertaken efforts related to DC energy efficiency, e.g. Mahdavi (2015), Bailey (2013), Johnston (2012), Bates et al. (2014), Zhang et al. (2014), Rath (2014), CleanTechnica (2013), Hammond (2013), Maxwell et al. (2015), Miller (2009), Sandia National Laboratories (2014), and Laros et al. (2018). The Green500 list encourages energy efficiency by ranking the 500 most energy efficient supercomputers in the world (Sharma et al., 2017; Feng and Cameron, 2007; Feng and Scogland, 2017). The Energy Efficiency HPC Working Group (EE HPC WG, 2018) works to improve HPC energy efficiency and has produced documents related to power measurement (EE HPC WG, 2017), liquid cooling (Martinez et al., 1979), and procurements (EE HPC WG, 2015).

A discussion of interactions between HPC DCs and electric suppliers and various approaches to energy efficiency is found in Bates et al. (2014). A "four-pillar framework for energy efficient HPC data centers" is presented in Wilde et al. (2014), and Schöne et al. (2014) discusses energy efficiency measurement and tuning on HPC platforms. A toolset to enable energy efficiency evaluations is presented in Shoukourian et al. (2008). Static and dynamic power management techniques are reviewed in Valentini et al. (2014). A number of authors address energy efficiency in the context of cloud computing, e.g. Lee and Zomaya (2012), Beloglazov et al. (2012), Kliazovich et al. (2014), Berl et al. (2010), Beloglazov and Buyya (2010), Miettinen and Nurminen (2005), and Buyya et al. (2010). "Energy-aware computing methods" for a variety of large-scale computing infrastructures and several unaddressed research questions are reviewed in Cai et al. (2011), and Ranganathan (2007) presents principles that can increase the energy efficiency of computing. Haaland et al. (2010) provides a framework for modeling DC cooling system behavior and predicting such behavior following changes to the cooling configuration. This work complements that in Haaland et al. (2010) by focusing on how node or server temperatures respond to changes in the DC's cooling configuration, its infrastructure, and the compute equipment it houses.

## 13.3   Background

High performance computing platforms, i.e., supercomputers, are composed of compute nodes or servers that are connected by an interconnect that enables the communication among nodes needed for large-scale computations. A compute node or server, referred to as a node in this work, has one or more processors, memory, and the networking components required for communication with other nodes. The nodes in a supercomputer are placed in racks, which are vertical cabinets that house nodes in horizontal rows, with each row containing multiple nodes in the HPC platforms considered here. The racks are arranged in aisles, with each aisle including multiple adjacent racks. Networking components are also housed in racks and may be included in racks that also contain nodes or in their own racks which are then interspersed with racks of nodes.

The HPC platforms discussed here are used primarily for large-scale parallel scientific calculations, e.g. a large-scale scientific simulation such as a climate simulation or a molecular dynamics simulation, that require many nodes for long periods of time. Table 13.1 provides details of the architectures of the platforms in the DC considered here.

As discussed in Section 13.4.1, if a node's operating temperature is sufficiently high it will automatically shut down in order to prevent temperature-related hardware damage. There is a computational cost to one or more nodes experiencing a temperature-related shutdown (or other crash or outage) during a large parallel scientific computation. This cost occurs because the remaining nodes performing the scientific computation must typically wait for the work being performed on the nodes that were shut down to be transferred to other nodes and completed before the entire calculation can be finished. This cost increases for applications that require greater numbers of nodes because then more nodes must wait for the results from the nodes that were shut down. Given the typical heavy demand for HPC platforms, this waiting time decreases the amount of scientific computation that can be performed on an HPC platform. Thus, preventing

**Table 13.1** Architectures of HPC platforms in the DC under consideration. Wolf was installed after certain changes to the DC cooling configuration had been made.

| Platform | Compute node processor(s) | Processors / compute node | Memory | Memory / compute node (GB) | # compute nodes | Network |
|---|---|---|---|---|---|---|
| Moonlight | 8-core Intel Xeon ES-2670 Nvidia Telsa M2090 | 2  2 | 1600 MHz DDR3 | 32 GB | 308 | Qlogic InfiniBand QDR Fat-Tree |
| Mustang | 12-core AMD Opteron 6176 | 2 | 1333 MHz DDR3 | 64 GB | 1600 | Mellanox InfiniBand QDR Fat-Tree |
| Pinto | 8-core Intel Xeon ES-2670 | 2 | 1600 MHz DDR3 | 32 GB | 154 | Qlogic InfiniBand QDR Fat-Tree |
| Wolf | 8-core Intel Xeon ES-2670 | 2 | 1866 MHz DDR3 | 64 GB | 616 | Qlogic InfiniBand QDR Fat-Tree |

temperature-related node shutdowns is a crucial consideration when making changes to improve energy efficiency.

In many DCs, the configurations of the compute resources and of the physical infrastructure are subject to frequent changes. These changes include a range of actions such as replacing/rearranging solid and perforated tiles, modifying power distribution and cabling, and adding or removing compute equipment. For example, during one six-month period for the DC considered here, three HPC platforms were decommissioned and removed from the DC and one new HPC platform was installed. In addition, modifications to cabling for networks and file systems and to power distribution to accomodate the new HPC platform were made. Communication about such changes is important because they can require the cooling configuration to be reevaluated and possibly reconfigured. Revision of the temperature monitoring infrastructure may also be needed so that it includes the collection of temperature measurements from the new equipment in the DC or temperature measurements that reflect the new layout of the DC.

Changes to a DC like these can make it difficult to separate the effects of changes to the DC cooling configuration on ambient and compute equipment temperatures from the effects of other changes to the DC (such as the addition or removal of HPC platforms) on ambient and compute equipment temperatures. Rigorous quantification of the effects of a change to the cooling configuration on hardware temperatures is straightforward if steady-state temperatures are recorded before modifying the cooling configuration and then following modifications the DC temperature is allowed to stabilize and any changes to compute equipment temperatures are assessed before further changes to the DC are made.

Sometimes, changes to a DC's cooling system will be undertaken during a time in which other changes to the DC are difficult to postpone so that it is not possible to suspend changes in order to estimate the effects of changes to the cooling configuration. In this case, the focus may instead be on using the temperature monitoring infrastructure to assess the compute hardware in the DC for elevated temperatures. If compute hardware operational temperatures are not approaching their limits with the current DC cooling and infrastructure configurations, then minor changes to the cooling configuration that are anticipated to modestly increase compute hardware operational temperatures can likely be undertaken without deleterious effects.

## 13.4    Approach to Improving DC Energy Efficiency

This section presents our approach to increasing DC energy efficiency. This includes a temperature monitoring infrastructure (temperature measurement collection, metrics for assessing measured temperatures and energy efficiency, statistical models for the temperature data, and graphical communication of results) and a process for improving a DC's energy efficiency. The steps in the development of the temperature monitoring infrastructure and the process for improving DC energy efficiency will likely be interspersed in time, and are so presented in the following sections. To illustrate the approach, highlights from its application to the LANL DC are discussed. Node temperatures were of greatest concern and thus are the focus in the sections below. Of course, the process applies to other compute equipment temperatures too.

### 13.4.1 Determine Hardware Temperature Limits and Metrics of Interest

Because it is important to prevent any changes from causing compute equipment in the DC to operate at temperatures that are too high, hardware operating temperature limits should be verified for all of the compute equipment in the DC. In addition, all equipment owners should be notified of the coming changes to the DC cooling configuration so they can be alert for any temperature-related issues.

As an example, Table 13.2 presents monitoring thresholds and factory limits for different compute equipment in the DC during the time period considered here. The support servers in the DC had the lowest maximum operating temperature, 35 °C. For the three HPC platforms that remained in the DC during the course of the cooling changes (Moonlight, Mustang, and Pinto) and a fourth HPC platform (Wolf) that was installed after the changes to the DC described in Section 13.5 had been completed, the monitoring thresholds and factory limits are for temperature measurements near the CPU sockets. The monitoring thresholds are set a little lower than the factory limits, with a warning issued when a node reports a temperature at or above its monitoring threshold warning value. When a node temperature reaches its monitoring threshold high value, it is taken out of service after it completes any job that is currently running on it and inspected for needed repairs. When a node reaches its highest factory limit value, it automatically shuts down.

Before making any changes to improve a DC's energy efficiency, metrics for quantifying the effects of given cooling, DC, and compute equipment configurations and changes to them should be defined. These can include metrics that measure energy efficiency and metrics that reflect the highest temperatures that compute hardware attain, which help ensure that all compute hardware is sufficiently cooled.

In addition, a code that is used as a temperature testing benchmark to estimate worst-case (highest) node temperatures is needed. A temperature testing benchmark code is necessary because the user demands on a system can vary with time, so that temperatures from user jobs during a short period of testing may not be representative of the highest temperatures user jobs will attain over longer time periods. Also, the use of a benchmark code facilitates testing during dedicated system time (DST) used for system maintenance when users cannot run jobs. Temperature testing during DST may be preferred because it should minimize the impact of testing on users. The benchmark

**Table 13.2** Compute hardware operating temperature limits in degrees Celsius.

| System | Monitoring threshold warning | Monitoring threshold high | Factory limit high | Factory limit critical |
|---|---|---|---|---|
| Support servers | — | — | Operating: 10 to 35 | |
| Moonlight | 89 | 95 | 90 | 100 |
| Mustang | 59 | 65 | 70 | 70 |
| Pinto | 89 | 95 | 90 | 100 |
| Panasas/NetApp | — | — | — | — |
| Tape media | — | — | 10 | 45 |
| Wolf | 89 | 95 | 90 | 100 |

should have input parameters that can be tuned to produce compute equipment temperatures that are a little warmer than the warmest user jobs.

For this study, High Performance Linpack (HPL; Petitet et al., 2010), a freely available software package that solves the equation $Ax = b$ for $x$, was used. High Performance Linpack can be used to test the temperature response of nodes in HPC platforms to demanding user jobs because it is computationally intensive and is highly configurable so it can be tuned to yield node temperatures a little higher than the highest temperatures from user jobs.

The temperature testing benchmark can be used to calculate one or more temperature metrics that are used to investigate and communicate the current state of compute equipment temperatures. The temperature metric(s) should reflect the need to prevent equipment from operating at temperatures that are too high.

The temperature metric for the nodes in the DC was intended to reflect the worst-case (highest) node temperature in the DC. The goal was that any heat-related failures, i.e. any nodes that reached a warning or high temperature threshold, would be the result of one or more underlying hardware issue(s) and not the result of changes intended to increase energy efficiency. The nodes in the HPC platforms could supply temperature measurements near their CPU sockets, and these temperature measurements were used for the temperature metric for HPC platform nodes.

Using HPL and the CPU temperature measurements, two temperature metrics were considered. The first provides a 95% upper prediction bound for the highest temperature each node would attain while running HPL for 24 hours. The second is platform-specific and provides a 95% upper simultaneous bound for the highest temperature one or more compute nodes would achieve if HPL were run for 24 hours on all of the compute nodes in a particular platform. This second bound is more stringent than the first because a 95% upper simultaneous bound is defined as the temperature for which there is only a 5% change that any one or more nodes would exceed it. Other metrics could be developed, and the metrics above could, of course, be used with different limits, e.g. 99% instead of 95%. Both metrics can be calculated using the results of the statistical modeling described in Section 13.4.8 and in Storlie et al. (2013). This chapter focuses on the first metric, referred to as Max24Hours since it provides a 95% upper bound for the maximum temperature each node would reach if it ran HPL for 24 hours, because it was used in practice for the changes to increase energy efficiency described here and provides information about variability in node temperatures within each platform.

### 13.4.2 Maximize Energy Efficiency of the DC before Changing its Cooling Configuration

Before making any changes to the cooling configuration, other energy efficiency measures should be implemented. For the work here, many optimizations to the DC had already been undertaken, including cold aisle containment, the installation of cold locks on floor tiles, the addition of building automation systems, and upgrades to the CRAC unit controllers.

Before implementing changes to the cooling configuration, some additional changes to improve energy efficiency were made. To guide this work, the TileFlow data center computational fluid dynamics modeling software simulation model (Innovative Research, 2014) was fitted to the DC. TileFlow uses the tile layout, the equipment in the DC and its airflow pattern (here, from the front to the back of a compute node), and the cooling configuration of the DC (here, the cubic feet per minute of supply air each CRAC unit can provide) to provide estimates of the RATs at the CRAC units, and of air flow, temperatures (inlet and outlet to the racks), and air pressure beneath the floor at different locations in the DC. Based on the results of the TileFlow model and standard tile placement guidelines from the American Society of Heating, Refrigerating, and Air-Conditioning Engineers (ASHRAE; American Society of Heating and Air-Conditioning Engineers, 2012) and the Uptime Institute (Pitt Turner et al., 2014), the configuration of solid and perforated floor tiles was changed to increase cooling and air flow in targeted locations in the DC and to increase the DC's overall efficiency.

### 13.4.3 Determine Excess Cooling Capacity

Before making changes to the DC cooling configuration, the presence of excess cooling capacity with the current cooling configuration should be determined.

As mentioned in Section 13.1, with HPC platforms that are air cooled, a minimum air flow is required for sufficient cooling of compute hardware. Thus, it is necessary to determine whether there is sufficient air flow to be able to make changes to the cooling configuration, e.g. turn off one or more CRAC units and/or raise the SAT. For the DC considered here, the results of the TileFlow model described in Section 13.4.2 suggested that there was sufficient air flow to make changes.

### 13.4.4 Instrument Temperature Data Collection

In order to monitor changes in compute equipment operating temperatures, temperature data collection must be instrumented. This includes deciding what temperature data to collect, how often to collect it, and how to manage it. For each temperature measurement it is important to include a timestamp with the date and time of the measurement; the compute equipment reporting the measurement, e.g. the HPC platform and node reporting the measurement; and, if multiple temperature readings are provided for a given piece of compute equipment, e.g. a node, the location within the piece of equipment reporting the measurement. While the most accurate method of determining operational compute hardware temperatures can be to collect temperature data from the hardware's temperature sensors, ambient temperature measurements in the DC can complement such data and provide temperature information for hardware with no temperature reporting capability. It is important that temperature data collection systems be configured to alert facilities and/or compute system personnel when sufficiently high temperatures are reached.

The compute nodes in the HPC platforms in the DC were able to report temperature measurements near their CPU sockets. As described in Section 13.4.1, prior to the cooling system changes, the temperature of each compute node was monitored, with specific actions taken when it exceeded certain thresholds. To support cooling configuration changes, the protocol was changed so that temperatures were collected every half hour from each node and stored in a Zenoss database (Zenoss, 2018).

For the remaining systems in the DC (support servers and disk and tape storage), hardware temperature data were not readily available. Instead, data from 54 Environet sensors (Geist, 2016) mounted on rack doors five feet from the floor were used to monitor the temperatures of those systems. The Environet sensor data provide a much coarser description of compute system temperatures than do temperature measurements from the compute hardware itself, both because there are not many Environet sensors per square foot and because they provide ambient air temperatures on rack doors rather than temperatures within compute systems themselves. Nonetheless, data from the Environet sensors can be used to monitor changes in ambient temperatures. Before the work described here was undertaken, Environet data were collected every 15 minutes. This collection frequency was increased to every minute before the initial temperature testing described in Section 13.4.7. After that, temperatures from the Environet sensors were collected every minute during the cooling configuration changes.

### 13.4.5 Characterize Operating Temperatures from User Jobs and Instrument Temperature Testing Benchmark

It is important to tune the input parameters of the code used as a temperature testing benchmark so that the maximum equilibrium temperatures compute equipment reach while running the benchmark are a little higher than the warmest temperatures attained during user jobs. This can be accomplished by first characterizing the temperatures of user jobs and then performing an experiment that varies relevant input parameters to the temperature testing benchmark code to find parameter settings that produce maximum equilibrium temperatures that are a little warmer than the warmest user jobs. The following two paragraphs detail how this was undertaken for HPL temperature testing on the Mustang platform. As explained in Section 13.6, this calibration of HPL temperatures to the warmest temperatures produced by user jobs was undertaken after the energy efficiency improvements described in Section 13.5 had been completed because the effect of the HPL input parameters values on node temperatures was not initially appreciated. In general, it is recommended that this calibration be performed *before* the initial characterization of node temperatures described in Section 13.4.7.

The investigation of node temperatures experienced during user jobs on Mustang proceeded as follows. Temperature data from all user jobs from a roughly four-month period were collected, and the maximum temperature that each node running a particular job attained was recorded. Based on the maximum temperatures recorded for each user, the 40 users whose jobs yielded the highest node temperatures were identified. The node temperatures from all of the jobs run by each of these 40 users were used to represent high-node-temperature jobs.

Next, an experiment was conducted in which HPL was run with 26 different combinations of the HPL input parameters $N$, the problem size or size of the matrix being inverted, and $NB$, the block size or amount of work each compute node performed in parallel to complete the HPL calculation. The 26 combinations included 25 combinations chosen using a Latin hypercube sample (McKay et al., 1979) in which values of $N \in 5000, \dots, 20000$ and values of $NB \in 1, \dots, 50$ were chosen, and previously used default HPL settings of $N = 12800$ and $NB = 1$, referred to as HPL1 below. All 26 HPL jobs were run on each of about 900 nodes during production use, i.e. times when users were able to run jobs on Mustang, because a DST period was not available for the testing. The spatial distribution of the Mustang nodes that ran these tests was close to uniform, meaning that they should provide a representative sample of hot and cold areas in Mustang and those in between. Node temperatures were found to be very sensitive to the $NB$ value, especially when $NB$ was small, but to have little sensitivity to the value of $N$.

Based on a comparison of the experimental results with the temperatures attained during high-node-temperature user jobs, an HPL benchmark with $N = 12800$ and $NB = 4$ was chosen for temperature testing on Mustang. A similar process could be used for other compute equipment.

### 13.4.6  Develop Method for Characterizing Compute Equipment Temperatures to Support Increasing Energy Efficiency

It is important to characterize compute equipment temperatures using the benchmark temperature testing code both before any changes are made to the cooling configuration and during the process of making changes to improve cooling efficiency. Hence, compute equipment temperatures may be characterized multiple times during the process of improving DC energy efficiency, and it can be important to minimize the impact of the temperature testing on users.

In order to decrease inconvenience to users, the temperature testing can be performed during DST or other times when the compute systems are not available to users. If that is not possible, with HPC platforms the impact of testing compute nodes during production use when the system is available for user jobs can be minimized by testing on otherwise idle nodes with no reservation, i.e. nodes that are not being used or held for an upcoming user job. While testing during production use has the potential to affect users, it offers several benefits. First, it provides more immediate notification of temperature changes and issues because it can occur in an ongoing manner rather than during DSTs, which may occur infrequently. Second, SATs can vary with time, and ongoing testing during production use helps to ensure that testing occurs at a range of SATs that includes the warmest SATs. A short period of testing during DST could occur by chance during a period of low SATs and hence underestimate node temperatures that might be obtained when SATs are higher.

For this work, minimizing the impact on users was crucial. Therefore, the desire was to use DSTs to characterize node temperatures for all of the HPC platforms in the DC that were not scheduled for decommissioning and removal. Each HPC platform had a different scheduled monthly DST period. To provide worst-case node temperatures, it was important to maximize the computational load on all of the HPC platforms in the DC during the temperature testing, and thus the heat generated by all of the HPC platforms in the DC. Therefore, the temperature testing protocol was to run HPL jobs on all nodes on the HPC platform that was on DST, and to run preemptible single-node HPL calculations on all otherwise idle nodes without a reservation on the HPC platforms that were not on DST. The preemptible HPL calculations that were performed on the HPC platforms that were not on DST were used to maximize the computational load on the HPC platforms not on DST while minimizing any impact on users.

During the testing, temperatures were recorded every minute from the HPC platform on DST. Temperatures were collected less frequently on the non-DST HPC platforms in order to lessen the impact of temperature collection on user jobs. HPC platform operators were advised of the HPL temperature testing so they could be alert for temperature-related issues.

The maintenance work required during some DSTs was sufficiently lengthy that performing HPL temperature testing during the DST was not possible. Because the ability to perform HPL testing during DST was uncertain, the protocol used in the LANL DC switched to performing HPL temperature testing during production use by running HPL temperature testing jobs on otherwise idle nodes with no reservation as described above.

### 13.4.7 Characterize Compute Equipment Temperatures before Changing the DC Cooling Configuration

Before making any changes to the cooling configuration, baseline temperature testing should be conducted using a temperature testing benchmark code that has been tuned to produce temperatures a little warmer than the warmest user jobs; see Sections 13.4.1 and 13.4.5. Such baseline testing provides information about compute equipment temperatures that can inform initial changes to the DC's cooling configuration. This testing can also be used to assess the amount of time required for the compute equipment to reach maximum equilibrium temperatures and whether starting temperatures affect maximum equilibrium temperatures. Determining the amount of time required to reach maximum equilibrium temperatures enables efficient future temperature testing that uses only the time required for accurate measurement of maximum equilibrium temperatures. If it is determined that starting temperature does affect maximum equilibrium temperature for certain compute equipment, worst-case temperature testing could be based on worst-case starting temperatures.

For the HPC platforms tested here, a compute node typically reached its maximum equilibrium temperature after about five minutes of running HPL, although to be conservative only temperature measurements taken after ten minutes of running HPL were assumed to be representative of a node's maximum equilibrium temperature. Starting node temperatures did not appear to affect their maximum equilibrium temperatures.

### 13.4.8 Develop Statistical Models and Graphical Summaries for Temperature Data

Statistical modeling results and graphical summaries can effectively communicate the outcomes of temperature testing, provide information that can guide future changes to improve a DC's energy efficiency, and enable identification of interesting features in compute equipment temperatures. The required statistical models and graphical methods can be developed after the baseline temperature characterization described in Section 13.4.7 has been completed, and then revised as necessary as changes are made to the DC cooling configuration, its infrastructure, and the compute resources it houses.

The statistical models developed for the node temperature data, explained in detail in Storlie et al. (2013), enabled estimation of the effects of cooling changes, of the dependency of one node's temperature on the temperatures of nearby nodes in the same HPC platform, of the dependency of a given node's current temperature on its past temperatures, and of the Max24Hours metric used in graphical communication of the temperature testing results.

A hierarchical model was used for the node temperature measurements. This model incorporated Normal spatial random effects associated with covariates that captured the cooling, DC infrastructure, and relevant node-specific conditions under which different temperature measurements were taken, a residual meta-Gaussian process (meta-GP;

Demarta and McNeil, 2005) that accounted for any additional spatio-temporal variation, and Normal measurement errors associated with each temperature measurement.

With the model, the temperature for a given node at a particular time depended on the configuration of the DC, its cooling infrastructure, and relevant node-specific conditions at the time the measurement was taken via a regression on covariates that captured those conditions. In this regression, each covariate had an associated coefficient vector that included an entry for each node. Each vector of regression coefficients followed a Gaussian Markov random field (GMRF) with a conditionally autoregressive mean that reflected the different ways nodes could be neighbors and the strengths of association between temperatures for different types of node neighbors. The model estimated spatial correlation in temperatures for neighboring nodes with three major categories of spatial relationships with each other: (1) nodes that were next to each other, (2) nodes that were above or below each other, and (3) nodes that were across an aisle from each other. As detailed in Storlie et al. (2013), the meaning of each of these three spatial relationships depended on the node and the geography of the HPC platform that housed it. As an example, Mustang nodes could be next to each other in three ways: they could be (1) next to each other within the same rack; (2) next to each other, but in different racks; or (3) next to each other, but with a rack of network components between them. Mustang nodes could further be above or below each other in two ways or across an aisle from each other in two ways, leading to seven types of Mustang node neighbors. With the GMRF mean structure, the mean of a particular regression coefficient for a given node depended on the regression coefficients of its neighboring nodes for that covariate and the type of neighbor each neighboring node was via a weighted average that reflected the strength of association between the temperature measurements of node neighbors of different types. The precision of the regression coefficient for a given node depended on the number of each of the different types of neighbors the node had and the previously mentioned weights that reflect the strengths of association between temperature measurements from nodes with different neighbor relationships. This model was chosen to allow for a variety of levels of dependence between the temperatures of nodes that were different types of neighbors, including the high correlations present between temperature measurements from certain types of neighbors described below. This model fitted the data better than a more common CAR model found in Reich et al. (2010), which led to much lower estimates of the higher correlations discussed below.

Early investigation indicated that the tail of the residual process would not be well approximated by a Normal distribution. Hence, the marginal distribution of the residual process for each node was assumed to be Normal for most values and to have a generalized Pareto distribution in its upper tail. A Gaussian process copula (Nelson, 2014; Sang and Gelfand, 2006) was developed that resulted in the desired meta-GP. Such meta-GP models do not account for asymptotic dependence of tail values (Frahm et al., 2005; Coles et al., 2013), and the node temperature data did not suggest that such asymptotic dependency existed. To ease the computation burden, it was assumed that the residual process was independent for different nodes, and this was found to be reasonable upon investigation of residuals after the node means, which varied spatially, were accounted for.

Diffuse priors were developed based on discussion with individuals on the cooling project team. A Markov chain Monte Carlo method was used to fit the model, and the resulting estimates were not found to be overly sensitive to the prior specification.

Both spatial and temporal correlation were apparent in the Mustang node temperature data. For Mustang, the highest estimated spatial correlations occurred between temperatures in nodes that were directly above, below, or beside each other, with estimated correlations of 0.87 or slightly higher. Estimated correlations between temperatures in nodes that were above, beside, or below each other with a shelf or the side of a rack between them were just under 0.85, while the estimated correlation between temperatures in nodes that were beside each other but separated by a rack of networking equipment was just under 0.76. The temperatures of nodes that were across an aisle from each other had much lower estimated correlations of about 0.56. The temperature measurements within a Mustang node were correlated over time, with an estimated correlation of 0.96 for temperatures that were a minute apart. See Section 13.4.10 for information about the effects of changing rack configurations on node temperatures.

Graphical communication of the modeling results can clearly and efficiently convey the current status of equipment temperatures to the team directing the cooling changes and can further facilitate the identification of hot spots and other important aspects of equipment temperatures. The graphical communication of results used for this work displayed the Max24Hours metric for all nodes in a given platform according to the HPC platform's physical layout. As an example, Figure 13.1 provides a 95% upper prediction bound for the highest temperature each Mustang node would achieve while running HPL1 for 24 hours before any changes were made to the DC cooling configuration. In Figure 13.1 the values for all nodes are substantially below the monitoring and factory limits for node temperatures.

### 13.4.9 Improve DC Cooling Efficiency and Recharacterize Compute Equipment Temperatures

If the results of the statistical modeling of the baseline temperature testing data suggest changes to the DC cooling configuration can be undertaken without deleterious effects, DC cooling efficiency improvements can be implemented. It is recommended that an energy efficiency improvement to the DC cooling configuration be followed by a period during which temperatures in the DC come to equilibrium before compute equipment temperatures are recharacterized. Temperature testing before each additional energy efficiency improvement is also recommended to ensure that temperatures are likely to stay within their operational limits following the improvement.

During the first eight months of work on the DC cooling configuration, four changes were made to the CRAC unit controls to increase SATs, four CRAC units were turned off, and some additional changes were made to the tile configuration in the DC. The process for making energy efficiency improvements to the DC's cooling configuration included implementing the change followed by temperature testing with HPL to assess whether additional changes to the cooling system were likely to yield temperatures that were too high. Temperature testing focused on Mustang because early testing indicated that it was the HPC platform that was most likely to experience high temperatures.

### 13.4.10 Use Temperature Monitoring Infrastructure to Conduct Pilot Studies of Possible Changes to DC or Compute Infrastructure

The temperature monitoring infrastructure developed in support of DC energy efficiency improvement can be leveraged to conduct pilot studies of certain changes

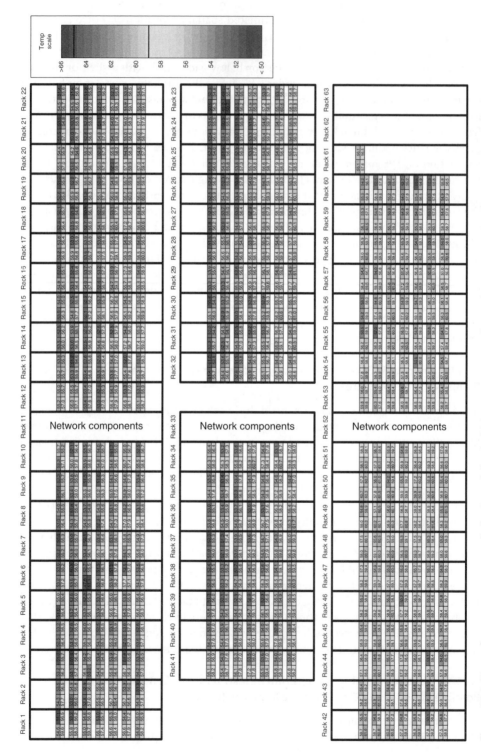

**Figure 13.1** Ninety-five percent upper prediction bounds for the highest temperature that each Mustang node would attain while running HPL1 for 24 hours (Max24Hours) before any changes were made to the DC cooling configuration. This figure from Storlie et al. (2017), published by Taylor and Francis Ltd. in the *Journal of the American Statistical Association* (http://www.tandfonline.com/toc/uasa20/current), reprinted by permission of Taylor and Francis Ltd. and the American Statistical Association. See color section.

intended to increase energy efficiency. This enables quantification of the effects of a change before it is implemented on a large scale, which can be expensive.

As an example, Mustang racks included both drive trays in slots that did not house hard drives and doors that could be removed. Removing drive trays and/or doors could lead to increased air flow to the nodes and hence cooler node temperatures, resulting in reduced need for cooling and increased energy efficiency.

An experiment was conducted on Mustang to more rigorously quantify the effects of removing rack doors and/or drive trays on node temperatures. Drive trays were removed from half of the Mustang racks (those with even numbers) before the experiment began. First, HPL1, the version of HPL with the default input parameters $N = 12800$ and $NB = 1$ that had been used in previous temperature characterization, was run on all Mustang nodes with all of the rack doors closed and then with the doors on the even racks open, where opening the rack doors was a proxy for the more labor-intensive act of removing them. Next, HPL with the input parameters $N = 12800$ and $NB = 4$ that were tuned to the warmest user jobs was run on all nodes with the doors on the even racks open. (Time constraints prevented further testing.) Revision of the statistical model described in Section 13.4.8 to account for the test conditions led to an average effect of removing drive trays on Mustang node temperatures of $-2.8\,°C$, with some node temperatures decreasing by as much as $5\,°C$. The effect of opening rack doors after the drive trays had been removed was more modest, roughly $-0.1\,°C$ on average and at most about $-0.9\,°C$. Because of these results, the drive trays in all odd racks were also removed, but rack doors were not removed.

### 13.4.11 Continue Temperature Monitoring After Changes Are Completed

The dynamic nature of DCs and user demands on them underscores the need for ongoing temperature monitoring even after all energy efficiency improvements have been completed. In particular, users may start using new codes or code versions that are more optimized than previous versions, both of which can lead to higher node temperatures and a need to revise the temperature testing benchmark. Modifications of the DC and the equipment it houses can also change cooling needs. Section 13.6 provides a case study that demonstrates the need for ongoing monitoring and the strengths of the approach presented here for determining root cause when nodes suddenly experience high temperatures.

## 13.5  Results

Our approach to increasing energy efficiency used the temperature monitoring infrastructure and process outlined in the previous section to successfully increase the DC's energy efficiency before changes were made to its cooling system, to determine that energy efficiency improvements made to the DC cooling configuration would likely not lead to compute equipment operating at temperatures that were too high, and then to make energy efficiency improvements to the DC's cooling configuration. Initially, the DC included 18 CRAC units supplying air at a temperature of $13\,°C$, with an average RAT of about $18.9\,°C$. Roughly six months later, 14 CRAC units were running, their average RAT had increased by a little less than $2\,°C$ degrees to about $20.7\,°C$, and their minimum RAT had increased by nearly 4.5 °C. Further, each CRAC unit was capable of cooling approximately 11 kW more than before the cooling configuration changes.

## 13.6 Application to Root Cause Determination

This section describes application of the temperature monitoring infrastructure to root cause determination when Mustang nodes suddenly experienced high temperatures, including temperatures that were close to or reached 70 °C, the temperature at which Mustang nodes automatically shut down. This occurred after the energy efficiency improvements to the cooling configuration described in Section 13.5 had been successfully implemented.

Just prior to the observation of high node temperatures on Mustang, two changes that could have affected the operating temperatures of Mustang nodes had occurred: (1) a new HPC platform named Wolf had been installed in the DC, and (2) a team within LANL's HPC Division began running more and longer HPL jobs on Mustang that used new HPL input parameter values tuned to investigate performance variability. This new HPL configuration is referred to as HPL2. Interest focused on how these two changes affected the operating temperatures of Mustang nodes and on identification of the root cause for the temperature-induced node shutdowns to ensure no future related issues.

It was first verified that node temperatures had in fact increased since Wolf was installed. Data from the temperature monitoring infrastructure showed that while maximum temperatures had increased, average node temperatures had not increased that much. Thus, while typical operating temperatures had not changed much, the nodes' highest operating temperatures had increased. Moreover, examination of the data revealed that HPL2 jobs experienced most of the temperature-related shutdowns. This suggested that the high Mustang node temperatures were the result of the new HPL2 jobs and not of changes to the DC's cooling configuration, the installation of Wolf, or other changes to the DC.

A statistically designed experiment was undertaken to confirm this hypothesis. This experiment and the statistical modeling of the resulting data enabled the cooling project team to quickly and efficiently understand the cause of Mustang's high node temperatures. Mustang was subjected to testing with HPL1 and with HPL2. The results confirmed that node temperatures while running HPL1 had not increased enough to explain the elevated node temperatures. Figure 13.2 provides the posterior mean effect of the deployment of Wolf and related to changes to the DC on node temperatures while running HPL1, with the unusually low temperatures observed in racks 44 and 58 discussed below. HPL2 was found to lead to node temperatures 8.5 °C degrees warmer on average than with HPL1, with a maximum increase of about 13 °C degrees for certain nodes; see Figure 13.3.

This information was crucial in responding to the increased Mustang node temperatures. Instead of increasing cooling to Mustang's nodes, which might have been the natural response given the history of changing the cooling configuration to increase energy efficiency, efforts instead focused on better management of HPL2 jobs.

The investigation of the high Mustang node temperatures further led to two additional studies, which were described previously. First, the high HPL2 temperatures motivated characterization of node temperatures attained during user jobs and investigation of

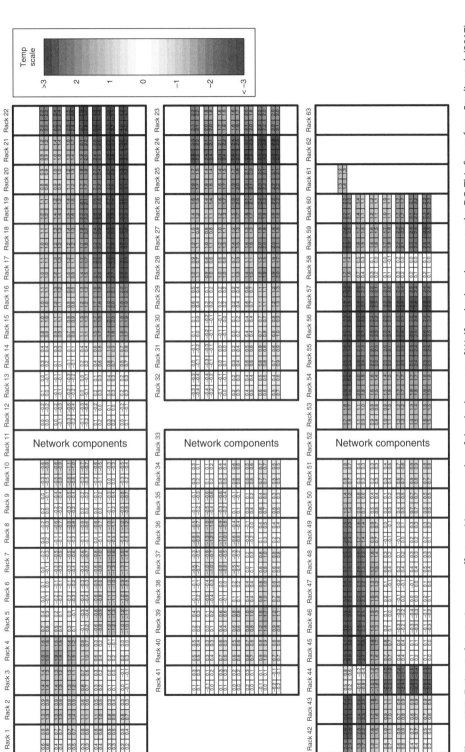

**Figure 13.2** Estimated posterior mean effect on Mustang nodes of the deployment of Wolf and other changes to the DC. This figure from Storlie et al. (2017), published by Taylor and Francis Ltd. in the *Journal of the American Statistical Association* (http://www.tandfonline.com/toc/uasa20/current), reprinted by permission of Taylor and Francis Ltd. and the American Statistical Association. See color section.

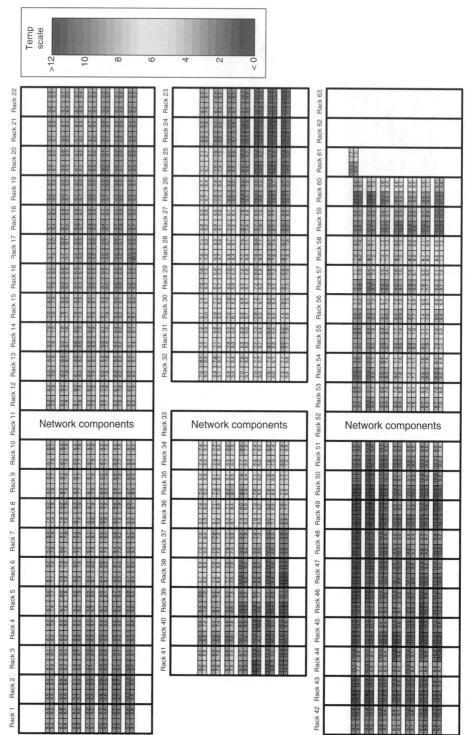

**Figure 13.3** Estimated posterior mean effect on Mustang nodes of running HPL2 instead of HPL1. This figure from Storlie et al. (2017), published by Taylor and Francis Ltd. in the *Journal of the American Statistical Association* (http://www.tandfonline.com/toc/uasa20/current), reprinted by permission of Taylor and Francis Ltd. and the American Statistical Association. See color section.

the relationship between the HPL input parameters and maximum equilibrium node temperatures while running HPL. As a result, the HPL input parameters were tuned to produce node temperatures a little higher than those produced by the warmest user jobs; see Section 13.4.5. While this process was performed after the changes described in Section 13.5 had been made to the DC cooling configuration, it is now a foundational step in our approach that should be completed before any temperature characterization studies are performed or any changes are made to the DC cooling configuration. Further, the input parameter values for the temperature testing benchmark code should be revised as needed when maximum operating node temperatures change as the result of changes to the cooling configuration or in user jobs. Second, before the experiment that investigated HPL1 and HPL2 was performed and in an effort to improve cooling, the doors on Mustang racks 44 and 58 had been removed and unnecessary drive trays installed in slots that did not house hard drives had been removed from rack 44. Figure 13.2 highlights that the nodes in these racks were estimated to have experienced much smaller increases in temperature than the nodes in nearby racks as a result of the installation of Wolf and other changes to the DC. Based on this observation, the experiment described in Section 13.4.10 was undertaken to quantify the effects of removing drive trays and doors from racks on node temperatures. The experiment used the temperature monitoring infrastructure to quantify these effects rigorously before any wide-scale implementation took place.

## 13.7 Summary

This work presents an approach to DC energy efficiency improvement that includes the development of a temperature monitoring infrastructure (temperature measurement collection, metrics for measured temperatures, statistical modeling of the data, and graphical communication of results) and a process for improving a DC's energy efficiency. A temperature monitoring infrastructure that uses a temperature testing benchmark that yields node temperatures a little higher than the highest temperatures likely to be observed during user jobs combined with statistical modeling permits estimation of the highest temperatures compute equipment is likely to experience under the current DC cooling, infrastructure, and compute equipment configurations. Graphical communication of the results using relevant temperature metrics facilitates their presentation and discussion of the results with individuals with varying backgrounds in statistics and discovery of interesting features in the temperature data. This, in turn, enables decisions about whether additional changes to improve energy efficiency that could lead to higher node temperatures are advisable.

With careful experimentation that includes periods when there are no changes to the DC or its cooling configuration, the effect of a particular change of interest could be estimated, allowing for prediction of likely changes to node temperatures following the specified change. In particular, the temperature monitoring infrastructure can be used to perform pilot studies of changes that may increase energy efficiency. Energy efficiency or temperature metrics can quantify the effects of any changes. This work used data from experiments conducted under field conditions in a production DC to demonstrate that removing unnecessary drive trays can result in decreased node temperatures so that cooling can be reduced.

This work further demonstrates that improvements to a DC's energy efficiency can be effectively undertaken even when the DC is subject to nearly ongoing changes that can affect the efficacy of a given cooling configuration. While the approach was described in the context of node temperatures because they were of greatest interest for the DC studied here, it could be applied to other compute equipment.

The temperature monitoring infrastructure can also provide early warning of increasing hardware temperatures. For this purpose, it is best to collect temperature data frequently. The temperature monitoring infrastructure used for this work leveraged existing temperature sensors near compute node CPUs and ambient temperature measurements from rack-mounted temperature sensors for systems unable to report temperature measurements. Temperature monitoring that incorporates temperature measurements and temperature operating limits from all systems and from hardware components in addition to the CPU is recommended to provide more complete monitoring of and information about the temperatures experienced by compute resources.

Plots of within-node temperatures while running HPL, such as those in Figure 13.1, suggest that node temperatures may vary substantially within an HPC platform. Such heterogeneity in node temperatures might be leveraged by thermal- or cooling-aware scheduling (Moore et al., 1999; Vanderster et al., 2013; Tang et al., 2007; Mukherjee et al., 2008; Ayoub et al., 2010; Banerjee et al., 2010; Wang et al., 2011, 2012; Liu et al., 2013; El Mehdi Diouri et al., 2013; Xu et al., 2013; Meng et al., 2012) to increase DC cooling efficiency. For example, more computationally intensive workloads could be placed on nodes that tend to run colder with a computationally intense workload.

## Acknowledgments

Many people contributed to the success this work. The authors thank Myra Branch, Gabriel De La Cruz, Michael Ferguson, Jennifer Green, Chung-Hsing Hsu, Craig Idler, Kathleen Kelly, Anthony Lopez, Cory Lueninghoener, Cindy Martin, Michael Mason, Eloy Romero Jr., Phillip Romero, Samuel Sanchez, Ben Santos, Ron Velarde, and David Walker, and apologize for any omissions from this list.

## References

American Society of Heating and Air-Conditioning Engineers (2012) *Thermal Guidelines for Data Processing Environments*, 3rd edn.

Ayoub, R., Sharifi, S., and Rosing, T.S. (2010) Gentlecool: Cooling aware proactive workload scheduling in multi-machine systems, in *Proceedings of the 2010 Conference on Design, Automation and Test in Europe*, pp. 295–298.

Bailey, A.M. (2013) LLNL data center consolidation initiative. http://svlg.org/wp-content/uploads/2013/11/Bailey-LLNL-presentation.pdf [accessed 22 January, 2018].

Banerjee, A., Mukherjee, T., Varsamopoulos, G., and Gupta, S.K. (2010) Cooling-aware and thermal-aware workload placement for green HPC data centers., in *Green Computing Conference*, pp. 245–256.

Bates, N., Ghatikar, G., Abdulla, G., Koenig, G.A., Bhalachandra, S., Sheikhalishahi, M., Patki, T., Rountree, B., and Poole, S. (2014) Electrical grid and supercomputing centers: An investigative analysis of emerging opportunities and challenges. *Informatik-Spektrum*, 38(2), pp. 111–127.

Beloglazov, A., Abawajy, J., and Buyya, R. (2012) Energy-aware resource allocation heuristics for efficient management of data centers for cloud computing. *Future Generation Computer Systems*, 28(5), 755–768.

Beloglazov, A. and Buyya, R. (2010) Energy efficient resource management in virtualized cloud data centers, in *Proceedings of the 2010 10th IEEE/ACM International Conference on Cluster, Cloud and Grid Computing*, pp. 826–831.

Berl, A., Gelenbe, E., Di Girolamo, M., Giuliani, G., De Meer, H., Dang, M.Q., and Pentikousis, K. (2010) Energy-efficient cloud computing. *The Computer Journal*, 53(7), 1045–1051.

Buyya, R., Beloglazov, A., and Abawajy, J. (2010) Energy-efficient management of data center resources for cloud computing: A vision, architectural elements, and open challenges. arXiv:1006.0308.

Cai, C., Wang, L., Khan, S.U., and Tao, J. (2011) Energy-aware high performance computing: A taxonomy study, in *Proceedings of the 2011 IEEE 17th International Conference on Parallel and Distributed Systems (ICPADS)*, pp. 953–958.

CleanTechnica (2013) NREL readies launch of world's most energy-efficient high-performance data center. http://cleantechnica.com/2013/03/12/nrel-readies-launch-of-worlds-most-energy-efficient-high-performance-data-center/ [accessed 22 January, 2018].

Coles, S.G., Heffernan, J., and Tawn, J. (2013) Dependence measures for extreme value analysis. *Extremes*, 2, 339–365.

Demarta, S. and McNeil, A.J. (2005) The *t* copula and related copulas. *International Statistical Review*, 73(1), 111–129.

EE HPC WG (2015) Energy efficiency considerations for HPC procurement documents: 2014. https://eehpcwg.llnl.gov/assets/aa_procurement_2014.pdf [accessed 22 January, 2018].

EE HPC WG (2017) Energy efficient high performance computing power measurement methodology. https://www.top500.org/static/media/uploads/methodology-2.0rc1.pdf [accessed 22 January, 2018].

EE HPC WG (2018) Energy Efficient High Performance Computing Working Group. eehpcwg.llnl.gov [accessed 22 January, 2018].

El Mehdi Diouri, M., Gluck, O., Lefevre, L., and Mignot, J.C. (2013) Your cluster is not power homogeneous: Take care when designing green schedulers!, in *Proceedings of the 2013 International Green Computing Conference (IGCC)*, pp. 1–10.

Feng, W.c. and Cameron, K.W. (2007) The Green500 List: Encouraging sustainable supercomputing. *Computer*, 40(12), 50–55.

Feng, W. and Scogland, T. (2017) The Green 500. green500.org [accessed 22 January, 2018].

Frahm, G., Junker, M., and Schmidt, R. (2005) Estimating the tail-dependence coefficient: Properties and pitfalls. *Insurance: Mathematics and Economics*, 37(1), 80–100.

Geist (2016) Environet facility. http://www.geistglobal.com/products/dcim/environet-facility [accessed 22 January, 2018].

Haaland, B., Min, W., Qian, P.Z., and Amemiya, Y. (2010) A statistical approach to thermal management of data centers under steady state and system perturbations. *Journal of the American Statistical Association*, 105(491), 1030–1041.

Hammond, S. (2013) Bytes and BTUs: Holistic approaches to data center energy efficiency. http://www.lanl.gov/conferences/salishan/salishan2013/2013 Hammond Salishan(2).pdf [accessed 22 January, 2018].

Innovative Research, Inc. (2018) Tileflow: Data center CFD modeling software. http://tileflow. com/tileflow/data-center-cfd-modeling-software-overview.html [accessed 22 January, 2018].

Johnston, D. (2014) Pilot water conservation project uses treated groundwater for cooling. https://www.llnl.gov/news/pilot-water-conservation-project-uses-treated-groundwater-cooling [accessed 22 January, 2018].

Kliazovich, D., Bouvry, P., and Khan, S.U. (2012) GreenCloud: A packet-level simulator of energy-aware cloud computing data centers. *The Journal of Supercomputing*, 62(3), 1263–1283.

Laros III,, J.H.L., DeBonis, D., Grant, R., Kelly, S.M., Levenhagen, M., Olivier, S., and Pedretti, K. (2014) *High Performance Computing – Power Application Programming Interface Specification Version 1*, Sandia Report SAND2014-17061. http://powerapi.sandia.gov/docs/PowerAPI_SAND.pdf [accessed 22 January, 2018].

Lawrence Berkeley National Laboratory (2018) Center of Expertise for Energy Efficiency in Data Centers. datacenters.lbl.gov [accessed 22 January, 2018].

Lee, Y.C. and Zomaya, A.Y. (2012) Energy efficient utilization of resources in cloud computing systems. *The Journal of Supercomputing*, 60(2), 268–280.

Liu, Z., Chen, Y., Bash, C., Wierman, A., Gmach, D., Wang, Z., Marwah, M., and Hyser, C. (2012) Renewable and cooling aware workload management for sustainable data centers. *ACM SIGMETRICS Performance Evaluation Review*, 40, 175–186.

Mahdavi, R. (2013) Energy efficiency opportunities in federal high performance computing data centers. http://energy.gov/sites/prod/files/2013/10/f3/dc_hpcc.pdf [accessed 22 January, 2018].

Manese, C. (2014) Facebook and Open Compute: Designing for Efficiency and Scale. https:// eehpcwg.llnl.gov/documents/conference/sc14/f0_manese_facebook_ sc14workshop.pdf [accessed 22 January 2018].

Martinez, D., Durbin, T., Ellsworth, M., and Tschudi, B. (2015) Systematic approach for universal commissioning plan for liquid-cooled systems. https://eehpcwg.llnl.gov/ assets/lcc1_liquid_cooling_commissioning.pdf [accessed 22 January, 2018].

Maxwell, D., Ezel, M., Donovan, M., Layton, C., and Becklehimer, J. (2014) Monitoring Cray cooling systems, in *Proceedings of the 2014 Cray Users Group*. https://cug.org/ proceedings/cug2014_proceedings/includes/files/pap181.pdf [accessed 22 January, 2018].

McKay, M.D., Beckman, R.J., and Conover, W.J. (1979) Comparison of three methods for selecting values of input variables in the analysis of output from a computer code. *Technometrics*, 21(2), 239–245.

Meng, J., McCauley, S., Kaplan, F., Leung, V.J., and Coskun, A.K. (2015) Simulation and optimization of HPC job allocation for jointly reducing communication and cooling costs. *Sustainable Computing: Informatics and Systems*, 6, 48–57.

Miettinen, A.P. and Nurminen, J.K. (2010) Energy efficiency of mobile clients in cloud computing, in *Proceedings of the 2nd USENIX Conference on Hot Topics in Cloud Computing*.

Miller, R. (2012) Oak Ridge: The frontier of supercomputing. http://www.datacenterknowledge.com/archives/2012/09/10/oak-ridge-the-frontier-of-supercomputing/ [accessed 22 January, 2018].

Moore, J.D., Chase, J.S., Ranganathan, P., and Sharma, R.K. (2005) Making scheduling "cool": Temperature-aware workload placement in data centers, in *USENIX Annual Technical Conference, General Track*, pp. 61–75.

Mukherjee, T., Banerjee, A., Varsamopoulos, G., Gupta, S.K., and Rungta, S. (2009) Spatio-temporal thermal-aware job scheduling to minimize energy consumption in virtualized heterogeneous data centers. *Computer Networks*, 53(17), 2888–2904.

Nelson, R. (1999) *An Introduction to Copulas*, New York: Springer-Verlag.

Petitet, A., Whaley, R.C., Dongarra, J., and Cleary, A. (2008) HPL – a portable implementation of the High-Performance Linpack benchmark for distributed-memory computers. http://netlib.org/benchmark/hpl/ [accessed 23 January, 2018].

Pitt Turner IV,, W., Klesner, K., and Orr, R. (2014) Implementing data center cooling best practices. https://journal.uptimeinstitute.com/implementing-data-center-cooling-best-practices/ [accessed 22 January, 2018].

Ranganathan, P. (2010) Recipe for efficiency: Principles of power-aware computing. *Communications of the ACM*, 53(4), 60–67.

Rath, J. (2014) NERSC flips switch on new Edison supercomputer. http://www.datacenterknowledge.com/archives/2014/01/31/nersc-flips-switch-new-edison-supercomputer/ [accessed 22 January, 2018].

Reich, B., Hodges, J., and Carlin, B. (2007) Spatial analyses of periodontal data using conditionally autoregressive priors having two classes of neighbor relations. *Journal of the American Statistical Association*, 102(477), 44–55.

Sandia National Laboratories (2014) Sandia National Laboratories High Performance Computing Power Application Programing Interface (API) Specification. powerapi.sandia.gov [accessed 22 January, 2018].

Sang, H. and Gelfand, A.E. (2010) Continuous spatial process models for spatial extreme values. *Journal of Agricultural, Biological, and Environmental Statistics*, 15(1), 49–65.

Schöne, R., Treibig, J., Dolz, M.F., Guillen, C., Navarrete, C., Knobloch, M., and Rountree, B. (2014) Tools and methods for measuring and tuning the energy efficiency of HPC systems. *Scientific Programming*, 22(4), 273–283.

Sharma, S., Hsu, C.H., and Feng, W.c. (2006) Making a case for a Green500 List, in *Proceedings of the 20th International Parallel and Distributed Processing Symposium (IPDPS 2006)*.

Shoukourian, H., Wilde, T., Auweter, A., and Bode, A. (2014) Monitoring power data: A first step towards a unified energy efficiency evaluation toolset for HPC data centers. *Environmental Modelling & Software*, 56, 13–26.

Storlie, C.B., Rust, W.N., Reich, B.J., Ticknor, L.O., Bonnie, A.M., Montoya, A.J., and Michalak, S.E. (2017) Spatiotemporal modeling of node temperatures in supercomputers. *Journal of the American Statistical Association*, 112(517), 92–108.

Tang, Q., Gupta, S.K., and Varsamopoulos, G. (2008) Energy-efficient thermal-aware task scheduling for homogeneous high-performance computing data centers: A cyber-physical approach. *IEEE Transactions on Parallel and Distributed Systems*, 19(11), 1458–1472.

US Department of Energy (2014) Better Buildings Challenge. https://betterbuildingssolutioncenter.energy.gov/challenge [accessed 22 January, 2018].

US Department of Energy (2017) Better Buildings Challenge: Data Centers. https://betterbuildingssolutioncenter.energy.gov/challenge/sector/data-centers [accessed 22 January, 2018]

Valentini, G.L. et al. (2013) An overview of energy efficiency techniques in cluster computing systems. *Cluster Computing*, 16(1), 3–15.

Vanderster, D.C., Baniasadi, A., and Dimopoulos, N.J. (2007) Exploiting task temperature profiling in temperature-aware task scheduling for computational clusters, in L. Choi, Y. Paek, and S. Cho (eds.) *Advances in Computer Systems Architecture*, pp. 175–185.

Wang, L., Khan, S.U., and Dayal, J. (2012) Thermal aware workload placement with task-temperature profiles in a data center. *The Journal of Supercomputing*, 61(3), 780–803.

Wang, L., Von Laszewski, G., Huang, F., Dayal, J., Frulani, T., and Fox, G. (2011) Task scheduling with ANN-based temperature prediction in a data center: A simulation-based study. *Engineering with Computers*, 27(4), 381–391.

Wilde, T., Auweter, A., and Shoukourian, H. (2014) The 4 pillar framework for energy efficient HPC data centers. *Computer Science – Research and Development*, 29(3–4), 241–251.

Xu, H., Feng, C., and Li, B. (2013) Temperature aware workload management in geo-distributed datacenters. *ACM SIGMETRICS Performance Evaluation Review*, 41, 373–374.

Zenoss Inc. (2018) Zenoss: The leader in software-defined IT operations. www.zenoss.com [accessed 22 January, 2018].

Zhang, Z., Lang, M., Pakin, S., and Fu, S. (2014) Trapped capacity: Scheduling under a power cap to maximize machine-room throughput, in *Proceedings of the 2nd International Workshop on Energy Efficient Supercomputing*, pp. 41–50.

# 14

# Agile Testing with User Data in Cloud and Edge Computing Environments

*Ron S. Kenett, Avi Harel, and Fabrizio Ruggeri*

## Synopsis

A primary concern of any software update is about risks of new ways for the users to make errors. This chapter presents a framework for mitigating the risks of user errors due to changes following software updates. The underlying methodology incorporates usability and user experience considerations in the design, testing, deployment, and operation of dynamic collaborative systems, so that the error-prone elements of the user interface are identified and eliminated. The core of this methodology is a means for ongoing analysis of the user's activity, suggesting corrective means to reduce the rate of user errors. The methodology incorporates statistical process control of program service indices, obtained by a decision support system for user interface design, in which the users are elements of the control loop. This methodology was previously applied to controlling the usability of web services and installing on service-oriented architecture platforms, and it is applicable to the agile test in production approach. The chapter extends the scope of this methodology to cloud and edge computing, proposing a way of controlling the performance of software as a service offerings.

## 14.1   Introduction

The traditional centralized model of cloud computing is undergoing a paradigm shift toward a distributed and federated model. The need for such a new computing model arises from cost/economic factors concerning data and service management. Future clouds need to respond to high heterogeneity across independent cloud systems, efficient and secure data exchange among clouds, and the ability to efficiently deploy microservices across federated systems. Components at the edge of the computing environment can be gateways, routers, and smart devices (such as smartphones and embedded devices equipped with sensors and actuators), accessible through well-defined application programming interfaces. In traditional clouds, an internet of things (IoT) application pushes/pulls data to/from the cloud through edge gateways. Context-aware microservices are deployed in both edge and cloud data centers and interconnected over secure communication systems that make active instances of microservices resilient to possible attacks or data stealing.

*Analytic Methods in Systems and Software Testing*, First Edition.
Edited by Ron S. Kenett, Fabrizio Ruggeri, and Frederick W. Faltin.

This chapter presents a method for assuring seamless testing of agile application updates, required in cloud and edge computing and in modern software as a service (SaaS) architectures. It also applies to testing in production methods. Testing in production (TiP) consists of live testing of web applications using live user and customer traffic, and is a test strategy that application developers increasingly integrate into their agile development methodology. It tests the quality of updates with live user traffic in a production environment, as opposed to synthesized data in a traditional test environment. By exposing new updates to a variety of real users, one can verify their functionality, performance, and value against the actual traffic volume and velocity (https://azure.microsoft.com/en-us/documentation/articles/app-service-web-test-in-production-get-start/).

### 14.1.1 Software as a Service and Microservices

Microservices and SaaS are a framework for dynamic integration of software programs stored in the cloud, provided by different vendors, to accomplish a user's operational goal (Alhamazani et al., 2015). Software as a service involves providers and users meeting on a service composition platform, and provides an open platform for integrating legacy, internal, and external software components in the form of services in a uniform and reusable approach. Software development of SaaS must enable flexible and dynamic service registration, discovery, matching, composition, binding, and reconfiguration. Software as a service, microservices, and cloud computing are considered a most promising computing paradigm for large-scale reuse, business agility support, and integration across heterogeneous environments (Bai et al., 2008; Kossman at al., 2010; Di Nitto et al., 2013; Khalaf et al., 2015; Levin et al., 2015).

### 14.1.2 The Quality of Edge Applications

Quality is often defined as the system capability to achieve customer satisfaction. The quality of systems is often defined by several attributes. For example, Garvin (1987) identified the eight most significant attributes predicting positive customer satisfaction, including performance, features, reliability, specification compatibility, life span, service, aesthetics, and perceived quality.

Quality attributes are the overall factors that affect run-time behavior, system design, and user experience. They represent areas of concern that have the potential for application-wide impact across layers and tiers. Some of these attributes are related to the overall system design, while others are specific to run-time, design-time, or user-centric issues. The extent to which the application possesses a desired combination of quality attributes such as usability, performance, reliability, and security indicates the success of the design and the overall quality of the software application (Microsoft online guide: https://msdn.microsoft.com/en-us/library/ee658094.aspx).

In this chapter we focus on dynamic aspects of system quality, namely those attributes which are most sensitive to design changes, mostly required in agile development. Mainly, we are concerned with those attributes that might change significantly after system deployment. In many applications, user experience is considered the most prominent attribute affecting customer satisfaction.

### 14.1.3   The User Experience

Typically, user interface designers focus on providing "cool and neat" layout, employing the newest widgets based on state-of-the-art technology. Many designers do not pay sufficient attention to the ways users may actually use these features. Skilled designers, who are aware of the user's needs and limitations, do their best to facilitate the user's workflow, yet they occasionally fail to anticipate the user's goals and behavior. Eventually, their designs often suffer from various usability limitations, such as poor readability, error-prone navigation controls, and attention-distracting features. Consequently, many users of such systems often feel inconvenienced by the various types of usability barriers (Nielsen, 1994; ISO, 1998; UPA, 2009; Harel et al., 2008a, 2008b).

Several measures of user experience have been proposed in the literature (e.g., Nielsen, 1994; UPA, 2009; Sauro, 2015). These measures typically depend on results from special purpose usability tests which are typically expensive and time consuming, and therefore inadequate for real-time control. In this chapter we focus on dynamic attributes, namely those attributes which are most sensitive to design changes.

### 14.1.4   The Program Service Index

A program service index (PSI) is a usability score for a system or any of its components that we wish to control. We apply here methods from cybernetics to trace these indices and to decide when a change in an index should denote a significant deviation that requires intervention. Cybernetics is the interdisciplinary study of the structure of complex systems, especially communication processes, control mechanisms, and feedback principles. Specifically, it considers closed loops, where action by the system in an environment causes some change in the environment. Such changes impact the way the system behaves, for example to achieve service goals (Tan et al., 2008; Emeakaroha et al. 2012; Leitner et al., 2012). Typical problems in this domain have included control of the software test process (how much and what additional effort is to be applied to achieve a desired quality objective under time/cost constraints?), software performance control (how best to adjust software parameters so that an optimal level of performance is maintained?), and control of the software development process (what is an optimal set of process variables required to achieve delivery objectives within cost/time constraints?). The approach introduced here is to apply sequential methods to monitor the user experience by comparing actual results to expected results and acting on the gaps.

A PSI is a measure of the user experience. It is closely related to the concept of "service level" (http://en.wikipedia.org/wiki/Service_level). Program service indices are defined at design time by service designers, and can be adjusted at operation time by the service managers. They should reflect the risks of poor service, and therefore they should measure the two main aspects of negative user experience: service performance and reliability. We can set distinct PSIs for service performance and service reliability, or we can set a shared index, for example by combining separate indices. In the first option, we run two parallel usability controls: one for performance control and the other for reliability control. In the second option, we run a single usability control combining both aspects. In many cases a service level agreement is set up as a percentile, say 95%. This implies, for example, that 95% of the time the PSI should not exceed a prespecified limit.

### 14.1.5 Usability Risks

When a service is offered, its features include various characteristics such as performance, reliability, security, and usability levels. Usability is a term used to denote the ease and reliability with which people can employ a system, a tool, or other human-made objects in order to achieve a particular goal. Usability failures attributed to human errors are typically a main factor in system failure. For example, 60%–80% of aircraft accidents are attributed to human error (Shappell and Wiegmann, 1996). Usability failures are not always identified correctly. For example, many calls for technical support about TV system malfunctions turn out to be the result of user error.

Operational risk management is focused on the identification, classification, and evaluation of risks of operational failures affecting systems and their stakeholders. For a comprehensive treatment of operational risks combining structured, quantitative data with unstructured data and semantic analysis, see Kenett and Raanan (2010).

Popular SaaS-based implementations enable users to choose from a list of services and integrate them in a way that suits the user's goal. It is assumed that the service seekers follow certain operational procedures to describe what service they need, and make no mistakes in the requirement specification. However, this flexibility introduces special usability concerns and provides too many opportunities for users to fail by setting the wrong sequencing or by disregarding mode constraints. Therefore, SaaS-based applications are particularly error prone, and the primary PSI suited mostly to SaaS architectures is operational reliability.

### 14.1.6 Understanding User Behavior

To understand how to enable seamless operation, software designers need to get diagnostic information about the particular sources of negative user experience. Common practices for getting this information are based on asking for user feedback and questionnaires about users' opinions. For a comprehensive treatment of surveys capturing customer opinions, with various statistical models, see Kenett and Salini (2011).

User feedback and opinion are, however, of low diagnostic value, because the users can inform mostly about their feelings, based on their limited experience of trying an application, but not about the *sources* of these feelings. This chapter is focused on the diagnosis of user behavior.

### 14.1.7 Modelling User Activity

User activity can be described using a Bayesian Markov model with states given by screen displays (or web pages). Two absorbing states are present (fulfillment of the task and unsuccessful exit), and our interest is in their probability. Transitions between one state and all the possible others are modeled as a multinomial distribution with probability $p_{ij}$ of going from state $i$ to state $j$.

A Dirichlet distribution is a natural, conjugate choice for the vector $\mathbf{p}_i = (p_{i1}, \ldots, p_{ik})$ denoting the probabilities of all possible transitions from state $i$. The choice of the hyperparameters is a tough, although typical, problem in Bayesian analysis, where the statistician has to transform experts' opinions into numbers. It is worth mentioning that the prior mean of $p_{ij}$ is equal to the corresponding hyperparameter, say $a_{ij}$, divided by the sum all of them. Prior variance can be computed as well, so that the expert's guess on each $p_{ij}$ can be used to match the prior mean, whereas variance can be adjusted (just multiplying all the hyperparameters by the same factor) so that it will be smaller or larger, depending on the strong or weak confidence, respectively, of the expert on his/her guess.

The activity log provides the number $n_{ij}$ of transitions from state $i$ to state $j$, and data can be combined with the Dirichlet prior via Bayes' theorem, so that a posterior distribution, again a Dirichlet one, is obtained. The posterior mean of $p_{ij}$, i.e. its Bayesian estimator under a squared loss function, is computed as before as the sum of the new hyperparameter (given by $a_{ij} + n_{ij}$) divided by the sum of all the updated hyperparameters. Based on the posterior distribution on the transition probabilities, operational reliability can therefore be analyzed by looking at the predictive probability of ending in the unsuccessful state, given a prespecified pattern, i.e. a sequence of transitions between screen displays. With such techniques, one can combine, in a decision support for user interface design (DSUID) system, expert opinions with data to model transitions between states, and describe cause-and-effect relationships.

For an approach to improving usability by providing online help with procedures based on past experience and modeling the transition between states, see Ben Gal (2007), Rios Insua et al. (2012), and West and Harrison (2009). For an introduction to Markov chains, see Karlin and Taylor (1975). Other complementary sources of information on usability are Chapter 11 in this book, by Ruggeri and Soyer, which deals with decision models for software testing, and Chapter 18 by Freeman et al. on testing defense systems.

### 14.1.8 Usability Assurance

In designing and controlling software systems, both software cybernetics and operational risk management are applicable. Recently, new methods have been proposed to help system designers prevent and protect the system from user errors (http://www .gordon-se.technion.ac.il/files/2012/05/Seattle-v3.pdf).

Usability assurance is a continuous effort spanning the whole system lifecycle: design, testing, deployment, and routine operation:

- At the *design* stage, human factors are considered in a user-centered design paradigm, to make sure that the user interface enables proper interaction with the system.
- At the *testing* stage, usability experts conduct usability testing with real users, doing typical tasks. The tests enable system designers to learn about the quality of the user interface and to identify usability bottlenecks.
- At the *deployment* stage, diagnostic usability reports using special statistics obtained from logs of user activity may reveal the usability barriers for users when doing real tasks in their real operational environment (Harel, 1999).
- At the *operational* stage, special service indices may reveal changes in the system usability due to exposure to new user profiles and to changes in the services or their availability. The original operational scenario might not be predictive of the evolving demands. After a change has been traced, diagnostic usability reports may be derived for the period following the change, enabling exploration of required changes in the service provision.

This chapter focuses on the operational stage, namely, on methods for tracking the changes in service indices. The framework presented in the next sections incorporates usability considerations in a system implementation, to ensure fluent and reliable operation of collaborative software programs. The underlying methodology enables adapting the user interface (UI) to changes in the service allocation, so that error-prone elements of the UI are identified and eliminated. Related versions of this model can be applied to TiP and edge computing with IoT devices.

### 14.1.9 Usability Verification

A more appropriate method for understanding the sources of negative user experience is by usability testing. The goal is to identify usability barriers, which prevent fluent and reliable system operation. The traditional method is by tests conducted by usability experts. At the initial testing stage the goal is to verify that user representatives can actually do their primary job, dictated by the system designers. At the deployment and operation stages the goal is to understand the behavior of the real users, doing their real tasks.

Typically, usability tests are expensive and require delay in implementing the design changes. Another limitation of traditional usability tests is that their face validity is low; a large number of users is required for getting reliable, comprehensive conclusions.

The method proposed here is based on an analysis of user activity obtained from log files. An initial attempt to extract usability information from logs of user activity was presented by Harel et al. (2008b). The approach there was that the logs of user activity do not include information about the user's intention, which was assumed to be essential for usability studies, and that sources of negative user experience can be deduced from the logs of user activity. The next subsection introduces models for usability control.

### 14.1.10 Decision Support for User Interface Design

In a previous work we proposed the method of DSUID for controlling the usability of web services based on statistical analysis of the user activity (Kenett et al., 2009). The term "decision Support for user interface design" refers to a methodology introduced by Harel et al. (2008a) for ensuring system usability. The methodology is applicable to the last two stages of the system development cycle described above: the deployment and operation stages. It employs a method for providing feedback to service managers about usability barriers, enabling them to adjust the resources and the interaction with users. The feedback is based on special statistical analysis of the user activity, revealing problematic patterns of user behavior.

### 14.1.11 Ongoing Usability Verification

Over time, systems are exposed to new user profiles. Also, the services may change, according to marketing demands. Additionally, services may be added or removed, and their availability may change by time of day, etc. The original operational scenario might not be predictive of the evolving demands, and there is a need to adapt the UI to these changes.

Harel et al. (2008b) and Kenett et al. (2009) presented a methodology for static usability diagnosis based on a seven-layer model. Common practices in software testing enable verification of the software behavior, as long as the users behave as expected. However, modern applications used in cloud and edge computing are dynamic (Tullis and Albert, 2008; Tullis and Tedesco, 2010; Parwekar, 2011; Abolfalzi et al., 2014; Bai and Rabara, 2015; Natu et al., 2016). When software changes might impact the user's orientation and behavior, other testing tools are required. Unfortunately, common practices in usability testing cannot handle agile changes.

This chapter expands the DSUID framework presented by Harel et al. (2008a) to account for effective online tracking of changes in usability and performance over time. We generalize the DSUID paradigm to a more general concept of cloud computing, and propose it as a general framework for agile analysis of user traffic data. The chapter lays out the foundations for agile testing by employing the DSUID paradigm in SaaS and microservices systems, in the context of cloud and edge computing.

## 14.2 Diagnostic Reports of the Quality of Service

Aggregate diagnostic reports provide essential feedback to service designers and managers. These are obtained by special statistical analysis of the records of user activity stored in log files on the servers.

Logs of user activity typically provide time stamps for all user actions. In addition, when server log files are available, they also include traces of image files and scripts used for the screen display. The time stamps of the additional files enable us to estimate three important time intervals:

- The time users wait until the beginning of file download is used as a measure of screen responsiveness.
- The download time is used as a measure of screen performance.
- The time from download completion to the user's request for the next screen, in which the user reads the screen content, but also does other things, some of them unrelated to the screen content.

The challenge is to decide, based on statistics of these time intervals, whether the users feel comfortable during the operation session; when do they feel that they wait too much for the screen download, and how do they feel about what they see on screen? In this section we provide details of a DSUID system with specific examples. For additional description of DSUID, including the design and analysis of system tests, see Kenett et al. (2009).

### 14.2.1 A Model of User Mental Activities

Figure 14.1 depicts a typical sequence of events. The chart demonstrates that the time stamps may be used to measure three time intervals:

- feedback time = time(download starts) – time(link to screen)
- download time = time(download finished) – time(download starts)
- response time = time(next link activation) – time(download finished).

The time variables with potential impact on the visit experience are:

- The screen download time, measured from the first screen view until the end of subsequent embedded files (images, JavaScripts, etc.).
- The elapsed time until the next user event, interpreted as reading time.
- The elapsed time since the previous user event, interpreted as the time until the user found the link to this screen.

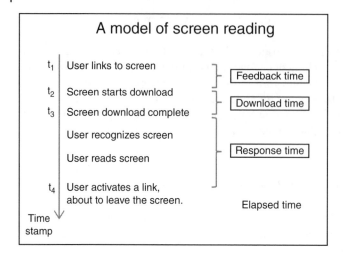

**Figure 14.1** Chart of user activities.

Other attributes obtained from the log files used for assessing the user experience include:

- text size – the number of text characters downloaded
- screen size – the number of characters downloaded, including images, animation, and video files.

### 14.2.2 Usability Problem Indicators (UPIs)

A UPI is a user event that suggests an instance of potential user difficulty. Typical UPIs are user actions reversing previous actions. Examples include: backward navigation, undo, cancel, selection from main menus, or help requests.

The model of user mental activities in system operation lists the most likely user reaction to exceptional situations. Based on this model, we can list the following indicators of user tolerance to design deficiencies:

- A user returning to the previous screen may indicate that the current screen was perceived as less relevant to the goal, compared to the previous screen.
- A user linking to a subsequent screen may indicate that the link was perceived as a potential bridge to a goal screen.
- A user activating a main menu may indicate that the user is still looking for the information, after not finding it in the current screen.
- User session termination may indicate that either the goal has been reached, or the overall user experience became negative.

Accordingly, besides session termination, other indicators of potential usability problem are the rates of navigation back to a previous screen and the user escaping from the screen to a main menu.

### 14.2.3 Usability Assessment

User experience is regarded as negative or positive according to whether or not a UPI was indicated. The expected relationships between the key variables are as follows:

- Screen load should directly affect the screen download time.
- The amount of text on screen should directly affect the screen reading time.
- The ratio of text size to screen size may be equal to 1 if the screen is pure HTML (has no graphics, video, etc.) or small (since images typically consume much more space than text).
- Screen download time should affect the visit experience. Too long a download time should result in a negative visit experience. Research indicates that too short a download time might also result in a negative visit experience; however, this might be a problem only for extremely good service (server performance).
- Screen seeking time may affect the visit experience. Too much seeking time might have a "last straw" effect, if the screen is not what the user expected.
- Screen reading time may have an effect on the visit experience. Users may stay on a screen because it includes lots of data (in this case study, the data is textual) or because the text they read is not easy to comprehend.

### 14.2.4   Statistical Time Analysis

To understand the user experience we need to know the user activity compared to the user expectation. Neither is available from the server log file, but can be estimated by appropriate processing. A DSUID system flagging possible usability design deficiencies requires a model of statistical manipulations of server log data. Assume now that we have a log of the user activity. How can we tell whether the service users encounter any difficulty in exploring a particular screen display (or a web display), and if so, what kind of difficulty do they experience, and what are the causes for this experience? We assume that the users are task driven, but we do not know if the users' goals are related to a specific service. Also, we have no way to tell if users know anything, a priori, about the service, and if they believe that the service is relevant to their goals, or if they have used it before. It may be that the users are simply exploring the service screens, or that they follow a procedure to accomplish a task. Yet, their behavior reflects their perceptions of the service, and estimates of their effort in subsequent investigation. How can we conclude that a time interval is acceptable to the users, is too short, or is too long? For example, consider an average screen download time of five seconds. Users may regard it as too lengthy, if they expect the screen to load quickly, for example, in response to a search request. However, five seconds may be quite acceptable if the user's goal is to learn or explore specific information, if they expect it to be related to their goal. Setting up a DSUID involves the integration of two types of information:

- Design deficiencies, which are common barriers to seamless operation, based on the first part of the model described above – the user's screen evaluation.
- Detectors of these design deficiencies, common indicators of possible barriers to seamless operation, based on the second part of the model – the user's reaction.

The way to conclude that the download time of a particular screen is too long is by a measure of potentially negative user experience, namely, the rate of session termination when on the particular screen. The diagnostic-oriented time analysis is by observing the correlation between the screen download time and the session termination rate. If the users are indifferent to the download time, then the exit rate should be invariant with respect to the screen download time. However, if the download time matters, then the exit rate should depend on the screen download time. When the screen download time is acceptable, most users may stay in the program, looking for additional information. However, when the download time is too long, more users might terminate the session, and go to the competitors. The longer the download time, the higher the exit rate.

### 14.2.5 Screen Relevance

Once we have an estimate of the task-related mental activities we can adapt the method for deciding about problematic performance to decide about problematic task-related mental activities. Users who feel that the information is irrelevant to their needs are more likely to respond quickly, and are more likely to go backward, or to select a new screen from the main menu. Therefore, the average time on screen over those users who navigated backwards or retried the main menu should be shorter than that of the average of all users. On the other hand, users who believe that the information is relevant to their needs, but do not understand the screen text very easily, are likely to spend more time than the average reading the screen content, and the average time on screen should be longer.

### 14.2.6 Screen Readability

Screen usability is a compound property, consisting of several factors. One factor, used to demonstrate the method, is screen readability. We demonstrate here the method for computing the confidence level of the actual screen readability. The model consists of one independent variable, which is the screen readability, and two dependent variables: the user's reaction time and the user's next action. The model assumes that:

1) The user's reaction time is correlated with the reading difficulties.
2) The rate of UPIs, such as backward navigation, is correlated with the reading difficulties.

Based on these assumptions, we expect that on average, the user reaction time for visits followed by a UPI will be higher than that for visits followed by another user action. Therefore, we apply a *t*-test to compare the user reaction time of two samples: one is of visits followed by a UPI and the other is of visits followed by another user action. The screen readability is determined by a *t*-test comparing the two samples. The time that users spend reading a screen depends on various perceptual attributes, including the screen relevance to their goals, the ease of reading and comprehending the information on screen, the ease of identifying desired hyperlinks, etc. Assume that for a particular screen, the average time on screen before backward navigation is significantly longer than the average time over all screen visits. Such a case may indicate a readability problem, due to design flaws, but it can also be due to good screen content, which encouraged users who spent a long time reading the screen to go back and re-examine the previous screen.

### 14.2.7 Statistical Decision

The way to conclude that the download time of a particular screen is too long is by a measure of potentially negative user experience, namely, the rate of session termination. To enable statistical analysis, we need to change our perspective on these variables, and consider how the download time depends on the exit behavior. We compare the download time of those visits that were successful with that of those visits that ended in session termination. If download time matters, we should expect that the average download time for those users who abandoned the program will be longer than the average for those who continued with the program. Otherwise, if the users are indifferent to the download time, we should expect that the download time of the two groups would not be significantly different. To decide about the significance of the usability barrier we compare the download time of two samples: one of screen visits that ended up in session termination and the other of all the other screen visits. The null hypothesis is that the two samples are of the same population. If the null hypothesis is rejected, we may conclude that the users' behavior depends on the download time: if the download time of the first sample exceeds that of the second sample, and the difference is statistically significant, then we may conclude that the download time of the particular screen is significantly too long. A simple two-tailed $t$-test is sufficient to decide whether the user behavior is sensitive to the screen download time.

## 14.3  Service Control

Usability control is the process of capturing instances of significant changes in the quality of service, analyzing the sources for these changes, and changing the UI to compensate for degradation in the service quality. The model is depicted in Figure 14.2, which shows that usability control is based on reports obtained by a usability analyzer. The control is employed based on four components:

- tracking user activity
- alerting about the need to apply corrective adjustments
- diagnosis: understanding user behavior
- intervention: applying corrective adjustments.

Once a service barrier has been identified, special means should be taken to remove it. Most often, the diagnosis leads to further investigation. For example, if system performance is poor, then the SaaS platform needs to determine the causes of the poor performance. On the other hand, if the diagnostic reports show that a particular UI control is error prone, a usability engineer should be asked to examine the conditions that led to the user errors. However, if these conditions cannot be prevented by design, then a user-controlled service facilitator, such as a pop-up dialog box, may be incorporated in the UI, enabling users to select from a list of options.

### 14.3.1  Redefining the Program Service Indices

A PSI is a usability score for the system or any of its components that we wish to control. In the original DSUID, the PSI was based on the concept of UPIs. For agile testing, we propose an extension of the PSI concept. We expand the approach to deal with dynamic systems and indicate how it can be applied to general SaaS architectures. The method, based on the multivariate approach, is by sharing the UPIs (Kenett and Zacks, 2014). We apply methods from cybernetics to decide when a change in an index should denote a significant deviation that requires intervention. Specifically, we apply the statistical

## Agile testing with users

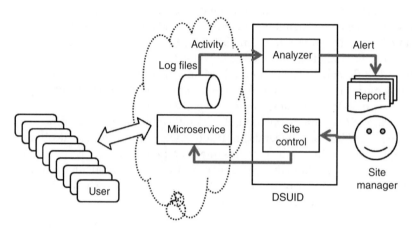

**Figure 14.2** Usability control.

process control (SPC) paradigm, applied previously in the general context of software cybernetics (Cai et al., 2004; Mathur, 2004), to decide when a change in a PSI requires intervention (Kenett and Zacks, 2014). In the context of usability control, we use SPC to decide if a PSI behaves normally by comparing actual values to reference values. Measuring the deviations from the references enables the system to identify undesired changes or drifts in the service indices.

The reference to the PSI used to describe normal behavior may be controlled by the users. For example, a user can define that the service reaction time should not exceed three seconds. The reference can also be set automatically according to prediction of future changes in the indices. The usability control can also be based on a dynamic linear model (DLM) where a Bayesian estimate of the current mean (BECM) is used to compute expected service indices. These tracking and change control methods are described in Zacks and Kenett (1994), Kenett and Baker (2010), and Kenett and Zacks (2014).

### 14.3.2 Tracking User Activity

Tracking user activity was traditionally used to effect software testing by employing the method of record/playback. Later, it was also proposed by Harel (1999) for usability validation. Tracking is also used in DSUID-based systems for capturing barriers to seamless operation. Tracking user activity is based on PSIs.

### 14.3.3 Applying SPC

The methods presented in this chapter are designed to solve the usability problem of cloud computing by integrating DSUID within an SPC framework.

Statistical process control is a method for identifying deviations from normal processing (Kenett and Zacks, 2014). Its origin is in manufacturing implementations, where statistical tools are used to observe the performance of production processes in order to identify and correct significant process deviations that may later result in rejected products. In a DSUID scheme we implement SPC and related tools for deciding when deviation of a PSI from its expected value indicates a possible barrier to fluent and reliable interaction.

### 14.3.4 A Dynamic Linear Model

Dynamic models represent a key method in Bayesian forecasting and modeling. In particular, DLMs with Gaussian-distributed components have found a plethora of applications and played a central role in time series analysis. Often their use has been suggested to model the evolution of complex systems over time with possible changes. Here the model is fitted to describe the evolution of various usability indices and to detect possible deviations from their behavior. In this sense, they can be used as a trigger to denote when an intervention is needed on the system to improve usability or to measure, on the positive side, when usability has improved.

### 14.3.5 Computing the Expected PSIs

In order to track usability changes, we need to continuously compute the service indices and to compare them to constraining values. The expected values should be estimated based on the history of the service indices. We propose a BECM for estimating the expected service indices.

Computing a PSI is an arithmetical operation defined as:

PSI = sum over the screen displays of [weight(SD) × WPUI(SD)] / number of SD,

where:

- SD stands for screen display.
- weight(SD) is a score in the range 0–1 denoting the importance of successful operation of the SD. This score reflects the application's marketing needs; for example, pages in a critical path to order submission in an e-commerce site should typically score high.
- WPUI(SD) is a score in the range 0–1 denoting the usability index of the SD. It can be defined as the product over several usability barriers to SD.

Observations from a continuous distribution characteristic of a process, such as a usability score, are recorded at discrete times $i = 1, 2, \ldots$; the observations, $X_i$, are determined by the value of a base average level, $\mu_0$, and an unbiased error, $\epsilon_i$. Possible system deviations at startup are modeled by having $\mu_0$ randomly determined according to a normal distribution with known mean $\mu_T$ (a specified target value) and a known variance $\sigma^2$. At some unknown point in time, a disturbance might change the process average level to a new value, $\mu_0 + Z$, where $Z$ models a change from the current level. The probability of having a disturbance between any two observations, taken at fixed operational time intervals, is considered a constant. Accordingly, we apply a model of at most one change point at an unknown location.

Once it is determined that a service has significantly degraded, we need to inform the service platform about the problem and its possible causes. The resulting report should include an indication of possible sources for the degradation and recommendations for how to compensate for them. Methods for diagnosis include requesting user feedback using a structured questionnaire and statistical analysis – for a modern treatment of customer surveys, see Kenett and Salini (2011). However, a most effective method is by

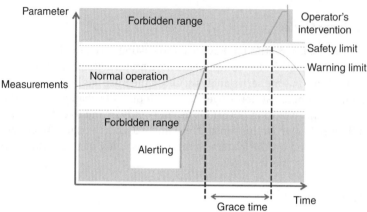

**Figure 14.3** Alerting.

a combination of the former methods, namely, by requesting user feedback at specific interaction points that were identified as error prone by statistical usability analysis. By prompting users to relate to particular interaction points, the service platform can get important information about the users' intentions. Then, it can compare the activity log to the users' intentions and figure out what the reasons were for the user error.

### 14.3.6 Alerting

Alerting is based on a measure of the user experience enabling the control of program usability in real time. The alerting method is based on that of the guide for resilience assurance proposed at http://resilience.har-el.com/Guide. It is depicted in Figure 14.3.

### 14.3.7 The Grace Time

Suppose that, in the context of system control, a service parameter needs to be changed due to usability issues. The procedure for parameter modification takes time, and process control should generate an expected event. The service control program continuously computes, by trend analysis, the predicted time for the PSI to exit the range of accepted values, and triggers the control accordingly. The predicted time may be computed by

[current PSI − PSI threshold] / change (PSI),

where change(PSI) is the rate of change (dPSI/dt) at the current PSI. The grace time should be set by the system administrator.

## 14.4 Implementation

Once it has been concluded that a service usability level has significantly degraded, we need to inform the SaaS platform about the new risk. The alert message should include an indication of the possible sources for the degradation and recommendations for how to compensate for them.

### 14.4.1    Statistical Analysis

In setting up a DSUID scheme we identify seven layers of data collection and data analytics. The statistical analysis used in a preliminary version of the DSUID framework is as follows.

The lowest layer, *user activity*, records significant user actions (involving screen changes or server-side processing). The second layer, *screen hit attributes*, consists of download time, processing time, and user response time. The third layer, *transition analysis*, produces statistics about transitions and repeated form submission (indicative of user difficulties in form filling). The fourth layer, *UPI identification*, provides indicators of possible operational difficulty, including:

- estimates for session termination by the time elapsed until the next user action (as no exit indication is recorded in the server log file)
- backward navigation
- transitions to main screens, interpreted as escaping the current subtask.

The fifth layer, *usage statistics*, consists of statistics such as:

- average introduction time
- average download time
- average time between repeated form submission
- average time on screen (indicating content-related behavior)
- average time on a previous screen (indicating ease of link finding).

In the sixth layer, *statistical decision*, for each of the screen attributes DSUID compares the statistics over the exceptional screen views to those over all the screen views. The null hypothesis is that (for each attribute) the statistics of both samples are the same. A simple two-tailed $t$-test can be used to reject it, and therefore to conclude that certain screen attributes are potentially problematic. A typical error level is 5%. In the seventh layer, *interpretation*, for each of the screen attributes DSUID provides a list of possible reasons for the difference between the statistics over the exceptional operation patterns and that over all the screen hits. However, it is the role of the usability analyst to decide which of the potential sources of user difficulties is applicable to the particular deficiency.

### 14.4.2    A Case Study

To illustrate this approach we applied the method on log data obtained from the site managers of a leading insurance company, www.555.co.il. The diagnostic reports obtained are available at http://www.ergolight.har-el.com/Sites/555/First/wtFrameSet .html.

## 14.5    Summary

Currently, customers and designers still do not have proper means to learn about user experience in real time. Web analytics tools, such as Google Analytics, provide valuable information to site owners, but these tools do not provide data to decide on the need for intervention. Workshops by the Usability Professionals' Association (UPA – now called UXPA) and others are now accessible on the web. In previous work we described a way to integrate usability quantification in SaaS. In this work we extended the methodology to general software development, when installed in the cloud. It is related to the concept of trust presented in Barreto et al. (2015), Butin et al. (2015), Dupont et al. (2015), and Lee et al. (2015). It needs to be adaptive, context sensitive, and situation aware, like other SaaS application characteristics. Specifically, this chapter presents a DSUID based on an SPC model for monitoring, collecting, and interpreting user data. We discuss several analytical models, including Markov chains, Bayesian networks, dynamic linear models, and the BECM procedure to analyze such data in a Bayesian framework.

Standard usability analysis is based on eliciting the opinion of experts. We introduce here a quantitative approach that allows combining expert prior assessment of the probability of failure/success when some patterns are followed by users with dynamically collected data. For a similar approach to setting up recommendation algorithms in web services, see Geczy et al. (2008) and Izuni et al. (2008). The Bayesian framework we introduce here provides for efficient and practical tracking and correction of usability aspects of microservices in service-oriented and SaaS architectures in general. This approach obviously needs to be expanded and further studied in specific software cybernetic applications. In particular, the PSI defined here should be calculated for sequences of log files, to learn how this index changes in real dynamic situations, such as in peak hours and during updates of the database of an IT system.

The chapter focuses on methodological aspects. Obviously, tools are needed to support such implementations. Providing more information on such tools is beyond the scope of this chapter. It is reasonable to propose that new tools should be designed, enabling real-time validation and control of the user's experience. Such tools may use a PSI for process control, in which the PSI is obtained from usability factors that may be calculated in future web analytics tools. In future studies, it may be possible to install software probes in code, similar to those installed today in many web analytic tools. Such probes will enable better distinction between users, and will enable enhancement of the indication of session termination.

## References

Abolfazli, S., Sanaei, Z., Ahmed, E., Gani, A., and Buyya, R. (2014) Cloud-based augmentation for mobile devices: Motivation, taxonomies, and open challenges. *IEEE Communications Surveys Tutorials*, 16(1), 337–368.

Alhamazani, K. Ranjan, R. Jayaraman, P., Mitra, K., Rabhi, F., Georgakopoulos, D., and Wang, L. (2015) Cross-layer multicloud real-time application QoS monitoring and benchmarking-as-a-service framework. *IEEE Transactions on Cloud Computing*, PP(99), 1–23.

Bai, T.D. and Rabara, S.A. (2015) Design and development of integrated, secured and intelligent architecture for Internet of Things and cloud computing, in *Future Internet of Things and Cloud (FiCloud), 3rd International Conference*, pp. 817–822.

Bai, X., Lee, S., Tsai. W., and Chen Y. (2008) Collaborative web services monitoring with active service broker, in *Proc. Computer Software and Applications Conference (COMPSAC'08)*, pp. 84–91.

Barreto, L. Celesti, A., Villari M., Fazio, M., and Puliafito, A. (2015) Authentication models for IoT clouds, in *International Symposium on Foundations of Open Source Intelligence and Security Informatics FOSINT-SI*. IEEE Computer Society.

Ben Gal, I. (2007) Bayesian networks, in F. Ruggeri, R.S. Kenett, and F. Faltin (eds.) *Encyclopedia of Statistics in Quality and Reliability*, Chichester: Wiley.

Butun, I., Kantarci, B., and Erol-Kantarci. M. (2015) Anomaly detection and privacy preservation in cloud-centric Internet of Things, in *Communication Workshop (ICCW) IEEE International Conference*, pp. 2610–2615.

Cai, K. Cangussu, J., Decarlo, R., and Mathur, A. (2004) An overview of software cybernetics, in *Proc. 11th Annual International Workshop on Software Technology and Engineering Practice*, pp. 77–86.

Di Nitto, E., da Silva, M., Ardagna, D., Casale, G., Craciun, C., Ferry, N., Muntes, V., and Solberg, A. (2013) Supporting the development and operation of multi-cloud applications: The MODAClouds approach, in *Proc. Intl. Symp. on Symbolic and Numeric Algorithms for Scientific Computing*, pp. 417–423.

Dupont, S., Lejeune, J., Alvares, F., and Ledoux, T. (2015) Experimental analysis on autonomic strategies for cloud elasticity, in *2015 International Conference on Cloud and Autonomic Computing*, pp. 81–92.

Emeakaroha, V.C., Ferreto, T.C., Netto, M., Brandic, I., and De Rose, C. (2012) CASViD: Application-level monitoring for SLA violation detection in clouds, in *Computer Software and Applications Conference (COMPSAC) IEEE 36th Annual Conference*, pp. 499–508.

Garvin, D.A. (1987) Competing on the eight dimensions of quality. *Harvard Business Review*, 65(6).

Geczy, P., Izumi, N., Akaho, S., and Hasida, K. (2008), Behaviorally founded recommendation algorithm for browsing assistance systems. *Annals of Information Systems*, 8, 317–334.

Harel, A. (1999) Automatic operation logging and usability validation, in *Proc. HCI International '99 (the 8th International Conference on Human–Computer Interaction)*, Munich, Germany.

Harel, A., Kenett R.S., and Ruggeri, F. (2008a) Decision support for user interface design: Usability diagnosis by time analysis of the user activity, in *Proc. Computer Software and Applications Conference (COMPSAC'08)*, pp. 836–840.

Harel, A., Kenett R.S., and Ruggeri, F. (2008b) Modeling web usability diagnostics on the basis of usage statistics, in W. Jank and G. Shmueli (eds.) *Statistical Methods in eCommerce Research*, Chichester: Wiley, pp. 131–172.

ISO (1998) *International Standards Organization Guidance on Usability*, ISO 9241-11.

Izumi, N., Geczy, P., Akaho S., and Hasida, K. (2008) Browsing assistance service for intranet information systems, in P. Perner (ed.) *Advances in Data Mining: Medial Applications, E-Commerce, Marketing, and Theoretical Aspects, ICDM 2008, Lecture Notes in Computer Science*, vol. 5077, Berlin: Springer Verlag.

Karlin, S. and Taylor, H.M. (1975) *A First Course in Stochastic Processes*, 2nd edn., New York: Academic Press.

Kenett, R.S. and Baker, E. (2010) *Process Improvement and CMMI for Systems and Software: Planning, Implementation, and Management*, Boca Raton, FL: Taylor and Francis, Auerbach Publications.

Kenett, R.S. and Raanan, Y. (2010) *Operational Risk Management: A Practical Approach to Intelligent Data Analysis*, Chichester: Wiley.

Kenett, R.S. and Salini, S. (2011) *Modern Analysis of Customer Surveys with Applications using R.* Chichester: Wiley.

Kenett, R.S. and Zacks, S. (2014) *Modern Industrial Statistics: With Applications in R, MINITAB and JMP*, 2nd edn. Chichester: Wiley.

Kenett, R.S., Harel, A., and Ruggeri, F. (2009) Controlling the usability of web services. *International Journal of Software Engineering and Knowledge Engineering*, 19(5), 627–651.

Khalaf. R., Slominski, A., and Muthusamy, V. (2015) Building a multi-tenant cloud service from legacy code with docker containers, in *IEEE International Conference on Cloud Engineering (IC2E)*, p. 18

Kossmann, T., Kraska, S., Loesing, S., Merkli, S., Mittal, R., and Pfaffhauser, F. (2010) Cloudy: A modular cloud storage system. *Proceedings of the VLDB Endowment*, 3(1–2), 1533–1536.

Lee, K., Kim, D., Ha, D., Rajput, U., and Oh. H. (2015) On security and privacy issues of fog computing supported Internet of Things environment, in *Network of the Future (NOF), 2015 6th International Conference*, pp. 1–3.

Leitner, P., Inzinger, P., Hummer, W., Satzger, B., and Dustdar, S. (2012) Application-level performance monitoring of cloud services based on the complex event processing paradigm, in *Proc. Intl. Conf. on Service-Oriented Computing and Applications*, pp. 1–8.

Levin, K., Barabash, Y., Ben-Itzhak, Y., Guenender, S., and Schour, L. (2015) Networking architecture for seamless cloud interoperability, in *IEEE 8th International Conference on Cloud Computing*, pp. 1021–1024.

Mathur, A.P. (2004) Software Cybernetics: Progress and Challenges, presentations given at the University of Paderborn, Paderborn, Germany. http://www.cs.purdue.edu/homes/apm/talks/Paderborn-talk.ppt [accessed 23 January, 2018].

Natu, M., Ghosh, R.K., Shyamsundar, R.K., and Ranjan, R. (2016) Holistic performance monitoring of hybrid clouds: Complexities and future directions. *IEEE Cloud Computing*, 3(1), 72–81.

Nielsen, J. (1994) *Usability Engineering*, Boston, MA: Academic Press Inc.

Parwekar, P. (2011) From Internet of Things towards cloud of things, in *Computer and Communication Technology (ICCCT), 2011 2nd International Conference on*, pp. 329–333.

Rios Insua, D., Ruggeri, F., and Wiper, M.P. (2012) *Bayesian Analysis of Stochastic Process Models*, Chichester: Wiley.

Sauro, J. (2015) www.measuringusability.com [accessed 20 October, 2015].

Shappell, S. and Wiegmann, D. (1996) U.S. naval aviation mishaps 1977–1992, *Aviation, Space, and Environmental Medicine*, 67, 65–69.

Tan, L., Chi, C., and Deng, J. (2008) Quantifying trust based on service level agreement for software as a service, in *Proc. Computer Software and Applications Conference (COMPSAC'08)*, pp. 116–119.

Tullis, T. and Albert, B. (2008) *Measuring the User Experience: Collecting Analyzing, and Presenting Usability Metrics*, Burlington, MA: Morgan Kaufmann.

Tullis, T. and Tedesco, D. (2010) *Online Usability Testing: Improving the User Experience through Automated Studies*, Burlington, MA Morgan Kaufmann.

UPA (2009) User Experience Professionals Association. uxpa.org [accessed 30 January, 2018].

West, M. and Harrison, P.J. (2009) *Bayesian Forecasting and Dynamic Models*, New York: Springer-Verlag.

Zacks. S. and Kenett, R.S. (1994) Process tracking of time series with change points, in J.P. Vilaplana and M.L. Puri (eds.) *Recent Advances in Statistics and Probability (Proc. 4th IMSIBAC)*, Utrecht: VSP International Science Publishers, pp. 155–171.

# 15

## Automated Software Testing

*Xiaoxu Diao, Manuel Rodriguez, Boyuan Li, and Carol Smidts*

## Synopsis

Automated software testing is a new trend in software engineering. It is a technology that allows the completion of test procedures with little or no human interaction from the initial test design to the final test execution. This technology benefits the testing process by increasing its efficiency and repeatability, reducing human error, and increasing the depth at which the analysis of the results is performed. In practice, model-based testing is a common methodology used to automate test procedures; it records information of interest to the test participants by utilizing a series of models to describe the system under test, the test requirements for verification and validation, application context, and other relevant knowledge. Building the models for testing requires cooperation between various activities such as modeling language selection, operational profile development, etc. Based on the generated models, the test cases can be carried out automatically to satisfy predefined test criteria such as statement coverage, branch coverage, functionality coverage, etc. Due to the complexity of software systems, the testing process would be impossible to complete because of the vast number of generated test cases. This problem is solved by reducing the number of test cases to an acceptable level through dedicated optimization algorithms. Test cases will execute on a testbed or a test environment which is capable of parsing test scripts, and collecting and reasoning about test results. In this chapter, we will detail the crucial phases of model-based automatic software testing and discuss existing challenges and solutions.

## 15.1 Introduction

As a critical phase in the system lifecycle, manual testing is currently encountering significant challenges. Due to the complexity of the systems under test (SUTs), the time expended on testing has increased dramatically. To combine a release with a subsequent release, the magnitude and complexity of the test effort will increase. However, manufacturers require incremental software releases that can provide tangible features to the customer within the given project timeline. Currently, organizations are turning to automated testing to verify the software while minimizing effort and schedules adequately.

A useful definition of automated testing, in general, is: "The management and performance of test activities, to include the development and execution of test cases

*Analytic Methods in Systems and Software Testing*, First Edition.
Edited by Ron S. Kenett, Fabrizio Ruggeri, and Frederick W. Faltin.
© 2018 John Wiley & Sons Ltd. Published 2018 by John Wiley & Sons Ltd.

so as to verify test requirements, using an automated test tool" (Dustin et al., 1999). During the development and integration stages, when reusable test cases may be run a great number of times, automated testing reduces the testing effort dramatically. In addition, the improved integration efficiency achieved by using an automated test tool for subsequent incremental software builds demonstrates the high value of automated testing. Although each new software build has to undergo a considerable number of verification processes, the test cases previously developed can be reused efficiently. Given the continual changes and additions to the requirements and software, tests can be generated automatically to ensure accuracy and stability through each build. For example, although controversies exist about automated testing, 72% of developers and researchers agree that automated software testing saves time and cost in comparison to manual testing with respect to rerunning tests, and 75% of them agree on the positive relation between automation and test coverage resulting in improved product quality (Petersen and Mantyla, 2012).

### 15.1.1 Glossary of Terms

In this chapter, we adopt the definitions of commonly used terms as follows:

*Operational Profile*: "A quantitative characterization of how a system will be used. The operational profile shows how to increase productivity and reliability and speed of development by allocating development resources to functions on the basis of use" (Musa, 1993).

*Test Case*: "A set of test inputs, execution conditions, and expected results developed for a particular objective, such as to exercise a particular program path or to verify compliance with a specific requirement" (IEEE, 2002).

*Test Case Generator*: "A software tool that accepts as input source code, test criteria, specifications, or data structure definitions; uses these inputs to generate test input data; and, sometimes, determines expected results" (IEEE, 2002).

*Test Coverage*: "The degree to which a given test or set of tests addresses all specified requirements for a given system or component" (IEEE, 2002).

*Test Criteria*: "The criteria that a system or component must meet in order to pass a given test" (IEEE, 2002).

*Test Effectiveness*: "The effectiveness of the test [suite] is determined by computing the fraction of the net requirements covered by the test suite (a set of test cases) generated using the technique" (Sinha and Smidts, 2006). Another definition considering structural coverage is proposed for white-box testing. Thus, test effectiveness can be defined as either

$$\text{test effectiveness} = \frac{\text{number of requirements covered}}{\text{total number of requirements in the application}}.$$

or

$$\text{test effectiveness} = \frac{\text{branches of structure covered}}{\text{total branches of structure in the application}}.$$

*Test Efficiency*: "The efficiency of the test suite is the measure that determines the effectiveness achieved per unit cost of developing the test suite" (Sinha and Smidts, 2006). Thus, test efficiency is calculated as:

$$\text{efficiency} = \frac{\text{effectiveness of the test suite}}{\text{cost to develop the test suite}}.$$

*Test Environment (Test Bed)*: "An environment containing the hardware, instrumentation, simulators, software tools, and other support elements needed to conduct a test" (IEEE, 2002).

*Test Method*: The method implementing a testing process, which usually includes the test data generation, test execution, and test result collection.

*Test Objective*: "An identified set of software features to be measured under specified conditions by comparing actual behavior with the required behavior described in the software documentation" (IEEE, 2002).

*Test Phase*: "The period of time in the software life cycle during which the components of a software product are evaluated and integrated, and the software product is evaluated to determine whether or not requirements have been satisfied" (IEEE, 2002).

*Test Plan*: "A document describing the scope, approach, resources, and schedule of intended test activities. It identifies test items, the features to be tested, the testing tasks, who will do each task, and any risks requiring contingency planning" (IEEE, 2002).

*Test Script*: "Detailed instructions for the set-up, execution, and evaluation of results for a given test case" (IEEE, 2002).

*Test Technology*: The techniques, including models, algorithms, and tools, involved in the testing process to solve a specific issue.

### 15.1.2 Model-Based Testing

As a methodology, MBT (model-based testing) is an automated testing process that leverages a series of models to generate, represent, and manage the resources sustaining the test. Models in MBT require raising the level of abstraction when contrasted with the code-centric view that currently dominates the software community. In accordance with the objectives of MBT, behavioral and structural models are treated as primary entities that can be used to generate test-supporting code automatically. Meanwhile, requirements and operational profile models are defined to produce test data or cases that satisfy the test objectives. The overarching process of MBT is displayed in Figure 15.1.

As shown in Figure 15.1, system usage information that defines the user operations and their occurrence is referred to for creating the operational profile (OP). The OP

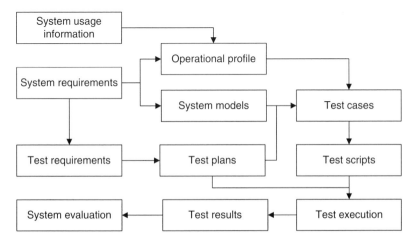

**Figure 15.1** Flowchart for the MBT process.

and system models derived from the system requirements are key reference points that can be used to generate test cases. The test cases are then turned into test scripts. Test scripts are executable representations that can be parsed by the test environments. In Figure 15.1, the test plans formulated using the test requirements contain information related to test execution such as the test objectives, the necessary resources, stopping criteria, etc. Driven by the test plans, the test scripts can be run by the test environment. Finally, the test results are collected to evaluate the significant properties of the SUT, such as correctness, performance, and reliability.

The goal of testing is typically to trigger the failure of the SUT to find defects. In black-box testing, test models are used to represent the interaction between the test automation environment and the SUT. In addition, some models contain test execution directives or test environment configurations to support the various test objectives. The models may take various forms: state diagrams, control flow graphs, interface control documents, etc. Usually, test modeling is completed at different levels of abstraction. When the SUT is developed using a purely model-driven approach, the testing model simply reuses the models used to support system development. The only aspect of the system being tested is the model transformation into code and not the system itself. When there is an important difference between the abstraction level of the test model and the SUT, there is a need to describe the transformation between abstract test cases and the concrete test scripts that will be implemented.

The test plans generated from test requirements contain the test objectives, which are usually referred to as the inputs for test case generation. In order to test a complex system, test objectives typically consist of a long sequence of user inputs with specific constraints. These test objectives are often described by natural languages and may be represented in a variety of ways in accordance with the variability of the objectives and the maturity with which the users employ the testing systems. Practically, the test objectives may be included in the model (model coverage), the implementation (code coverage), user experience (operational profile, use cases), or any combination of the above (Hartman et al., 2007). However, in order to automate the test generation, the test objectives are required to be described in formal formats that can be recognized by generation algorithms.

In summary, there are four important issues in the field of model-based software testing: (1) the modeling language used to describe the SUTs, (2) development of the operational profile, (3) automated test case generation, and (4) test execution supported by tools. In this chapter, each issue is discussed in detail. Section 15.2 discusses system modeling for testing and includes an introduction to current modeling languages and how to select the correct language. Section 15.3 introduces the development of the operational profile. Then, the automated test case generation process is discussed in Section 15.4, which includes a discussion of the test generation strategies and corresponding optimization algorithms. Finally, the necessary methods and tools for test execution are introduced in Section 15.5. Section 15.6 summarizes the chapter.

## 15.2 Modeling Languages for Testing

Automated testing involves methodologies to assist testers in acquiring information about the SUTs. The leap from traditional scripted testing to MBT is one of the most promising improvements to support moving from manual to automatic test execution.

In accordance with MBT, testers will first construct a series of models to describe the critical information necessary for preparing the test. During model construction, a few distinct options can be considered to build the models efficiently. One such option is to use one of the main languages that are widely used by developers for system design; another is to develop a language specific to the testers for defining the test objectives.

Choosing an appropriate language for building test models depends on the particular system's lifecycle phase considered, because the resources and methodologies available are distinct and the test objectives vary per phase. For example, in the requirements phase, design models are used in testing to expose potential reliability or safety problems that may increase the risk of system failures. In the development phase, unit and module tests based on structural models can discover defects in the software structure or functions that exist in the source code. In the system integration phase, verification and validation tests referring to OP can provide evidence of system performance properties and allow reliability evaluations. Criteria driving the selection between modeling languages options should synthetically consider aspects related to the test objectives, available facilities, and restrictions on time and human resources. This section discusses the language selection criteria and assesses the advantages and disadvantages of each option.

## 15.2.1 Dominant Modeling Languages for Testing

### 15.2.1.1 Design Languages Used for Testing

A design language is a language used to describe the design of a system. Modeling languages such as UML (OMG, 2010a), SysML (OMG, 2010b), AADL (Feiler et al., 2006), etc. are considered industry standards for system design and development. UML is widely used to describe the structures and behaviors of hardware and software systems during the phases of design, development, and integration. To cover the gap between models for development and testing, the UML Testing Profile (OMG, 2013) was developed from experiences gathered from former UML model-based testing. By reusing the language's elements from the design models, testing models can be organized and generated more efficiently and precisely because it is not necessary for any additional effort to be made in model construction or validation. A well-defined design model facilitates the resolution of specification ambiguities and synchronizes the understanding of SUT requirements. As an example, Figure 15.2 demonstrates UML design models for an automated teller machine (ATM). The class diagram shows the static structure of the software. The use case diagram, the state diagram, and the sequence diagram illustrate the dynamic interactions between the participants.

However, building models using design languages is a difficult process for testers as they need to learn and master the design languages and the supporting tools. Often, models developed using design languages are so difficult to understand that it becomes infeasible or too error prone for an untrained tester to establish a test plan based on the models. Developers are required to interpret the model and specify system information to bridge the information gap between the tester and the developer. This process consumes additional human resources as a result of the developers needing to attach several unnecessary interpretations to the system models.

### 15.2.1.2 Test-Specific Languages

Test-specific languages can ease the development of test models because testers do not need to learn the design languages whose primary purpose is to support development

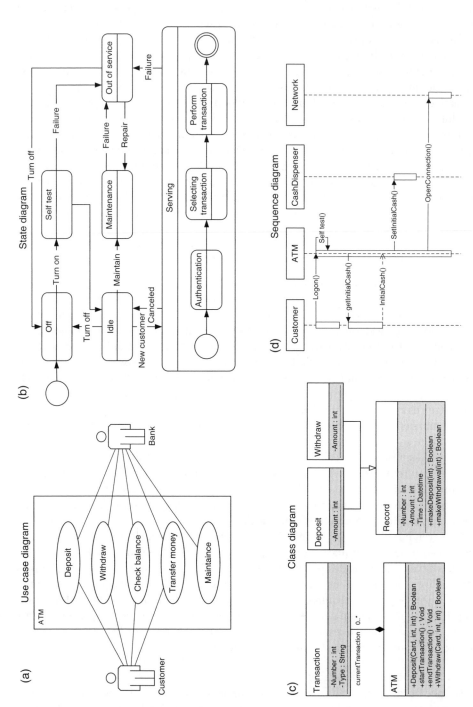

**Figure 15.2** Example UML diagrams of an ATM.

```
module ATM_Test {
  modulepar { // Module parameter Definition
    float maxDurOfTC_Par := 1.8; // Parameter with default value
    card_Type validCard_Par;
    integer validPin_Par, validAmount_Par;
  }
  external const float TestExecutionTime; // External constant
  import from Network_Interface all; // Import of all definitions
  group Test_Behaviour_Definitions { // Group definition
    testcase Withdrawal ... { ...// Test case for the withdraw transactions
    }
    testcase Authentication ... { ...// Test case for authentication procedure
    }
    testcase Deposit ... { ...// Test case for the deposit transactions
    }
  } // end group
} // end module
```

**Figure 15.3** A TTCN-3 example of ATM test.

activities and not testing activities. TTCN-3 (Grabowski et al., 2003) is a widespread modeling language dedicated to describing test specifications and test cases; it is not just an executable scripting language for test tools, but also provides a representation of system-level structures. In many cases, test-specific languages can be used for model-based testing in conjunction with a higher-level specification language that can parsed by a source-to-source compiler. Figure 15.3 is a brief example of the ATM system mentioned in Figure 15.2.

The modeling process for TTCN-3 can be implemented automatically, as many studies have determined the mapping relations between TTCN-3 and design languages (Schieferdecker et al., 2003; Zander et al., 2005). These studies have allowed for the full automation of the translation from TTCN-3 to other design languages.

### 15.2.2 Domain-Specific Languages for Testing

In several industrial domains, such as aerospace (Gaeta and Czarnecki, 2015), nuclear (Yoshikawa and Nakagawa, 2016), and network computing (Bergmayr et al., 2014), commercial tools for testing are widely used during system design, development, and verification. However, in the domains where no commercial tools are commonly available for system modeling (e.g. cryptography testing) or where special test requirements should be tested (e.g. typed-related properties in applications), testers usually prefer to develop a toolset to implement the testing activities. In this case, a domain-specific language is required to represent the test requirements that are difficult to describe using standard modeling languages. The use of domain-specific languages involves development of:

- a predefined representation for the semantics and the syntax of the modeling language
- a test design tool for modeling the tests using the domain-specific language
- a model checker for verifying the correctness and consistency of the models
- a translator (or a compiler) to interpret the models for the test execution environment.

For convenience, testers typically tailor a mature modeling language to create and represent the model for a specific domain. Most modeling languages provide the facilities to specialize and extend the model elements from existing specifications. For example, UML provides a standard extension mechanism called "profiles" to support domain-specific modeling. Profiles are collections of stereotypes, tagged values, and constraints usually defined using OCL (Object Constraint Language) expressions. As an example, Botella et al. (2013) defines a specific testing language, named UML4MBT, to represent the behavior model of a control algorithm to be tested. However, many problems arise when using profiles, because profiles that support integrating aspects of the problem domain with the test models lack the ability to hide unnecessary details that increase the complexity of test models (Hartman et al., 2007). In other words, although some constraints can be defined to hide redundant information in UML models, such as classes, methods, and attributes, other information like test data remains and complicates the test models. Meanwhile, an additional problem arises when the domain is hard to define using a UML profile. In practice, defining the stereotypes and tags is not sufficient to describe a testing environment in many domains.

Similar to TTCN-3, other domain-specific languages can also be translated into a generic modeling language. This translation can be carried out by an adapter which parses the semantic and syntax of the domain-specific language and reorganizes the language elements into a generic model. There exist tool-specific adapters for certain domains. However, the effort required to adapt a domain-specific language to a generic language should still be considered if a generic tool is chosen to implement tests in which the SUTs are modeled by a domain-specific language.

Table 15.1 lists some typical studies in which domain-specific languages were developed and utilized to achieve particular test objectives. Some of the studies provide prototypical tools for checking model consistency or translating models into a generic modeling language.

Model construction is the prerequisite of MBT. The accuracy of system models significantly affects the correctness and effectiveness of the following test activities. As an infrastructure for system modeling, the language and its supporting tools are of high importance to system testers and researchers. Future trends in modeling language development should be concerned with improvements in describing capability, accuracy, and portability.

## 15.3 Operational Profile Development

An OP is a quantitative characterization of how a software system will be used (Musa, 1993), and can be seen as a model of the user's inputs. It consists of the set of operations that a system is designed to perform and their probabilities of occurrence. An OP allows one to obtain reliability estimates by testing a software system as it will be used in the field. Reference to an OP also increases productivity and shortens development time by allocating resources on the basis of use. As stated by Musa: "Using an operational profile to guide testing ensures that if testing is terminated and the software is shipped because of imperative schedule constraints, the most-used operations will have received the most testing and the reliability level will be the maximum that is practically achievable for the given test time."

**Table 15.1** Domain-specific languages used for testing.

| Language studies | Description |
| --- | --- |
| Qualifying cryptographic components (Botella et al., 2013) | A subset of UML, termed UML4MBT, is developed, including a series of UML diagrams to represent the algorithm using behavior modeling. In parallel, a specific test purpose language is defined to allow the test engineer an understanding of the testing objectives. |
| MOS for functional test generation (Enoiu et al., 2013) | A function block diagram (FBD) modeling language aims to support model-based and search-based testing methods to verify the quality requirements of system components. The FBD models are transformed automatically into timed automata models supporting the timing aspects of the system behavior. |
| HOTTest for testing database (Sinha et al., 2006) | A strongly domain-specific language is used to extract typed-related system invariants that can be related to the properties of applications. Based on the Haskell functional programming language, the test model is translated into an extended finite state machine from which the actual tests are generated. |
| Environment MBT of Real-time Embedded Software (Iqbal et al., 2012) | An extension of MARTE, a UML real-time profile is developed to model the environment of real-time embedded software. Black-box test cases are generated to achieve high state coverage using a heuristic algorithm. |
| Routing Protocol Testing of Mobile Ad hoc Networks (Oancea et al., 2016) | A modeling language is studied based on the Gamma formalism relying on a chemical reaction metaphor. By mapping the network nodes into molecules, this modeling language has the capability of handling the characteristics of data transfer in mobile networks. |
| Testing for Context-Aware Applications (Yu et al., 2016) | A bigraphical model is developed to describe the structures and behaviors of context-aware applications that automatically adapt to a changing environment. |

### 15.3.1 Introduction

There are five important steps involved in the testing process related to an OP: (1) development of the OP model, including model analysis and validation; (2) derivation of OP probabilities; (3) generation of test cases; (4) test planning (including the design of pass, fail, and stop criteria); and (5) test execution. An OP model consists of a representation of how the software system will be used. Representations based on tree-like models, Markov chains, or statecharts are commonly used. The core of an OP model typically consists of many events that represent the use of the software system (such as mouse clicks or data input, customer/user actors, configuration alternatives, and software functions) together with their associated occurrence probabilities. These probabilities can be determined in different ways by employing techniques built on the analysis of historical data, experimental data, or expert judgment.

Operational profile models are commonly evaluated and validated using assumptions and test constraints. Test cases can be generated with a higher or lower degree of automation from OP models by employing different techniques (random walk models, analytical formulae, etc.) and strategies (with respect to probability levels, test cost, model coverage, etc.).

## 15.3.2 Operational Profile Development

Since the origins of the concept of the OP (Adams, 1984; Musa, 1993; Whittaker and Thomason, 1994), several research efforts have focused on its development and application in the software testing field. Smidts et al. (2010) further investigated OP development and classified OPs into detailed categories. They provided analyses oriented towards the development of future extensions and improvements of theories and techniques that involve OP model development. Major research areas in this topic include the definition of profiles, the extension of classical reliability prediction theories, and the application of OPs to emerging technologies.

### 15.3.2.1 Operational Profile Definition

Musa splits the OP into several profiles. A profile is defined as a set of disjoint (only one can occur at a time) alternatives with the probability that each will occur. The profiles that Musa defined are as follows: customer profile, user profile, system-mode profile, functional profile, and operational profile. A customer is defined as the group, person, or institution that is acquiring the system, and a customer group as a set of customers that will use the system in the same way. A user is the person, group, or institution that employs, not acquires, the system, while a user group is a set of users who will employ the system in the same way. A system mode is a set of functions or operations that are grouped for convenience in analyzing execution behavior (e.g., environmental conditions, architectural structure, criticality, user experience, hardware component, etc.). An operational profile must be developed for each system mode. A function is a task or part of the overall work to be done by the system, sometimes in a particular environment, as viewed by the system engineer and high-level designer. The functional profile includes the notions of explicit vs. implicit profiles, initial vs. final function lists, and environmental variables. An operation represents a task being accomplished by the system, sometimes in a particular environment, as viewed by the people who will run the system (also as viewed by testers, who try to put themselves in this position). The operational profile (i.e., the profile defined in regard to operations) includes the notions of runs, input space, and partitions. The operational profile is used to select operations in test cases in accordance with their occurrence probabilities, taking into account run categories and specific run types.

Another early definition of the OP was provided by Juhlin (1992). The OP is divided into a configuration profile (the way customers set up their system) and a usage profile (the way in which customers use the system). The configuration profile defines the physical and logical configuration that makes up the OP, and is the environment in which one or more usage profiles may take place. The usage profile defines the uses, tasks, and activities of the product by its users, including the rate, volume, and/or frequency of that usage. The rationale for this decomposition is the reduction of the broad variation in customer usage into a set of test environments that can be implemented as part of a commercial product development/test process.

The OP defined by Gittens, Lutfiyya, and Bauer (2004) consists of a process profile, a structural profile, and a data profile. The process profile is the same as Musa's operational profile. The structural profile represents the structure of the computer system on which the application is running, as well as the configuration or structure of the actual application that is being used. It introduces the concept of a measurable quantity for the attributes of data structures (e.g., arrays, lists, windows, abstract data types, etc.). The data profile summarizes the values of the inputs to the SUT from one or more users.

Huang et al. (2007) defined a fault injection profile. While the OP quantifies the occurrence probabilities of valid software inputs, the fault injection profile focuses on invalid software inputs. Through fault injection testing, such an OP allows for estimating the unreliability of the software due to failures (i.e., the invalid inputs) originating in the underlying executive platform as a consequence of the operational usage of the computer system. The fault injection profile is defined as the probability that a particular system device is affected by a specific fault type at a given time, and is supported by a tool framework for analyzing and simulating semiconductor wear-out (due to operational usage) and abnormal circuit behavior.

Smidts et al. (2014) collected and analyzed previous work and categorized OP into detailed classes. The classes were organized as a taxonomy composed of common features (e.g., profiles, structure, and scenarios), software boundaries (which define the scope of the OP), dependencies (such as requiring knowledge of the code or developed for a particular field of interest), and development (which specifies when and how an OP is developed).

### 15.3.2.2  Operational Profile Derivation

Research on OP derivation facilitates the automation of an OP development process. Many significant efforts have been made to derive OPs from software engineering specifications and artifacts that are commonly available in software development projects.

Runeson and Regnell (1998) focused on the transformation of use case models (typically employed in requirements engineering for modeling the usage of an intended system) into OP testing models for reliability engineering. Two derivation approaches were presented: transformation (where rules are defined to shape and refine a use case into an OP) and extension (where a use case is extended with OP-like data and attributes to enable test case definition). In the transformation approach, a mapping is done from actors into customers/users, services into system modes, use cases into functions, scenarios into function inputs, and system actions into operations.

The derivation of OPs from well-known, standardized engineering models that are employed in industry is a promising research venue. Challenges include the systematic use of automated means and tools, and the validation of the correctness and completeness of the derived OP with respect to the original models and operating environments.

### 15.3.2.3  OP Development Case Studies

In this subsection, we review the application of the OP concept to testing for component-based development, synchronous reactive systems, and communication protocols.

Hamlet, Mason, and Woit (2000) developed a foundational theory of software system reliability based on components. The theory described how to make component measurements that are independent of operational profiles, and how to incorporate the overall system-level operational profile into the system reliability calculations. Although a component developer cannot know how the component will be used and so cannot certify it for an arbitrary use, the component developer can still perform a certification based on results that a component buyer can factor in later via the usage information. The component developer is supposed to provide profile mapping information in a standardized way in the form of data sheets, one for component reliability and another for profile transformation.

Lutess is an SUT-like tool for testing synchronous reactive software in the telecommunication industry and other industrial contexts (Du Bousquet et al., 1998). In a recent application, it was extended to testing multimodal interactive systems. The OP is based on Markov chain usage models for which automaton states synthesize not only the input state that was issued immediately prior to it but also the past history of the software usage. Lutess also includes a random simulator of the environment, which produces input data that are valid with respect to the environment description.

Popovic and Kovacevic (2007) applied robustness testing to the class of communication and network protocol implementations based on the State design pattern and Java programming environment. The usage model is based on hidden finite state machines (FSMs), which correspond to FSMs extended with error states and transitions that do not appear in the original FSM model. The usage model is called the stress operational profile, and it models the behavior of the protocol implementation under test when it is exposed to both valid and invalid messages. The invalid messages are either syntactically or semantically faulty messages.

### 15.3.3 OP Applications in Software Testing

In this section, software testing used in conjunction with OPs is divided into several common testing categories such as statistical usage testing, software reliability growth testing with OPs, input domain testing, etc. The research literature identified is classified in Table 15.2 and elaborated in the following subsections.

#### 15.3.3.1 Statistical Usage Testing

Statistical usage testing approaches are characterized by (1) a usage model (such as an OP) that represents the expected operational use of the software system from which test cases are extracted, (2) a testing environment that simulates the actual operational environment, and (3) an approach to analyze test data and obtain reliability estimates (Walton et al., 1995). The statistical usage testing category is divided into single-profile OP and multiple-profile OP. An OP can contain more than one profile and therefore can be classified as a single-profile OP or a multiple-profile OP. We further divide the single-profile OP category into papers that "focus on the usage model" and papers with a "generic focus on reliability testing."

A usage model characterizes the operational use of a software system in terms of its intended use and environment. The typical structure of usage models for testing is a Markov chain, although other structures such as Harel statecharts or probabilistic event flow graphs are also used. The usage model consists of a graph with states and transitions. Between states, the transitions are labeled with the occurrence probability. There are two phases of development for a usage model: the structural phase and the statistical phase. The structural phase concerns the definition of states and transitions by focusing on model granularity and correctness with respect to specifications. The statistical phase assigns probabilities to each transition. The set of transition probabilities defines a probability distribution over the input domain of the software system (Poore et al.,

**Table 15.2** Classification of literature on testing domains that involve the use of an OP.

| | | | |
|---|---|---|---|
| **Statistical usage testing** | Single-profile OP | Focus on usage model | Avritzer and Weyuker (1995), Du Bousquet, Ouabdesselam, and Richier (1998), Du Bousquet et al. (1999), Brooks and Memon (2007), Dulz and Zhen (2003), El-far and Whittaker (2001), Özekici and Soyer (2001), Pant, Franklin, and Everett (1994), Popovic and Kovacevic (2007), Walton, Poore, and Trammell (1995), Amrita and Yadav (2015) |
| | | Generic focus on reliability testing | Avritzer and Larson (1993), Brown and Lipow (1975), Madani et al. (2005), Prowell (2003), Sayre (1999), Hartmann (2016) |
| | Multiple-profile OP | | Arora, Misra, and Kumre (2005), Kuball and May (2007), Li and Malaiya (1994), Musa (1992), Özekici, Altínel, and Özçelikyürek (2000), Runeson and Wohlin (1993) |
| **Software reliability growth testing with OP** | | | Arora, Misra, and Kumre (2005), Horgan and Mathur (1996), Özekici, Altínel, and Özçelikyürek (2000), Özekici, Altínel, and Angün (2001), Runeson and Wohlin (1993), Ali-Shahid and Sulaiman (2015), Hartmann (2016) |
| **Input domain testing** | Random testing | | Hamlet (2006), Mankefors-Christiernin and Boklund (2004), Ouabdesselam and Parissis (1995), Woit (1994) |
| | Partition testing | | Brown and Lipow (1975), Chen (1998), Li and Malaiya (1994), Vagoun (1996) |
| | Mixed | | Frankl, Hamlet, and Littlewood (1998), Sayre and Poore (2000), Amrita and Yadav (2015) |
| **Fault seeding testing** | | | Mankefors-Christiernin and Boklund (2004), Whittaker and Voas (2000) |
| **Fault injection with OP** | | | Huang et al. (2007), Popovic and Kovacevic (2007) |

2000). Transition probabilities can typically be classified as informed, uninformed, or intended. Model constraints are also used to shape usage assumptions and probability distributions.

Markov chains provide simulation capabilities and are supported by a number of analysis tools. Test quality can be improved based on information such as the expected test case length or the estimated number of test cases required to cover all states. A discriminant (also called a Kullback–Liebler information number) has been proposed to terminate testing when using Markov chain models. The discriminant approaches zero when the usage calculated from test experiments resembles the expected usage (Prowell, 2003). This information can also help develop test plans that require optimal resource allocation (e.g., test case number, test sequence length, etc.) before testing to reach reliability goals.

Several different approaches may be used to generate and select test cases. Typically, a test case corresponds to a path or sequence of states generated by randomly walking through a state-based usage model. This is done in accordance with the model's transition probabilities, i.e., the probability weights of the outgoing transitions from a current state. Other, less representative strategies from a statistical viewpoint include selecting the highest-probability items, which allow for optimization of test resources, and selecting the most-likely-to-fail items, which allow for optimization of fault exposure (Juhlin 1992). Amrita and Yadav (2015) proposed an OP-based approach for software test case allocation using fuzzy logic. For special cases – including mandatory tests, importance tests, and infrequent but critical operation tests – alternative strategies may be required.

### 15.3.3.2 Software Reliability Growth Testing

Software reliability growth models (SRGMs) are applied in the software testing phase to estimate software reliability, and are traditionally concerned with the relationships of a number of testing measurements, such as the cumulative number of failures experienced during testing, the time interval between failures, or the test duration. They rely on the basic assumptions that when a fault is detected during testing it is removed, and that the rate of introduction of new faults is less than that of fault removal through debugging. The software test phase is thus viewed as a process in which the reliability of the software increases as the testing progresses.

Software reliability is defined as "the ability of a system or component to perform its required functions under stated conditions for a specified period of time" (IEEE, 2002). Software reliability is estimated using data collected from the testing process, including the total number of failures, the time between failures, and other metrics. The common approach is to feed this data into software reliability models. A wide range of models supporting different complexities and characteristics are available. Several classifications have been proposed based on different viewpoints, such as failure mechanisms, statistical assumptions, failure latency, and lifecycle phase. A comprehensive classification is provided by Smidts et al. (2002) by considering three complementary axes: lifecycle, structure, and information. Among the existing models applicable to testing, software reliability growth models are the most commonly used for estimating reliability from testing data. The second most common are architectural models based on a modular representation of the software system. Various architectural models have been explicitly developed to estimate system reliability from Markov chain usage models. These include the models proposed by Whittaker and Thomason (1994), Sayre (1999), Hamlet, Mason and Woit (2000), and Dulz and Zhen (2003). Ali-Shahid and Sulaiman (2015) proposed a framework based on a software operational and testing profile that aims to provide better guidance to software testers when testing under delivery pressure. The framework enhanced reliability by assigning probabilistic priorities to the testing mechanism. An e-billing application was used as a case study to generate test suites. Hartmann (2016) proposed a statistical model that captures the relation between faults and failures. Using the model, the effects of using operational profiles for reliability growth testing is analyzed.

A large number of SRGMs have been published over the last 30 years. Many OP-based reliability testing approaches rely on the application of SRGMs to estimate reliability. However, very few of these approaches explicitly consider the integration of an analytical OP model into an extended SRGM. Such an extension leads to a more accurate representation of the operational reliability of the system (Pasquini et al. 1996). The use of OP-extended SRGMs comes, however, at the expense of having to collect extra data during the testing process. On some occasions, this might require the use of more advanced software instrumentation and monitoring techniques to obtain the required measurements, depending on the level of detail of the OP parameters integrated into an SRGM (e.g., collection of measurements that enable the establishment of a relationship between a software fault and a failed operation). Some authors go into deeper details and consider factors related to the software code (Bishop, 2002; Bishop and Bloomfield, 1996).

### 15.3.3.3 Input Domain Testing

Input domain testing relies on sampling techniques applied to the software input space to select test cases and to calculate the overall probability of failure of the software system. Traditionally, input domain models do not involve the removal of faults when they are detected during testing, so this technique has commonly been employed for validation purposes (i.e., the software system is accepted or rejected by the customer based on the calculated reliability). In general, whether the operational profile of the software system is accurately represented or not will depend on the approach chosen for the sampling of software inputs.

Input domain testing can be divided into random testing, partition testing, and stochastic testing according to the inputs of the SUTs. In random testing (also referred to as probabilistic or statistical testing), the software inputs can be randomly sampled based on a probability distribution representing the expected field use. In partition testing, the software input space is partitioned into equivalence classes. Depending on the size and selection probability of the equivalence classes, partition testing might become equivalent to random testing. Stochastic testing is characterized by a stochastic model of the input population that defines the structure of how the software system is stimulated by its environment and a sampling approach that produces test cases from the model.

### 15.3.3.4 Fault Seeding Testing

Fault seeding is a software engineering technique used to measure test coverage. A number of known faults, called "seeded faults," are randomly added to the source code of a software system. During testing, both seeded faults and inherent faults are detected and removed. After testing, the number of faults that remain can be estimated from the number of seeded faults and inherent faults discovered. Some effort has been made to consider the operational conditions in fault-seeding approaches, especially in regard to testing and seeding profiles (Mankefors-Christiernin and Boklund, 2004; Whittaker and Voas, 2000).

### 15.3.3.5   Fault Injection with an OP

Fault injection is a software testing technique that aims at improving the efficiency of testing by using faults as part of the test patterns. Fault injection techniques allow for accelerating the occurrence of software failures and for testing the error handling paths of the software that otherwise would be rarely covered. When injected faults represent external, abnormal events (such as invalid software inputs or hardware failures that impact the software), fault injection techniques help improve the representativeness and completeness of the testing profile (and likewise the OP) with respect to exceptional operational conditions. This application of fault injection techniques belongs to the category of input domain testing techniques.

A research challenge for applying fault injection techniques with an OP is the ability to provide field probabilities for observed conditions. The work described by Huang et al. (2011) is oriented towards this end. The methodology aims at developing a fault injection profile that consists of an OP for use in fault injection techniques. The fault injection profile defines the set of abnormal environmental events that impact the software and their probabilities of occurrence. The authors focused on transient failures (or "single event upsets") and on permanent failures due to operational usage of hardware devices. The method to calculate OP probabilities is supported by analytical and simulation tools that reproduce environmental phenomena and underlying semiconductor physics responsible for operational circuit failures that lead to software faults.

### 15.3.4   Test Results Issues

Automated testing with an OP relies on the ability of the OP to deliver accurate testing results. Sensitivity analysis techniques help determine the extent to which testing results are affected by inaccuracies of the OP.

Software reliability depends on the operational conditions, which may change with time or not be known with certainty. Likewise, an important question is how sensitive a software reliability estimate is to the changes, errors, or uncertainties in an OP. This question has mostly been addressed from an SRGM perspective. Three examples, Musa (1994), Chen, Mathur, and Rego (1994), and Pasquini, Crespo, and Matrella (1996), are discussed next.

Musa (1994) analyzes the relative error in failure intensity and in the OP for a single operation. The author applies an analytical approach for sensitivity analysis and considers two SRGMs: the "Musa basic execution time" model and the "Musa–Okumoto logarithmic" model. Musa defines an OP as the set of occurrence probabilities of the software functions. The sum of the occurrence probabilities is equal to one, and an error in one occurrence probability causes countervailing errors in other occurrence probabilities. Based on this observation, Musa suggests multiple errors in the OP will also tend to have countervailing (rather than cumulative) effects on the failure intensity of software. He concludes that reliability models are robust with respect to errors in the estimation of the OP. He quantifies this robustness, and states that errors in the estimation of the OP occurrence probabilities can be five times the percent error that is acceptable for measuring failure intensity.

Chen, Mathur, and Rego (1994) investigate the sensitivity of reliability estimates to errors in the OP through simulation. Using a reliability estimation tool called TEERSE, the authors model a software program as a graph of arcs and nodes with associated transition probabilities. Errors in the estimation of the OP are then simulated by modifying the arc transition probabilities. The effects in the reliability estimates are then investigated with respect to various SRGMS, such as Musa–Okumoto, Goel–Okumoto, coverage-enhanced Musa–Okumoto, and coverage-enhanced Goel–Okumoto. The authors conclude that (1) inaccuracies in OP estimates may result in significant errors in reliability estimates, and (2) reliability models that are more robust to OP errors are required.

Pasquini, Crespo, and Matrella (1996) discuss an empirical approach based on a case study aimed at evaluating the sensitivity of SRGMs to errors in the estimation of an OP. The characteristics of the case study allow for measurements of the actual reliability growth of the software (by using Nelson's reliability model) and its comparison with the estimates provided by several SRGMs. Measurements and comparisons are repeated for different OPs, which provide information on the effect of possible errors in the OP estimation. Three SRGMs – Littlewood and Verrall Bayesian quadratic, geometric, and Brooks and Motley Poisson – offered an acceptable fit to the available failure data and were used for the evaluation. The OP was modeled using a graph with (1) nodes that represent functions, and (2) transitions that represent the control flow between functions. The arc probabilities were determined through interviews with program users. The software is a European Space Agency application that calculates parameters for an array of antennas. The results show that the predictive accuracy of the model is not heavily affected by errors in the estimation of the OP. However, the type of experiment (a single case study without replication) limits the applicability of the results to different development projects and reliability estimation approaches. Also, a strong relationship is shown to exist between the number of executed test cases and the sensitivity of models to errors in the estimation of the OP. This relationship may explain the apparent contradictions between the results of Chen, Mathur, and Rego (1994) and Musa (1994).

## 15.4 Test Case Generation

A test case is a set of data that verifies whether an SUT's behavior performs as expected. The data in a test case usually includes inputs, execution conditions, and expected results. This information is generated through a specific algorithm which ensures that the generated test case fulfills predefined test requirements. During the process of generation, the system models and OPs support the generation algorithm in acquiring structural and usage information of the SUTs. In practice, the system validation test is designed to prove that the system functions are coherent with the system requirements. This type of test refers to the OP models to generate the test cases which activate as many functionalities as possible. Otherwise, the system verification tests that aim to discover defects in a software program should create test cases which can activate a maximum number of statements or branches in the program. In this section, we will consider different strategies for test case generation based on the system models and OPs. Also, we will introduce optimization algorithms to overcome the state explosion problem.

### 15.4.1 Generation Criteria

The criteria for test case generation are related to the test objectives and requirements. Testing can be classified as either structure based or specification based. A structure-based test depends on the software inner structure, statements, and execution paths to produce the test data and expected responses. This type of test is also denoted as white-box testing and usually utilizes models of control flow and data flow. Therefore, the generation criteria for structure-based testing depend on the structural coverage. On the other hand, specification-based testing is usually denoted as black-box testing or functional testing in which the OP and mapping relationship between inputs and outputs are used to create the test cases. In this case, the generation criteria are formulated for fulfilling the functionality coverage. In addition, the number and type of faults can be seen as the criterion to evaluate the test cases. The following details studies conducted on formulating and assessing the generation criteria mentioned above.

#### 15.4.1.1 Structural Coverage Criteria

The criterion for structural coverage depends on the structural model which contains nodes and links to describe the subcomponents and their relationships in a software system. The subcomponents can be the statements or routines in source code or the functional modules in a complex program.

For example, Zhu, Hall, and May (1997) described the criteria for statement, branch, and path coverage. These criteria are based on the program-based models generated from the analysis of source code. The authors analyzed the time spent on the generation process under each criterion and suggested the principles for selecting an adequate criterion to perform the test efficiently. Frankl and Weiss (1993) experimentally compared the effectiveness of the data-flow and branch-flow criteria. They referred to the error-exposing ability of each criterion for estimating their suitability. They used all-uses test cases for data-flow criteria and all-edges cases for branch-flow criteria. According to the results of their experiments, all-uses cases performed better than all-edges cases.

An FSM is a high-level structural model that is frequently referred to for test case generation. Lee and Yannakakis (1996) summarized several coverage criteria related to FSM models and discussed five important FSM testing issues, including final state determination of a test, state identification, state verification, conformance testing, and machine identification. Ntafos (1988) evaluated several types of structure-based criteria via the number of test cases generated to discover a predefined problem.

#### 15.4.1.2 Data Coverage Criteria

Data coverage criteria are used to select a series of values from the input domain of the SUTs to satisfy the test requirements. These criteria split the input space into several equivalence classes and select one of them to perform the test. Therefore, these criteria are typically combined with data boundary analysis or domain analysis.

For example, Kosmatov et al. (2004) introduced data coverage criteria based on a partitioning method. They formalized a heuristic of boundary testing to obtain the hierarchy of the boundary. Weyuker and Jeng (1991) analyzed the aspects that impact the effectiveness of data coverage criteria. The authors also systematically compared data coverage criteria with random testing to assess the efficiency quantitatively. Vagoun (1996) studied the effectiveness of coverage criteria related to the number of test cases, and proposed an approach using input domain partitioning to improve the test performance. He demonstrated how partitioning could decrease the number of required cases. His improvement depended on the internal state variables accessed by user-level functions. The study investigated the parameters impacting the probability of defect detection through simulation. These include test case size, defect type, the number of partitions, and the ratio of read/write access permissions.

### 15.4.1.3 Functional Coverage Criteria

Functional coverage criteria are based on system requirements in which the functionality is divided into several functional modules. Each module can be implemented by an individual software program. These criteria are typically used for system validation which provides the evidence to prove that the behaviors of SUTs are consistent with user requirements.

For measuring and improving the test effectiveness, Kapfhammer and Soffa (2007) provided a metric considering the necessary time of test execution to assess the criteria. Walkinshaw et al. (2010) proposed an experiment-based method to improve test efficiency by referring to the test results concerning system behaviors. Li and Malaiya (1994) analyzed the effects of input profile selection on software testing using the concept of a "fault detectability profile." They demonstrated that the optimal input profile for defect removal depends on factors such as the OP and the defect detectability profile of the software. They stated that coverage testing can be effective in practice if coverage metrics are chosen to meet reliability requirements for different modules or systems. They concluded that to achieve highly reliable software, test inputs must be uniformly selected among different input domains and tests must be conducted according to the software's OP. If testing efforts are limited due to cost or schedule constraints, it is only necessary to test the highly used input domains. Finally, the authors concluded that an OP is required for accurate determination of operational reliability.

### 15.4.1.4 Fault-Based Criteria

Fault-based criteria are employed when the test aims to verify the efficiency of fault tolerance or to detect specific types of faults. The test process is usually assisted with fault injection (e.g. program mutation) to implement certain faults. Due to the diversity of faults, fault-based criteria focus on a set of particular faults and leverage control-flow, data-flow, or other structural models to generate the test cases.

In practice, Paradkar et al. (2005) studied the criteria of mutation testing for fault detection. Their method was based on an improved state machine model to generate the test cases, which contains a set of boundary variables of system states and operations. Their results showed that the method that employs mutation and state sequences has better effectiveness than only state-based criteria for fault detection.

To evaluate the effectiveness of fault-based criteria, Li, Praphamontripong, and Offutt (2009) made a comparison between fault-based and structural-based criteria by observing the number of detected faults and the number of required cases for discovering faults. The results proved that mutation testing is more efficient at discovering faults than random testing during unit testing. Andrews et al. (2006) performed a series of experiments to assess fault-based criteria by generating all possible mutants related to a set of mutation operators. They applied a large number of mutants to an average-sized industrial program to investigate the time required to satisfy both fault-based and structure-based criteria. The investigation concentrated on the issues related to the relationships between fault detection, test suite size, and control/data flow coverage.

### 15.4.1.5 Random Testing Criteria

Random testing criteria are conventional testing criteria first proposed in 1994 (Hamlet, 1994). In random testing, test cases are selected randomly from the input domain. The primary target of the random test is to maximize the coverage of the structure or test requirements by drawing the test inputs randomly. Although the random method is a low-efficiency algorithm for test coverage, it is easy and cheap to implement and is suitable for solving most test case optimization problems.

A considerable amount of work has been done to improve the efficiency of random testing. For instance, mirror adaptive random testing was proposed to avoid the coverage repetition of test cases (Chen et al., 2004). Mayer (2005) proposed a lattice distribution based method with linear runtime. In 2007, Gerlich, Gerlich, and Boll (2007) improved random testing by considering the structural information of the SUTs. However, the improvements were limited due to the absence of detailed information related to the SUTs.

## 15.4.2 Test Case Optimization

The state explosion problem is a significant obstacle to the feasibility of testing that attempts to cover the structures or functionalities of the components in a complex system. An exponential increase in the number of required test cases arises from this attempt. Therefore, the test generation procedure should apply an effective test selection algorithm to reduce the number of redundant test cases. As an example, the test cases of specification-based testing typically correspond to operating sequences that can be generated by randomly walking through a function-based usage model. This is done in accordance with the model's transition probabilities (i.e., the probability weights of the outgoing transitions from the current state). Other, less representative strategies from a statistical viewpoint include selecting the highest-probability items (which allows for optimization of test resources) and selecting the most-likely-to-fail items (which allows for optimization of fault exposure) (Juhlin, 1992). For special cases – including mandatory tests, importance tests, and infrequent but critical operation tests – alternative strategies may be required.

### 15.4.2.1 Genetic Algorithms

A genetic algorithm (GA) is a heuristic method for solving both constrained and unconstrained optimization problems based on a natural selection process mimicking biological evolution (Holland, 1975). Jones, Sthamer, and Eyres (1996) utilized genetic algorithms to generate test cases for the first time. In their research, the chromosomes in the GA represented the branches or edges in a structural model. The variants of the chromosomes exhibited the different combinations of the branches. A generation of the chromosomes described an iteration of the algorithm that could create particular sequences of branches. These sequences were considered as an instance of the test cases. The primary objective of the GA was to carry out a generation that contained chromosomes (the test case equivalent) that maximize the coverage of test requirements. Jones et al. provided the GA method for full branch coverage and created an acceptable number of iterations of the conditional loop. They also proposed a fitness function for evaluating the quality of the test data. They proved that the quality of the test data produced was higher than the data generated from random testing.

Moreover, Pargas, Harrold, and Peck (1999) improved the test generation process by combining GAs with control dependence graphs. The improvement reduced the time spent on analyzing the source code and provided higher coverage of the code statements and branches compared to random testing. Girgis (2005) involved data flow analysis in the optimization process, focusing on the interactions between variable definitions (def) and references (use) in a program. The author achieved a def–use path coverage and provided an enhanced generation algorithm with respect to the number of iterations, coverage ratio, and the number of test cases. This method was later improved by Ghiduk, Harrold, and Girgis (2007), where the all-uses were covered and the ability of the test cases to cover the test requirements was ranked. This advance overcame the problem of losing valuable information of usage nodes. Pinto and Vergilio (2010) proposed a multi-objective fitness function to optimize the integration of coverage, test time, and the capability of failure exploration. Aleti and Grunske (2015) enhanced the algorithm's efficiency via an adaptive framework which uses feedback from the optimization process to adjust parameter values of the algorithm. Sharma, Sabharwal, and Sibal (2013) summarized the efforts undertaken to improve GAs in the context of software testing techniques.

Although GAs improve the effectiveness of the generation process compared to random testing, the improvement mostly depends on the diversity of the chromosomes, which means the effectiveness is related to the number of candidate test cases waiting for selection. A paucity of chromosomes may decrease the efficiency of finding the optimal results.

### 15.4.2.2 Ant Colony Optimization

Ant colony optimization (ACO) was originally proposed to address the traveling salesman problem (Dorigo and Gambardella, 1997). It is a heuristic optimization algorithm that finds an optimal path by emulating the behavior of ants in the process of finding food, and utilizes a substance, termed a *pheromone*, for communication of information. McMinn and Holcombe (2003) applied this algorithm to solve the testing state space explosion problem. They improved the data dependency analysis that identifies program statements responsible for state transitions. Meanwhile, Doerner and Gutjahr (2003) leveraged ACO to select test paths balancing coverage and testing costs. Carino and Andrews (2016) applied ACO to graphical user interface (GUI) testing. They also compared the efficiency of fault detection between the conventional ACO and an improved version involving Q-learning, a behavioral reinforcement algorithm. Suri (2012) surveyed 21 studies on improvements of ACO for test case optimization. The author compared these studies with respect to convergence speed, input domain, redundant path, and flexibility.

In practice, ACO is a promising technique for creating event sequences to solve the state explosion problem. The incremental process used to construct solutions allows for a straightforward incorporation of data dependency procedures to identify the possible transition from one program node to another. Nevertheless, a shortcoming of ACO is the slow convergence ratio and missing the global optimal results because the calculation result usually falls into a local optimum.

### 15.4.2.3 Particle Swarm Optimization

Particle swarm optimization (PSO) is a method for identifying optima in continuous nonlinear functions (Kennedy and Eberhart, 1995) which utilizes a series of particles that fly through the problem hyperspace with given velocities. According to a user-defined fitness function, the best position of the particles and their neighborhood are derived through adjustments of the particles' velocities. Windisch, Wappler, and Wegener (2007) used this optimization to achieve high structural code coverage for software testing. They integrated PSO into an automatic test case generation technology, termed evolutionary structural testing, to optimize the branch coverage problem. Khin, YoungSik, and Jong (2008) utilized PSO to prioritize the test cases automatically based on software units which were modified during software maintenance. The goal is to prioritize the test cases into an optimal order so that the high-priority test cases can be selected in the regression testing process. Nayak and Mohapatra (2010) applied this method to data flow testing. They observed that PSO outperforms GA in 100% def–use coverage. Jiang et al. (2014) applied PSO to generate test cases for path coverage. They evaluated the coverage rate and average iteration times of PSO and compared these evaluations to GA, ACO, and other optimization algorithms. The comparison showed that PSO performed more efficiently on path coverage than the other algorithms.

Particle swarm optimization is fast and robust in solving nonlinear problems. However, the key issue of PSO is the design of the fitness function. Inappropriate fitness functions will increase the computational workload, especially for complex software systems.

## 15.5 Test Execution

Test execution is the process of running test cases on a test device or environment which connects to the SUT through various types of I/O interfaces. The test execution process is constrained by the test strategies that include the stopping criteria to verify whether the test results are sufficient to fulfill the test requirements. In this section, the critical factors involved in the test execution phase are detailed.

### 15.5.1 Testing Methods

Testing methods provide mechanisms and technologies for implementing the test execution process. The testing process is usually achieved by the toolchains for designing, developing, and controlling the test procedures. In test design, testers are responsible for formulating the test requirements, which define the expectation of the upcoming tests. The test requirements are the reference points allowing the testers to formulate the test strategies and create test cases. These test cases are interpreted by test tools which execute the SUTs and collect the responses. Finally, this data is organized as test results and reported to the testers for further analysis. Figure 15.4 briefly illustrates this process.

The standardized procedure for automated testing can be applied to both hardware and software domains. For instance, Huang et al. (2005) introduced a test method for validating composite web services. In this approach, the authors generate the test cases via a model checker which handles the control flow models. The testing environment is a web browser which communicates with an online shopping service.

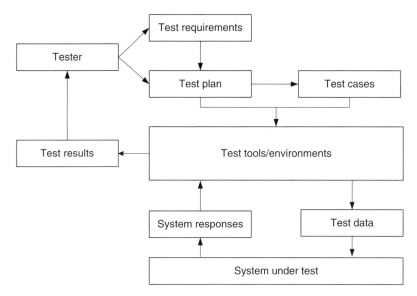

**Figure 15.4** Automated testing process.

Amalfitano et al. (2012) studied the test method for GUI applications in smartphones. This method automatically extracts and records the sequences of actions by dynamically executing the applications. They developed the MobiGUITAR (Mobile GUI Testing Framework) conceptual framework, which is implemented in a toolchain that executes on Android (Amalfitano et al., 2014). MobiGUITAR models the state of the application's GUI, which helps more accurately model the mobile application's state-sensitive behavior. Ma et al. (2016) developed a toolset for automated testing to solve the fragmentation problem of the Android system.

### 15.5.2   Testing Tools/Environments

Testing tools and environments are the physical infrastructures that execute the tests. These tools simulate the SUT's working environment or surroundings and automatically execute the generated test cases under predefined conditions. Table 15.3 lists popular commercial tools organized by testing category.

In Table 15.3, prevailing commercial tools used for software modeling are listed. These modeling tools provide the capability of transforming general models into field-specific models via interfaces and addons. Next, a list of unit testing tools and source code analyzers is introduced. These tools can be seen as white-box testing facilities to perform test activities on software routines, variables, and classes. Source code analyzers are capable of evaluating the properties of the system such as complexity, reliability, maintainability, and so on. Furthermore, Table 15.3 lists several popular system testing tools used for black-box testing. By collecting and assessing the system or module's responses, these tools are employed for verifying the system performance and functional correctness of SUTs. Lastly, a list of test management tools is provided. The test management tools involved in a testing process are utilized to organize the test resources, including human, time, device, and materials.

**Table 15.3** Examples of popular automated testing tools and environments.

| Categories | Inventory |
|---|---|
| System and OP modeling | • IBM Rational Rhapsody (IBM, 2016a)<br>• Sparx Systems Enterprise Architect (Sparx, 2016)<br>• innoSlate (InnoSlate, 2016)<br>• MaTeLo (Dulz and Zhen, 2003) |
| Unit test and source code analysis | • McCabe Visual Quality Toolset (McCabe Software, 2016)<br>• iSight++ by Integrisoft Inc.<br>• IBM Rational LogiScope (IBM, 2016d)<br>• DevPartner Studio by Micro Focus (Micro Focus, 2016) |
| Module and system test | • Insure++ by Parasoft Inc. (Parasoft, 2016)<br>• IBM Rational Robot (IBM, 2016c)<br>• WinRunner by HP (HP, 2016) |
| Test management | • Test Director by Mercury Inc. (Mercury, 2016)<br>• Test Expert by Siemens (Siemens, 2016)<br>• IBM Rational Quality Manager (IBM, 2016b) |

## 15.6   Summary

Automated software testing is a novel test methodology which notably improves the effectiveness and efficiency of the testing process. This methodology consists of several technologies such as system modeling, test case generation, and automatic testing methods. In automated software testing, human error is replaced by processes implemented by instruments that are running in compliance with predefined strategies. This allows testers to concentrate on the work of test planning and results analysis.

This chapter introduced the definition and crucial phases of automated software testing including software system modeling, operational profile development, test case generation, and automatic test execution. The trends observed in methodologies and technologies relevant to automated software testing were discussed. Derived from the model-driven test method, a series of system modeling approaches were discussed along with the corresponding research concerned with modeling language selection. Then, the focus turned to test case generation methods for both black-box and white-box testing. The development of operational profiles was also discussed. Recommendations on methods and processes for operational profile development were given, and a description of specific applications of operational profiles under various testing conditions was detailed. The sensitivity problems related to OP-based tests were also discussed. By utilizing operational profiles, test cases can be generated through test criteria for the coverage of test requirements in an efficient way. The number of test cases generated can be reduced via an optimization algorithm. The advantages and disadvantages of modern optimization algorithms for practical applications were then discussed. Finally, the execution phase of the test was introduced, including automatic testing methods supported by testing tools or environments. Also, we listed several typical commercial tools for unit testing, system testing, and test management.

# References

Adams, E.N. (1984) Optimizing preventive service of software products. *IBM Journal of Research and Development*, 28(January), 2–14.

Aleti, A. and Grunske, L. (2015) Test data generation with a Kalman filter-based adaptive genetic algorithm. *Journal of Systems and Software*, 103, 343–352.

Ali-Shahid, M.M. and Sulaiman, S. (2015) Improving reliability using software operational profile and testing profile, in *Proc. 2nd International Conference on Computer, Communications, and Control Technology*, pp. 384–388.

Amalfitano, D. et al. (2014) MobiGUITAR: Automated model-based testing of mobile apps. *IEEE Software*, 32(5), 53–59.

Amalfitano, D. et al., 2012. Using GUI ripping for automated testing of Android applications, in *Proc. 27th IEEE/ACM International Conference on Automated Software Engineering*, pp. 258–261.

Amrita  and Yadav, D.K. (2015) Operational profile based software test case, in *2nd International Conference on Computing for Sustainable Global Development (INDIACom)*, pp. 1775–1779.

Andrews, J.H. et al. (2006) Using mutation analysis for assessing and comparing testing coverage criteria. *IEEE Transactions on Software Engineering*, 32(8), 608–624.

Arora, S., Misra, R.B., and Kumre, V.M. (2005) Software reliability improvement through operational profile driven testing, in *Proc. Reliability and Maintainability Symposium*, pp. 621–627.

Avritzer, A. and Larson, B. (1993) Load testing software using deterministic state testing. *ACM SIGSOFT Software Engineering Notes*, 18(3), 82–88.

Avritzer, A. and Weyuker, E.J. (1995) The automatic generation of load test suites and the assessment of the resulting software. *IEEE Transactions on Software Engineering*, 21(9), 705–716.

Bergmayr, A. et al. (2014) UML-based cloud application modeling with libraries, profiles, and templates. *CEUR Workshop Proceedings*, 1242(317859), 56–65.

Bishop, P.G. (2002) TEST PROFILE: Rescaling reliability bounds for a new operational profile. *ACM SIGSOFT Software Engineering Notes*, 27(4), 180.

Bishop, P.G. and Bloomfield, R.E., 1996. A conservative theory for long-term reliability growth prediction. *IEEE Transactions on Reliability*, 45(4), 550–560.

Botella, J. et al. (2013) Model-based testing of cryptographic components: Lessons learned from experience, in *Proceedings of IEEE 6th International Conference on Software Testing, Verification and Validation*, pp. 192–201.

Brooks, P.A. and Memon, A.M. (2007) Automated GUI testing guided by usage profiles, in *Proceedings of the 22nd IEEE/ACM International Conference on Automated Software Engineering*, p. 333.

Brown, J.R. and Lipow, M. (1975) Testing for software reliability, in *Proceedings of the International Conference on Reliable Software*, pp. 518–527.

Carino, S. and Andrews, J.H. (2016) Dynamically testing GUIs using ant colony optimization, in *Proceedings 30th IEEE/ACM International Conference on Automated Software Engineering*, pp. 138–148.

Chen, M.-H., Mathur, A.P., and Rego, V. (1994) A case study to investigate sensitivity of reliability estimates to errors in operational profile, in *Proceedings of 1994 IEEE International Symposium on Software Reliability Engineering*, pp. 276–281.

Chen, T.Y. et al. (2004) Adaptive random testing through dynamic partitioning, in *Proceedings of Fourth International Conference on Quality Software*.

Chen, Y. (1998) Modelling software operational reliability via input domain-based reliability growth model, in *28th Annual Symposium on Fault-Tolerant Computing*.

Doerner, K. and Gutjahr, W.J. (2003) Extracting test sequences from a Markov software usage model by ACO, in *Proc. Genetic and Evolutionary Computation Conference 2003*, pp. 2465–2476.

Dorigo, M. and Gambardella, L.M. (1997) Ant colony system: A cooperative learning approach to the traveling salesman problem. *IEEE Transactions on Evolutionary Computation*, 1(1), 53–66.

Du Bousquet, L., Ouabdesselam, F., and Richier, J.-L. (1998) Expressing and implementing operational profiles for reactive software validation. *International Journal of Educational Research*, 222–230.

Du Bousquet, L., Ouabdesselam, F., Richier, J.-L., and Zuanon, N. (1999) Lutess: A specification-driven testing environment for synchronous software, in *Proceedings of the 1999 IEEE International Conference on Software Engineering*, pp. 267–276.

Dulz, W. and Zhen, F. (2003) MaTeLo: Statistical usage testing by annotated sequence diagrams, Markov chains and TTCN-3, in *Proceedings of the International Conference on Quality Software*, pp. 336–342.

Dustin, E., Rashka, J., and Paul, J. (1999) *Automated Software Testing: Introduction, Management, and Performance*, Boston, MA: Addison Wesley.

El-far, I.K. and Whittaker, J.A. (2001) Model-based software testing. *Encyclopedia of Software Engineering*.

Enoiu, E.P. et al. (2013) MOS: An integrated model-based and search-based testing tool for function block diagrams, in *Proceedings of the 1st International Workshop on Combining Modelling and Search-Based Software Engineering*, pp. 55–60.

Feiler, P.H., Gluch, D.P., and Hudak, J.J. (2006) *The Architecture Analysis & Design Language (AADL): An Introduction*. CMU/SEI-2006-TN-011. Pittsburgh, PA: Carnegie Mellon University.

Frankl, P.G., Hamlet, R.G., and Littlewood, B. (1998) Evaluating testing methods by delivered reliability. *IEEE Transactions on Software Engineering*, 24(8), 586–601.

Frankl, P.G. and Weiss, S.N. (1993) An experimental comparison of the effectiveness of branch testing and data flow testing. *IEEE Transactions on Software Engineering*, 19(8), 774–787.

Gaeta, J.P. and Czarnecki, K. (2015) Modeling aerospace systems product lines in SysML. *International Software Product Line Conference*, pp. 293–302.

Gerlich, R., Gerlich, R., and Boll, T. (2007). Random testing: From the classical approach to a global view and full test automation, in *Proceedings of the 2nd International Workshop on Random Testing: Co-located with the 22Nd IEEE/ACM International Conference on Automated Software Engineering (ASE 2007)*, pp. 30–37.

Ghiduk, A.S., Harrold, M.J., and Girgis, M.R. (2007) Using genetic algorithms to aid test-data generation for data-flow coverage, in *Proceedings of the Asia-Pacific Software Engineering Conference, APSEC*, pp. 41–48.

Girgis, M.R. (2005) Automatic test data generation for data flow testing using a genetic algorithm. *Computer*, 11(6), 898–915.

Gittens, M., Lutfiyya, H., and Bauer, M. (2004) An extended operational profile model, in *Proceedings 15th International Symposium on Software Reliability Engineering*, pp. 314–325.

Grabowski, J. et al. (2003) An introduction to the testing and test control notation (TTCN-3). *Computer Networks*, 42(3), 375–403.

Hamlet, D. (2006) When only random testing will do, in *Proceedings of the 1st International Workshop on Random Testing, RT'06*, pp. 1–9.

Hamlet, D., Mason, D., and Woit, D. (2000) Theory of system reliability based on components, in *Third International Workshop on Component-Based Software Engineering*, pp. 1–9.

Hamlet, R. (1994) Random testing. *Encyclopedia of software Engineering*.

Hartman, A., Katara, M., and Olvovsky, S. (2007) Choosing a test modeling language: A survey, in *Proceedings, Hardware and Software, Verification and Testing*, pp. 204–218.

Hartmann, H. (2016) A statistical analysis of operational profile driven testing, in *Proceedings IEEE International Conference on Software Quality, Reliability and Security Companion (QRS-C)*, pp. 109–116.

Holland, J.H. (1975) *Adaptation in Natural and Artificial Systems: An Introductory Analysis with Applications to Biology, Control, and Artificial Intelligence*. Ann Arbor, MA: University of Michigan Press.

Horgan, J.R. and Mathur, A.P. (1996) Software testing and reliability, in M. Lyu (ed.) *The Handbook of Software Reliability Engineering*, New York: IEEE Computer Society Press, pp. 531–565.

HP (2016) ALM Application Lifecycle Management. https://software.microfocus.com/en-us/products/application-lifecycle-management/overview [accessed 23 January, 2018].

Huang, B. et al. (2007) On the development of fault injection profiles. *Proceedings of the Annual Reliability and Maintainability Symposium, RAMS*, pp. 226–231.

Huang, B. et al. (2011) Hardware error likelihood induced by the operation of software. *IEEE Transactions on Reliability*, 60(3), 622–639.

Huang, H. et al. (2005) Automated model checking and testing for composite web services. *Proceedings of the Eighth IEEE International Symposium on Object-Oriented Real-Time Distributed Computing, ISORC 2005*, pp. 300–307.

IBM (2016a) IBM Rational Rhapsody family. http://www-03.ibm.com/software/products/en/ratirhapfami [accessed 23 January, 2018].

IBM (2016b) IBM Rational Quality Manager. http://www-03.ibm.com/software/products/en/ratiqualmana [accessed 23 January, 2018].

IBM (2016c) IBM Rational Robot. https://www-01.ibm.com/software/in/awdtools/tester/robot/ [accessed 20 October, 2010].

IBM (2016d) IBM Rational Logiscope. http://www-01.ibm.com/common/ssi/cgi-bin/ssialias?infotype=OC&subtype=NA&htmlfid=897/ENUS5724-X69&appname=totalstorage [accessed 23 January, 2018].

IEEE (2002) *Glossary of Software Engineering Terminology (Reaffirmed 2002)*, IEEE Standard 610.12-1990. New York: Institute of Electrical and Electronics Engineers.

InnoSlate (2016) Systems Engineering Tools. www.innoslate.com [accessed 4 October, 2016].

Iqbal, M.Z., Arcuri, A., and Briand, L. (2012) Empirical investigation of search algorithms for environment model-based testing of real-time embedded software, in *Proceedings of the 2012 International Symposium on Software Testing and Analysis*, pp. 199–209.

Jiang, S. et al. (2014) An approach for test data generation using program slicing and particle swarm optimization. *Neural Computing and Applications*, 25(7–8), 2047–2055.

Jones, B.F., Sthamer, H.H., and Eyres, D.E. (1996) Automatic structural testing using genetic algorithms. *The Software Engineering Journal*, 11, 299–306.

Juhlin, B.D.D. (1992) Implementing operational profiles to measure system reliability, in *Proceedings ISSRE'02*, pp. 286–295.

Kapfhammer, G.M. and Soffa, M.L. (2007) Using coverage effectiveness to evaluate test suite prioritizations, in *Proceedings of the 1st ACM International Workshop on Empirical Assessment of Software Engineering Languages and Technologies held in conjunction with the 22nd IEEE/ACM International Conference on Automated Software Engineering (ASE)*, pp. 19–20.

Kennedy, J. and Eberhart, R. (1995) Particle swarm optimization, in *Proceedings of the IEEE International Conference on Neural Networks*, pp. 1942–1948.

Khin, H.S.H., YoungSik, C., and Jong, S.P. (2008) Applying particle swarm optimization to prioritizing test cases for embedded real time software retesting, in *Proceedings of the 8th IEEE International Conference on Computer and Information Technology Workshops*, pp. 527–532.

Kosmatov, N., Legeard, B., Peureux, F., and Utting, M. (2004) Boundary coverage criteria for test generation from formal models, in *15th IEEE International Symposium on Software Reliability Engineering*, pp. 139–150.

Kuball, S. and May, J.H.R. (2007) A discussion of statistical testing on a safety-related application. *Proceedings of the Institution of Mechanical Engineers, Part O: Journal of Risk and Reliability*, 221(2), 121–132.

Lee, D. and Yannakakis, M. (1996) Principles and methods of testing finite state machines: A survey. *Proceedings of the IEEE*, 84(8), 1090–1123.

Li, N. and Malaiya, Y.K. (1994) On input profile selection for software testing, in *Proceedings of the 1994 IEEE International Symposium on Software Reliability Engineering*, 196–205.

Li, N., Praphamontripong, U., and Offutt, J. (2009) An experimental comparison of four unit test criteria: Mutation, edge-pair, all-uses and prime path coverage, in *2009 IEEE International Conference on Software Testing, Verification, and Validation Workshops*, pp. 220–229.

Ma, X. et al. (2016) An automated testing platform for mobile applications, in *2016 IEEE International Conference on Software Quality, Reliability and Security Companion (QRS-C)*, pp. 159–162.

Madani, L. et al. (2005) Synchronous testing of multimodal systems: An operational profile-based approach. *Proceedings of the International Symposium on Software Reliability Engineering, ISSRE*, pp. 325–334.

Mankefors-Christiernin, S. and Boklund, A. (2004) Multiple profile evaluation using a single test suite in random testing, in *Proceedings of the 15th International Symposium on Software Reliability Engineering*, pp. 283–294.

Mayer, J. (2005) Lattice-based adaptive random testing, in *Proceedings of the 20th IEEE/ACM International Conference on Automated Software Engineering - ASE '05*, p. 333.

McCabe Software (2016) Software quality, testing, and security analysis. www.mccabe.com [accessed 24 January 2018].

McMinn, P. and Holcombe, M. (2003) The state problem for evolutionary testing, in *Proceedings of the Conference on Genetic and Evolutionary Computation, GECCO 2003*, pp. 2488–2498.

Mercury (2016) Mercury Quality Center – TestDirector. http://www.gillogley.com/testing_mercury.shtml [accessed 24 January, 2018].

Micro Focus (2016) Automated testing solution – DevPartner. https://www.microfocus.com/products/devpartner/ [accessed 24 January, 2018].

Musa, J.D. (1992) The operational profile in software reliability engineering: An overview, in *Proceedings, Third International Symposium on Software Reliability Engineering*, pp. 140–154.

Musa, J.D. (1993) Operational profiles in software-reliability engineering. *IEEE Software*, 10(2), 14–32.

Musa, J.D. (1994) Sensitivity of field failure intensity to operational profile errors, in *Proceedings of the 5th International Symposium on Software Reliability Engineering*, pp. 334–337.

Nayak, N. and Mohapatra, D.P. (2010) Automatic test data generation for data flow testing using particle swarm optimization, in *Proceedings of the International Conference on Contemporary Computing*, pp. 1–12.

Ntafos, S.C. (1988) A comparison of some structural testing strategies. *IEEE Transactions on Software Engineering*, 14(6), 868–874.

Oancea, D.-G., Pura, M.L., and Morogan, L. (2016) Using gamma formalism for model-based testing for the routing protocols for mobile ad hoc networks, in *Proceedings of the 2016 IEEE International Conference on Communications (COMM)*, pp. 177–180.

OMG (2010a) OMG Unified Modeling Language$^{TM}$ (OMG UML), superstructure v.2.3. *InformatikSpektrum*, 21(May), p. 758.

OMG (2010b) OMG Systems Modeling Language (OMG SysML$^{TM}$) v.1.2. *Source*, June, p. 260.

OMG (2013) UML Testing Profile (UTP) specification version 1.2. http://www.omg.org/spec/UTP/1.2/About-UTP/ [accessed 24 January, 2018].

Ouabdesselam, F. and Parissis, I. (1995) Constructing operational profiles for synchronous critical software. *Proceedings of Sixth International Symposium on Software Reliability Engineering. ISSRE'95*, pp. 286–293.

Özekici, S., Altínel, I.K., and Angün, E. (2001) A general software testing model involving operational profiles. *Probability in the Engineering and Informational Sciences*, 15(4), 519–533.

Özekici, S., Altínel, I.K., and Özçelikyürek, S. (2000) Testing of software with an operational profile. *Naval Research Logistics*, 47(8), 620–634.

Özekici, S. and Soyer, R. (2001) Bayesian testing strategies for software with an operational profile. *Naval Research Logistics*, 48(8), 747–763.

Pant, H., Franklin, P., and Everett, W. (1994) A structured approach to improving software-reliability using operational profiles, in *Proceedings of the Annual Reliability and Maintainability Symposium*, pp. 142–146.

Paradkar, A. et al. (2005) Case studies on fault detection effectiveness of model based test generation techniques, in *Proceedings of the First International Workshop on Advances in Model-Based Testing – A-MOST '05*, pp. 1–7.

Parasoft (2016) Parasoft Insure++. https://www.parasoft.com/products/insure/ [accessed 24 January, 2018].

Pargas, R.P., Harrold, M.J., and Peck, R. (1999) Test-data generation using genetic algorithms. *Software Testing, Verification and Reliability*, 9(4), 263–282.

Pasquini, A., Crespo, A.N., and Matrella, P. (1996) Sensitivity of reliability-growth models to operational profile errors vs testing accuracy. *IEEE Transactions on Reliability*, 45(4), 531–540.

Petersen, K. and Mantyla, M.V. (2012) Benefits and limitations of automated software testing: Systematic literature review and practitioner survey, in *Proceedings of the 7th International Workshop on Automation of Software Test (AST)*, pp. 36–42.

Pinto, G.H.L. and Vergilio, S.R. (2010) A multi-objective genetic algorithm to test data generation, in *Proceedings of the 22nd IEEE International Conference on Tools with Artificial Intelligence (ICTAI)*, pp. 0–5.

Poore, J.H., Walton, G.H., and Whittaker, J.A. (2000) A constraint-based approach to the representation of software usage models. *Information and Software Technology*, 42(12), 825–833.

Popovic, M. and Kovacevic, J. (2007) A statistical approach to model-based robustness testing, in *Proceedings of the International Symposium and Workshop on Engineering of Computer Based Systems*, pp. 485–494.

Prowell, S.J. (2003) JUMBL: A tool for model-based statistical testing, in *Proceedings of the 36th Annual Hawaii International Conference on System Sciences, HICSS 2003*.

Runeson, P. and Regnell, B. (1998) Derivation of an integrated operational profile and use case model, in *Proceedings of the International Symposium on Software Reliability Engineering, ISSRE*, pp. 70–79.

Runeson, P. and Wohlin, C. (1993) Statistical usage testing for software reliability certification and control, in *Proceedings EuroSTAR '93*, pp. 309–323.

Sayre, K.D. (1999) *Improved Techniques for Software Testing Based on Markov Chain Usage Models*. PhD thesis, The University of Tennessee.

Sayre, K. and Poore, J.H. (2000) Partition testing with usage models. *Information and Software Technology*, 42(12), 845–850.

Schieferdecker, I. et al. (2003) The UML 2.0 Testing Profile and its relation to TTCN-3, in *Proceedings of the 15th IFIP International Conference on Testing of Communicating Systems (TestCom2003)*, pp. 79–94.

Sharma, C., Sabharwal, S., and Sibal, R. (2013) A survey on software testing techniques using genetic algorithm. *International Journal of Computer Science Issues*, 10(1), 381–393.

Siemens (2016) Test Expert. https://www.plm.automation.siemens.com/en_us/products/ tecnomatix/electronics-manufacturing/test-expert.shtml [accessed 24 January, 2018].

Sinha, A. et al. (2006) HOTTest: A model-based test design technique for enhanced testing of domain-specific applications. *ACM Transactions on Software Engineering Methodologies*, 15(3), 242–278.

Sinha, A. and Smidts, C. (2006) An experimental evaluation of a higher-ordered-typed-functional specification-based test-generation technique. *Empirical Software Engineering*, 11(2), 173–202.

Smidts, C. et al. (2002) Software reliability models. *Encyclopedia of Software Engineering*.

Smidts, C. et al. (2010) Operational profile testing. *Encyclopedia of Software Engineering*.

Smidts, C. et al. (2014) Software testing with an operational profile. *ACM Computing Surveys*, 46(3), 1–39.

Sparx (2016) Enterprise Architect. http://www.sparxsystems.com/products/ea/ [accessed 4 October, 2016].

Suri, B. (2012) Literature survey of ant colony optimization in software testing, in *Proceedings of the Sixth International Conference on Software Engineering (CONSEG)*.

Vagoun, T. (1996) Input domain partitioning in software testing, in *Proceedings of the 29th Hawaii International Conference on System Sciences*, p. 261.

Walkinshaw, N. et al. (2010) Increasing functional coverage by inductive testing: A case study. In *Proceedings of the IFIP International Conference on Testing Software and Systems*, pp. 126–141.

Walton, G.H., Poore, J.H., and Trammell, C.J. (1995) Statistical testing of software based on a usage model. *Software – Practice and Experience*, 25(January), pp. 97–108.

Weyuker, E.J. and Jeng, B. (1991) Analyzing partition testing strategies. *IEEE Transactions on Software Engineering*, 17(7), 703–711.

Whittaker, J.A., and Thomason, M.G. (1994) A Markov chain model for statistical software testing. *IEEE Transactions on Software Engineering*, 20(10), 812–824.

Whittaker, J.A. and Voas, J. (2000) Toward a more reliable theory of software reliability. *Computer*, 33(12), 36–42.

Windisch, A., Wappler, S., and Wegener, J. (2007) Applying particle swarm optimization to software testing, in *Proceedings of the 9th Annual Conference on Genetic and Evolutionary Computation – GECCO '07*, p. 1121.

Woit, D. (1994) *Operational Profile Specification, Test Case Generation, and Reliability Estimation for Modules*. PhD dissertation, Queen's University.

Yoshikawa, H. and Nakagawa, T. (2016) Development of plant DiD Risk Monitor system for NPPs by utilizing UML modelling technology. *IFAC-PapersOnLine*, 49(19), 397–402.

Yu, L., Tsai, W.-T., and Perrone, G. (2016) Testing context-aware applications based on bigraphical modeling. *IEEE Transactions on Reliability*, 65(3), 1584–1611.

Zander, J. et al. (2005) From U2TP models to executable tests with TTCN-3: an approach to model driven testing, in *Proceedings of the IFIP International Conference on Testing of Communicating Systems*, pp. 289–303.

Zhu, H., Hall, P.A.V., and May, J.H.R. (1997) Software unit test coverage and adequacy. *ACM Computing Surveys*, 29(4), 366–427.

# 16

## Dynamic Test Case Selection in Continuous Integration: Test Result Analysis using the Eiffel Framework

*Daniel Ståhl and Jan Bosch*

## Synopsis

The popular agile practices of continuous integration and delivery stress the rapid and frequent production of release candidates and evaluation of those release candidates, respectively. Particularly in the case of very large software systems and highly variable systems, these aspirations can come into direct conflict with the need for both thorough and extensive testing of the system in order to build the highest possible confidence in the release candidate. There are multiple strategies to mitigate this conflict, from throwing more resources at the problem to avoiding end-to-end scenario tests in favor of lower-level unit or component tests. Selecting the most valuable tests to execute at any given time, however, plays a critical role in this context: repeating the same static test scope over and over again is a waste that large development projects can ill afford. While a number of alternatives for dynamic test case selection exist – alternatives that may be used interchangeably or even in tandem – many require analysis of large quantities of in situ real-time data in the form of trace links. Generating and analyzing such data is a recognized challenge in industry. In this chapter we investigate one approach to the problem, based on the Eiffel framework for continuous integration and delivery.

## 16.1 Introduction

Dynamic test case selection means selecting which tests to execute at a given time, *dynamically at that time*, rather than from predefined static lists. It also implies performing that selection somewhat intelligently – blind random selection might be considered dynamic, but is arguably not overly helpful. Consequently, what we mean by dynamic selection is this *intelligent* selection, designed to serve some specific purpose.

There are many such purposes which may be served, not least in a continuous integration and delivery context. Continuous integration has been shown to be difficult to scale (Roberts, 2004; Rogers, 2004). One problem in the continuous integration and delivery of very large systems is that the test scopes of such systems can be both broad and time consuming – often much longer than the couple of hours beyond which some will argue that the practice is not even feasible (Beck, 2000). At the same time, others state that a cornerstone of continuous integration practice is that all tests must pass (Duvall et al., 2007), which is clearly problematic.

*Analytic Methods in Systems and Software Testing*, First Edition.
Edited by Ron S. Kenett, Fabrizio Ruggeri, and Frederick W. Faltin.
© 2018 John Wiley & Sons Ltd. Published 2018 by John Wiley & Sons Ltd.

This is particularly the case in certain segments of the industry. While in a generic cloud environment the problem can to a certain extent be solved, or at least mitigated, by throwing more inexpensive hardware at it, large embedded software systems developed for specialized bespoke hardware do not have that option. Examples of this, studied by us in previous work (Ståhl and Bosch, 2014, 2016b), include telecommunication networks, road vehicles, and aircraft. The problem is further exacerbated by the high degree of customizability and the large number of variants of these products, where it is no longer even clear what 100% of tests passed actually means. All tests passing in *all* product variants – of which there may be many thousands – is clearly not feasible (not to mention verifying all requirements).

The conclusion from this is that there is reason to carefully consider which tests to execute when, the better to maximize coverage and confidence in the software, while minimizing the time and cost required; the larger the test scope and the more expensive and/or scarce the test equipment, the greater the reason for doing so.

That being said, there are multiple ways to seek that optimization, and they are not mutually exclusive. The practice of minimizing high-level scenario tests in favor of low-level unit or component tests – essentially pushing test value down through the "test pyramid" (Cohn, 2010) – is often highlighted, particularly in agile circles (Fowler, 2012). Once everything has been pushed as far down as it can be pushed, however, one is still left with the high-level (and expensive) tests that remain, and the need to decide which ones to execute.

There are a number of options for such selection. One may wish to prioritize tests that have not been executed for a long time, tests verifying recently implemented or changed requirements, recently failed tests, tests that have not recently been executed in a certain configuration, tests with a low estimated cost of execution (Huang et al., 2012), tests that tend to fail *uniquely* as opposed to failing in clusters together with other tests, tests that tend to fail when certain parts of the source code are modified, and so on. They all have one thing in common, however: they require real-time traceability of not only which tests were executed when and for how long, but also items under test, requirements, source changes, and test environments.

Such traceability capabilities require advanced tool support, yet traceability is a domain where the industry is struggling, with an identified lack of infrastructure and tooling-oriented solutions (Cleland-Huang et al., 2014), and particularly tooling "fully integrated with the software development tool chain" (Rempel et al., 2013), with few studies on industry practice (Mäder et al., 2009). Against this background, we will discuss the open source continuous integration and delivery framework Eiffel, originally developed by Ericsson to address these challenges.

## 16.2 The Eiffel Framework

Providing a wide portfolio of products which constitute part of the critical infrastructure of modern society, Ericsson must not only meet strict regulatory and legal demands, but also live up to demanding non-functional requirements, ensuring, e.g., high availability and robustness. With an ambitious continuous integration and delivery agenda, the company has faced the dual challenge of making these practices scale to the considerable size of its product development – many of its products requiring thousands of engineers to develop – and to not only preserve but also improve the traceability of its development efforts.

In response to this challenge, and finding no satisfactory commercially available alternatives – particularly considering its very heterogeneous development environment – Ericsson created its own enterprise continuous integration and delivery framework, called Eiffel. Originally developed in 2013 and now licensed and available as open source (https://github.com/Ericsson/eiffel), Eiffel affords both scalability and traceability by emitting real-time events reporting on the behavior of the continuous integration and delivery system. Whenever something of interest occurs – a piece of source code was changed, a new composition was defined, a test was started, a test was finished, a new product version was published, etc. – a message describing the event is formed and broadcast globally. Each such event further contains references to other events, constituting *trace links* to semantically related engineering artifacts. For instance, a test may thus identify its item under test, which in turn identifies the composition it was built from, which references an included source change, which finally links to a requirement implemented by that change. A more elaborate Eiffel event graph example is shown in Figure 16.1.

By listening to and analyzing these events, Ericsson has managed to address both scalability and the traceability challenges outlined above. Scalability, because each of the globally broadcast events serves as an extension point where a particular continuous integration and delivery system may be hooked into by others in the company interested in the communicated information. That way, differences in tooling, equipment, technology, processes. or geographic location are abstracted away, enabling a decentralized approach to building very large yet performant systems. Traceability, because when persistently stored the graph formed by these events and their semantic references allows a great number of engineering questions to be answered; in the very simple example above, questions such as whether the requirement has been verified, which versions of the product the software change has been integrated into, or, conversely, which software changes and requirement implementations have been added in any one version of that product.

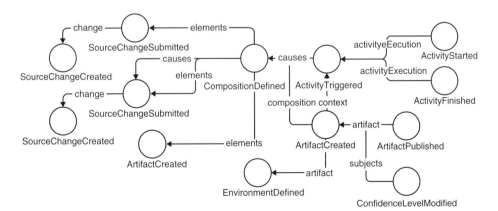

**Figure 16.1** A simple example of an Eiffel event graph. Event names have been abbreviated: the full names of all Eiffel events are of the form *Eiffel…Event*, e.g. *EiffelArtifactCreatedEvent*.

**Table 16.1** Comparison of traceability data collection procedures in 2011 and 2015.

|  | 2011 | 2015 |
| --- | --- | --- |
| Data acquisition method | Asking engineers by mail and phone | Web-based database interface |
| Data acquisition lead time | 15 days | 10 minutes |
| Integration lead time | Varies, estimated average of 30 days | Varies, up to 11 hours |
| Integration frequency | Varies, 0–2 times a month | Once every 5 days |

It should be pointed out that this is done in real time – not by asking colleagues, making phone calls, or by managing spreadsheets, but by database queries. This constitutes a crucial difference to traditional approaches to traceability, which tend to be manual and/or *ex post facto* (Asuncion et al., 2010). Consequently, as found in previous work (Ståhl and Bosch, 2016a), the improvement in traceability effectiveness in projects after the adoption of the Eiffel framework is significant.

Table 16.1 shows the results of an experiment conducted in one such project. Before adopting Eiffel, the tracing of which components used which versions of their dependencies was completely manual and tracked via multiple non-centralized spreadsheets. Consequently, any attempt at collating this information into a coherent overview was also a manual process relying on mail and phone queries, taking weeks and occurring at irregular intervals. Using Eiffel, however, the same data is continuously gathered in minutes using simple database queries. While this example shows the importance of effective and conducive tooling for content traceability – which changes, work items, and requirements have been included in a given version of the system and vice versa – the same applies to test traceability as well. We argue that this radical improvement in traceability practice is a game changer that truly enables dynamic test case selection: in a continuous integration and delivery context one is simply forced to decide the test scope based not on mail conversations but on database queries.

This is particularly the case in large organizations developing large systems. To exemplify, a study of multiple industry cases (Ståhl et al., 2016) reveals that the larger the developing organization, the larger the average size of commits (see Figure 16.2). When, as in one of the studied cases, 40–50 changes averaging nearly 3000 lines of code are committed every day, manual analysis to determine the test scope of those changes is simply not an option.

## 16.3 Test Case Selection Strategies

As discussed in Section 16.1, there are a number of options available with regards to the criteria by which to dynamically select test cases for execution. Regardless of which option one chooses, the rules governing the selection in a particular scenario need to be carefully described in a structured fashion; we term this description a *test case selection strategy* to emphasize the difference from traditional static collections of test cases often referred to as test suites or test campaigns.

To be clear, a test case selection strategy may combine one or more methods of selection, including static identification of test cases. Consequently, such a strategy may in

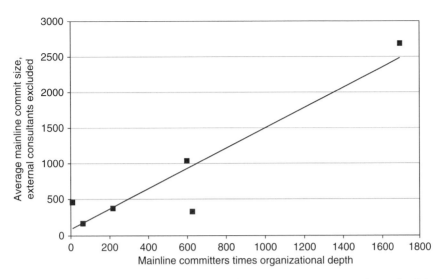

**Figure 16.2** Plot of organizational size (number of mainline committers times hierarchical depth) versus average commit size, excluding commits by external consultants, in six industry cases.

its simplest form be equivalent to a traditional test suite, but may also be much more advanced. To exemplify, it may dictate that tests A, B, and C shall be included, as well as tests tagged as "smoke-test" and any scenario tests which have failed in any of the last five executions.

Entering this type of logic into a single selection strategy description affords a high degree of flexibility to test leaders and test managers: it constitutes a single point of control where they can adjust the testing behavior of the continuous integration and delivery system. Perhaps more importantly, however, it can serve as a vital bulkhead between the separate concerns of continuous integration and delivery job configuration and test management. As we study implementations of continuous integration and delivery practice in the industry, not only do we frequently see static test scopes which remain unchanged for years at a time, but we see them woven into hundreds of, e.g., Jenkins job configurations where they are tangled into build scripts, environment management, triggering logic, etc., causing great difficulties for non-expert users to control and maintain the system.

## 16.4 Automated vs. Manual Tests

In the paradigm of continuous integration and delivery, considerable emphasis is placed on the automation of tests, and rightfully so. Achieving the speed, frequency, and consistency required to produce and evaluate release candidates at the rapid pace these practices call for mandates automation wherever automation is feasible. In our experience, both as practitioners and as researchers, we find that there is still room for manual testing, however; not the repetitious rote testing often seen in traditional development methods to verify functionality, but testing in areas where computers are not (yet) a match for human judgment. Such areas include advanced human–machine interfaces and exploratory testing. To exemplify, in previous work we have studied the continuous integration and delivery system of jet fighter aircraft development (Ståhl and Bosch, 2016b) where the ability of controls and feedback systems – not only visual, but also tactile – to aid the pilot is ultimately determined by the pilot's subjective perception of them. As for exploratory testing, there is great value in letting a knowledgeable human do their utmost to explore weaknesses and try to break the system any way they can. That being said, such manual testing activities arguably do not belong on the critical path of the software production pipeline, where they may increase lead times and cause delays, but as parallel, complementary activities.

Regardless of why, where, or how one performs manual tests, however, from a traceability point of view it is crucial that manual test results are as well documented as automated ones, and preferably documented in the same way so that a single, unified view of test results, requirements verification, and progress can be achieved. All too often we witness that not only are manually planned, conducted, documented, and tracked test projects treated as completely separate from and irreconcilable with automated tests, but automated tests of diverse test frameworks are also reported, stored, and analyzed independently. This results in multiple unrelated views on product quality and maturity – views that project managers, product owners, and release managers must take into account to form an overview.

One advantage of the Eiffel framework is that it creates a layer of abstraction on top of this divergence. While clearly identifying the executed test case and the environment it was executed in, it makes no difference between types of tests, test frameworks, or indeed whether it was manually conducted or not (with the caveat that the execution method is recorded so that it may be filtered on in subsequent queries, if relevant). This is not only important from a traceability point of view and a prerequisite for non-trivial dynamic test case selection, as will be discussed in Section 16.5, but going back to the ability of the framework to not only document but also *drive* the continuous integration and delivery system this agnosticism forms a bridge between human and computer agents in that system: here it is entirely feasible for an automated activity (such as further testing, or a build job) to be triggered as a consequence of manual activities.

## 16.5  Test Case Selection Based on Eiffel Data

In previous sections we have discussed the need for dynamic test case selection and how it requires traceability while touching upon selection strategies and handling of manual and automated tests on a conceptual level. We have also introduced the Eiffel framework and looked at its ability to afford that traceability. Now let us investigate on a very concrete level how such test selection may be carried out, based on data provided by the Eiffel framework.

In Section 16.1 we listed several examples of methods for test case selection. We suggest that all of these may favorably be achieved through analysis of Eiffel events and their relationships. To demonstrate this, we will look at two of these methods in greater detail.

**Selecting tests that tend to fail when certain parts of the source code are modified** requires a historical record of test executions mapped to source code changes, which may be generated from *EiffelTestCaseFinishedEvent* (TCFE), *EiffelTestCaseStartedEvent* (TCSE), *EiffelArtifactCreatedEvent* (ACE), *EiffelCompositionDefinedEvent* (CDE), and *EiffelSourceChangeSubmittedEvent* (SCSE). Figure 16.3 shows how TCFE references

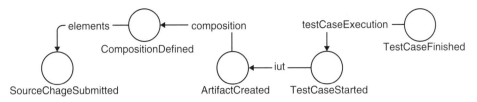

**Figure 16.3** Eiffel events required for selecting tests that tend to fail when certain parts of the source code are modified.

TCSE via its *testCaseExecution* link, whereupon TCSE references ACE via the *iut* (item under test) link, which in turn references CDE via its *composition* link, which references any number of SCSEs via *elements*. Traversing this event graph allows test executions to be connected to source code changes. The TCFE contains the verdict of the test case execution, TCSE identifies the test case, and SCSE points to the relevant source code revision (e.g. a Git commit). Analyzing a sufficient set of source code changes and resulting test executions, it is thus possible to map changes to particular parts of the software (e.g. individual files) and failure rates of subsequent test executions. This information can then be used to prioritize tests likely to fail, and/or adjusting the test scope for each individual change as it affects more or less error-prone areas of the software.

**Selecting tests that have not recently been executed in a certain configuration** may similarly be done based on analysis of *EiffelTestCaseStartedEvent* (TCSE) and *EiffelEnvironmentDefinedEvent* (EDE), where TCSE references EDE via its *environment* link. The latter event describes a specific environment in greater or lesser detail – depending on the technology domain and the need for detail, such a description may consist of anything from, e.g., a Docker image to a network topology or to the length of the cables used to connect the equipment. By querying for EDEs matching certain criteria and then selecting any TCSEs referencing those events, a list of test case executions in matching environments can be built. As TCSE identifies the test case which was executed, as well as a time stamp, a list of test cases sorted by the time they were last executed in a matching environment can be compiled.

The remainder of the use cases listed in Section 16.1 can be addressed in a similar way.

It should be noted that when analyzing historical test records it is imperative that one distinguishes between what one *intended* to do, and what was *actually* done – in terms of which tests were executed, but particularly with regards to the environment in which it was done. In other words, linking a test execution to the test request, including environment constraints (in Eiffel terminology, the test execution recipe) may be useful, but the much more important link is to a snapshot of the environment where the test was truly executed. This is why the Eiffel framework clearly distinguishes between these, and lets *EiffelTestCaseStartedEvent* link to both of them with explicitly different semantics.

## 16.6   Test Case Atomicity

In any dynamic test case selection scheme, the smallest selectable entity is one which is atomic in the sense that it can be executed in isolation, independently of other test cases which may or may not have preceded it. In practice, it is not uncommon to see test cases implemented in suites, with explicit or implicit dependencies on the particular order of execution. This poses a severe impediment to any attempt to dynamically select test cases: no longer can the individual test cases be selected to optimize for a desired outcome, but instead one must select entire suites containing those test cases.

We argue that dynamic test case selection may still be feasible in such a situation, but that its efficacy is severely reduced.

## 16.7 Summary

In this chapter we have described how the open source continuous integration and delivery framework Eiffel was developed by Ericsson to address the challenges of scalability and traceability. Furthermore, we have discussed the dynamic selection of test cases as a method to reduce time and resource usage of, particularly, continuous delivery testing. We have then posited that the traceability data generated by Eiffel can in fact be used to great effect to facilitate a wide range of dynamic test selection methods, and have shown through examples how this can be achieved.

We believe that the possibilities outlined in this chapter serve as opportunities for further research, particularly into empirical validation of the ability of the Eiffel framework to satisfy the traceability requirements of dynamic test case selection, with regards to functionality as well as performance.

## References

Asuncion, H.U., Asuncion, A.U., and Taylor, R.N. (2010) Software traceability with topic modeling, in *Proceedings of the 32nd ACM/IEEE International Conference on Software Engineering*, pp. 95–104.

Beck, K. (2000) *Extreme Programming Explained: Embrace Change*. Reading, MA: Addison-Wesley Professional.

Cleland-Huang, J., Gotel, O.C., Huffman Hayes, J., Mäder, P., and Zisman, A. (2014) Software traceability: Trends and future directions, in *Proceedings of the Conference on the Future of Software Engineering*, pp. 55–69.

Cohn, M. (2010) *Succeeding with Agile: Software Development Using Scrum*. London: Pearson Education.

Duvall, P., Matyas, S., and Glover, A. (2007) *Continuous Integration: Improving Software Quality and Reducing Risk*. Reading, MA: Addison-Wesley.

Fowler, M. (2012) Test Pyramid. http://martinfowler.com/bliki/TestPyramid.html [accessed 24 May, 2016].

Huang, Y.C., Peng, K.L., and Huang, C.Y. (2012) A history-based cost-cognizant test case prioritization technique in regression testing. *Journal of Systems and Software*, 85(3), 626–637.

Mäder, P., Gotel, O., and Philippow, I. (2009) Motivation matters in the traceability trenches, in *Proceedings of the 17th IEEE International Requirements Engineering Conference*, pp. 143–148.

Rempel, P., Mäder, P., and Kuschke, T. (2013) An empirical study on project-specific traceability strategies, in *Proceedings of the 21st IEEE International Requirements Engineering Conference*, pp. 195–204.

Roberts, M. (2004) Enterprise continuous integration using binary dependencies, in *Proceedings of the International Conference on Extreme Programming and Agile Processes in Software Engineering*, pp. 194–201.

Rogers, R.O. (2004) Scaling continuous integration, in *Proceedings of the International Conference on Extreme Programming and Agile Processes in Software Engineering*, pp. 68–76.

Ståhl, D. and Bosch, J. (2014) Automated software integration flows in industry: A multiple-case study, in *Proceedings of the 36th International Conference on Software Engineering*, pp. 54–63.

Ståhl, D. and Bosch, J. (2016a) Continuous integration and delivery traceability in industry: Needs and practices. Submitted to and accepted by SEAA 2016. Not yet published.

Ståhl, D. and Bosch, J. (2016b) Industry application of continuous integration modeling: A multiple-case study, in *Proceedings of the 38th International Conference on Software Engineering*, pp. 270–279.

Ståhl, D., Mårtensson, T., and Bosch, J. (2016) The continuity of continuous integration correlations and consequences. In review.

**17**

# An Automated Regression Testing Framework for a Hadoop-Based Entity Resolution System

*Daniel Pullen, Pei Wang, Joshua R. Johnson, and John R. Talburt*

## Synopsis

Software development is increasingly moving toward distributed processing environments such as Hadoop MapReduce and Spark in order to exploit the power of parallel processing. At the same time, distributed processing environments present a new set of challenges for the software development life cycle. This is particularly true for application testing. Because distributed platforms are often designed to optimize performance by dynamically reconfiguring themselves at run time, creating and maintaining standard test cases can be difficult. Even creating a single event log or statistical summary is not that simple. This chapter presents a case study describing how these challenges were addressed in the development and deployment of a regression testing framework on the Hadoop MapReduce platform. Even though the framework was specifically built to support a large-scale entity resolution application, the design principles can be applied to almost any type of distributed processing application.

## 17.1 Introduction

An ongoing component of software development is testing. Though the duration and timing vary, testing is a critical component of software development that is included in most software development methodologies. Though there are many different types of software testing, this chapter focuses on regression testing. This specific type of testing is designed to ensure that a previously developed and tested piece of software continues to behave in a manner that is conformant with the software requirements that it was written against.

In a traditional development environment, there are many tools available that help support regression testing processes. Three main categories of interest are:

- unit testing
- functional testing
- system testing.

Unit testing is a software testing method that focuses on individual units or pieces of the code. This is considered a white-box type of testing. In other words, the creator

*Analytic Methods in Systems and Software Testing*, First Edition.
Edited by Ron S. Kenett, Fabrizio Ruggeri, and Frederick W. Faltin.
© 2018 John Wiley & Sons Ltd. Published 2018 by John Wiley & Sons Ltd.

of the unit test understands how the code functions. In object-oriented programming languages, unit tests are implemented at the class level. In procedural programming, they may be implemented at the level of a module or an individual function. In development environments that are constrained on resources, the developers that write the code often also write and implement the unit tests. In Java, a common unit testing framework is JUnit. It allows for easy integration into the source code and build environment. Various unit testing frameworks are available for different programming languages. To provide the greatest benefit, unit testing should be implemented early in the development process.

Functional testing is a black-box type of testing. This type of testing is used to check that the software in question performs specific functions in the manner that is described in the software requirements. These tests are created without a comprehensive understanding of the underlying code. The goal of this type of testing is conformance to standards based on functions available in the underlying code. Often, this type of testing is performed by a quality assurance team.

System testing is the big picture perspective of testing. Unlike the previous two categories, these tests are performed on an entire system. The focus of this type of testing is how the software in question behaves with regard to the software requirements when all of the components are integrated together and performing as a whole. In reality, regression testing is a component of system testing as a whole. However, it is still important to understand the nuanced differences between the three categories discussed here.

In the context of Hadoop distributed processing, unit tests cover an essential component of regression testing. However, they come up short in challenging some areas. Rather than focus on the existing and readily available tools for unit testing, this chapter will discuss how to handle some of the shortcomings of unit testing when moving to the Hadoop MapReduce programming paradigm.

MapReduce is an implementation of a programing model for processing large data sets in a network of distributed processors (Talburt and Zhou, 2015). Hadoop MapReduce allows programmers to massively scale applications across hundreds or thousands of processors in a Hadoop cluster. One of the benefits of Hadoop is its scalability of processing using many small, interconnected processors instead of one very large processor. MapReduce handles records as key–value pairs.

MapReduce comprises five primary components:

- Map
- partition
- comparison
- Reduce
- output writer.

The Map component works on individual records (key–value pairs) to prepare for aggregate record processing to be done in the reduce component. Code in the map component performs tasks such as filtering, sorting, and hashing on individual records.

The partition component divides the records into groups for processing in the reducer component. Grouping is based on the key of the key–value pair.

After partition, the comparison component sorts the key–value pairs in the partition according to a user-specified comparison function. Each set of key–value pairs with the same key forms a partition group.

The Reduce component operates on individual partition groups (key–value pairs with the same key). A reducer may count records in a group, find the maximum value in the group, or perform more complex calculations across the records in a group. Reducers apply algorithms to the groups of records that share the same key.

The output writer component writes the output of the reducer to stable storage.

## 17.2   Challenges of Hadoop Regression Testing

Hadoop applications require a paradigm shift in their design and development. With this shift on the development side, there are changes and new challenges in the area of regression testing. Some of these challenges stem from the limitations of distributed environments; others are caused by the massive increase in the volume of data processed by Hadoop applications. Although it is not necessary to directly handle all of these challenges, it is important to understand their impact.

### 17.2.1   Limitations of Distributed Environment

Hadoop is, by design, a distributed computing framework. The many advantages it offers come with a list of disadvantages that put a greater burden on designers and developers. Some of the disadvantages extend to regression testing. The limitations discussed here are primarily focused at MapReduce specifically. A non-definitive list of major limitations of Hadoop MapReduce are:

- batch processing
- difficult or impossible to implement some algorithms
- non-shared memory
- Map and Reduce tasks run in isolation.

Fortunately, not all of these are "deal breakers" for regression testing. Most regression testing frameworks are batch processing by design. Additionally, the ability to implement some algorithms tends to limit the type of software that can be developed for the platform but not the type of software you can test. This limits the real focus to the third and fourth items as the primary challenges for regression testing. The impact of these two items on testing is interrelated.

It is important to understand that these tasks run across many different machines that transmit data between each other across a local area network. One machine, or node, in a Hadoop cluster cannot directly access the memory of another machine. Furthermore, the Map and Reduce tasks cannot directly communicate with each other. The communication that occurs between the machines must follow what is provided by the MapReduce programming model. Except for some small exceptions, data can be transmitted at three different times during MapReduce jobs:

- after the input is read and transmitted to a Map task
- after the Map task, during the partition and shuffle steps
- after the Reduce task when it is transmitted to the output writer.

For each of these, there is a guarantee that the data will transmit between tasks but not a guarantee it will transmit between machines (except for data replication on the Hadoop distributed file system, HDFS).

The developer also has limited control over the destination of data. This means the developer cannot always predict the Map or Reduce task (or machine on the cluster) that will process a given piece of data. Furthermore, exercising the limited control over the destination task for a piece of data can require additional work such as creating custom partition and sort classes for a particular MapReduce job. These limitations together encompass the isolated nature of Map and Reduce tasks.

To overcome these limitations, a couple of features can benefit regression testing initiatives. First, Hadoop MapReduce provides a robust logging interface using Log4j. For testing purposes, this allows a developer to accumulate a fair amount of information accessible for each of the tasks that run and the Hadoop job itself. Hadoop counters are also of particular interest. Counters are used internally by MapReduce to track a large amount of metadata about a job. Examples of built-in counters include:

- total volume of data processed in tasks
- total number of records processed tasks
- amount of CPU time consumed
- amount of RAM utilized.

There are many more than this, but this gives an idea of some of the data already captured by MapReduce. In addition to the already provided counters, the developer can create custom counters that are accessible from the Map and Reduce tasks through the Context class. This allows the developer to iterate a counter from the Map and Reduce tasks. There is a limitation to the number of counters a developer can use, configurable through the mapreduce.job.counters.limit property in the mapred-site.xml configuration file. The default limit is 120. Since a large number of counters puts a strain on the resource manager, developers should be conservative in their use of custom counters.

### 17.2.2 Data Volume

Testing Hadoop applications can be simplified by focusing on smaller data sets. Unfortunately, experience has shown that not all regressions or bugs can be identified by small data sets. Testing with large-volume datasets can be particularly challenging for a couple of reasons. The ability to measure the accuracy of the results can be difficult. This is discussed in more detail below. The other challenge is the amount of time it takes regression tests to run. Planning needs to be undertaken to consider the trade-offs between the delay created while waiting on regression testing to complete and the cost of allocating time on larger Hadoop clusters or even dedicated testing clusters. Many organizations already accommodate development and production clusters. For those organizations, the decision is easier. It is not mandatory to test Hadoop applications at scale. However, it is important to identify the best trade-off for the needs of your business.

### 17.2.3 Measuring Accuracy of Results

One challenge of Hadoop regression testing is measuring the accuracy of the results. It is impractical to do manual measurements with large-volume data sets. Using an automatic way to measure the accuracy, a benchmark data set is needed. How can a benchmark be automatically generated? There are two solutions:

- procedural generation of the data sets
- replication of manually validated benchmark data.

The first solution needs a data generator to automatically generate the data sets based on the request. This approach can control the variation of the data sets and the volume of the data sets. The benchmark will have very high accuracy.

The second solution has several steps:

- Pick up a small subset from the data, though picking the subset is a challenge. It can be picked by random sampling, systematic sampling, stratified sampling, or cluster sampling.
- Use the subset data as input to run the software.
- Manually review the results; create the benchmark subset.
- Replicate the subset to the required volume.

#### 17.2.3.1 Dependent on Application

It is imperative that testing initiatives are planned and understood in the context of a specific application. Though the challenges outlined here are identified as generically as possible, the authors of this chapter identified these challenges based on the testing of an entity resolution system that was designed for and implemented in MapReduce. Some of these challenges may not be an issue for other applications. Furthermore, other challenges may arise that are not detailed here.

## 17.3 The HiPER Entity Resolution System

This chapter describes the regression testing framework for the HiPER (High-Performance Entity Resolution) system, which runs in a MapReduce environment. The regression resting is designed to accommodate the limitations of MapReduce environments as defined in Section 17.2, based on the software system as described in this section.

### 17.3.1 Overview

HiPER is a big data software product developed by Black Oak Analytics, Inc., and was specifically designed to run in the Hadoop MapReduce environment to scale entity resolution for large-scale data applications. The HiPER entity resolution engine supports both deterministic matching and probabilistic matching, transitive linking, and asserted linking. A key feature of HiPER is that its entire configuration, including input reference layouts, entities, match rules, and blocking, is interpreted at run time through user-defined XML scripts.

The key elements of the HiPER design philosophy are:

- designed specifically for the Hadoop distributed computing platform
- full-context computing
- keep all of the original input data
- on-the-fly data preparation to reduce dependency on external preprocessing
- generate extensive metadata for process explainability and traceability
- many points of system configurability to adapt to different use cases without modifying source code – single code base, multiple configurations
- run in any environment – the cloud, dedicated cluster, single server using VM cluster – all with the same code base

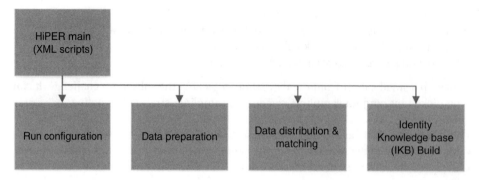

**Figure 17.1** HiPER functional components.

- extensible Java classes for adding plug-and-play comparators and data preparation functions
- UTF-8 character compliant
- built-in support for language localization
- support for clerical review and correction.

The HiPER functional components (see Figure 17.1) are:

- HiPER Main – Reads and interprets XML configuration scripts.
- Run Configuration – Sets the mode of operation for the requested configuration.
- Data Preparation – Optional pre-ER processes such as parsing and standardization.
- Data Distribution and Matching – Blocks the data and matches reference within blocks.
- Identity Knowledgebase (IKB) Build – Creates and stores identity structures and other outputs such as link index, run report, run log, run statistics, and clerical review exceptions.

### 17.3.2 Entity Resolution

Entity resolution (ER) is the process of determining whether two references to real-world objects in an information system are referring to the same object or to different objects (Talburt, 2011). The term *entity* describes the real-world object, a person, place, or thing, and the term *resolution* is used because ER is fundamentally a decision process to answer (resolve) the question: Are the references to the same or to different entities? Although the ER process is defined between pairs of references, it can be systematically and successively applied to a larger set of references to aggregate all of the references to same object into subsets or clusters. Viewed in this larger context, ER is also defined as "the process of identifying and merging records judged to represent the same real-world entity" (Benjelloun et al., 2009).

Entities are described in terms of their characteristics, called *attributes*. The values of these attributes provide information about a specific entity. Identity attributes are those that when taken together distinguish one entity from another. Identity attributes for people are things like name, address, date of birth, or fingerprint – the kinds of things often asked for to identify a person requesting a driver's license or hospital admission. For a product, identity attributes might be model number, size, manufacturer, or universal product code.

A *reference* is a collection of attribute values for a specific entity. When two references are to the same entity, they are sometimes said to *co-refer* (Chen et al., 2009) or to be *matching references* (Benjelloun et al., 2009). However, for reasons that will become clear, the term *equivalent references* will be used throughout this text to describe references to the same entity.

### 17.3.3 Entity Identity Information Life Cycle Management

Entity identity information management (EIIM) is the collection and management of identity information with the goal of sustaining entity identity integrity (Talburt and Zhou, 2015). Entity identity integrity requires that each entity must be represented in the system once, and only once, and distinct entities must have distinct representations in the system (Maydanchik, 2007). Entity identity integrity is a fundamental requirement for master data management (MDM) systems.

Entity identity information management is an ongoing process that combines ER and data structures representing the identity of an entity into specific operational configurations (EIIM configurations). When these configurations are executed in concert, they work together to maintain the entity identity integrity of master data over time. Entity identity information management is not limited to MDM – it can be applied to other types of systems and data as diverse as reference data management systems, referent tracking systems (Chen et al., 2013), and social media (Mahata and Talburt, 2014).

Identity information is a collection of attribute–value pairs that describe the characteristics of an entity – characteristics that serve to distinguish one entity from another. For example, a student name attribute with a value such as "Mary Doe" would be identity information. However, because there may be other students with the same name, additional identity information such as date of birth or home address may be required to fully disambiguate one student from another.

Although ER is necessary for effective MDM, it is not sufficient to manage the life cycle of identity information. Entity identity information management is an extension of ER in two dimensions: knowledge management and time. The knowledge management aspect of EIIM relates to the need to create, store, and maintain identity information. The knowledge structure created to represent a master data object is called an *entity identity structure* (EIS).

The time aspect of EIIM is to assure that an entity under management in the MDM system is consistently labeled with the same, unique identifier from process to process. This is only possible through EIS, which stores the identity information of the entity along with its identifier so that both are available to future processes. Persistent entity identifiers are not inherently part of ER. At any given point in time, the only goal of an ER process is to correctly classify a set of entity references into clusters, where all of the references in a given cluster reference the same entity. If these clusters are labeled, then the cluster label can serve as the identifier of the entity. Without also storing and carrying forward the identity information, the cluster identifiers assigned in a future process may be different.

## 17.4   Entity Identity Information Life Cycle and HiPER Run Modes

In the field of information management it has long been recognized that information has a life cycle. Information is not static, it changes over time. Several models of information life cycle management have been developed. English (1999) formulated a five-phase information life cycle model of plan, acquire, maintain, dispose, and apply, adapted from a generalized resource management model. McGilvray (2008) later extended the model by adding a "store and share" phase and naming it the POSMAD life cycle model, an acronym for

- plan for information
- obtain the information
- store and share the information
- maintain and manage the information
- apply the information to accomplish your goals
- dispose of the information as it is no longer needed.

The CSRUD model was developed specifically to address life cycle entity identity information (Talburt and Zhou, 2015). It adapts the CRUD (create, read, update, and delete) model used for many years by database developers to address identity management specific issues. The five phases of CSRUD are:

- Capture – The initial creation of EIS for the system. Capture occurs when an MDM system is first installed. However, there is almost always some form of MDM, either in a dedicated system or an internal ad hoc system, that must be migrated into the new system.
- Store and Share – The storage of EIS in a persistent format such as a database or flat-file format.
- Resolution and Retrieve – The actual use of the MDM information in which transactions with master data identifying information are compared (resolved) against the EIS in order to determine their identity. When an entity reference in a transaction is determined to be associated with a particular EIS, the EIS identifier is added to the transaction. For this reason, the process is sometimes called "link append" because the EIS identifier added to the transaction is used to link together transactions for the same entity.
- Update – Adding new EIS related to new entities and updating previously created EIS with new information. The update process can be either automated or manual. Manual updates are often used to correct false positive and false negative errors introduced by the automated update process.
- Dispose – Retiring EIS from the system, which can be for two reasons. The first is the case where the EIS is correct but is no longer active or relevant. The second is in the correction of false negative errors where two or more EIS are merged into a single EIS.

Because HiPER was designed to support the entity identity information life cycle, most of its run modes correspond to one of these phases.

### 17.4.1   Identity Capture Only

The identity Capture Only run mode of HiPER supports the capture phase of the CSRUD model. Figure 17.2 illustrates the major inputs and outputs of the Capture Only run mode.

The purpose of the Capture Only run mode is to create the initial IKB of EISs that serves as the system identity memory. Each EIS preserves the identity information related to a particular entity. In addition, each EIS is assigned a unique identifier (EID) that is used to represent that identity. By storing these EIDs along with the identify information, the IKB provides a mechanism for assigning this same identifier to the same entity over time, i.e. creates a persistent entity identifier.

The Link Index output is simply a two-column table that shows for each input reference processed, the identifier of the EIS (identity) into which the reference was linked. The Link Index allows the client system (the system providing the input references) to append these identifiers to the input references to facilitate further processing. For example, if the input is a set of customer references, then the Link Index shows which references were to the same customers, i.e. the references linked by the same EID.

The Review Indicators output is an exceptions report used in the assertion process described later. Review indicators are notes to the system operators that certain matches have a low level of confidence and could be sources of error. These reports are reviewed by domain experts who use them to investigate possible errors in the IKB.

#### 17.4.1.1   Entity Identity Structures

HiPER follows the model of full-context computing in which all input data are saved. HiPER collects and saves a copy of the identity information from each entity reference it processes. Even in cases where certain data items are parsed or standardized in preparation for matching, e.g. a combined address field, the original data is kept in the EIS. Each EIS also contains metadata items generated by the Capture Only process such as creation date, identifier of the run in which it was created or last updated, identifiers for rules that brought a particular reference into the EIS, and assertion metadata.

#### 17.4.1.2   Identity Knowledgebase

The HiPER IKB is the aggregate of all the EISs that exist in the system at any given time. Because HiPER runs in the HDFS, the IKB itself has a distributed file structure. The base

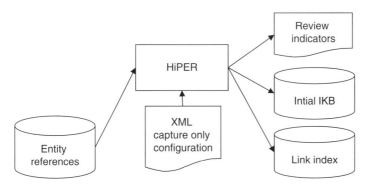

**Figure 17.2** Identity Capture Only configuration of HiPER.

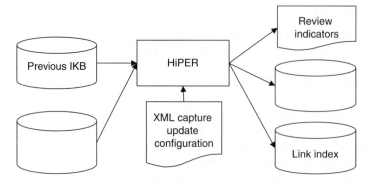

**Figure 17.3** Identity Capture Update configuration of HiPER.

table comprises the references that have been processed along with any reference-level metadata. Another table stores EIS-level metadata, and yet a third table stores IKB-level metadata.

### 17.4.2 Identity Capture Update

The identity Capture Update run mode of HiPER supports the update phase of the CSRUD model. Figure 17.3 illustrates the major inputs and outputs of the HiPER Capture Update run mode.

The purpose of the Capture Update run mode is to update the IKB with new identity information. The Capture Update run mode requires two inputs: a file of entity references and a copy of the IKB to be updated with new information. During the Capture Update process HiPER makes decisions for each input reference. If the entity reference belongs to (matches) one and only one EIS in the IKB, then the identity information in the reference is added to the EIS, and the EID of the reference in the Link Index is the EID of the EIS into which it was merged. If the entity reference is determined not to match any EIS already in the IKB, then it creates a new EIS with a new EID, and that EID is assigned to the reference in the Link Index.

There is another situation in which the input entity reference matches two or more EISs in the IKB: if the reference and all matching EISs are merged into a single EIS. The EID of the surviving EIS is the same as the EID for oldest EIS in the merge set. The EIDs of the other EISs become inactive. It is for this reason that the EID are said to be "persistent" rather than "permanent," because the EID can change when these merge situations arise.

### 17.4.3 Identity Resolution

The Identity Resolution run mode of HiPER supports the resolve and retrieve phase of the CSRUD model. Figure 17.4 illustrates the major inputs and outputs of the HiPER Identity Resolution run mode.

The purpose of the Identity Resolution run mode is to allow client systems to obtain EIDs for entity references, but without updating the IKB. The Identity Resolution run mode requires two inputs: the current IKB (current identity knowledge) and a file of references to be identified. Because the references do not update (only read) the IKB, the only output is the Link Index. For each input entity reference, the system decides if the

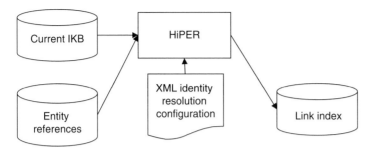

**Figure 17.4** Identity Resolution configuration of HiPER.

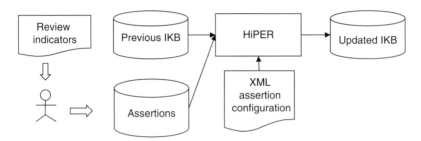

**Figure 17.5** Assertion configuration of HiPER.

reference matches an existing IKB or not. If it matches, then the reference is assigned the EID of the matching EIS in the Link Index, otherwise the reference is assigned a default EID value that tells the client system that no match was found.

### 17.4.4 Assertion

The Assertion run mode of HiPER supports the update phase of the CSRUD model. Figure 17.5 illustrates the major inputs and outputs of the HiPER Assertion run mode.

The purpose of the Assertion run mode is to correct errors in the IKB made by the automated match rules used in Capture Only and Capture Update processing. The transactions read by HiPER to correct these errors are called *assertions*. Assertion transactions are created manually by operators following the instructions of domain experts who have examined the clerical review exceptions reports. Based on these reports, experts determine what should be the true state of certain incorrectly formed EISs in the IKB. Assertions bypass the matching rules to directly access and adjust the IKB structure. The automated rules in both the Capture Only and the Capture Update modes can produce two kinds of errors: a false positive error and a false negative error.

A false positive error occurs when the match rules merge references to different entities into the same EIS. The correction of this error is to split the EIS into two or more EISs that correctly group the references. The assertion transactions that correct the false positive error identify the references that should be moved out of a particular EIS to form a new EIS. In addition to splitting the EIS, the assertion transactions also embed metadata into the EIS that prevent them from being merged together again by the match rules in future Capture Only or Capture Update runs.

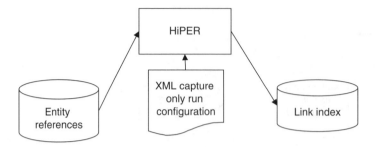

**Figure 17.6** Merge–Purge configuration of HiPER.

A false negative error occurs when the match rules fail to merge two references into the same EIS. The assertion transaction to correct a false negative simply identifies the two EISs to be merged.

In order to make the clerical review and assertion process more efficient, HiPER also supports two other types of assertion: true positive and true negative. When an expert determines that EISs appearing in the clerical review report are actually correct, then the EIS can be asserted as correctly configured. These assertions do not change the structure of the IKB, but they do insert metadata into the EIS to prevent them from being reported again for clerical review unless and until their state has changed. In this way, the effort expended in clerical review can be reduced over time.

### 17.4.5 Merge–Purge

The Merge–Purge run mode of HiPER is a special configuration used to support data cleansing as part of an extract–transform–load (ETL) process. Merge–Purge processing is not part of the identity information management life cycle, it is one-time operation to deduplicate references prior to further ETL operations. Figure 17.6 illustrates the major inputs and outputs of the Merge–Purge run mode.

The only input is the file of entity references and the only output is the Link Index. No IKB is produced or modified in the Merge–Purge run mode. The client providing the input references can use the Link Index to determine which references are for the same entity (duplicate references).

## 17.5 The HiPER Regression Testing Framework

The HiPER Testing Framework (HTF; see Figure 17.7) is a Java program designed to run on a virtual machine with HiPER already installed in a Hadoop environment. The HTF runs regression testing based on the configuration of a test suite XML script and test case XML scripts.

The Testing Framework inputs include:

- source data
- all required HiPER XML scripts
- HiPER JAR file
- test suite XML script
- test case XML script(s)

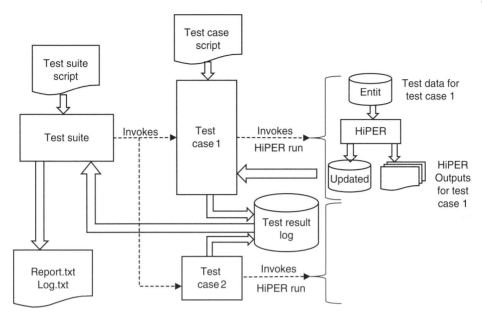

**Figure 17.7** Test Framework architecture.

- library of HTF Java resources
- HTF JAR file.

The HTF's primary outputs are a log file and a report file; additionally, any outputs created by test case runs are either stored in Hadoop or can be set to be output to directories in the HiPER suite directory. All pieces of the HTF are modularized, so input sources, HiPER JAR files, HiPER scripts, test case scripts, etc. can be replaced or modified as needed to adapt test suites to new use cases.

Multiple test suites can be configured to test a variety of use cases, and test the entire HiPER system for stability and accuracy. The purpose of the framework is to provide regression testing. Every time a new version of HiPER is released, it can be run through multiple test suites and be tested through hundreds of test cases, each representing a different configuration of HiPER.

### 17.5.1 Design Principles of the Framework

The Testing Framework was designed to be modular, to test by comparing output results to expected results, to have multiple access points for verification, and to include sufficient logs and reports. Because testing software in MapReduce environments has limitations based on the available data, the HTF was designed to validate values produced by Log4J and the output writer.

### 17.5.2 Modular Design

The Testing Framework is set up to include test suites with test cases. Each test suite contains test cases and their associated files. The test cases are independent of each other (with a couple of exceptions, discussed in Section 1.5.6) and can be added, subtracted, or changed without affecting each other. One of the disadvantages of the modular design is that in a test suite with hundreds of test cases, if any of the script validation is changed in the HiPER system, each individual test case script must be changed. The Testing Framework now includes a Python tool that can update/change each of the scripts for testers, should the need arise to make the same change to all of the scripts manually.

While a test suite includes many (hundreds) of test case scripts, each used to test specific run settings, the test cases, taken as a whole, test much larger use cases. For example, a Capture Only run tests HiPER's creation of a Link Index and IKB; a subsequent Capture Update run takes the IKB output from the previous Capture Only run and updates the IKB with new input data. Another Capture Only run can take the IKB created by the previous Capture Update run and run it again with new input data, thus testing the Capture Update, Update use case. In this manner, multiple test cases test the larger functionality of the system. These larger functionality tests are broken up into multiple test suites, each designed to test specific business use cases as part of the EIIM.

### 17.5.2.1 Test Suite

Test suites refer to the inputs necessary to regression test a specific set of test cases, often in testing a business use case. Specifically, a test suite refers to a test suite XML script, Testing Framework JAR file, HiPER JAR file, workspace folder, test case folder, and results folder. Each test suite includes specific test cases, each defined by a test case XML script. When instantiated, the test suite runs HiPER, based on its source files, for each test case script, in the order the scripts are listed in the test suite script. The results for the runs are output to the results folder, which includes logs and reports. The log includes run-time data, and the reports include validations based on manually input values. Test suites can have the HiPER and test case components replaced, modified, updated, or interchanged with other test suites.

An example test suite script is shown below:

```
<TestSuiteScript SuiteName="TestSuite">
<TestCaseWorkspace>/home/hadoop/TestSuiteA/HiPERWorkspace/</TestCaseWorkspace>
<HiPERWorkspace>/home/hadoop/TestSuiteA/TestCaseWorkspace/</HiPERWorkspace>
<HadoopPath>/home/hadoop/hadoop/bin/hadoop</HadoopPath>
<HiPERJar>HiPER_v12345.jar</HiPERJar>
<TestResultLog>/home/hadoop/TestSuiteA/TestResultLog</TestResultLog>
<TestCases>
     <TestCase>TestCaseScriptA.xml</TestCase>
     <TestCase>TestCaseScriptB.xml</TestCase>
     <TestCase>TestCaseScriptC.xml</TestCase>
</TestCases>
</TestSuiteScript>
```

### 17.5.2.2 Test Case

Every test case is defined by an XML script that includes the test case name, a description, the output statistics to be pulled for that test case, and the expected values. The test cases each compare specific values from the output of the HiPER run for that specific test case. The test case output statistics are chosen from a predefined list that are captured for each test case run from various HiPER outputs.

An example test case script is shown below:

```
<TestCaseScript CaseName="CaptureOnly" HiperRunScript="CaptureOnlyRunScript.xml">
<Compare Label="Reference Counts">
    <Output>Link.ReferenceCount</Output>
    <Output>IKB.ReferenceCount</Output>
    <Expected>2866</Expected>
 </Compare>
 <Compare Label="Cluster Counts">
    <Output>Link.IdentityCount</Output>
    <Output>IKB.IdentityCount</Output>
    <Expected>718</Expected>
 </Compare>
</TestCaseScript>
```

For example, the test case could pull the reference counts from the Link file (which includes the HiPER IDs) and the knowledgebase. Those results are then compared to a manually entered value, and if the results match the expectation the result of the test case comparison is True; if they do not match then the comparison is determined to be False. All of the compared values are printed in the final test suite report.

### 17.5.3    Output-Oriented Testing Approach

The HiPER Testing Framework runs pass/fail reports based on the results of each test case, which is a single run of HiPER (see Figure 17.8). The pass/fail determination is made by taking parsed outputs from a run of HiPER and comparing them with the expected results. For some test cases, the expected results are easy to calculate.

For example, if we know that out of five references, the result should be two EISs, one of three references and one of two references, it is easy to calculate that the Link results should include the IdentityCount as 2. However, in more complex chains of test cases, where later test cases rely on the results of earlier test cases, calculating the expected values can be more difficult.

#### 17.5.3.1    Parsing HiPER Outputs

For each test case, HiPER runs produce five outputs that can be used by each test case to make comparisons for the test suite report, the Link file, IKB, statistics, change report, and transitive closure statistics. These outputs are parsed during the run of each test case, and recorded in the test suite report. Every test case output is added to the report and verified as pass/fail based on the entered expected value.

#### 17.5.3.2    Reuse of Test Sets and Verified Results

The current HiPER test suites are designed to allow for results to be built upon by adding test cases that use existing test sets and verified results, for increasingly complex test cases. Because test cases (and related folders) can be added across test suites, test sets can be reused in future suites. Once results of new test cases have been verified as true, those test cases can be added to the list of reusable test sets.

### 17.5.4    Currently Defined Access Points for HiPER Verification

The following HiPER outputs have values that can be compared for the pass/fail comparisons: Link file, IKB, statistics file, change report, and transitive closure statistics file:

- Link file: Includes the EID and reference ID produced during the run.
- IKB: Identities from the previous run.

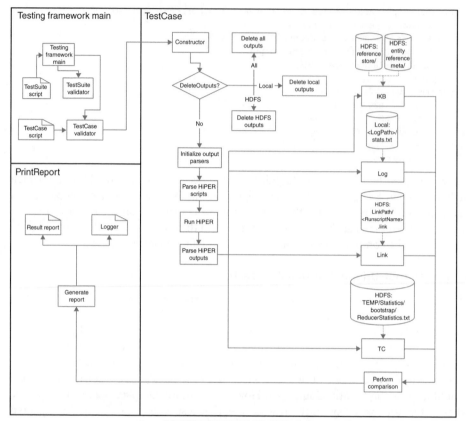

© BLACK OAK ANALYTICS, INC, 2016 – *HiPER Testing Framework (HTF) v3.0*

**Figure 17.8** HiPER Testing Framework detailed diagram.

- Statistics: Includes summary, EIS, index, and timing statistics.
- Change report: Includes statistics related to EISs such as input EIS, output EIS, updated EIS, not updated EIS, merged EIS, and new EIS.

### 17.5.4.1 Counts

The Link, IKB, and statistics files can provide reference counts and EIS counts.

The change report can provide the output EIS, input EIS, updated EIS, not updated EIS, merged EIS, and new EIS counts. For validation purposes, output EIS = input EIS + new EIS and input EIS = updated EIS + not updated EIS.

### 17.5.5 Logging and Reporting

The log report (see Figure 17.9) includes the number of test cases processed, the time started, the time the runs were completed, and the time the test suite finished processing. The report includes the counts compared for each test case, along with pass/fail results and the difference between the expected value and the actual value.

```
HiPER Testing Framework 3.0

Test Case Results:

MergePurge                              Pass
    Reference Counts                        Fail
        Detailed Failure Report:
        Link.ReferenceCount                         -1

    Cluster Counts                          Fail
        Detailed Failure Report:
        Link.IdentityCount                          -1

CaptureOnly                             Pass
    Reference Counts                        Fail
        Detailed Failure Report:
        IKB.ReferenceCount                          -1
        Link.ReferenceCount                         -1

    Cluster Counts                          Fail
        Detailed Failure Report:
        IKB.IdentityCount                           -1
        Link.IdentityCount                          -1
```

**Figure 17.9** Framework example report outputs.

## 17.5.6 Test Case Examples

The basic test case examples include Merge–Purge, Identity Capture Only, Identity Capture Update, Identity Resolution, and Assertion, as described in Section 17.4.

### 17.5.6.1 Standalone Test Cases

Some test cases are standalone, and have the input files needed to run the test cases included in the test suite HiPER workspace. Typically, these are test cases such as permutations of Merge–Purge and Identity Capture Only runs, because these run configurations run directly from input sources. However, cases such as Identity Capture Update and Identity Resolution require an initial run of Identity Capture Only, to create a Link file and IKB, so these test cases (and permutations) use previous HiPER run outputs as the inputs.

### 17.5.6.2 Interdependent Test Cases

Because Identity Capture Update and Identity Resolution test cases require Link files and IKBs, the test cases are designed so previous Identity Capture Only test case outputs are used as inputs by these test cases. One result is that if the input for the source run test case is changed, each dependent test case will propagate changed output results. Using this incremental approach to designing test cases, test suites are able to test large groups of interdependent test cases without having to manually validate all of the steps.

One of the issues that had to be solved for the Testing Framework is how to use HiPER identity structure identifiers in test data. HiPER structure identifiers are generated on demand using a time-dependent hashing algorithm, and this can cause a problem for test cases that need to reference these structure identifiers. At the moment a new identity structure is created, the hashing algorithm uses several pieces of data, including the system time, to generate a new identifier. Using time as part of the input helps ensure that the new identifier will be unique. In a production application this is a desirable feature. Once an identity structure is created and assigned a unique identifier, the same identifier value will persist as long as the identity structure is maintained in the client's system. However, the dynamic generation of identity structure identifiers creates a problem when trying to build test scenarios that use these identifiers.

As an example, consider a scenario to test whether two HiPER structures merged by an assertion transaction have been correctly merged. The first step in the test scenario is to build a HiPER IKB from a set of test references using a Capture Only process. The second step is an Assertion process in which two structures from the IKB are merged by a structure-to-structure assertion transaction. The problem is that the HiPER identifiers generated in the first step will always be different each time the process is run, even though the same references are input as test data each time. Thus, the HiPER identifiers referenced in the structure-to-structure assertion transaction cannot be known prior to the completion of the Capture Only process. Having to manually insert the correct identifiers into the second process each time the test is executed is impractical and defeats the purpose of having an automated testing framework.

The solution to the problem was to create a new test function that extracts the HiPER identifier based on a particular reference in the structure and dynamically generates an assertion transaction of the desired type. This is possible because each reference processed by HiPER must have a unique record identifier. Once a reference is created for inclusion in a test dataset, its identifier will not change.

The functionality was added in the form of a new < GenerateAssertions > child element of the < TestCaseScript > element. The generate assertions element is used to obtain information from an IKB generated in a previous < TestCaseScript > of the same < TestSuiteScript>. The < GenerateAssertions > element directs the generation of particular assertion transactions that reference HiPER identifiers in the previously generated IKB. The < GenerateAssertions > element provides a way for the < TestCaseScript > generating the assertions to dynamically find and reference valid HiPER structure identifier values at run time.

The process relies on knowing the identifiers of references in the IKB. The process flow of a < TestCaseScript > using < GenerateAssertions > is shown in Figure 17.10.

The < GenerateAssertions > step will generate HiPER assertions of a given type (true positive, true negative, structure-to-structure, and structure split) that reconfigure one or more HiPER identity structures in an IKB created in a previous < TestCaseScript>. The < GenerateAssertions > defines the

- type(s) of assertions to be generated
- dependent (previous) < TestCaseScript > that generated the IKB to which the generated assertions are to be applied
- name of the IKB created in the dependent < TestCaseScript > .

The < GenerateAssertions > is a child element of < TestCaseScript > and follows the pattern shown below:

```
<TestCaseScript CaseName="NameOfCase"   HiPERRunScript="NameOfRunScript>
 < GenerateAssertions   DependentTestCase = "NameOfTestCase"   IKBName = "NameOfIKB">
  <Assertions Type ="TypeOfAssertion">
   <Assert>Value1, Value2, Value3, ...</Assert>
```

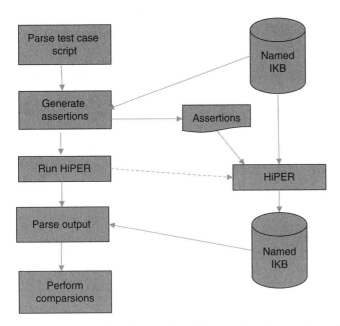

**Figure 17.10** Logic flow for <TestCaseScripts> using <GenerateAssertions>.

```
   <Assert>...
  </Assertions>
  <Assertions Type = ...
   <Assert>...
    ...
  </Assertions>
 </GenerateAssertions >
 ...
</TestCaseScript>
```

The script allows for more than one assertion type to be tested in the same test case. HiPER currently supports four types of assertions. There are two correction assertion types, structure split ("StrSplit") to correct false positive IKB configurations, and structure-to-structure ("StrToStr") to correct false negative IKB configurations. In addition there are two confirmation assertion types, true positive ("TP") to label a structure having been verified as correctly linked, and true negative ("TN") to label a set of structures as having been verified as correctly not linked. Even though the confirmation assertions do not alter the configuration of IKB structures, they do insert metadata tags that prevent the same structures from being called out for clerical review again after they have already been verified as correct, unless and until they are altered by some other process. Their purpose is to save time and effort in the clerical review process.

Each of the four assertion types requires a different pattern of parameters as the < Assert > character content. For example, generation of a structure-to-structure transaction requires the following < Assert > pattern:

```
<Assert>AssertID, *RefID, GroupNbr, UserName</Assert>
```

Here, "AssertID" is a unique value to identify the assertion being generated; "*RefID" is a valid record identifier in the IKB, the asterisk (*) prefix indicating that the RefID is a placeholder for the identifier of the HiPER identity structure containing the reference; and "GroupNbr" is a link value that connects this < Assert > element with other < Assert > elements sharing the same "GroupNbr" value. All of the identity structures sharing the same "GroupNbr" link are to be merged. Therefore, an < Assertions > element of Type="StrToStr" must enclose at least two < Assert > elements. The "UserName" is inserted into the metadata of the merged structure to identify the reviewer who made the merge decision.

The < Assert > pattern for structure-split assertions is similar:

```
<Assert>AssertID, RefID1, *RefID2, GroupNbr, UserName</Assert>
```

In this case, "RefID1" identifies a reference that should be removed from the structure "*RefID2" to create a separate structure. The number of child < Assert > elements in an < Assertions > element of type "StrSplit" depends on how many references need to be removed. In addition, "GroupNbr" indicates how the removed references should be grouped in new structures. The following example illustrates an actual test case for structure-split assertions:

```
<TestSuiteScript>
 <TestCaseScript CaseName="IdentityCapture" HiperRunScript="CaptureRunScript.xml">
 . . . .
 </TestCaseScript>
 <TestCaseScript CaseName="StructureSplit" HiperRunScript="AssertionRunScript.xml">
  <GenerateAssertions DependentTestCase=IdentityCapture IKBName="CaptureOutput">
   <Assertion Type ="StrSplit">
    <Assert>SPA.1, 0001.A915657, *0001.A915657, 1, PWang</Assert>
    <Assert>SPA.2, 0001.B712583, *0001.A915657, 1, PWang </Assert>
    <Assert>SPA.3, 0001.C265894, *0001.A915657, 2, PWang </Assert>
    <Assert>SPA.4, 0001.D456879, *0001.A915657, 2, PWang </Assert>
   </Assertion>
  </GenerateAssertions >
  <Compare Label="Reference Counts">
   <Output>Link.ReferenceCount</Output>
   <Output>IKB.ReferenceCount</Output>
   <Expected>143</Expected>
  </Compare>
  <Compare Label="Cluster Counts">
   <Output>Link.IdentityCount</Output>
   <Output>IKB.IdentityCount</Output>
   <Expected>13</Expected>
  </Compare>
 </TestCaseScript>
</TestSuiteScript>
```

In this example, four references are removed from the structure containing reference "0001.A915657" into two new structures. References "0001.A915657" and "0001.B712583" are to be removed and placed together into a new structure (indicated by both have group number "1"). References "0001.C265894" and "0001.D456879" will be placed together into yet a different structure (indicated by the group number "2").

**Table 17.1** Inventory of test cases.

| Run configuration | Blocking | Rule | Hash | Grouping | Parser | Block indexing | Index |
|---|---|---|---|---|---|---|---|
| Merge–Purge | Standard | Null | None | Yes | Address | None | Index |
| Merge–Purge | Standard | Null | None | Yes | Address | None | RefID |
| Merge–Purge | Standard | Null | None | Yes | No | None | Index |
| Merge–Purge | Standard | Null | None | Yes | No | None | RefID |
| Merge–Purge | Standard | Null | None | No | Address | None | Index |
| Merge–Purge | Standard | Null | None | No | Address | None | RefID |
| Merge–Purge | Standard | Null | None | No | No | None | Index |
| Merge–Purge | Standard | Boolean | None | No | No | None | Index |
| Merge–Purge | Standard | Null | None | No | No | None | RefID |
| Merge–Purge | Standard | Boolean | None | No | No | None | RefID |
| CaptureOnly | Standard | Null | None | Yes | Address | None | Index |
| CaptureOnly | Standard | Null | None | Yes | Address | None | RefID |
| CaptureOnly | Standard | Null | None | Yes | No | None | Index |
| CaptureOnly | Standard | Null | None | Yes | No | None | RefID |
| CaptureOnly | Standard | Null | None | No | Address | None | Index |
| CaptureOnly | Standard | Null | None | No | Address | None | RefID |
| CaptureOnly | Standard | Null | None | No | No | None | Index |
| CaptureOnly | Standard | Boolean | None | No | No | None | Index |
| CaptureOnly | Standard | Null | Soundex | No | No | None | Index |
| CaptureOnly | Standard | Null | None | No | No | None | RefID |
| CaptureOnly | Standard | Boolean | None | No | No | None | RefID |
| CaptureOnly | Standard | Null | Soundex | No | No | None | RefID |

The < Compare > elements that follow then validate whether the expected reference counts and structure counts (IdentityCount) are present in the output after the assertion was applied.

### 17.5.7 Inventory of Test Cases

HiPER test cases are permutations of run settings including the run configuration, the blocking in relationship to ER, the rules used, the hash functions used, the rule grouping used, the parser functions, the block indexing, and whether the indexing is based on the index key or reference ID, as shown in Table 17.1.

### 17.5.8 Versioning

Versioning is controlled at the Testing Framework level, test suite level, and test case level:

- The Testing Framework is versioned through a code repository (with bug tracker).
- The test suites are versioned based on the name of the test suite and the test suite XML script, which is versioned by date.
- The test cases are versioned by the test case script name.

### 17.5.9   Documentation

The Testing Framework has four levels of documentation:

- Unit testing (JUnit).
- Javadoc source documentation.
- XML script documentation, which includes descriptions of each test suite, test case, and associated scripts. These descriptions are used to identify the position of the individual script in the process. For example, test cases dependent on the outputs of previously run test cases have those dependencies recorded in the XML script descriptions.
- Reference guide, which describes the functionality of the XML scripts and the requirements for configuring a run of the HTF.

## 17.6   Future Work

For the system outlined in this chapter, the addition of explicit test case dependencies is an important next step. In the current Testing Framework, the test cases are executed in the order defined in the test suite configuration. This is very hard to manage if there are a large number of test cases. It could be easier if the user could designate dependencies for a test case and allow the system to determine the order of processing. In addition to the ease of configuration, it would allow the system to make determinations at run time that could reduce the overall processing time of the regression testing suite. Currently, parallel processing of test cases would require separate test suites that have been manually separated by an end user. If dependencies are specified at the test case level, the system could automatically determine which test cases can be processed in parallel.

Another function that should be given additional consideration in the future is how to generate different benchmark data sets. The current Testing Framework can only process data that the user has explicitly created and fed into the system. If the Testing Framework could generate its own data based on the features that are being tested, it could dramatically reduce the amount of time required to implement new test cases.

## 17.7   Conclusion

Despite the challenges of testing applications running on the Hadoop MapReduce platform, the creation of a Hadoop-based testing framework has proved to be well worth the effort. This is especially true during the early stages of HiPER development where code modification happened quite frequently, sometimes daily. The time saved by having a robust automated testing framework has been immense.

In addition to testing for internal development and release control, the testing framework has also proved to be useful as a software validation tool. Because of the increasing regulation and risk around sensitive data, many HiPER installations are behind client firewalls and running on different Hadoop configurations. The ability to perform a complete regression test in the client's environment not only informs the HiPER development team, it helps build client confidence in the product. In some cases, clients require the validation of software installed on their internal systems as part of their due diligence process.

# References

Benjelloun, O., Garcia-Molina, H., Menestrina, D., Su, Q., Whang, S.E., and Widom, J. (2009) Swoosh: A generic approach to entity resolution. *The VLDB Journal* 18(1), 255–276.

Chen, C., Hanna, J., Talburt, J.R., Brochhausen, M., and Hogan, W.R. (2013) A demonstration of entity identity information management applied to demographic data in a reference tracking system. *Proceedings of the International Conference on Biomedical Ontology (ICBO 2013)*, pp. 136–137.

Chen, Z., Kalashnikov, D., and Mehotra, S. (2009) Exploiting context analysis for combining multiple entity resolution systems, in *Proceedings: ACM SIGMOD '09 Conference.*

English, L. (1999) *Improving Data Warehouse and Business Information Quality: Methods for Reducing Costs and Increasing Profits.* Chichester: Wiley.

Mahata, D. and Talburt, J.R. (2014) A framework for collecting and managing entity identity information from social media. *Proceedings of the 19th MIT International Conference on Information Quality*, pp. 216–233.

Maydanchik, A. (2007) *Data Quality Assessment.* Westfield, NJ: Technics Publications.

McGilvray, D. (2008). *Executing Data Quality Projects: Ten Steps to Quality Data and Trusted Information.* Burlington, MA: Morgan Kaufmann.

Talburt, J.R. (2011) *Entity Resolution and Information Quality.* Burlington, MA: Morgan Kaufmann.

Talburt, J.R. and Zhou, Y. (2015) *Entity Information Life Cycle for Big Data.* Burlington, MA: Morgan Kaufmann.

**Part IV**

**Testing Applications**

# 18

## Testing Defense Systems

*Laura J. Freeman, Thomas Johnson, Matthew Avery, V. Bram Lillard, and Justace Clutter*

## Synopsis

Defense systems are complex and contain a variety of software and hardware components. They consistently push the limits of scientific understanding, synergizing new and unique interfaces between software and hardware. They range from the traditional military system, such as fighter aircraft, to systems less likely to be associated with defense, like business and information technology systems. The complex, multifunctional nature of defense systems, along with the wide variety of system types, demands a structured but flexible analytical process for testing systems. Additionally, rigorous testing must ensure that representative users can effectively operate the system in a variety of environments and mission scenarios. This chapter highlights the core statistical methodologies that have proven useful in testing defense systems. Case studies illustrate the value of using statistical techniques in the design of tests and analysis of the resulting data.

## 18.1    Introduction to Defense System Testing

Testing and the evaluation of test outcomes is a critical aspect of the acquisition process for all defense systems. The types of systems that undergo testing in the United States Department of Defense (DoD) are extremely diverse and include systems such as submarines, aircraft carriers, fighter aircraft, cargo aircraft, radar, radar jammers, transport trucks, armored vehicles, chemical agent detectors, payroll systems, medical information systems, and identification cards. The test and evaluation process provides valuable information throughout the acquisition process.

Testing is an essential aspect of the systems engineering process. From system design to the consumer's decisions to purchase the product, testing and the resulting analyses can provide insight and inform decisions. Conducting tests and evaluating the results are essential elements of developing quality products that deliver the performance required by the end user.

*Analytic Methods in Systems and Software Testing*, First Edition.
Edited by Ron S. Kenett, Fabrizio Ruggeri, and Frederick W. Faltin.
© 2018 John Wiley & Sons Ltd. Published 2018 by John Wiley & Sons Ltd.

A scientific approach to testing provides a structured, defensible process that ensures the right amount of testing is conducted to answer key questions. A scientifically planned test requires the input and expertise of all stakeholders including project management, engineering and scientific expertise, and statistical expertise.

### 18.1.1 Operational Testing Overview

Defense system testing is commonly divided into phases: contractor testing, developmental testing, live fire testing, and operational testing.

During contractor testing, the contractor developing the system's design tests to compare design prototypes, characterize critical component capabilities, and verify that the system can meet requirements, among dozens of other objectives.

Developmental testing evaluates system capabilities using prototypes of critical subsystems and prototypes of full systems. Recent efforts in developmental testing within the DoD have focused on characterizing system performance, assessing reliability, ensuring interoperability with other systems, and understanding cybersecurity vulnerabilities. Developmental testing can be conducted at the component or subsystem level, or by using the full system. Developmental testing can occur in the laboratory, at test ranges, or under operational conditions.

Live fire testing evaluates the vulnerability of the systems to potential threats and, in the case of weapon systems, the lethality of the system against potential targets. Because of the cost and safety limitations associated with detonating live weapons, live fire testing typically has limited sample sizes and relies heavily on design analyses, modeling and simulation, and existing data from combat.

In this chapter we focus on statistical design methods for operational testing. Operational testing is the final test event in the DoD acquisition process and acts as a form of acceptance testing by the government. It is required by law before a program can proceed to full rate production or the system can be fielded.

Operational testing is conducted with operationally representative users on production representative systems. Operators and maintainers use the system in an operationally realistic environment to conduct operational missions. The two primary areas of evaluation from an operational test are operational effectiveness and suitability. Operational effectiveness captures the overall degree of mission accomplishment or success of a unit equipped with a system. Operational suitability is the degree to which a system can be satisfactorily used in the field, with consideration given to: availability, compatibility, transportability, interoperability, reliability, maintainability, safety, human factors, wartime usage rates, manpower supportability, logistics supportability, documentation, and training requirements.

For those unfamiliar with the DoD acquisition system, the National Research Council (1998) publication *Statistics, Testing, and Defense Acquisition: New Approaches and Methodological Improvements* provides a detailed overview of the process, useful terminology, and a more detailed history. In this chapter, we focus on the unique statistical challenges of designing operational tests, many of which can be attributed to the process, but some of which are inherent to the complexity of the systems and the missions system operators must complete.

## 18.1.2 Unique Statistical Challenges of Operational Testing

The statistical design and analysis techniques used in operational tests must be diverse because DoD systems are purchased and employed to address a diverse set of missions. It would be impossible to cover all possible methods used to test complex systems in one chapter of a textbook, especially considering the plethora of test objectives, including analyses of mission capability, performance, reliability, maintainability, human factors, logistics, etc. However, there are common themes that emerge across defense testing. This chapter focuses on those common themes and their statistical implications.

### 18.1.2.1 Multidimensional Evaluation

The analysis of complex systems requires the measurement of many different types of variables. System performance is important, but how users interact with a system is also critical in determining whether units can use a system to complete operational missions. Reliability and maintainability measure the likelihood that a system will be available for use when called upon by operators. All of these variables, along with countless others, must be collected during testing. Additionally, designing a test that focuses solely on system performance might miss the larger context of the mission. Therefore, it is essential to think about the full operational context when designing tests.

### 18.1.2.2 Complex System/Complex Operational Space

Not only do test outcomes (dependent variables) span a multidimensional space, the inputs (independent variables) also cover a range of conditions. Frequently, defense systems are designed to be used in multiple missions and each of those missions might also cover a complex operating space. Being able to cover the full operating environment efficiently is a core challenge of planning a defensible operational test. As a result, this chapter focuses on the experimental design process for complex systems.

### 18.1.2.3 Testing is Expensive

The space across which we wish to understand mission capability is complex, and often timelines are short and test points are expensive. This typically limits sample sizes in an operational test to a few dozen missions. Additionally, the cost of testing in an operational environment requires that any proposed test be scrutinized to ensure that all data collected provide necessary and useful information in the most efficient manner. Gauging the right amount of testing is more than simply determining the number of test points; equally important is the placement of those points across the operational envelope. The placement of the points is the most important aspect of determining whether the testing will be adequate to support the goals of the analysis.

### 18.1.2.4 Humans are Part of the System

Systems on their own do not accomplish missions, humans equipped with systems accomplish missions. This also means that we must consider human–system interactions in evaluating system effectiveness and suitability. Therefore, tracking those human interactions through both behavior metrics and surveys is a critical component of operational testing.

### 18.1.2.5 Focus on Timeliness, Accuracy

For many systems, variables that capture timeliness and accuracy are useful for assessing whether the system is performing as desired. These variables capture whether a new system allows operators to perform a task better and/or faster than a previous system. However, these variables often have left-skewed distributions. Statistical methods for analyzing this data must account for this appropriately.

### 18.1.2.6 Focus on Prediction

Typically, it is interesting to know whether a factor considered in the test design (say conducting missions during the day versus at night) has an impact on performance, but a more meaningful analysis for decision makers and system operators informs them of how those factors translate into future performance. They need to understand how good or bad system performance might be based on the specifics of missions that they might encounter. Statistically, this means that hypothesis tests on model coefficients are not in themselves of interest. Instead, the focus is on translating model results to predictions of future mission performance with associated levels of precision.

In this chapter we summarize commonly used techniques in defense system testing and specific challenges imposed by the nature of defense system testing. Case studies illustrate the application of the methods.

### 18.1.3 The Statistical Science of Testing Complex Systems

As every system is different, the engineering and scientific expertise will vary based on the system or process being tested. However, statistics provides us with concrete methods and tools for generating the knowledge sought in testing. These statistical methods are often independent of the application, although the appropriate methods are influenced by the characteristics of the data. The same statistical techniques that are used to compare a generic drug to a brand-name counterpart can be used to compare the accuracy of a new guidance system in an air-to-ground missile to the original guidance system.

Design of experiments (DOE) is a structured and purposeful methodology for test planning. In the complex operational mission space, it is often only feasible to test a limited number of points, and DOE provides the method for selecting those points from among the many possibilities. Montgomery (2008) provides a comprehensive resource for planning experiments, though most defense system tests are more complex than the examples in Montgomery's text.

After the test is complete, there are many statistical methods to choose from for analyzing the resulting data. Empirical models produce objective conclusions based on the observed data. Parametric regression models maximize information gained from test data, while non-parametric methods provide a robust assessment of the data free from model assumptions. Bayesian methods provide avenues for integrating additional sources of information.

The analysis model should reflect the observed data and not the planning process. In designing tests we often assume a statistical model for the analysis. However, there is no limitation requiring the use of that model in the analysis. Often, qualities of the data observed (e.g., skewness, lurking variables, etc.) lead us to employ different analysis methods than originally planned; this is completely acceptable and can better inform test design on similar products and processes in the future.

## 18.2    Design of Experiments for Defense System Testing

### 18.2.1    DoD Historical Perspectives

Historically, the DoD has not always employed methods like DOE for test planning. Common test design approaches for operational testing in the past included specialized/singular combat scenarios, changing one test condition at a time, conducting case studies, and avoiding control over test conditions (termed "free play"). These approaches fail to ensure that testing is both efficient and able to characterize operational mission capabilities across the full range of operating conditions. Moreover, for complex systems, performance often depends on interactions (commonly second order) of variables, and these historical test strategies are inadequate to support estimation of interactions.

The National Research Council (1998) reviewed operational test strategies of the time and concluded that "[c]urrent practices in defense testing and evaluation do not take full advantage of the benefits available from the use of state-of-the-art statistical methodology," and that "[s]tate-of-the-art methods for experimental design should be routinely used in designing operational tests."

The broad application of DOE to operational test and evaluation is a relatively new development. In 2009, the Service Operational Test Agencies (OTAs), in collaboration with the Director, Operational Test and Evaluation (DOT&E), endorsed the use of DOE methods in DoD testing. In 2010, DOT&E outlined clear requirements for using DOE to plan operational tests. In 2012, the Deputy Assistant Secretary of Defense for Developmental Test and Evaluation (DASD DT&E) endorsed a scientific approach to testing by including DOE methods in the Scientific Test and Analysis Techniques (STAT) T&E Implementation Plan. Over the last few years, the testing community has developed best practices in applying the methods.

### 18.2.2    Overview

Experimental design is a defensible methodology for deciding what data should be collected to answer the experiment (test) objectives. Notably, the first chapter of Montgomery (2008) focuses on the process. The essential elements of an experimental design process are:

1) Identify the questions to be answered, also known as the goals or objectives of the test.
2) Identify the quantitative metrics, also known as response variables or dependent variables, that will be measured to address the key questions.
3) Identify the factors that affect the response variables. Also known as independent variables, these factors frame the broad categories of test conditions that affect the outcome of the test. Identify the levels for each factor. The levels represent various subcategories between which analysts and engineers expect test outcomes to vary significantly. When performance is expected to vary linearly, two levels are used. Identifying non-linear performance requires three or more levels.
4) Identify applicable test design techniques. Examples include factorial designs, response surface methodology, and combinatorial designs. The applicable test design method depends on the question, metrics, types of factors (numeric or categorical), and available test resources. Identify which combinations of factors

and levels will be addressed in each test period. If the test is to be a "one-shot" test with no follow-up planned, then a more robust test may be required. If testing can be sequential in nature, then smaller screening experiments aimed at determining the most important factors should precede more in-depth investigations. Determine how much testing is enough by using relevant statistical measures (e.g., power, prediction variance, correlation in the factor space, etc.).

5) Conduct testing.
6) Analyze data.
7) Draw conclusions.

Employing DOE in this holistic sense ensures efficient and adequate testing, and also aids in determining how much testing is sufficient. As a result, we can determine the optimal allocation of constrained test resources and provide an analytical trade-space for test planning.

### 18.2.3 Statistical Test Objectives

A clear test objective is the first step in planning a defensible experimental design and is essential in deciding how much testing is enough to meet the stated objectives. Table 18.1 summarizes common general classes of test goals, commonly associated test phases, and potentially useful experimental design strategies for achieving the specified goal.

Experimental design is a rich scientific methodology, containing many design types. The specific tool that is employed depends on the question to be answered (step 1). The questions and objectives can change as the system under test matures. The choice of DOE technique (step 4) should reflect the objective. Table 18.1 lists several common objectives and the corresponding designs one might select to satisfy the corresponding objective. This list is intended to show the breadth of tools that are included in DOE,

**Table 18.1** General classes of test goals.

| Test objective | Potentially useful experimental designs |
| --- | --- |
| *Characterize* performance across an operational envelope and determine whether a system meets requirements across a variety of operational conditions | Response surface designs, optimal designs, factorial designs, fractional factorial designs |
| *Compare* two or more systems across a variety of conditions | Factorial or fractional factorial designs, matched pairs optimal designs |
| *Screen* for important factors driving performance | Factorial or fractional factorial designs |
| *Test for problem* cases that degrade system performance | Combinatorial designs, orthogonal arrays, space-filling designs |
| *Optimize* system performance with respect to a set of conditions and inform system design | Response surface designs, optimal designs |
| *Predict* performance, reliability, or material properties at use conditions | Response surface designs, optimal designs, accelerated life tests |
| *Improve* system reliability or performance by determining robust system configurations | Response surface designs, Taguchi designs (robust parameter designs), orthogonal arrays |

but is far from exhaustive. The *NIST Engineering Statistics Handbook* discusses several goals of tests including prediction, characterization, and optimization (NIST, 2013).

### 18.2.3.1 Characterize

Operational testing should be adequate to *characterize* system capabilities and shortfalls across all relevant operating conditions. Such full characterization ensures that fielding decisions are made with a clear understanding of system performance, since it is not cost effective to field weapon systems that do not work or provide no clear improvement over existing systems. Full characterization also enables testers to inform the warfighters, whose lives may depend upon these systems, about what these systems can and cannot do. It is important to note that if we are able to characterize performance with sufficient precision across a variety of conditions, then we are also able to determine whether the system meets a specified requirement at a similar level of precision across those same conditions. Multiple classes of test designs may be useful when characterization is the primary test goal, including factorial and fractional-factorial designs, response surface designs, and optimal test designs. The appropriate test design will depend on the complexity of the operational envelope and expected performance variation across the operational envelope. Some conditions (levels of the factors) might be difficult to obtain, making some test designs more suitable (e.g., optimal over factorial). In most cases, covering arrays and combinatorial test designs are inappropriate for characterization because they provide low (or no) power for detecting differences in performance across the operational envelope. Power is an important measure when the test goal is characterization.

### 18.2.3.2 Compare

Direct comparison between two or more systems is a common operational test goal. A variety of test designs are useful for comparing multiple systems. The best comparisons can be made using a matched design where the systems (or processes) being compared are subjected to the same tests across all conditions. This approach controls for unwanted variability in the comparison. While it is always important to consider human factors in designing tests, it is especially important in comparison designs. When using a within-subject design that has users perform the same tasks on both systems, one must be sure to control for order effects. When using a between-subject design, where the same users do not use both systems, ensure enough testing is conducted to have high power for comparing the systems. Power for detecting performance differences among systems is particularly important for comparison tests. Low-power tests could result in an inability to draw conclusions about differences in performance between systems after the testing is completed.

### 18.2.3.3 Screen

Screening is an important test goal prior to operational testing. An important part of an integrated test process is early identification of key factors that affect system performance. Identifying these key factors and screening out unimportant factors is essential to constructing a defensible operational test. Factorial designs and fractional factorial designs are extremely useful tools in screening for important factors. When the number of factors and levels under consideration is large, optimal designs and highly fractionated factorials can also be useful.

### 18.2.3.4 Problem Cases

All operational testing seeks to identify problems in performance. However, test designs specifically geared for quickly finding problem cases are unique to situations where the outcome of the test is deterministic. These experimental designs are common for software testing and are covered extensively in Chapter 5 of this book. Phadke (1995) provides an overview of orthogonal arrays for problem identification. Combinatorial tests based on orthogonal arrays provide an efficient methodology for covering use cases such that faults caused by certain combinations of factors (two-way, three-way, etc.) can be detected quickly. Kuhn, Lei, and Kacker (2008) and Kuhn et al. (2009) provide additional detail about combinatorial testing. Because the system's performance and test outcome is not stochastic, statistical power is not meaningful. Rather, the strength of the design is defined by the number of factor combinations (two-way, three-way, etc.) covered. Since the outcome is deterministic, a fault will be caused by a specific combination of factors, so a test that includes as many combinations as possible in as few runs as possible is desired.

Testing for problems is not limited to determinist outcomes. Often, when one develops a test plan consisting of the most stressing cases, a goal of the test is to look for problems. While this testing approach may find problems, overall it is a risky design strategy because it fails to provide understanding of system performance across a range of conditions. It typically results in confounding between factors, and limited ability to determine the cause of failures.

### 18.2.3.5 Optimize

Process optimization is not a common test goal of operational testing. However, it is useful in system design and manufacturing. Additionally, at the system level it can be useful in the development of tactics, techniques, and procedures (TTPs). Myers, Montgomery, and Anderson-Cook (2009) is a comprehensive resource on response surface methodologies for process optimization.

### 18.2.3.6 Predict

Two general classes of prediction are interpolation and extrapolation. In operational testing we often wish to predict performance in areas within the design space tested. In these cases an experimental design that provides flexible modeling options and low prediction variance is preferred. Such designs include response surface designs and optimal designs. The other class of prediction is based on extrapolation. Extrapolation is inherently riskier than interpolation. For example, many weapon and sensor systems become less accurate when used against targets at long ranges than at short ranges, but the relationship is not necessarily linear. If test data is only collected using targets at short ranges, an analyst may be tempted to use the resulting model to estimate accuracy against long-range targets as well. This approach could fail if, for example, accuracy degrades at an increasing rate as targets get further and further away. When the relationship between the factor being extrapolated and the response variable being modeled is not well understood, extrapolation should be avoided. In cases where those relationships are well understood, extrapolation may provide useful information that would otherwise be inaccessible. Accelerated life tests for predicting reliability at use conditions are a type of extrapolation. In general, models used in operational testing must rigorously be validated and accredited using live test data collected across the full range of operational conditions.

### 18.2.3.7 Improve

Improve (unlike optimize) refers to tests that are specifically designed to make processes or systems robust to uncontrollable conditions. These types of experiments are used in designing systems to ensure robust performance across all operating conditions. Additionally, these designs are useful in design for reliability efforts. In these experiments the tester controls both controllable factors (that is, factors that can be controlled in the manufacturing process) and uncontrollable factors (often referred to as noise factors, e.g. humidity, operating conditions, etc.). The goal of the test is to determine the settings of the controllable factors that result in robust performance across all levels of the noise factors. This test goal is important, but typically arises during the manufacturing process, rather than in operational testing where characterizing system performance across all conditions is the priority. Taguchi designs (robust parameter designs) were originally developed to meet the *improve* test objective. Phadke (1995) provides an introduction to these design methods. In the Taguchi thinking, interactions and statistical power are not important because the goal of the test is only to find the most robust setting of the controllable factors. However, the research community has identified many improvements over traditional Taguchi designs based on orthogonal arrays. Myers, Khuri, and Vining (1992) show how the response surface methodology can be used to achieve similar objectives.

### 18.2.4 Common Design Methods

*Full factorial* designs include at least two factors and examine all possible combinations of each level. This allows the evaluator to determine the impact of each factor as well as whether one factor influences the impact of another factor on the test outcome. They are highly efficient and informative designs, though potentially prohibitively costly when many factors are involved (e.g., more than four in most operational test cases). For this reason, factorial designs are more common in developmental than operational testing scenarios, or whenever there are relatively few factors in the design.

*Fractional factorial* designs are a variation of full factorial designs that do not require all combinations of factor levels to be included in the test, and include only a fraction of the test points that would be in a full factorial. By trading off the ability to estimate high-order interaction effects, fractional factorial designs can achieve large reductions in test points over full factorials. Fractional factorial designs are a popular choice for defense system testing due to the large numbers of factors.

*Response surface designs* are a collection of designs that spread test points to collect data throughout the experimental region such that a more detailed model of the pattern of responses can be ascertained. They are used to locate where in the design space (or under what conditions) responses are optimal, and often to "map" a system's performance across a variety of conditions. This often involves the inclusion of higher-order effects that can estimate curves instead of monotonic linear effects. These designs also allow estimates of lack of fit and experimental error by adding center points, replications, and axial runs to factorial or fractional factorial base designs.

*Optimal designs* are defined in terms of some optimality criteria typically related to minimizing prediction variance or model parameter standard errors. Based on a researcher-specified model and a fixed sample size, software algorithms identify the optimal test points to satisfy the chosen criteria. Optimal designs are most useful when the number of test points is constrained to preclude a factorial design.

Myers, Montgomery, and Anderson-Cook (2009) provide an overview of common optimality criteria. Several methods exist for optimizing the test point coverage in optimal designs; these include, but are not limited to, D-, I-, and G-optimal criteria. D-optimal designs minimize the overall variance of the parameter estimates while also not letting the covariance between the parameter estimates get too large. I-optimal designs minimize the prediction variance. G-optimal designs minimize the *maximum* prediction error over the design space rather than the average prediction error. While many other design criteria have been proposed in the literature, D-, I-, and G-optimal designs are perhaps the most popular and are available options in most statistical software packages capable of generating optimal designs.

Jones and Nachtsheim (2011) developed *definitive screening designs* to estimate main effect and quadratic effects for a relatively large number of factors using a small design. These designs focus on quantitative variables and therefore have not gained much traction yet in operational testing.

*Experimental campaign sequential designs* (often simply referred to as sequential designs) stand in contrast to "one-shot" designs and employ a series or sequence of smaller tests. The goal is to learn from one test and modify subsequent tests based on this information. Results from preliminary tests may lead the evaluator to drop or add factors, modify the levels of a factor, or to the creation of more precise response variable measures.

*Point-by-point sequential design* is a form of sequential design used commonly in ballistic resistance testing. These designs sequentially (often after each data point) estimate the probability that a projectile will perforate a surface as a function of factors such as velocity and impact angle. Evaluators employing these tests commonly look for the velocity at which a projectile has a specific probability (e.g., 50% probability) of penetrating a surface. Johnson et al. (2014) provide a summary of several different sequential designs.

### 18.2.5 Statistical Model Supported (Model Resolution)

The statistical model supported by the test design is a primary consideration in determining test adequacy that is often overlooked. The statistical model is very important as it provides a snapshot of the knowledge gained about the behavior of the response across the operational envelope. The following types of statistical models are useful in thinking about test adequacy:

- First-order models allow for the estimation of main effects only (shifts in the mean for categorical factors or linear relationships for continuous factors).
- Second-order models allow for the estimation of main effects, two-way interaction effects, and quadratic terms for continuous factors.

Model complexity can extend to any order. Additionally, partial-order models are possible; one example of a reduced second-order model might contain main effects and two-way interactions, but not quadratic terms. Larger-order models result in more flexible modeling, allowing for a closer fit to the observed data. However, in operational testing, when the goal is to characterize performance, second-order models tend to be adequate for describing major changes in performance across the operational envelope due to the principle of *sparsity of effects* (Myers et al., 2009), which states that most systems are dominated by a few main effects and low-order interaction effects. If the test goal is to screen for important factors, a lower-order model may be appropriate. For prediction of a complex response surface, higher-order models may be necessary.

For two-level full and fractional factorial experiments, the order of the statistical model is often discussed in terms of its design "resolution." A design with greater resolution can accommodate higher-order model terms than a design with lower resolution. The lower the resolution, the more terms in the model are confounded with other terms, making the cause of observed performance differences amongst the different test conditions difficult to resolve. Resolution III, IV, and V designs are particularly important because they address second-order models, which are common in operational testing. Definitions of these designs are shown below:

- Resolution III designs: Main effects may be indistinguishable from some two-factor interactions.
- Resolution IV designs: All main effects can be estimated independently but some two-factor interactions may be indistinguishable from other two-factor interactions.
- Resolution V designs: All main effects and two-factor interactions can be estimated independently from each other.

### 18.2.6 Evaluation of Design Properties: Statistical Measures of Merit

A statistical measure of merit quantifies the quality of a specific aspect of a designed experiment. The cost of testing in defense applications requires that all test designs are defensible and statistical measures of merit provide the tools to quantify the quality of designs and compare between designs. No single statistical measure of merit, or group of measures, can completely characterize the quality of an experimental design. Certain qualities are difficult to quantify, such as the selection of factors or responses. In these situations, decisions are based on engineering and operator expertise. Table 18.2 summarizes many different statistical measures of merit and their utilities. The appropriateness of a given measure in assessing the quality of a design is dependent on the goal of the test and the experimental design methodology used.

#### 18.2.6.1 Confidence

In an operational test we typically have one or more research questions in mind. "Is the new system more survivable than the old system?" or "Does the reliability of the system meet the threshold?" or, most importantly, "How does system performance vary across the operational envelope?" When answering these questions with statistical rigor, the concept of confidence is crucial. Confidence is best understood in the context of a specific hypothesis test. For example, consider a new information technology system designed to process repair requests quickly. To test whether this system has adequate performance across a range of operationally relevant request sizes, we use the following hypotheses:

$H_0$: Request size does *not* significantly impact the time to process requests
$H_a$: Request size does significantly impact the time to process requests

**Table 18.2** Statistical measures of merit.

| Statistical measure of merit | Experimental design utility | Goal |
| --- | --- | --- |
| Confidence | The true negative rate (versus the corresponding risk, the false positive rate). Quantifies the likelihood in concluding that a factor has no effect on the response variable when it really has no effect. | Maximize |
| Power | The true positive rate (versus the corresponding risk, the false negative rate). Quantifies the likelihood in concluding a factor has an effect on the response variable when it really does. | Maximize |
| Correlation coefficients | The degree of linear relationship between individual factors. | Minimize correlation between factors |
| Variance inflation factor (VIF) | A one-number summary describing the degree of collinearity with other factors in the model (provides less detail then the individual correlation coefficients). | 1.0 is ideal, aim for less than 5.0 |
| Scaled prediction variance | Gives the variance (i.e., precision) of the model prediction at a specified location in the design space (operational envelope). | Balance over regions of interest |
| Fraction of design space | Summarizes the scaled prediction variance across the entire design space (operational envelope). | Keep close to constant (horizontal line) for a large fraction of the design space |

Note that if processing time varies across the operational envelope, we will *reject* this null hypothesis. While not specifically stated in the hypothesis, if we reject the null hypothesis we also will be able to estimate the effect that the request size has on processing time and compare that to requirements across the operational envelope. The confidence level of our test dictates how much we believe our result in the case we reject. For example, if it turns out that the request size does not impact the processing time for operationally representative file sizes, but we reject the null hypothesis, the result of the test is a false positive or Type I error. A desired Type I error rate of $\alpha$ can be achieved by using a test with $100(1 - \alpha)\%$ confidence.

However, this doesn't necessarily mean we should always choose a high level of confidence for our test. The trade-off for a high level of confidence is that we will fail to reject the null hypothesis more often, even when we should reject. Suppose now that the request size does significantly impact processing, but only by a small percentage over the range of operationally relevant file sizes. Without conducting a large test, it is likely that we will fail to reject the null hypothesis, resulting in a false negative or Type II error.

### 18.2.6.2 Power

One of the most direct ways to assess a test plan is answering the question, "Will this test plan tell me what I want to know?" In statistical analysis, a standard way to answer that question is by finding the power of the test plan. Power is the probability of finding a statistically significant relationship between the response variable and a particular factor.

Recall that "statistically significant" depends on the choice of confidence level for your test; a smaller $\alpha$ makes it less likely that you will reject your null hypothesis and therefore decreases your power. The opposite is true for relatively larger values of $\alpha$. Power in the above example reflects the probability of finding a statistically significant difference between the processing times due to the request size when a difference truly exists. The difference in time to process between a large and small request is the effect size. Larger effect sizes are easier to detect, so all else being equal, your power will be higher if you have a larger effect size.

At the same time, if there is a lot of noise in the data, it becomes harder to identify these effects. When we execute ten requests, we will get ten different values for processing time, even if we have the same request size. This run-to-run variability (independent of our experimental factors) is called random error or noise. The more "noisy" the data, the less power we will have. We use the signal-to-noise ratio (SNR) to summarize these competing elements of power; it is calculated by taking the effect size and dividing by the standard deviation of the random error in the data. Larger SNRs result in higher power.

Another important consideration in calculating power is the sample size. In general, larger tests result in higher power for detecting significant relationships between experimental factors and the response variable. However, not all data points are equal. Careful consideration must be placed on selecting the points across the design region for maximizing power for the most important factors. The point placement across the operational envelope, the order in which runs were conducted, and the factor combinations selected all have a substantial impact on power. A general rule for optimizing power through experimental design is to balance the test across factor combinations. If we are trying to determine if there is a statistically significant difference in time to process between large and small requests, and we will have 20 runs, it is generally best to make sure that ten of those runs are small requests and the other ten are large requests. Rarely is a test design this simple (only a single, two-level factor), and as designs get more complex (additional factors, factors with more than two levels, restrictions such as blocking) it becomes more difficult to find an optimal experimental design. Software packages can be used to generate efficient designs and give you power estimates.

In summary, power is a function of the statistical confidence level, the effect size of interest, the variability in the outcomes, and the number and placement of test points. It not only describes the risk in concluding that a factor does not have an effect on the response variable when it really does (Type II error), but also is directly related to the precision we will have in reporting results. This precision, which is related to the effect size, is key in the determination of test adequacy; without a measure of the precision we expect to obtain in the data analysis, we have no way of determining if the test will accurately characterize system performance across the operational envelope.

### 18.2.6.3 Collinearity

When designing an experiment, collinearity describes the degree of linear relationship between two or more factors. A well-designed experiment minimizes the amount of collinearity between factors. Two or more factors are considered collinear if they move together linearly (as one increases, so does the other). Analysis of data containing highly collinear factors can be misleading, confusing, and imprecise. Variances of coefficient estimates become greatly inflated (making the precision of the test worse) when factors

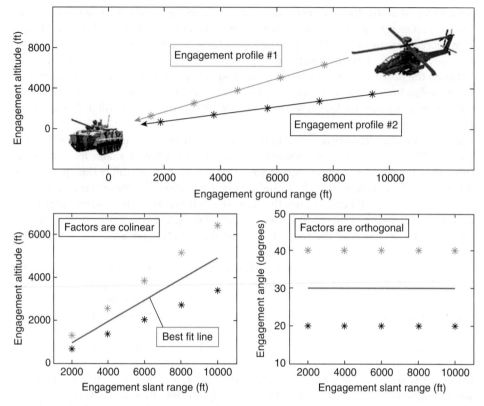

**Figure 18.1** Engagement profiles and factor plots.

are highly collinear, leading to inflated non-significant $p$-values (Type II errors). Additionally, collinearity can lead to false positives. When a response is regressed on two highly collinear factors, an analysis of variance might report that both factors are significant. Yet, if only one factor is included in the model, the analysis of variance may indicate that the factor is not significant. Finally, using a model containing highly collinear factors to extrapolate or interpolate between design points can yield estimates with large uncertainty.

Figure 18.1 shows an example from operational testing in the use of multiple factors to describe the geometric location of an aircraft. An Apache helicopter engages an enemy tank at ten locations along two different profiles. The response variable is weapon accuracy, while the factors are slant range and altitude. The slant range is said to be collinear with altitude because there is a near-linear relationship between the two. The fit line, shown in the bottom left of Figure 18.1, shows a positive linear relationship between the two factors as demonstrated by its positive slope; hence, altitude and slant range are, to some degree, collinear.

While selecting a different flight profile could mitigate the collinearity between slant range and altitude, the factors are mathematically related. Therefore, we cannot completely eliminate the collinearity by adjusting the flight profile. What we can do to break the collinearity is replace altitude with engagement angle, as shown in the bottom right of Figure 18.1. Using the same ten points as before, the bottom right of Figure 18.1 shows that we can create a similar experiment where the factors are not collinear. Notice that the best-fit line is horizontal, indicating that there is no linear relationship between factors; that is, the factors are *orthogonal*.

Correlation coefficients and VIFs are useful for assessing the collinearity of a design. Both of these measures of merit are calculated and monitored prior to executing an experiment. They are functions of the number of runs, the factors and levels in an experiment, and how those factors vary from run to run. They are not a function of the data collected from the test. They serve as a tool for establishing the merit of an experiment and can be used to compare DOEs.

The correlation coefficient is bounded between negative one and one, and summarizes the linear dependence of two independent variables. This can be a confusing concept to those used to thinking of correlation as a summary of a relationship between an independent and a dependent variable. However, as shown above, design choices often result in correlations between factors. Good designs minimize this correlation between independent variables.

Variance inflation factors provide a one-number summary description of collinearity for each model term. For an experiment with multiple factors, the VIF associated with the $i$th factor reflects the increase in the variance of the estimated coefficient for that factor compared to if the factors were orthogonal, and is defined as $VIF_i = \frac{1}{1-R_i^2}$, where $R_i^2$ is the coefficient of determination of a regression model where the $i$th factor is treated as a response variable in the model with all of the other factors. $VIF_i$ can range from one to infinity. Values equal to one imply orthogonality, while values greater than one indicate a degree of collinearity between factors. The square root of the variance inflation factor indicates how much larger the standard error is (and therefore, how much larger the confidence intervals will be), compared to a factor that is uncorrelated with the other factors. As a rule of thumb, values greater than five suggest that collinearity may be unduly influencing coefficient estimates.

### 18.2.6.4 Scaled Prediction Variance

In addition to coefficient estimation, one key reason we conduct experiments is to predict the future performance of a system. *Prediction variance* describes the error involved with making a prediction using a regression model. Consider an operational test that consists of $N$ runs and $k$ factors. The corresponding first-order regression model is $y = X\beta + \varepsilon$. The response variable $y$ is of size $[N \times 1]$, $X$ is the $[N \times k]$ design matrix, $\beta$ is a $[k \times 1]$ vector of coefficients, and $\varepsilon$ is an $[N \times 1]$ vector of random errors that has $E(\varepsilon) = 0$ and $Var(\varepsilon) = \sigma^2 I_N$. The predicted value at any point in the design space is $\hat{y} = x_0'\hat{\beta}$, where $\hat{\beta} = (X'X)^{-1}X'y$ is the maximum likelihood estimator of $\beta$ and $x_0$ is the vector corresponding to the prediction point in the design space, such as $x_0 = \begin{bmatrix} 1 & x_1 & x_2 & \dots & x_k \end{bmatrix}$. The prediction variance at any point in the operational envelope, $x_0$, is defined as $V = \sigma^2 x_0'(X'X)^{-1}x_0$. Thus, prediction variance is a function of the designed experiment $(X)$, the location in the design space where the prediction is made $(x_0)$, and the overall variance is the response $(\sigma^2)$. Since $\sigma^2$ is unknown, prediction variance can be difficult to use for evaluating the merit of an experimental design. Scaled prediction variance (SPV), on the other hand, normalizes the prediction variance by $\sigma^2$ so that SPV is a function of $N$, $X$, and $x_0$, that is

$$\text{SPV} = \frac{NV}{\sigma^2} = N x_0'(X'X)^{-1}x_0.$$

**Table 18.3** Two candidate experiments.

| | Design A | | Design B | |
|---|---|---|---|---|
| Run number | Muzzle velocity | Range to target | Muzzle velocity | Range to target |
| 1 | −1 | −1 | −1 | −1 |
| 2 | 1 | −1 | 1 | −1 |
| 3 | −1 | 1 | −1 | 1 |
| 4 | 1 | 1 | 1 | 1 |
| 5 | −1 | 0 | 0 | 1 |
| 6 | 1 | 0 | 1 | 0 |
| 7 | 0 | −1 | 0 | 0 |
| 8 | 0 | 1 | 0 | 0 |
| 9 | 0 | 0 | 0 | 0 |
| 10 | 0 | 0 | 0 | 0 |

The benefit of SPV is that it can be used to evaluate a designed experiment prior to running the test and collecting data. Multiple designed experiments can be postulated for a single test event and compared using SPV, allowing the best design to be selected. When assessing a design in this way, it is important to consider the full range of values each factor can take. For categorical factors, this is just a matter of considering prediction at each level of the relevant factors. For continuous variables, graphical methods such as contour plots are available.

To get a better understanding of SPV, consider a notional penetration test example. In this example, only ten shots are available to characterize the penetration depth of a small arms munition impacting an armor plate on a light combat vehicle as a function of muzzle velocity and range to target. Additionally, the test team plans to fit a second-order regression model and has two candidate experimental designs, as shown in Table 18.3. The levels of muzzle velocity and range are expressed in normalized units between minus one and one. The SPV contour plots for each experiment are shown in Figure 18.2.

Careful inspection of Figure 18.2 shows that while Design B has a larger region with the minimal SPV (less than 3.5), a greater portion of the design space for Design A has an SPV less than 4.0. Meanwhile, the SPV near the extremes is much greater for Design B. Based on these observations, Design A is preferred for the limited ten shots.

In this example, it is reasonable to plot SPV because it is a two-dimensional problem. In cases where there are more than two factors, detailed plots are not straight forward. *Fraction of design space* (FDS) plots show the cumulative distribution of the SPV across the operational envelope (design space). An FDS shows the proportion of the design space with SPV less than or equal to a given value. For the previous example, Figure 18.3 shows the FDS for Designs A and B. This chart shows that nearly 80 percent of the Design A space has an SPV below 4.0, while roughly 55 percent of the Design B region has an SPV below 4.0. From this chart it is clear that the Army should choose Design A.

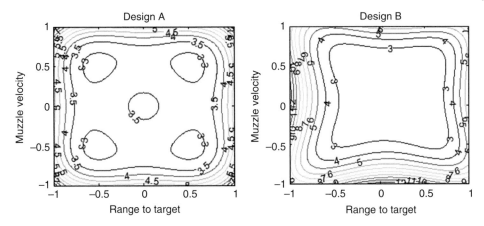

**Figure 18.2** Scaled prediction variance for two candidate experimental designs.

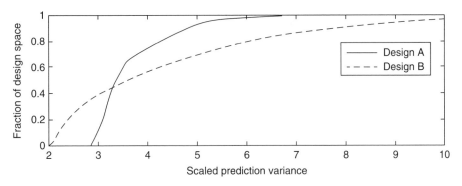

**Figure 18.3** FDS graph for candidate experimental designs.

### 18.2.7 Best Practices for Using DOE in Operational Testing

As DOE has become standard practice for DoD operational tests, several best practices have been identified. First and foremost, it is essential to apply experimental design concepts early in the test planning process. Programs successful in using experimental design establish test planning working groups that include all stakeholders, including the program manager, the requirements representative, developmental and operational testers, and subject matter experts in experimental design and analysis. All of these stakeholders are necessary to identify the key elements of test planning, including goals, response variables, factors, and analysis requirements. Ideally, any test strategy should be iterative in nature, accumulating evidence of system performance before and during operational testing.

Several specific analytical best practices have proven useful in ensuring efficient and effective testing while reducing the test resources required. These best practices include the following:

- Where possible, use continuous metrics as the primary measures of system performance as opposed to pass/fail probability-based metrics. Using continuous metrics has been shown to reduce test resource requirements by at least 30 to 50 percent for the same level of information. Cohen (1983) and Hamada (2002) provide examples and discussion on the advantages of continuous measurements in estimating conformance probabilities. However, many DoD requirements are specified in terms of probabilities. Even in these cases, testers should seek to recast requirements as continuous metrics.

- Similarly, use continuous factors when possible to cover the operational envelope. Identifying these continuous factors, or casting operational conditions in a continuous manner, enables the use of response surface design techniques specifically available for continuous factors. Using these techniques will also afford test efficiencies and provide more information-rich test results.
- Use sequential experimentation approaches to reduce required test resources in each test phase, while developing a comprehensive view of system performance.
- When employing DOE for test planning, focus on the factor-by-factor power calculations, rather than a single "roll-up" power estimate. Historically, test planners have used one-sample hypothesis calculations to directly evaluate a single requirement, even in cases where multiple conditions were evaluated in testing. This approach is inadequate for ensuring performance is well characterized in a test.
- Test goals should not be limited to verifying a single narrowly defined requirement in a static condition. Rather, testing should aim to characterize performance of the unit when it is equipped with the system across all feasible and operationally realistic conditions.
- Include all relevant factors (cast as continuous where possible) in the test design. By selecting relevant test factors and forcing purposeful control of those factors we can ensure that the operational test covers conditions the system will encounter once fielded. Leveraging developmental test data is essential for narrowing the list of relevant factors and mitigating the risk of excluding important factors. Omitting known important factors from the test design results in holes in our knowledge of system performance. When resources are highly constrained we should leverage advanced design techniques coupled with developmental testing to ensure we can incorporate as many factors as possible in the test design.

### 18.2.8 Survey Design and Human Factors

A key aspect of operational testing is observing the quality of human–system interactions and their impact on mission accomplishment. Operators are a critical component of military systems. Hardware and software alone cannot accomplish missions. Systems that are too complex for operators to use compromise mission success by inducing system failures. Problems that arise because of poor interface design force the Services to invest in lengthy and expensive training programs to mitigate problems. It is critical to evaluate the usability of military systems as well as the workload, fatigue, and frustration that operators experience while employing the system. Surveys are often the only means to evaluate these issues, and proper scientific survey design must be done to ensure that the data collected to evaluate the quality of human–system interactions are valid and reliable.

Surveys capitalize on the thoughts and experience of the system operators to derive essential information for system evaluation. However, their use in operational testing has not always reflected the best practices of the human factors community. The resulting data have had limited utility in evaluations. Data from well-written surveys are useful for (a) diagnosing why certain performance goals were not met (e.g., training, system design), and (b) empirically measuring human–system integration (HSI) components such as workload and usability. Workload and usability ratings can also form the basis of a robust comparison between new and legacy systems. In nearly every case, data from well-written and well-administered surveys aid the evaluator in assessing effectiveness and suitability.

One of the most common survey mistakes is the inclusion of questions that ask whether the user thought the system's performance was effective, accurate, timely, or precise enough to complete the mission. Accurate measurement of performance, effectiveness, and situation awareness requires knowledge of ground truth for the test, which operators and maintainers typically do not have. Surveys are measures of thoughts that are highly affected by context and are therefore relative, whereas requirements and performance are absolute, and are better measured by the tester.

The following are some of the best practices, highlighted by the survey community, that we should consider when writing and administering surveys:

- Neutrality in the questions: The goal of the survey is to obtain the respondent's thoughts. Phrasing questions in a manner that leads a respondent towards the tester's opinions will reduce the likelihood that the respondent provides unbiased answers.
- Knowledge liability: Do not ask questions the respondent cannot answer (e.g., did the system provide accurate tracking information?).
- User friendly: Reduce the respondent's effort by making questions brief and clear. Also, make sure that the order of the questions is logical to the respondent.
- Singularity: Address only one topic in a question.
- Minimal length: The perceived length of a survey and the actual time it takes to complete it affects data accuracy. Ask the minimum number of questions needed for the goal of the test.
- Confidentiality: When respondents believe that their data will be kept confidential, they are more likely to provide their true thoughts. Names and other personally identifiable information should be kept separate from the actual survey.

## 18.3 Statistical Analyses for Complex Systems

### 18.3.1 Overview

Rigorous statistical analysis methodologies are crucial in testing complex military systems. A sound experimental design is less beneficial if we fail to employ the appropriate corresponding analysis techniques. In the past, analysis has been driven primarily by assessing summary requirements:

- The system should detect 90% of targets at 50 miles.
- The minimum detection range should be at least 50 miles.
- The minimum detection range for moving targets at night should be at least 40 miles.

Because these requirements do not include the operational conditions or provide an overly narrow specification of a limited set of conditions they were historically interpreted as on average across the operational space (first two examples) or drove testing in a very specific set of conditions (the third example), while excluding much of the relevant operational environment.

A better approach is to use experimental design and regression techniques that characterize performance across the full operational space. Approaches like regression and linear models allow us to identify conditions that have an impact on system performance and determine the sets of conditions under which the system's requirements are met. This is achieved without requiring excessive replication under any one condition. These approaches maximize the information gained from each data point, resulting in efficient tests and defensible analyses.

## 18.3.2 Linear Models

Linear models provide an analysis framework for both continuous and categorical independent variables. The general form of the linear model is:

$$y = X\beta + \epsilon,$$

where $\epsilon \sim N(0, \sigma^2 I)$, $X$ is the model matrix of size $n \times p$, $n$ is the number of runs, $p$ is the number of model parameters, $y$ is the $n \times 1$ response variable of interest, and $\beta$ is the vector of model parameters of size $p \times 1$. The model is termed linear because it is linear in the model parameters ($\beta$) and not necessarily in the variables. Therefore, the $X$ matrix can contain quadratic terms or higher-order polynomials.

Estimation of the linear model is done using maximum likelihood estimation, resulting in:

$$\hat{\beta} = (X' \cdot X)^{-1} \cdot X' \cdot y,$$

where $\hat{\beta}$ are the estimated model coefficients. Many textbooks have discussed the general linear model, so we will focus only on aspects for defense system testing here. Searle and Gruber (2016) provide a comprehensive overview of linear models. In this section we highlight some aspects of modeling that are frequently used in defense analyses.

### 18.3.2.1 Dummy Variables/Variable Coding

One technique used frequently in defense analyses is dummy variable coding. When testing defense systems the variables are often categorical in nature. Examples include:

- Time of day categorized into day or night. This factor is often included in operational test designs to ensure that systems can be operated during both the day and night without any degradation in performance. In some systems, (e.g., infrared sensors) it might be worthwhile to capture illumination level continuously, but for many systems we are more worried about human factors considerations and day/night is adequate to capture any variability in outcomes.
- Vehicle types. Often when testing tracking systems we want to ensure they can track different types of vehicles successfully (e.g., tanks, trucks, boats). While it might be possible to capture the elements that differ between these vehicle types continuously,

it may require more variables than types of vehicles, resulting in the categorical variable being more efficient. Additionally, these continuous variables may not result in orthogonal independent variables, introducing correlation in the design space.

- Presence/absence variables for defensive techniques, such as countermeasures and jamming.
- Operating mode. Often operators have the ability to select different operating modes that tune system settings. For example, a tracking system might have operating modes optimized for detection in a littoral environment, versus mountains, versus desert. The operator can only select one of the three modes, but we want a test design that covers all three.

Dummy variables allow for any categorical factor with $k$ levels to be recoded as a $k-1$ indicator variables and used in regression analysis. For example, for a vehicle type variable we can define

$$x_1 = \begin{cases} 1 \text{ if wheeled,} \\ 0 \text{ if tracked,} \end{cases}$$

where $x_1$ represents the second column in $X$, and the first column is a vector of ones for the intercept of the model. If there were actually three vehicle types of interest (tracked, wheeled, boat) then we could make two dummy variables to include them in the regression:

$$x_1 = \begin{cases} 1 \text{ if wheeled,} \\ 0 \text{ if tracked,} \\ 0 \quad \text{ if boat,} \end{cases}$$

$$x_2 = \begin{cases} 0 \text{ if wheeled,} \\ 1 \text{ if tracked,} \\ 0 \quad \text{ if boat.} \end{cases}$$

If both $x_1$ and $x_2$ equal zero for a given run, then a boat was the vehicle used in that run. Using dummy variable coding we can capture both continuous and categorical factors. However, it is important to make sure the coefficients are properly interpreted. Additionally, one would not include interaction variables between levels of dummy variables. Focusing on model predictions instead of coefficient interpretations often helps in this regard.

### 18.3.2.2 Prediction

Another aspect of analysis that deserves additional discussion is the importance of prediction using linear models. In academic texts, much of the interpretation of linear models focuses on the interpretation and testing of model coefficients. In a defense context, decision makers and operators are often most concerned with a characterization of how good/bad performance might be across a range of operational conditions and how that performance compares to threshold requirements. Figure 18.4 shows an example of a system characterization that focuses on the time to find an initial target using a regression analysis. In this example, the Apache helicopter was tasked to find targets

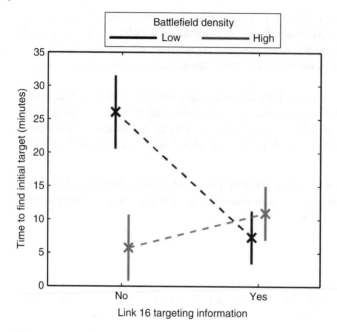

**Figure 18.4** Prediction plot for time to find initial target (minutes).

in both low- and high-density combat scenarios. One goal of the test was to determine if adding Link 16 (a military tactical data network) to the helicopter improved performance. Additionally, the test sought to characterize search capability across a variety of operating conditions. The factors considered in the test design were:

- time of day (day/night)
- battlefield density (low/high)
- Link 16 data (yes/no).

The regression analysis revealed that both battlefield density and Link 16 information as well as their interaction resulted in different timelines to find initial targets. Figure 18.4 shows the predicted time to find, which could easily be compared to a requirement by decision makers.

We emphasize the importance of prediction here, because while straightforward for a linear model, correctly estimating confidence intervals can be challenging for more complex models. For a linear model the predicted values are given by:

$$\hat{y} = X\hat{\beta},$$

where $\hat{\beta}$ are the estimated model coefficients and $\hat{y}$ are the predicted expected values; $X$ is now the prediction matrix, containing the values of the predictor variables where predictions are required. Confidence intervals can be constructed for any realization of the input variables, $x_0$, for $\hat{y}$ using:

$$\hat{y}(x_0) \pm t_{\alpha/2, N-p} \widehat{SE}(\hat{y}(x_0)),$$

Where $\widehat{SE}(\hat{y}(x_0))$ is the estimated standard error of the prediction at $x_0$ and equal to:

$$\widehat{SE}(\hat{y}(x_0)) = \sqrt{\frac{y'y - b'X'y}{(N-p)}(x_0' \cdot (X' \cdot X)^{-1} \cdot x_0)}.$$

The confidence interval on the expected predicted value at $x_0$ is notably different from a prediction interval which provides a confidence interval on a single future observation $(y_0)$. The prediction interval and corresponding standard error are given by:

$$\hat{y}_0(x_0) \pm t_{\frac{\alpha}{2}, N-p} \widehat{SE}(\hat{y}_0(x_0)),$$

$$\widehat{SE}(\hat{y}_0(x_0)) = \sqrt{\frac{y'y - b'X'y}{(N-p)}(1 + x_0' \cdot (X' \cdot X)^{-1} \cdot x_0)}.$$

### 18.3.3 Lognormal Transformation

One of the primary limiting assumptions for linear models in defense testing is that the response variable is normally distributed. An effective defense system typically improves the accuracy and/or timeliness of processes or missions. Therefore, variables that focus on time and distance typically reflect well the outcome of operational missions. Examples of variables that might follow this pattern include times to repair, detection times, detection ranges, miss distances, and target location errors. While these variables are continuous, they are inherently right skewed because of the lower bound at zero. Arguably, in the Apache analysis a lognormal distribution would have better represented the data.

The lognormal distribution often provides a quick and easy solution for these cases due to its close relationship with the normal distribution. If the random variable $T$ is lognormally distributed, then the random variable $Y = \log(T)$ is normally distributed. Figure 18.5 illustrates this concept visually. In Figure 18.5, the notional mission completion times, $t_i$, on the left follow a lognormal distribution with parameters $\mu = 2$ and $\sigma = 0.5$, applying a log transformation to the data on the right, $\log(t_i)$, which follow a normal distribution with the parameters $\mu = 2$ and $\sigma = 0.5$.

Because the transformed data follow a normal distribution we know that the expected value of the transformed data confidence intervals can be constructed

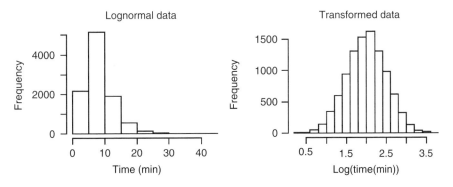

**Figure 18.5** Illustration of lognormal transformation on data.

using the *t*-distribution. However, no one is interested in the expected value of the mission completion times in log-scale time. In order to compare the expected value to the requirement we must transform back to the original time units. The mean and confidence intervals in the original data units, however, cannot be constructed via a simple exponentiation of the mean and confidence intervals of the transformed data (i.e. the mean of the lognormal data is not found by $\exp(\mu = 2)$). This common mistake has been made countless times in the analysis of skewed data. The correct expression for the mean in the original data units is estimated by:

$$\exp\left(\mu + \frac{\sigma^2}{2}\right)$$

and a confidence interval for this quantity can be constructed using the multivariate delta method (i.e., propagation of error) to estimate the variance (see the next section for a derivation of this variance). Alternatively, if the median is of interest it can be calculated using $\exp(\mu)$.

### 18.3.3.1 Multivariate Delta Method

To illustrate the multivariate delta method for a simple lognormal regression, let $T$ be the vector of mission completion times, and $y = \log(T)$. A simple linear regression model using the previous linear model notation is:

$$X = \begin{bmatrix} \bar{1} & x \end{bmatrix}, \quad \beta = \begin{bmatrix} \beta_0 \\ \beta_1 \end{bmatrix},$$

where $\bar{1}$ is a vector of ones, used to estimate the intercept, $x$ is a vector of a single independent variable, $\beta_0$ is the intercept, and $\beta_1$ is the slope of the regression equation relating $x$ to log mission completion time.

We can use maximum likelihood estimation to calculate the estimates of $\hat{\beta}_0$, $\hat{\beta}_1$, and $\sigma^2$ in a linear model framework. Additionally, we will need the variance–covariance matrix for the parameter estimates from the linear model. To construct confidence intervals around the expected mean mission completion time at a given value of the independent variable, $x_0$, let $X_0 = \begin{bmatrix} 1 & x_0 \end{bmatrix}$. The value we are seeking is then:

$$E(T_0) = \exp\left(X_0^T \beta + \sigma^2/2\right) = \exp\left(\beta_0 + \beta_1 x_0 + \sigma^2/2\right).$$

It is not straightforward to calculate a confidence interval on this quantity. The multivariate delta method provides one solution that results in a closed form expression for the confidence interval on the mean time at $x_0$.

The multivariate delta method states that if one is interested in a function of maximum likelihood estimates $g(\theta)$ where $\theta$ is a vector of coefficients estimated using maximum likelihood estimation resulting in $\hat{\theta}$, then the function $g(\hat{\theta})$ is approximately normally distributed with mean $g(\theta)$ and variance–covariance matrix:

$$\Sigma_{\hat{g}} = \left[\frac{\partial g(\theta)}{\partial \theta}\right]^T \Sigma_{\hat{\theta}} \left[\frac{\partial g(\theta)}{\partial \theta}\right],$$

where $\Sigma_{\hat{\theta}}$ is the variance–covariance matrix for the original coefficients.

For our particular function of interest, $\exp\left(\beta_0 + \beta_1 r_0 + \sigma^2/2\right)$, we can calculate the vector of partial derivatives to be:

$$
\left[\frac{\partial \mathbf{g}(\theta)}{\partial \theta}\right] = \begin{bmatrix} \frac{\partial T_0}{\partial \beta_0} \\ \frac{\partial T_0}{\partial \beta_1} \\ \frac{\partial T_0}{\partial \sigma} \end{bmatrix} = \begin{bmatrix} \exp\left(\beta_0 + \beta_1 x_0 + \sigma^2/2\right) \\ x_0 \exp\left(\beta_0 + \beta_1 x_0 + \sigma^2/2\right) \\ \sigma \exp\left(\beta_0 + \beta_1 x_0 + \sigma^2/2\right) \end{bmatrix}.
$$

Now all that remains is to do the matrix multiplication and to produce an estimate of the variance of expected miss distance. Notice that the dimension of the above vector is $3 \times 1$ for this example and the variance–covariance matrix is a $3 \times 3$ matrix, resulting in one value of variance for the miss distance. Once the matrix math is complete, confidence intervals can be constructed using Wald's method assuming a normal distribution:

$$
(1 - \alpha)\% \text{ CI}: \exp\left(\beta_0 + \beta_1 r_0 + \sigma^2/2\right) \pm z_{1-\alpha/2}\Sigma_{\hat{g}},
$$

where $z_{1-\alpha/2}$ is the critical value of the normal distribution.

Notice that any function could be used here. Often for lognormal data people like to provide percentile estimates (e.g., the 50th percentile, or median, is often of interest). Confidence intervals for that expression can be calculated in a similar fashion. These confidence intervals tend to result in wider confidence intervals than the actual coverage would dictate. Bootstrapping methods can provide better coverage probabilities. Efron and Tibshirani (1994) provide a comprehensive overview of bootstrapping methods.

### 18.3.4 Discussion of Complex Analysis Methods

The combination of linear models with dummy variables and log transformations can be used to solve many operational test analysis problems. They are easy to implement and provide straightforward approaches to prediction. Because they are parametric in nature, they are useful for the smaller data sets that often result from operational testing. However, they can lack the overall model complexity to represent the data closely.

One common response variable used in operational testing that immediately violates the assumptions of a linear model is binary (pass/fail) data. These measures are desirable because they are easy to measure. Additionally, for some systems, there may not be easily captured continuous variables, or they may not directly translate into mission outcomes. For example, when testing a torpedo it might seem logical to capture miss distance as a continuous measure of torpedo effectiveness. However, torpedoes have point detonation fuses, meaning they must directly impact their target to have any effectiveness; a narrow miss is equivalent operationally to a wide miss, making miss distance an inadequate response variable to capture torpedo effectiveness.

Generalized linear models (GLM) provide an analysis solution for these more complex modeling needs. Developed by Nelder and Wedderburn (1972), generalized linear models generalize the classical linear model by breaking the model into three parts – the random component, the systematic components, and the link function between the two. The random component allows for response variables that follow any distribution in the exponential family. The systematic component is the linear predictor that consists of a linear function of the independent variables. The link function relates the mean of the distribution (random component) to the linear predictor. Logistic regression is one common type of generalized linear model.

Mixed models are another complex methodology that may prove very useful to operational testing but have not been widely implemented. In operational testing, observations occur within the context of missions and the missions often include clusters of observations. Mixed models contain both fixed and random effects. The random effect (unlike the random component of a GLM) is useful in accounting for mission variability and the fact that the missions executed are a sample of all possible missions. A *mixed* model containing *fixed* or *population-averaged* effects can be used to address both systematic variation across the missions due to the factors selected and *random* or *subject-specific* effects which address within-mission variation. Mixed models for continuous normal outcomes were first presented by Laird and Ware (1982), and these models appear extensively in the literature. In non-normal data situations, these mixed models are commonly referred to as *generalized linear mixed models*, and Myers et al. (2010) discuss applications of these models to engineering and industrial data.

## 18.4  Case Studies of Design and Analysis in Operational Testing

The following case studies provide representative illustrations of the designs and statistical techniques used in operational testing. The first example focuses on an experimental design for a new air-to-ground weapon and highlights the challenge of covering a complex operational space. The second example shows the full process, from planning to analysis, for a sonar software upgrade. The final example shows the evaluation of a counter-fire radar, but omits the design discussion.

It is worth noting that all of these designs are "one-shot" experiments as opposed to sequential experiments. It has long been a goal of DoD testing to conduct integrated testing, where data from earlier test phases could be used to either augment or inform later testing, but that is often unachievable due to competing test objectives and limited resources. Dickinson et al. (2015) show the value of considering multiple phases of testing in the reliability analysis of the Stryker family of vehicles.

### 18.4.1  New Air-to-Ground Weapon Design Case Study

A new air-to-ground weapon ready for operational testing has three different methods of finding and targeting ground enemies. It is designed to launch from several different types of aircraft and to find both fixed and moving targets. There are three targeting methods: coordinate attack, laser attack, and new attack mode. Coordinate attack uses a fixed set of GPS coordinates to guide the weapon to the target. Laser attack uses laser designation from either the aircraft or a ground-based source to find the target. The new attack uses both millimeter wave and infrared targeting methods. Of the three targeting technologies, coordinate attack and laser attack are well understood and have been implemented on many legacy systems. The new attack methodology, however, is new technology for this weapon.

The goal of the operational test is to characterize the weapon's ability to find, fix, target, track, and engage a variety of operationally realistic targets. The test team identified several response variables of interest in assessing the system, including track accuracy, target location error, and miss distance of the weapon upon firing. One aspect of developing operational test programs is that the designs more often than not do not reflect a single response variable; instead, many response variables are typically necessary to fully characterize the system. Even in a system like an air-to-ground missile which has a fairly specific mission, several variables are necessary to fully understand the operational effectiveness of the system.

The design strategy breaks the test into three designs, one for each attack mode that focuses solely on the factors that are expected to affect weapon performance for that attack mode. The three designs are then stacked together and matched with variables from an overarching design. The overarching design ensures coverage across a range of launch conditions. While these overarching factors are not expected to impact performance, they all represent operationally realistic conditions that operators might encounter. The blocking methodology used in the overarching design ensures that attack mode's designs are not confounded with the factors in the overarching design.

In all of the designs the primary response variable is miss distance. Miss distance provides continuous information about end-to-end weapon engagement capability, but it may be measured differently for each of the attack modes. For example, miss distance for coordinate attack is expected to be the difference between the hit location and the GPS coordinates provided to the weapon; in the new attack mode, miss distance may be the difference between the target centroid and the hit location.

### 18.4.1.1 Overarching Design

Table 18.4 summarizes the factors and levels in the overarching design strategy. The first three factors (airspeed, altitude, range) represent the acceptable launch region for the weapon. Time of day was included to ensure that test points would be collected both during the day and at night. This is an example of a factor that is actually not expected to impact the test outcome, but was included as a precaution. Previous system tests have shown that sometimes lighting considerations can lead to potentially catastrophic events due to human factors considerations, for example displays are not clearly visible during the day (due to glare).

The five overarching factors in Table 18.4 were used to generate a 32-run full factorial, which was then partially replicated based on the required sizes for the test design for each of the attack modes for a total of 50 runs.

Figure 18.6 provides the power analysis for the overarching design for identifying the difference between these factors that span all three attack modes. The power calculations

**Table 18.4** Overarching open air test – test design summary.

| Factor | Levels |
| --- | --- |
| Release range | Far, Near |
| Release altitude | High, Low |
| Release airspeed | Fast, Slow |
| In-flight target update (IFTU) rate | Fast, Slow |
| Time of day (TOD) | Day, Night |

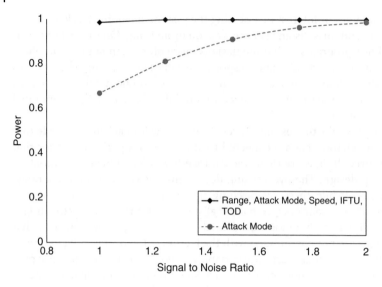

**Figure 18.6** Overarching design power at 80% confidence level.

**Table 18.5** GPS coordinate attack – test design summary.

| Factor | Levels |
| --- | --- |
| GPS | Degraded, Full |
| Impact angle | High, Low |

are provided for generic signal to noise ratios (SNRs). The SNR reflects the difference we wish to detect (signal) due to a change in the factor relative to the unexplained variability (noise). Generally, a value of 2 is considered a large effect size, while an SNR of 1 is a relatively small effect size. It should be noted that these power calculations assume independence from all of the factors nested in the attack mode designs, which may not be reasonable. Therefore, the power calculations in Figure 18.6 do not provide a strong understanding of test adequacy; rather, they provide an indication of the coverage of the space. It is notable that if differences exist in performance between attack modes, the overarching design provides high power for detecting those differences, shown by the attack model curve in Figure 18.6.

### 18.4.1.2 GPS Coordinate Attack

Coordinate attack is a well-understood legacy capability. Additionally, as many historical tests have shown, very few factors impact the accuracy of a coordinate attack. Table 18.5 shows the factors and design approach for coordinate attack. A replicated full factorial design with eight runs was selected because this is a relatively simple targeting mode that requires few factors to characterize performance across the employment space. Additionally, since this capability is the legacy capability, we are only interested in detecting large changes in performance.

Figure 18.7 shows the power analysis for the main effects in the coordinate attack design. The second-order interaction also has the same power because this is a full

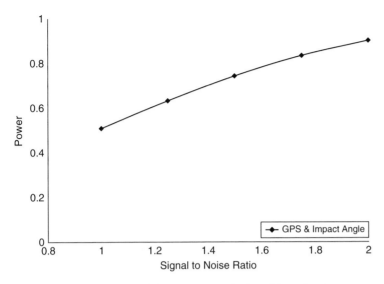

**Figure 18.7** Coordinate attack design power at 80% confidence level.

**Table 18.6** Laser attack – test design summary.

| Factor | All levels |
| --- | --- |
| Target aspect | Head/Tail, 90° |
| Laser source | Self, Other |
| Target speed | High, Low |

factorial design. The eight-run design will be able to detect only large changes in performance across the factor space. However, this is acceptable due to the relatively straightforward nature of the coordinate attack targeting mode.

### 18.4.1.3 Laser Attack

Table 18.6 summarizes the anticipated factors for laser attack and the design. Historical analyses of laser-guided tests have shown that the lasing source, target speed, and target aspect are the primary factors that impact the effectiveness of a laser-guided bomb. Again, a relatively small eight-run full factorial design was selected because we only wish to determine large changes in performance for this straightforward targeting method.

Since there was some concern from the test team that time of day could impact the weapon's ability to acquire the laser spot, the laser attack design was matched with the overarching design in a way that allows for the estimation of the effects of time of day. Figure 18.8 shows the power calculations for the main effects for the laser attack design factors (all the same) and time of day. Again, the relatively low power for this design is acceptable because we only wish to determine large changes in system performance for the laser targeting method. The power is slightly lower for time of day due to small correlations between time of day and the other factors.

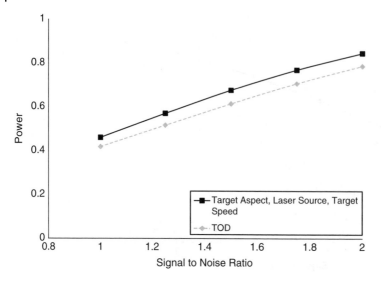

**Figure 18.8** Laser attack design power at 80% confidence level.

**Table 18.7** New attack mode – test design summary.

| Factor | All levels |
| --- | --- |
| Target type | Wheeled, Tracked |
| Target speed | Stopped, Fast, Slow |
| Target aspect | Head/Tail, 90° |
| Infrared countermeasures (IR CM) | No, Yes |
| Millimeter wave countermeasures (MMW CM) | No, Yes |
| Confusers | Yes, None |

#### 18.4.1.4   New Attack Mode

The new attack is the most complex attack mode and the primary reason one would employ new weapons compared to a less expensive legacy laser-guided bomb or GPS-guided bomb. The new attack uses both millimeter wave and infrared targeting methods. Table 18.7 describes the factors that could impact the performance of the new attack mode.

A 34-run D-optimal test design supports all the main effects and all two-way interactions for the design factors. Additionally, the update rate and TOD could be important in normal attack performance. Figure 18.9 shows the power analysis for all main effects. The power is high for all main effects. However, to estimate all of the main effects and two-factor interactions for the full design space would require a design with at least 45 points. Instead, the test team selected a design that minimizes correlations between two-factor interactions.

Figure 18.10 shows the correlations between all factors and their two-way interactions. Using this design will allow the analysis to pull the most important factors from all possible two-factor interactions. It is possible that if we learn that factors are not important

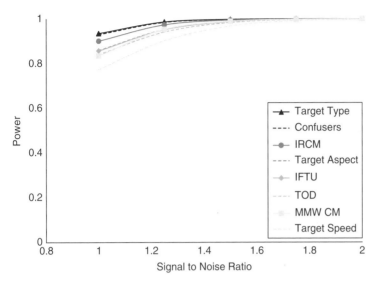

**Figure 18.9** Normal attack design power at the 80% confidence level.

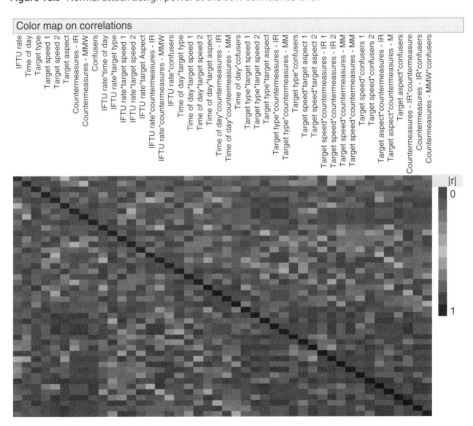

**Figure 18.10** Color map of correlations for all factors and two-way interactions for normal attack design.

in earlier testing we can better optimize the test design to address the most important factors.

This illustrative example of using DOE in operational testing shows how different design techniques can be used to address multiple test objectives. Arguably, three separate designs could have been constructed, but breaking the overarching design into blocks ensures that the full operational space is covered across the three different attack modes. Power analysis provides a methodology for showing that 50 missions will be adequate to fully characterize the new attack mode as well as provide limited information on the two legacy modes.

### 18.4.2 Operational Testing Using Statistically Designed Laboratory Tests

Acoustic-Rapid COTS Insertion (A-RCI) is the Navy's newest submarine sonar processing system. It provides hardware and software to process data from the submarine's sonar arrays and display those data to the sonar operators. The project uses a spiral development model to procure new, commercial off-the-shelf (COTS) computing hardware every two years. Buying new computing hardware over time capitalizes on the ever-decreasing cost of computing hardware and ensures an acceptable balance between obsolescent and modern hardware is maintained. To take advantage of the ever-improving processing power from hardware upgrades, a new version of A-RCI software, termed an Advanced Processing Build (APB), is developed every other year and incorporates feedback from Fleet users, fixes bugs discovered in previous versions, and adds new algorithms developed by industry and academia.

The primary role for A-RCI is to manage the large amount of information coming from the sonar arrays and display it so that the operator can make sense of it. To understand the scale of the operator's problem, consider that a *Virginia*-class submarine uses six sonar arrays for submarine searches, each providing information on all bearings, multiple elevation angles, and a range of frequencies. The sonar operators must monitor this multidimensional search space constantly, and it is impossible to display all of the information simultaneously. A-RCI provides displays and automation to help the operator manage this search space and help detect contacts as quickly as possible.

The Navy's primary metric used to evaluate A-RCI performance in the anti-submarine warfare mission is the median time it takes for an operator to detect a submarine contact, once that submarine's signal becomes available for display on sonar system screens ($\Delta T$). This measure quantifies A-RCI's role in the detection process. The goal of A-RCI processing improvements is to minimize the time needed to find target signatures.

At-sea tests of A-RCI consist of two submarines searching for each other in a specified area. Although this technique provides an operationally realistic environment, it suffers from several drawbacks. Most notably, at-sea testing has never been able to show a statistically significant improvement in A-RCI performance over the course of a decade, during which time many software and hardware upgrades were fielded to the Fleet. A comparison has been impossible because two software versions are never compared in the same at-sea event, and the results of a test can depend on target and environmental characteristics that are impossible to control. Additionally, at-sea testing uses a single target and a single operational environment, which limits the assessment of performance of the new APB to only a small portion of the operational envelope. Finally, the cost and variability of at-sea testing has resulted in poor quantification of APB performance (wide error bars) in the conditions tested.

**Table 18.8** Factors and levels used in the OIL testing analysis.

| Factor | Levels | Hypothesized effect |
|---|---|---|
| Target type | A, B | Different submarine types exhibit different acoustic signatures. Type A has more discrete tonal information than Type B. |
| Array type | A, B | Array type A typically detects targets at longer ranges, which would be expected to generate larger $\Delta T$s. |
| Target noise | Loud, Quiet | Loud targets are detected at longer ranges, which could lead to longer $\Delta T$s. Conversely, loud targets typically have more discrete tonal information and are easier to identify, which could result in shorter $\Delta T$s. |
| APB version | APB-09, APB-11 | The primary goal of the test was to compare the latest version of the sonar system, APB-11, with the previous version, APB-09. |
| Operator proficiency | 1 to 20 | More proficient operators will detect a submarine more quickly. The numeric scale was developed by the Naval Undersea Warfare Center and is based on an operator's experience with the A-RCI system. |

To address the shortcomings of A-RCI at-sea testing, we use operator-in-the-loop (OIL) laboratory testing. In this method, a fleet operator sits at a laboratory mockup of the sonar system. The laboratory then plays back a recorded encounter between two submarines, and the operator declares when he has detected the threat submarine. The laboratory allows the same encounter to be replayed on different versions of the software, which perfectly controls for environmental and target variability; the only difference between the two presentations is the software used to process the data. The primary limitation of the laboratory testing is that it only allows for a single array to be processed at one time. Therefore, the sonar array to be processed needs to be a controlled test factor, whereas in real operations all arrays operate simultaneously.

Table 18.8 shows the factors considered in the experimental design. The primary goal of the test was to compare the latest version of the sonar system, denoted APB-11, with the previous version, APB-09. To better characterize the systems, the test used operators of varying proficiency and controlled for characteristics of the target and the array being used.

Figure 18.11 shows the original design proposed for this operational testing. The design is a 120-run factorial design with strategic replication. While not apparent in Figure 18.11, the design is a split-plot design. A "run" consists of a single operator viewing a single recorded encounter, and a "Null" run is one in which no target is present. The split-plot structure was used to limit the number of changes of the APB version, as each software change required approximately 12 hours. The large amount of replication was built into the design to account for the fact that operator proficiency was not explicitly controlled. Instead, operators were chosen at random and their proficiency was recorded during the events, which ensured a balanced distribution of proficiencies. Each operator reviewed up to six tapes, including a blank tape to check for false alarm rate. Finally, the Navy desired to focus the testing on APB-11, which resulted in the asymmetric test design shown; while this was not optimal for

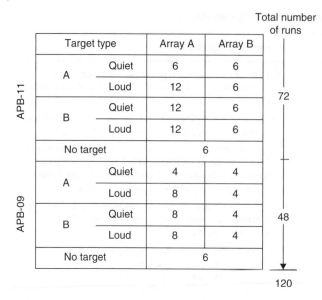

**Figure 18.11** Operator-in-the-loop test design matrix.

determining whether a significant APB difference existed, it did provide a more precise (tighter confidence intervals) understanding of the performance of APB-11.

Figure 18.12 shows the raw results of the test. Each panel shows the results for a recorded encounter, with APB-09 results on the left and APB-11 results on the right. The blue dots are detection times, and the red dots indicate runs in which the operator never detected the target before the recording finished. The location of the red dot indicates how long the target was on tape and not detected. A close inspection of Figure 18.12 reveals that the test points do not match the experimental design. It is not uncommon for test plans to change during operational testing where unanticipated execution challenges can arise. In this case, operator availability was not uniformly distributed across operator proficiency levels. Therefore, instead of preserving the original design, testers assigned operators to tapes to ensure a balance of proficiency levels across other factors.

Figure 18.12 shows the results by both experimental design bin and individual cut of tape used in that bin. The advantage to examining the results by recording is that recordings control all aspects of the encounter; the environment and target are exactly the same for each playback, so any difference in performance is due to either operator proficiency or the capability of the processing system. The test was well balanced in terms of operator proficiency, so any observed differences are most likely due to the processing system. In general, APB-11 exhibited improved performance in almost all of the recorded encounters; in each panel, the dots are generally lower for APB-11 than they are for APB-09. Therefore, even without statistical analysis, APB-11 appears to be an improvement over APB-09. Such a limited analysis does not, however, make use of all the available information; APB-11 appears to be better, but the improvement varies with recording and it is unclear why. The test was designed to determine which of the controlled factors affect A-RCI performance, and for that a more rigorous statistical analysis is necessary.

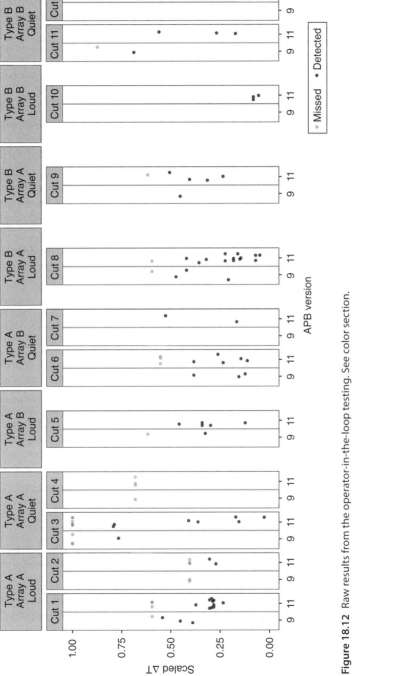

**Figure 18.12** Raw results from the operator-in-the-loop testing. See color section.

A regression analysis allows us to better understand how the controlled factors affected A-RCI performance. Our analysis, however, must account for missed detections. We treated them as censored data points, a useful methodology from reliability analysis (see Meeker and Escobar, 1998). In the case of a failed detection, we assumed that the operator would have detected the contact if given enough time. We assumed that the data followed a lognormal distribution, in which the probability of observing a detection time $x$ is the following:

$$P(x|\mu, \sigma) = \frac{1}{x\sigma\sqrt{2\pi}} e^{-\frac{(\ln(x)-\mu)^2}{2\sigma^2}}.$$

Here, $\mu$ is related to the median of the distribution, and $\sigma$ is a measure of its spread. Making this assumption allowed us to incorporate the missed detections using standard censored data analysis techniques.

Although there is no a priori reason why the data should follow a lognormal distribution, our initial assumption was well supported by the data. Figure 18.13 shows the empirical cumulative distribution function of the data, along with a lognormal fit and corresponding confidence region on the lognormal fit. The dashed lines represent the confidence region on the non-parametric fit. The data appear to be well described by a lognormal distribution.

After performing model selection we settled on the following model, which minimized Akaike's information criterion (AIC):

$$t \sim \text{lognormal}(\mu, \sigma).$$

$$\sigma = \text{constant}.$$

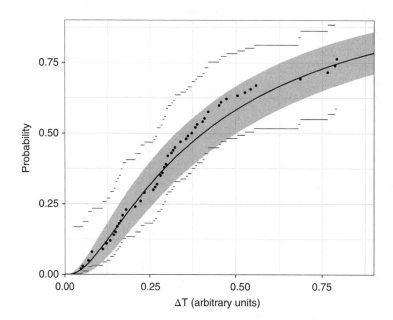

**Figure 18.13** Empirical cumulative distribution of the OIL data.

$$\mu = \beta_0 + \beta_1 \, (\text{OP}) + \beta_2(\text{APB}) + \beta_3(\text{target}) + \beta_4(\text{noise}) + \beta_5(\text{array})$$
$$+ \beta_6(\text{target} \times \text{noise}) + \beta_7(\text{target} \times \text{array}) + \beta_8(\text{noise} \times \text{array})$$
$$+ \beta_9(\text{target} \times \text{noise} \times \text{array}).$$

Table 18.9 shows the results of the final fit and describes the qualitative behavior of the coefficients. We found that the median detection time depends on the factors considered in the design, and that the $\sigma$ parameter was constant. All of the first-order effects were highly significant. APB-11 is significantly better than APB-09. Also notable is the fact that APB had no interaction with the other factors, which means that APB-11 produced an improvent regardless of the other factors. It did not matter whether the target was loud or quiet, or of hl type SSN or SSK; switching from APB-09 to APB-11 reduced the median detection time.

Figure 18.14 shows the results of the model fit (black dots, with 80% confidence intervals shown as vertical lines), along with the actual median detection times in each group (black) and the raw detection times (light gray and gray, as before). The model predictions generally agree with the median in each bin. There is, however, notable disagreement between the data median a the model prediction for one bin: quiet, type A targets with array type B in APB-09. The da median in this case is based on only three data points and is therefore highly variable, making it a poor estimator of the true performance in that bin. We believe the model estimate predicts t performance that would be observed if additional runs were conducted with APB-09.

The lognormal regression analysis provides several benefits over the naïve analysis based solely on individual recordings. First, differences in performance are now attributable to operationally relevant factors, such as target type, array type, etc. In contrast to the naïve analysis by recording, the statistical analysis shows that APB-11 outperforms APB-09 across all conditions. Second, our analysis allows us to extrapolate to areas where the data are limited. A few of the experimental configurations presented

**Table 18.9** Results of the model fit to the data. APB-11 provides a statistically significant improvement.

| Term | Value[†] | Description of the effect |
|---|---|---|
| $\beta_1$ (operator experience level) | $-0.074 \pm 0.041$ | Increased operator proficiency results in shorter detection times. An increase in proficiency of one unit reduces median detection time by 7%. |
| $\beta_2$ (APB) | $0.307 \pm 0.129$ | Detection time is shorter for APB-11. |
| $\beta_3$ (target) | $0.359 \pm 0.126$ | Detection time is shorter for one target. |
| $\beta_4$ (noise) | $-0.324 \pm 0.125$ | Detection time is shorter for loud targets. |
| $\beta_5$ (array) | $0.347 \pm 0.125$ | Detection time is shorter for the Type B array. |
| $\beta_6$ (target × noise) | $0.186 \pm 0.126$ | The third-order interaction is marginally significant, so all second-order interactions nested within the third-order interaction were retained to preserve model hierarchy. |
| $\beta_7$ (target × array) | $0.011 \pm 0.125$ | |
| $\beta_8$ (noise × array) | $0.021 \pm 0.126$ | |
| $\beta_9$ (target × noise × array) | $-0.180 \pm 0.125$ | |

[†]Confidence interval is an 80% Wald interval.

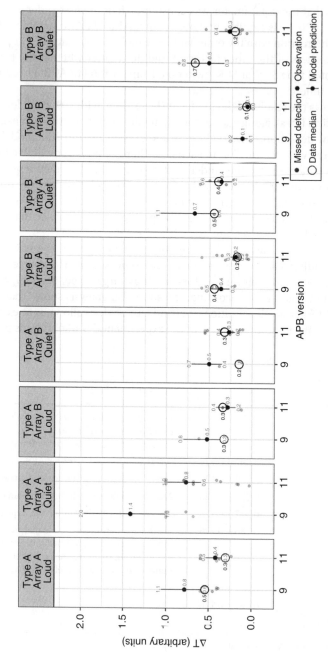

**Figure 18.14** Model predictions (see figure legend), along with the median detection time observed in each bin.

in Figure 18.14 do not have an observed data median for comparison with the model prediction, either because there were few data points or there was an excess of censored values. An analysis using a simpler technique would not have been able to estimate performance in regions where the data were inadequate to produce an estimate of performance.

### 18.4.3 Counterfire Radar Statistical Analysis

The performance of combat systems can be affected by a wide variety of operating conditions, threat types, system operating modes, and other physical factors. In this case study we look at how different statistical analyses can be used to summarize complex system behavior. The AN/TPQ-53 Counterfire Radar shown in Figure 18.15 is a ground-based radar designed to detect incoming mortar, artillery, and rocket projectiles; predict impact locations; and locate the threat gun geographically. Threat location information allows US forces to return fire on the enemy location, and impact location information can be used to provide warnings to friendly troops. The Q-53 is the next generation of counterfire radar, replacing the currently fielded radars.

The Q-53 has a variety of operating modes designed to help optimize its search. The 360° mode searches for projectiles in all directions around the radar, while 90° search modes can be used to search for threats at longer ranges in a specific sector. In addition, the 90° mode has two sub-modes. In the 90° normal mode, the radar searches a 90°

**Figure 18.15** Soldiers emplacing the Q-53 radar during operational testing.

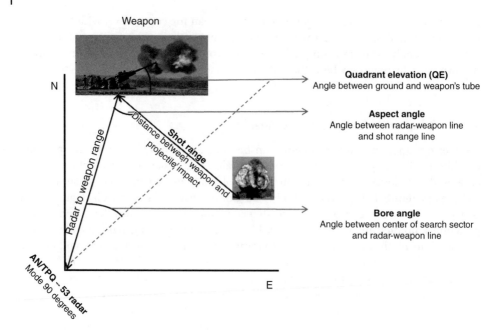

**Figure 18.16** Example fire mission including relevant geometric factors impacting Q-53 system performance.

sector out to 60 kilometers, while in the 90° short-range optimized mode (SROM), the radar focuses on short-range threats, sacrificing some performance at longer ranges.

In addition to the various operating modes, the Q-53 radar's performance can vary depending on the characteristics of an incoming projectile's trajectory and geometry relative to the radar's position. Determining how much the radar's performance varies across all these factors is essential to inform users of the capabilities and limitations of this system. Figure 18.16 outlines a standard fire mission for the Q-53. During a threat fire mission, the threat will fire projectiles at a target inside the search area of the Q-53. Figure 18.16 shows a Q-53 operating in a 90° mode, so its search sector is limited to the area within the black bars. The Q-53 must detect the projectile's trajectory and then estimate the threat's position. The specific geometry of the scenario will impact the Q-53's ability to track the projectile. Relevant factors include radar weapon range (the distance between the Q-53 and the weapon firing the projectile), quadrant elevation (the angle of the projectile's trajectory relative to the horizon), and shot range (the distance between the weapon and its target). When operating in 90° modes, the angle between the center of the radar's sector and the projectile's trajectory (bore angle) may also impact performance.

Two metrics best answer the key questions about system performance:

1) Can the Q-53 detect projectiles with high probability?
2) Can the Q-53 locate a projectile's origin with sufficient accuracy to provide an action-able counterfire grid location?

The operational test of the Q-53 replicated typical combat missions as much as possible given test constraints. Four radars (two battalions) observed shots fired from a variety of weapons. Each battalion decided how to employ the radar, within given test parameters, based on intelligence reports provided by the test team. Test personnel fired US and threat weapons throughout four 72-hour test phases. During a single threat fire mission, test personnel fired projectiles (typically ten) from a single location using the same gun parameters, simulating a typical engagement that a Q-53 battalion might encounter in a combat scenario. During a volley fire mission, test personnel fired projectiles from three weapons at the same time. Volley fire is a common technique used to increase the number of rounds hitting the target in a fire mission. Since the radar did not move during these missions, all of the factors in Figure 18.16 were held constant during each threat fire mission. Many missions were observed by two radars, so a single threat fire mission could be detected by two radars. Testers fired 2873 projectiles, which resulted in 323 usable fire missions.

Figure 18.17 shows the raw probability of detection data. Each point represents a fire mission, with the size of the point determined by the number of shots taken in the fire mission, ranging from a single shot to as many as 20 projectiles. The percentage of those shots detected by the Q-53 counterfire radar is shown on the $y$-axis. The colors of the points show the munition, with different operating modes and fire rates separated across the $x$-axis.

As Figure 18.17 shows, there is substantial variability in probability of hit across different combinations of operating mode, munition, and rate of fire. There are geometric differences between operating modes, complicating the definition of a shot's geometry. Bore angle is the angle between the weapon and the center of the radar's search sector. In 360° mode, there is no angular center and therefore no bore angle. As a result, the 90° modes must be analyzed separately from the 360° modes to ensure that bore angle is properly taken into account. Additionally, the data are heavily imbalanced. The choice of the 90° operating mode was left to the battalions. They quickly learned that most of the threat missions were within SROM capabilities, so 90° Normal was used substantially less than 90° SROM. There are substantially fewer volley-fire shots than single-fire shots. Furthermore, many of the geometric factors were confounded with each other because of limited available firing points on the test range. As often happens in operational testing, the Q-53 test conditions resulted in imbalanced, correlated data. The challenges in analyzing these types of data are best addressed using statistical regression models.

When characterizing system performance, it is important to account for all factors that impact system performance. A logistic regression analysis was the natural analysis model choice considering the complex nature of the problem, allowing us to identify which factors were driving performance and to generate estimates of probability of detection for all combinations of factors. Most importantly, this approach identified the effect of each factor, after accounting for the others, helping determine which factors have the largest impact on performance. The general logistic regression equation is

$$\log\left(\frac{p}{1-p}\right) = \beta_0 + \beta_1 x_1 + \ldots + \beta_p x_p.$$

In our case, $p$ is the probability of detection and the $x_i$ and $\beta_i$ represent the factors and coefficients, respectively. This approach relates the log of the odds ratio of probability of detection to the various factors that impact the probability of detection. Unlike a more naïve approach that looks at factors one at a time, this method allows us to attribute changes in probability of detection to specific factors. Importantly, this also allows us

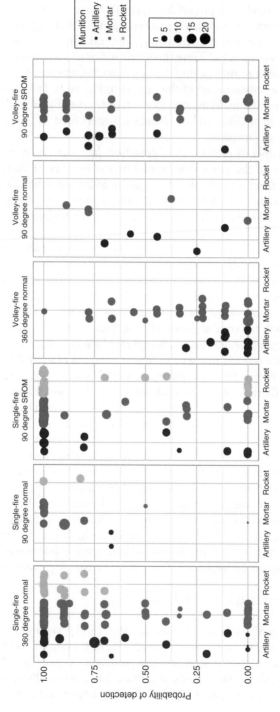

**Figure 18.17** Detection probabilities for 323 fire missions conducted during the Q-53 IOT&E. See color section.

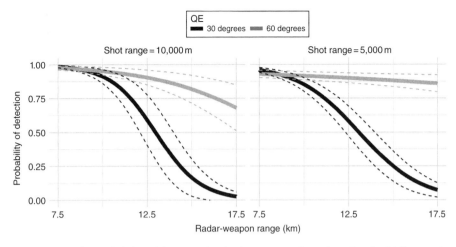

**Figure 18.18** The probability of detection for the Q-53 counterfire radar using the 360° operating mode against single-fired artillery.

to identify which of our considered factors are not driving performance. Such factors can be eliminated from the statistical model, simplifying the final expression without surrendering its explanatory power.

The logistic regression model, once determined from the data, showed that, in addition to projectile type, operating mode, and rate of fire, radar weapon range, quadrant elevation (QE), aspect angle, and shot range had an impact on system performance. Figure 18.18 shows how the probability of detection for artillery projectiles changes as the distance between the weapon and the Q-53 counterfire radar increases when the system is in the 360° operating mode observing single-fire artillery engagements. The data also revealed that radar–weapon range and QE had large impacts on Q-53's ability to detect incoming projectiles. These factors are linked to the time the projectile travels through the radar search sector. High-arcing shots (larger values for QE) are easier to see than shots with more shallow trajectories that stay closer to the ground (low QE) and are more likely to be masked by terrain. Shots with trajectories exposing larger cross-sections of the projectile to the radar (smaller aspect angles) were easier for the Q-53 to detect, although the data showed this factor to be less important than radar weapon range and QE.

The logistic regression approach also allows us to analyze the impacts of these factors simultaneously and observe how they interact. In Figure 18.18, as the radar–weapon range increases, the probability of detection drops sharply around 12 km for shots with shallow shot trajectories (QE = 30°, shown with the black lines). For the shots with more arc (QE = 60°, shown with gray lines), the Q-53 is still able to detect with high probability at longer ranges. While these factors have large effects, other factors such as aspect angle have relatively minor effects on probability to detect. Comparing the left and right panels of Figure 18.18, we can see that a 30° change in aspect angle results in a small change in probability of detection.

In addition to detecting incoming projectiles, the Q-53 counterfire radar will also estimate the location from which the detected projectiles were fired. The radar tracks the projectile through most of its flight and then backtracks the trajectory to estimate the threat's location (the point of origin of the trajectory). The distance between the true point of origin and the location estimated by the Q-53 is referred to as target location error (TLE). For this analysis, a single target location estimate was calculated for each fire mission, since all projectiles from a fire mission originated from the same location. As a result, there are fewer data for the TLE analysis than the probability of detection analysis.

Figure 18.19 shows quantile plots of TLEs for the 360° operating mode, broken down by munition type. The lines represent what the distribution would look like if the data were normal. The chart on the left shows the raw data. The data do not fall along the straight lines, indicating that the underlying distribution does not conform to the normal distribution. The data are skewed to the right, with many large TLEs in excess of what would be expected for normal data. The figure on the right shows the same data on the log scale; the data fall much closer to the lines, indicating that the log scale is more appropriate.

As a result, the TLE data were analyzed using a lognormal regression. This approach allows us to take the skewness of the data into account so that the fit has the same characteristics as the data. Figure 18.20 shows the results, with the figure on the left showing TLE for mortars and the figure on the right showing TLE for artillery and rockets. The green lines show the system's requirements, and the black lines show the estimated median TLE along with 80% confidence intervals. While TLE for mortars showed substantial variability, the large number of mortar fire missions allows us to make precise estimates of the median TLE. The analysis revealed that the estimated median TLE tends to increase (get worse) as radar weapon range increases. While the Q-53 is more accurate at estimating a mortar's location than the location of artillery and rocket weapons, the requirements for artillery and rockets were less stringent.

Physical factors related to the shot's geometry, as well as threat and operating mode, impact Q-53 performance. Understanding the effects of these factors helps commanders in the field choose the best operating mode for the system, allowing them the best chance of detecting incoming projectiles and locating their origins accurately for a counterfire response. The statistical regression techniques used to analyze the data identified which factors affected system performance and quantified their impact and practical significance for soldiers employing this system.

## 18.5 Summary

The methodologies of design of experiments and the corresponding statistical analyses provide a framework for testing a variety of complex military systems. However, examples in existing literature that show how these methods apply to military systems are far from widespread. This chapter provided an overview of the process of designing experiments for military systems with operational users in an operational environment. An important distinction of operational testing is that often the data collected are not exactly the same as the planned data. The analysis model should reflect the observed data and not the planning process. The sonar testing case study provided an example of how the data collected can deviate from the design. In other cases the operating environment results in completely uncontrolled data collection. Kenett and Nguyen (2017) discuss the process for evaluating the information quality of unplanned experiments or experiments that change during execution. The analysis should reflect the data collected, and that data must also be reviewed to ensure it reflects the original goals of the

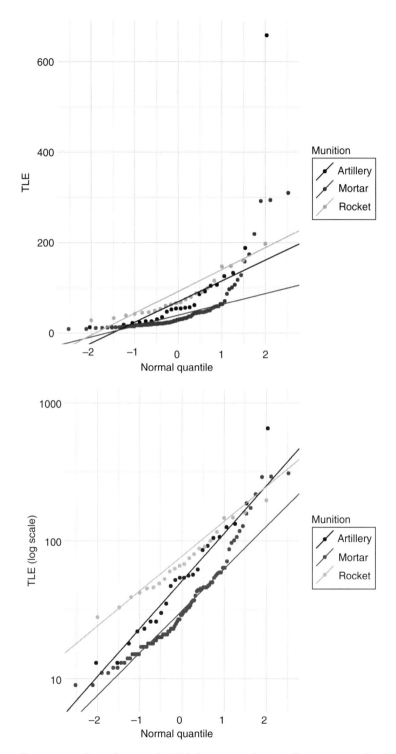

**Figure 18.19** Quantile–quantile (QQ) plots are used to visually assess normality.

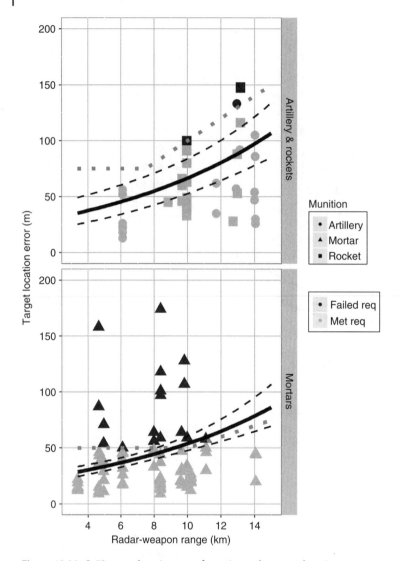

**Figure 18.20** Q-53 target location error for estimated weapon locations.

experiment and caveated appropriately if limitations exist. In the case of the new sonar software testing, the impact of the new software ultimately was larger than expected, resulting in a larger detectable difference. Therefore the lower replication than planned with imbalance across conditions was ultimately acceptable.

This chapter also summarized some essential parametric statistical analyses used in the analysis of defense systems. Because full system tests in the operational environment are often very expensive, parametric analysis provides the most information for limited sample sizes. However, it is important to ensure that the parametric model reflects the qualities of the data observed (e.g., skewness).

Three case studies illustrated different design and analysis approaches for a bomb, a sonar software upgrade, and a radar.

# References

Cohen, J. (1983) The cost of dichotomization. *Applied Psychological Measurement*, 7(3), 249–253.

Dickinson, R.M., Freeman, L.J., Simpson, B.A., and Wilson, A.G. (2015) Statistical methods for combining information: Stryker family of vehicles reliability case study. *Journal of Quality Technology*, 47(4), 400.

Efron, B. and Tibshirani, R.J. (1994) *An Introduction to the Bootstrap*. Boca Raton, FL: CRC Press.

Hamada, M. (2002) The advantages of continuous measurements over pass/fail data. *Quality Engineering*, 15(2), 253–258.

Johnson, T.H., Freeman, L., Hester, J., and Bell, J.L. (2014) A comparison of ballistic resistance testing techniques in the Department of Defense. *IEEE Access*, 2, 1442–1455.

Jones, B. and Nachtsheim, C.J. (2011) A class of three-level designs for definitive screening in the presence of second-order effects. *Journal of Quality Technology*, 43(1), 1.

Kenett, R.S. and Nguyen, N. (2017) New frontiers in evaluating experimental conditions for enhanced information quality. To appear in *Quality Progress*.

Kuhn, R., Lei, Y., and Kacker, R. (2008) Practical combinatorial testing: Beyond pairwise. *IT Professional*, 10(3).

Kuhn, R., Kacker, R., Lei, Y., and Hunter, J. (2009) Combinatorial software testing. *Computer*, 42(8).

Laird, N.M. and Ware, J.H. (1982) Random-effects models for longitudinal data. *Biometrics*, 38(4), 963–974.

Meeker, W.Q. and Escobar, L.A. (1998) *Statistical Methods for Reliability Data*. Chichester: Wiley.

Montgomery, D.C. (2008) *Design and Analysis of Experiments*. Chichester: Wiley.

Myers, R.H., Khuri, A.I., and Vining, G. (1992) Response surface alternatives to the Taguchi robust parameter design approach. *The American Statistician*, 46(2), 131–139.

Myers, R.H., Montgomery, D.C., and Anderson-Cook, C. (2009) *Response Surface Methodology: Process and Product Optimization Using Designed Experiments*. Chichester: Wiley.

Myers, R.J., Montgomery, D.C., Vining, G.G., and Robinson, T.J. (2010) *Generalized Linear Models with Applications in Engineering and the Sciences*, 2nd edn. Chichester: Wiley.

National Research Council (1998) *Statistics, Testing, and Defense Acquisition: New Approaches and Methodological Improvements*. Washington, DC: National Academies Press.

Nelder, J.A. and Wedderburn, R.W.M. (1972) Generalized linear models. *Journal of the Royal Statistical Society. Series A (General)*, 135(3), 370–384.

NIST (2013) NIST Engineering Statistics Handbook. http://www.itl.nist.gov/div898/handbook/ [accessed 24 January 2018].

Phadke, M.S. (1995) *Quality Engineering Using Robust Design*. Upper Saddle River, NJ: Prentice Hall.

Searle, S.R. and Gruber, M.H. (2016) *Linear Models*. Chichester: Wiley.

# 19

## A Search-Based Approach to Geographical Data Generation for Testing Location-Based Services

*Xiaoying Bai, Kejia Hou, Jun Huang, and Mingli Yu*

## Synopsis

Location-Based Service (LBS) has been a key infrastructure service for mobile applications like navigation and recommendation. Once hosted on a cloud platform, the service can support a large number of location-related queries from diversified devices and applications. The quality of the service, such as correctness, precision, and timeliness of the location information, is critical to those dependent applications and the ecosystem of mobile software built around it. Testing of the open services is thus necessary. However, LBS testing is challenging due to the enormous input domain of geographical areas. It is impossible to exhaustively enumerate all the possible inputs, and the corresponding expected outputs as well. How to design test inputs to achieve a certain level of confidence? How to validate test results when there are a large number of test cases? As manual testing is expensive, requiring extensive time to develop and computation resources to execute, how to devise an automatic approach enhancing test efficiency and effectiveness?

To cope with the difficulties, this research proposes a testing framework to support automatic LBS testing from two aspects: geographical test data generation and query result verification, taking geocoding and reverse geocoding as illustrative LBS functions. To generate geographical data, a simulated-annealing algorithm is designed to formulate the problem as an iterative optimization search process. It is based on a defect clustering heuristic assumption. That is, defects tend to cluster in certain areas. Hence, areas with detected defects deserve more effort in the follow-up testing. A Bayes classifier is used to predict the defect probability of geographical areas, and to guide geolocation data generation in the search process. To automate the validation of query results, it applies a voting mechanism to compare results across different platforms providing the same LBS functions. Experiments were exercised on four LBS platforms. The results showed considerable improvements in defect detection with reduced cost.

## 19.1 Introduction

The mobile cloud provides infrastructure services for diversified mobile devices with scalable resources like computing and storage. With the expansion of the mobile cloud,

*Analytic Methods in Systems and Software Testing*, First Edition.
Edited by Ron S. Kenett, Fabrizio Ruggeri, and Frederick W. Faltin.
© 2018 John Wiley & Sons Ltd. Published 2018 by John Wiley & Sons Ltd.

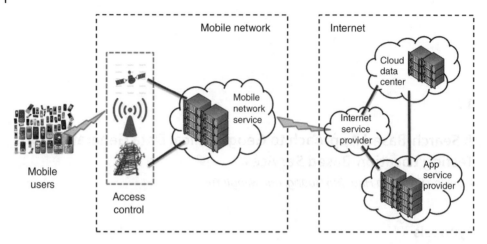

**Figure 19.1** Mobile cloud architecture overview.

location awareness has been an enabling technique for a variety of mobile applications (Rao and Minakakis, 2003; Kupper, 2005). The service is deployed on a cloud platform to support various clients querying geographical locations, finding points of interest, navigating, and encoding and decoding geolocation data. Figure 19.1 shows an overview of the mobile cloud architecture. A geographical position in terms of longitude and latitude is obtained from the mobile network, and passed to the cloud services through the internet.

LBS can be a basic built-in feature of almost all mobile applications such as transportation, entertainment, and healthcare for the given physical location. For example, with LBS support, an instant message application can support users by sharing instant locations with partners in real time, and entertainment software can recommend restaurants and hotels close to users' locations. LBS is thus considered as the "killer application" of mobile commerce (Dhar and Varshney, 2011).

Many platforms have been built to provide LBS as open APIs (application programming interfaces). For example, Baidu, Tencent, and Amap are three widely used LBS platforms in China. Each of the vendors provides two sets of APIs, with one for the web and the other for Android mobile Apps. There may be considerable differences between the services provided by different providers, and between two versions of the services provided by the same provider. For example, they could support location search at different scales and precisions, such as the query results are returned in a rectangular or circular area, and accurate to different granularity of administrative regions like city or district. In addition, the quality of location awareness of mobile applications is different in terms of accuracy, completeness, and timeliness (Rao and Minakakis, 2003; Kupper, 2005). In a preliminary experiment, 1000 geolocation data items were generated randomly and fed to the LBS APIs from four vendors to test their services against each other. There were around 10% inconsistencies in the returned values between any pair of platform services, which indicated potential errors in the services. Hence, it is critical to validate the correctness, precision, and timeliness of LBS services.

Research has been conducted on LBS quality assurance from various perspectives, including service content, service usage, hardware device, and communication network. For example, Li and Longley (2006) proposed a monitoring-based approach where an

LBS environment was built for studying the behavior and interactions between environments, individuals, and mobile devices. Software was developed to capture and analyze the spatial-temporal data of users' geographical positions (or administrative locations) within the test environment, including movement, information access, and participant observations.

Complementary to the log-based analysis approach, testing is still one of the important techniques for detecting and locating software defects using simulated inputs which are carefully designed with a high potential to trigger abnormal software behavior. However, test generation is difficult due to the large input domain. It is too expensive, even impossible, to enumerate all the positions. It is necessary to enhance test effectiveness with carefully selected inputs. According to G. Myers' classic definition, "[a] good test case is one that has a high probability of finding an as-yet undiscovered error. A successful test is one that uncovers an as-yet undiscovered error" (Myers, 1979). Testing is thus an intelligent process to productively and economically find a set of defect-sensitive test cases.

In this research, we formulate test input generation as an optimized search problem and propose an enhanced simulated-annealing algorithm to optimize the search for geographic locations. The objective function is defined with a defect clustering heuristic; that is, defects tend to aggregate in an area. If a defect is detected at some position, its neighboring positions have a high probability of containing defects. Hence, subsequent test data could be generated by exploring the neighboring area.

As shown in Figure 19.2, we have designed an automatic testing architecture to support an iterative process of defect-driven search. A feedback mechanism is introduced to adapt the search objective function based on previous test results and defect predictions. Test results are validated using a voting mechanism, by which location queries are forwarded to a set of LBS platforms with the same functionality. The returned results are collected and compared against each other. A difference between any pair of returned results indicates a potential error that needs further investigation. A Bayes classifier is

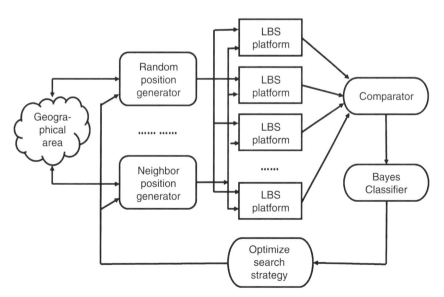

**Figure 19.2** Approach overview.

used to evaluate defect detection probabilities for the physical positions in a given geographic area, and to guide the optimization of search strategies.

The rest of the chapter is organized as follows: Section 19.2 introduces the background of search-based testing. Section 19.3 proposes the method and algorithm of LBS testing by simulated annealing. Section 19.4 presents experiments on four open LBS platforms to illustrate the improvements in test productivity and defect detection efficiency.

## 19.2   Search-Based Testing

Search-based testing (SBT) can be seen as an application of the generalized search-based software engineering (SBSE) research that has gained increasing attentions in recent years (Harman, 2007). Many software activities can be formulated as optimization problems, such as allocating resources to development tasks, locating a bug in a large number of suspicious program blocks, and so on. To deal with large-scale optimization problems in software engineering, SBSE applies metaheuristic search techniques to find near-optimal or good-enough solutions.

For software testing, heuristic search techniques offer promising solutions to optimizing test case generation. Given an item of software under test, SBT is the optimization process to find a set of test cases that meet test adequacy criteria. There exist various adequacy criteria, thus the optimization objective. For example, for structure-based test generation, the optimization objective is defined by structure coverage such as condition/decision coverage or path coverage, while for fault-based testing, the objective function can be defined by the coverage of various fault patterns.

Various search-based algorithms have been investigated in different contexts, such as hill climbing, tabu search, and generic algorithms. In this research, simulated annealing (SA) is used for the optimization process (Kirkpatrick, 1984). SA is originated from the chemical process of annealing in materials. It simulates the cooling process and system energy changes until it reaches a steady state. The basic idea of SA is to simulate the physical annealing processes. When an object is heated to a certain temperature, its molecules move freely in the state space $D$. As the temperature drops, the molecules gradually stay in different states. The molecules are rearranged in a certain structure when it reaches the lowest temperature. Statistically, the probability that molecules stay at state $r$ at temperature $T$ satisfies the Boltzmann distribution.

Test generation by SA has gained more and more attentions in recent years, with the increasing adoption of SBSE (Cohen et al., 2003; Harman, 2007; Kossler and Kumar, 2008). Tracey, Clark, and Mander (1998) applied SA in automated program analysis. Preliminary experiments showed encouraging results that optimization can greatly enhance the effectiveness and efficiency of test data generation. Zhai, Jiang, and Chan (2014) proposed a set of location-centric metrics for prioritizing test cases for regression test selection. The input-guided metrics measure the randomness of test inputs which are defined by location sequences. The POI (point of interest) aware metrics measure the central tendency of the outputs in correlation to the corresponding test inputs. In this way, they guide the selection of regression test cases to enhance test effectiveness. (McMinn, 2011) summarized various application areas and fitness functions of SA testing, such as temporal testing, functional testing, and structural testing. Simulated annealing has many advantages in solving optimization problems, such as high quality, robustness of initial values, and ease of implementation. It also has some limitations, such as high  initial temperature, slow cooling rate, low termination temperature, and enough sampling at each temperature, resulting in a long optimization process. Hence, in practice, a basic SA process is usually adapted to different problems with specially designed parameters. Key to the optimization technique is the definition of a problem-specific fitness function. In this research, the energy function of the SA algorithm is calculated by defect prediction using a naive Bayes classifier.

## 19.3 LBS Testing by Improved Simulated Annealing

### 19.3.1 Problem Statement

This research takes the geocoding and reverse-geocoding services as representative cases for designing search-based geographic data generation for testing LBS systems. To provide information for given geographic locations, the services usually depend on an internal encoding database system which associates a physical position (in terms of latitude and longitude) with an administrative location, in terms of city, district, and so on. The transformation process from physical position to administrative location is called geocoding, and the reverse process is called reverse-geocoding. Geocoding is critical to the quality of information services. For example, a navigation system may fail to find the route if a location is missing from the database, or an entertainment system may recommend a distant restaurant due to low location precision, or out-of-date information if the restaurant has been relocated.

Suppose that $S^D = \{s_1, s_2, \ldots, s_m\}$ is a set of geographical positions within an area $D$. The scope of $D$ is defined by its range of latitude, $D.rangLat$, and longitude, $D.rangLng$; that is, $D.rangLat = [Lat_l, Lat_h]$ and $D.rangLng = [Lng_l, Lng_h]$, where $(Lat_l, Lng_l)$ are the lower bounds and $(Lat_h, Lng_h)$ are the upper bounds of the given area. Let $s_i.lat$ be the latitude and $s_i.lng$ be the longitude of a position $s_i$. Then, $\forall s_i \in S^D$, $Lat_l \leq s_i.lat \leq Lat_h$ and $Lng_l \leq s_i.lng \leq Lng_h$.

Suppose that $L = \{l_1, l_2, \ldots, l_n\}$ is a set of LBS platforms, such as Baidu and Amap. Each $l_i$ provides the same information services for given positions. Let $R(l_i, s_j)$ denote the returned results of a platform service $l_i$ for an input of position $s_j$, and $R(L, s_j) = \{R(l_1, s_j), R(l_2, s_j), \ldots, R(l_n, s_j)\}$. A potential defect exists if, for a test input $s_j$, $\exists l_k, l_m \in L$ such that $R(l_k, s_j) \neq R(l_m, s_j)$. That is, there exist inconsistences in the responses of different services for the same inputs.

Test generation is an iterative search process to find a sequence of test inputs $\langle s_1, s_2, \ldots, s_k \rangle$ to detect potential defects in $L$. The optimal algorithm is based on a defect clustering assumption. This heuristic says that defects tend to be clustered in a section, so sections with detected errors have high probabilities of containing more errors. Thus, historical test results can be used to predict software error proneness, and test cases need to be designed and exercised on the error-prone sections. Regarding LBS systems, suppose that $Defect(D)$ is the number of defects detected in the area $D$. Given two areas, $D_1$ and $D_2$, if $Defect(D_1) > Defect(D_2)$ after a test, then in the next iteration, $D_1$ needs to be tested more and earlier than $D_2$.

### 19.3.2 The Search Algorithm

The generation of location test data is formulated as an SA optimization search algorithm. Simulated annealing is used to solve complex optimization problems based on the idea of neighborhood search. It is basically a stochastic optimization algorithm, based on a Monte Carlo iterative solution. The core of the algorithm is moving selection, that is, the selection of fitness values and new state. The search process switches from one state to another with decreasing "temperature" until it reaches the optimized solution at the lowest "temperature." At each step, it probabilistically moves between the current state and a neighboring state, which is heuristically selected. A typical definition of the probability $P$ is $P = \exp(-\frac{\delta}{t})$, where $\delta$ is the difference of the objective function between the neighboring state and the current state, and $t$ is the parameter representing temperature in the annealing process. It can be proved that, using the Metropolis principle for accepting or discarding a new state, if the temperature drops slowly enough, the probability of finding a global optimum solution is close to 1.

The SA algorithm is controlled by a set of parameters that are critical to the effectiveness and efficiency of the search process. For example, given an initial temperature $T$, if $T$ is high, the process of cooling down will be slow, which is time consuming, but could find a global optimal solution. On the contrary, if $T$ is low, the cooling and search process will be fast, but may miss a global optimum to find local acceptable solutions. In general, SA control parameters include the initial and terminating temperatures, cool-down rate, and the number of iterations.

The area $D$ in which to generate geographic position inputs is divided into $n \times m$ subareas, and is defined as a matrix $[D_{i,j}]$ where $D_{i,j}$ ($1 \leq i \leq n$ and $1 \leq j \leq m$) is a subarea within $D$. Searching for positions in $D$ as test inputs is a zooming-in process which progressively limits the scope of the search area, following the cooling schedule, until it satisfies the optimization goal. Figure 19.3 illustrates the search process: Figures 19.3(a) and (b) show the search through the subareas of $D$, while Figures 19.3(c) and (d) zoom into one subarea, $D_{22}$ and $D_{33}$, respectively. In Figure 19.3(a), based on Bayesian analysis, $D_{22}$ is ranked with the highest probability to detect errors, and is exercised first to generate test inputs within it. In Figure 19.3(b), it exits $D_{22}$, reranks the subareas, and $D_{33}$ is selected to be explored next. The process repeats until it reaches the coverage criteria for the subareas. It then gets into the selected subareas, and further explores them at a finer granularity. In Figure 19.3(c), two subareas of $D_{22}$, $D_{11}^{22}$ and $D_{13}^{22}$, are selected one after another, while in Figure 19.3(d) two subareas of $D_{33}$, $D_{12}^{33}$ and $D_{23}^{33}$, are selected in sequence. A sequence of test suites is generated through the iterations. At each iteration, a set of test cases, called a test suite, is generated randomly. Let $TC = \{TC_1, TC_2, \ldots, TC_n\}$ be the sequence of test suites, and $TC_i = \{s_1, s_2, \ldots, s_m\}$ ($s_k \in S^D$) be the test cases (geographic positions) generated in a test suite $TC_i$.

In order to estimate the defect distribution and guide the search directions, at each iteration a Bayes classifier is introduced to predict the error proneness of positions in the district based on previous test results. In the training process, a set of positions $S^D = \{s_i\}$ are randomly generated within $D$ to cover all the subareas evenly. That is, $|S^{D_{1,1}}| = |S^{D_{1,2}}| = \cdots = |S^{D_{n,m}}|$, where $|S^{D_{i,j}}|$ is the number of position data items generated within subarea $D_{i,j}$.

A characteristic function Hit() is defined to estimate potential errors for testing with $S^D = \{s_i\}$ on the set of target LBS platforms $L$, as follows. It is defined as the number of positions that trigger differences in the execution results.

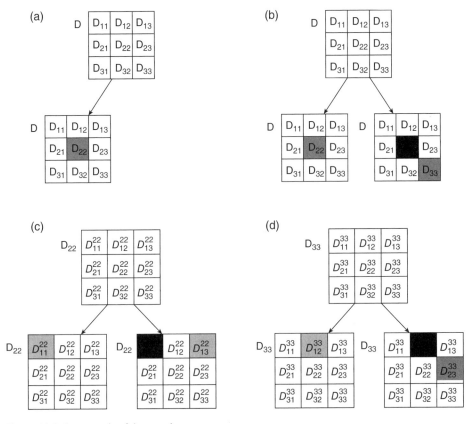

**Figure 19.3** An example of the search process.

$$
\text{Hit}(s_i, L) = \begin{cases} 1 & \text{if } \exists\ l_k, l_m \in L \text{such that } R(l_k, s_i) \neq R(l_m, s_i), \\ 0 & \text{otherwise;} \end{cases}
$$

$$
\text{Hit}(S^D, L) = \sum_{s_i \in S^D} \text{Hit}(s_i, L).
$$

The error proneness of $D_{i,j}$ is evaluated by the posterior probability for a potential error detected by $s_k \in S^D$ in an area $D_{i,j}$, $P(\text{Hit}(s_k, L) = 1|D_{i,j})$, using a naive Bayes classifier.

$$
P(D_{i,j}|\text{Hit}(s_k, L) = 1) = \frac{\text{Hit}(S^{D_{i,j}}, L)}{\text{Hit}(S^D, L)},
$$

$$
P(\text{Hit}(s_k, L) = 1) = \frac{\text{Hit}(S^D, L)}{|S^D|},
$$

$$
P(D_{i,j}) = \frac{|S^{D_{i,j}}|}{|S^D|},
$$

---

**Algorithm 19.1** Location data generation by simulated annealing.

---

**Input:**

    The district $D$;

    Initial temperature $t_0$;

    Termination temperature $t_t$;

    Cool down rate *CoolingFactor*;

    The maximum number of data samples within each subarea $Threshold^{D_{state}}$;

    The maximum number of all data samples $Threshold^{All}$;

**Output:**

    Generated test data;

1:  $T = t_0$;

2:  Divide $D$ into $n \times m$ subareas $\{D_{ij}\}$;

3:  Generate the training data sets *data*;

4:  Train BayesClassifier with *data* to initialize the estimated error proneness of each subarea;

5:  Initialize the test data set for each subarea $TC[D_{ij}]$;

6:  Initialize the defect number of each subarea $Defect[D_{ij}]$;

7:  **while** $(T > t_t)$ and $(TC[].length < Threshold^{All})$ **do**

8:    Pick the subarea $D_{state}$ in $\{D_{ij}\}$ with the highest estimated error proneness;

9:    **while** $(T > t_t)$ and $(TC[D_{state}].length < Threshold^{D_{state}})$ **do**

10:     Cool down $T = T \times CoolingFactor$;

11:     Randomly generate data $e_{new}$ within the sub-area $D_{state}$ ;

12:     Test with $e_{new}$;

13:     Collect results and calculate $Hit(S^{D_{state}}, L)$;

14:     Calculate energy function $Energy(D_{state})$;

15:     Calculate energy changes in successive iterations $\Delta Energy$;

16:     **if** $(\Delta Energy < 0)$ and $(\exp[\frac{\Delta Energy}{T}] < AP)$ **then**

17:       break;

18:     **end if**

19:    **end while**

20: **end while**

---

$$P(Hit(s_k, L) = 1 | D_{i,j}) = \frac{P(D_{i,j} | Hit(s_k, L) = 1) \times P(Hit(s_k, L) = 1)}{|P(D_{i,j})|}.$$

The optimization objective is searching for the positions with the highest probability of detecting defects. Accordingly, we define the energy using the error proneness predicted by the naive Bayes classifier as follows:

$$Energy_{D_{state}} = Hit(s_{k+1}, L) \times P(Hit(s_k, L) = 1 | D_{state}).$$

At each iteration, $Energy_n$ is recalculated based on the execution results of $TC_n$ for the selected subarea $D_{state}$. When the energy decreases in two successive test suites ($\Delta Energy < 0$), it accepts $D_{state}$ as the current state with some probability $AP$. It then jumps out of $D_{state}$, and picks up the next subarea of highest error proneness, that is, $P(\text{Hit}(s_k, L) = 1|D'_{state}) = \max(P(\text{Hit}(s_k), L) = 1|D)$. The algorithm is described in Algorithm 19.1.

## 19.4 Experiments and Results

### 19.4.1 Experiment Setup

The experiments are exercised on the reverse geocoding functions of four widely used LBS platforms: Amap Android, Tencent Android, Baidu Android, and Baidu Web. The APIs provide equivalent functional services, but differ in their protocols and data formats. For example, Baidu Android API is asynchronous, while Amap Android and Tencent Android are synchronous. Adapters are constructed to transform the generated test data to the various API formats, and to collect the returned results. To avoid business sensitivity, the specific vendor names are hidden and replaced by the code names API1, API2, API3, and API4.

An area $D$ is chosen with $D.rangeLat = [35, 53]$ and $D.rangeLng = [110, 150]$, which is of reasonable size and covers various landscapes including cities, mountains, and forests. $D$ is further divided into $2 \times 2$ subareas.

To balance between search efficiency and solution optimization, we set up the parameters to produce good enough test cases with acceptable cost. With a set of preliminary exercises, the parameters of the SA algorithm are set as follows:

- The initial temperature $t_0$ is 250.
- The termination temperature $t_t$ is 0.05.
- The training size is 100.
- The cooling factor is 0.98.

The results are analyzed from two perspectives: defect detection capabilities ($DD$), and test efficiency ($TE$).

$DD$ measures the number of potential defects detected by the set of generated geographic data. A potential defect is identified by a difference between a pair of APIs. Regarding the reverse geocoding services, three scenarios exist for potential defects:

Type1: One service returns a null value while another returns an administrative location.
Type2: The administrative locations returned from two services are different.
Type3: The administrative locations returned from two services are reported to different precisions.

$DD$ gives the statistics of potential defects of different types between any pair of the selected APIs. $TE$ measures the goodness of test generation. It is defined by the percentage of the number of test data choices that detect potential defects over the total number of generated tests:

$$TE = \frac{|S^{\text{Diff}}|}{|S|} \times 100\%,$$

**Table 19.1** Detected differences between pairs of APIs.

|        | API1 | API2 | API3 | API4 |
|--------|------|------|------|------|
| API1   | —    | 87   | 591  | 193  |
| API2   | 87   | —    | 588  | 196  |
| API3   | 591  | 588  | —    | 591  |
| API4   | 193  | 196  | 591  | —    |

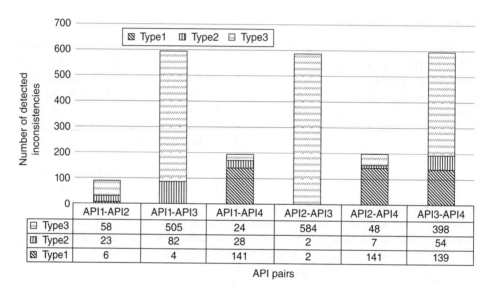

| | API1-API2 | API1-API3 | API1-API4 | API2-API3 | API2-API4 | API3-API4 |
|---|---|---|---|---|---|---|
| ▨ Type3 | 58 | 505 | 24 | 584 | 48 | 398 |
| ▥ Type2 | 23 | 82 | 28 | 2 | 7 | 54 |
| ▨ Type1 | 6 | 4 | 141 | 2 | 141 | 139 |

API pairs

**Figure 19.4** The number of detected differences of each type for each API pair.

where $S = \{s_i\}$ is the set of all generated test cases, with $|S|$ the number of generated test cases, and $S^{Diff} = \{s_j\}$ is the set of test cases which detect API differences; that is, $\forall s_j \in S^{Diff}$, $s_j \in S$ and $Hit(s_j, L) = 1$. $|S^{Diff}|$ is the size of $S^{Diff}$.

Random testing is exercised as the baseline for evaluating the proposed approach, which randomly samples data in the given area.

### 19.4.2 Defect Detection

Altogether, 730 position data items are generated with 100 for training. Considerable API differences are detected, which are potential defects or improvements of the LBS platforms. Table 19.1 shows the detected differences between any pair of APIs. API3 is identified to be the most error prone, with over 60% tests inconsistent with the others. Figure 19.4 shows the number of detected differences of each type for each API. Type3 is identified as the most common fault.

Figure 19.5 compares SA and random methods of identifying the differences between any pair of APIs, and of different types. The results show that with the same size of tests, SA can identify a lot more potential defects than random methods.

(a)

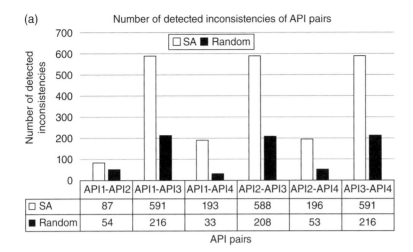

Number of detected inconsistencies of API pairs

| | API1-API2 | API1-API3 | API1-API4 | API2-API3 | API2-API4 | API3-API4 |
|---|---|---|---|---|---|---|
| □ SA | 87 | 591 | 193 | 588 | 196 | 591 |
| ■ Random | 54 | 216 | 33 | 208 | 53 | 216 |

API pairs

(b)

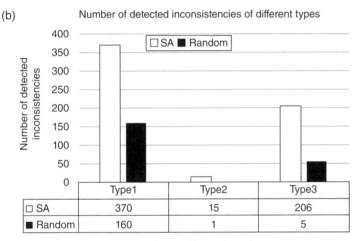

Number of detected inconsistencies of different types

| | Type1 | Type2 | Type3 |
|---|---|---|---|
| □ SA | 370 | 15 | 206 |
| ■ Random | 160 | 1 | 5 |

Types of inconsistencies

**Figure 19.5** Comparison between SA and random algorithms.

### 19.4.3 Test Effectiveness

The efficiency of SA and random methods are calculated below, showing that with the same size of tests (630 generated test data items), tests generated by SA can identify considerably more potential defects than those randomly generated. Hence, SA is much more effective than the random method.

$$TE_{SA} = \frac{370 + 206 + 16}{630} \times 100\% = 93.97\%,$$

$$TE_{Random} = \frac{160 + 55 + 1}{630} \times 100\% = 34.29\%.$$

## 19.5 Summary

With the rapid development of mobile cloud, mobile services have become an integrated part of people's daily life and work. Hence, service quality, such as availability, reliability, and security, is of great importance. Testing is one of the major techniques for quality assurance. Taking location-based services as a case study, this research has investigated the method of geographical test data generation. A simulated-annealing algorithm is proposed based on the assumption of LBS fault models, with the optimization objective to enhance the probability of potential defect detection. To guide an effective search process, the energy in the annealing algorithm is defined by the predicted probability of defect detection using a naive Bayes classifier. Experiments are exercised on real mobile service platforms, which show encouraging results that the proposed methods can enhance test effectiveness and defect detection capabilities with a tolerable time cost.

## References

Cohen, M.B., Colbourn, C.J., and Ling, A.C.H. (2003) Augmenting simulated annealing to build interaction test suites, in *Proceedings of International Symposium on Software Reliability Engineering (ISSRE)*, pp. 394–405.

Dhar, S. and Varshney, U. (2011) Challenges and business models for mobile location-based services and advertising. *Communications of the ACM*, 54(5), 121–128.

Harman, M. (2007) The current state and future of search based software engineering, in *Proceedings of Future of Software Engineering (FOSE)*, pp. 342–357.

Kirkpatrick, S. (1984) Optimization by simulated annealing: Quantitative studies. *Journal of Statistical Physics*, 34(5–6), 975–986.

Kossler, W. and Kumar, N. (2008) An adaptive test for the two-sample location problem based on $U$-statistics. *Communications in Statistics – Simulation and Computation*, 37(7), 1329–1346.

Kupper, A. (2005) *Location-Based Services: Fundamentals and Operation*. Chichester: Wiley.

Li, C. and Longley, P. (2006) A test environment for location-based services applications. *Transactions in GIS*, 10(1), 43–61.

McMinn, P. (2011) Search-based software testing: Past, present and future, in *Proceedings of IEEE International Conference on Software Testing, Verification, and Validation Workshops (ICSTW)*, pp. 153–163.

Myers, G. (1979) *The Art of Software Testing*. Chichester: Wiley.

Rao, B. and Minakakis, L. (2003) Evolution of mobile location-based services. *Communications of the ACM*, 46(12), 61–65.

Tracey, N., Clark, J., and Mander, K. (1998) Automated program flaw finding using simulated annealing. *ACM SIGSOFT Software Engineering Notes*, 23(2), 73–81.

Zhai, K., Jiang, B., and Chan, W.K. (2014) Prioritizing test cases for regression testing of location-based services: Metrics, techniques, and case study. *IEEE Transactions on Services Computing*, 7(1), 54–67.

# 20

# Analytics in Testing Communication Systems

*Gilli Shama*

## Synopsis

Customer experience is the heart of communication services. Everyone, at almost any minute of the day, is communicating. When you pick up a phone, surf the web, watch TV, or use a home security appliance connected to a central service, you are using a communication service – and you have no tolerance for errors. Communication systems require high quality, which is why the telecom and media industry captures 16% of the testing services market. This chapter will explore communication service providers' testing metrics and data analytics to support decision making processes from C-level through test management, and ending with the testers. We begin with a discussion and examples of both metrics to report on end of release results and metrics for monitoring the progress of testing. This will be followed by examples of using analytics to predict testing progress trends. Finally, we will end with an example from an analytics-based system that supports automation of smart test design. These three types of examples provide an extensive review of the high emerging potential of test data analytics in the world of communication service providers at all working levels. We conclude with a discussion and summary.

## 20.1    Introduction

### 20.1.1    Understanding Communication Service Providers' Testing Requirements

The communication service provider (CSP) industry (also known as "telecom and media," or just "telecom") has unique characteristics in comparison to other industries which need to be taken into consideration when carrying out testing activities. In this discussion, we will focus on analytics in typical acceptance testing for CSP software (analytics in special testing areas such as performance testing, security testing, network testing, digital testing, equipment testing, and data testing are outside the current scope).

*Analytic Methods in Systems and Software Testing*, First Edition.
Edited by Ron S. Kenett, Fabrizio Ruggeri, and Frederick W. Faltin.
© 2018 John Wiley & Sons Ltd. Published 2018 by John Wiley & Sons Ltd.

Below are five common CSP characteristics that are most relevant to typical software acceptance testing in the telecom and media industry:

1) The CSP industry is a cautious industry, requiring *high quality results.* CSPs are expected to provide high-quality communication services. Communication is used for many needs from entertainment through essential services for everyday life, and is also a must in emergency situations. The CSP industry is a cautious industry, similar to financial services, medical device and pharmaceutical manufacturers, and mission-critical government agencies, as identified in the Custom Standish Advisory (2013) testing survey.

2) CSP organizations suffer from both *complex IT architectures* and a demand to *reduce costs.* Many CSPs are large companies, and are also early adopters of IT. A tier-1 communication service provider has hundreds of systems, interacting with each other, including both legacy systems and new digital applications. The cost of testing CSP software is stretched. On the one hand, complex CSP architectures make system integration testing essential, and high in effort. On the other hand, the testing effort for existing systems is a maintenance expense, and as such is subject to *cost reduction* pressures. The size and complexity of CSPs combined with cautious requirements make the telecom and media industry one of the top three sectors in testing market services, holding on to 16% of the testing market (Raviart, 2016).

3) The CSP industry – as an entertainment provider – is a very *competitive market,* requiring fast time to market (TTM). Time to market is the time from a new idea for a service until this service is implemented and available. High competition means that service providers require fast TTM to launch services and content quickly, before the competition. Within the IT chain, testing is a process and phase which is often perceived as a bottleneck in releasing a new product to production, even when working in agile methodologies. The business impact of shortening the testing bottleneck period, even by one day, can be translated into millions of dollars.

4) The CSP industry is set to support *business processes.* There are many common processes in the industry under a standard business process framework (ETOM by the telecom standards forum, TMF), with variations between the different service providers.

5) The CSP industry has *common development methodologies.* Over the last decade, the IT departments of CSPs underwent large and long transformation projects, followed by many releases (ongoing maintenance projects). These types of projects, while rare, are especially required when transforming many connected systems, merging CSPs, opening new lines of business, or setting up IT for a new CSP. Transformation projects provide many data points and components over time and are well suited for applying predictive analytics. *Agile* methodologies are becoming more common to support the flexible structure of CSP software projects. The expectation for fast time to market forces development teams to work in agile (continuous delivery) mode. In agile delivery, code is developed in small parts, which when ready are delivered to acceptance testing. In such circumstances, acceptance testing is performed almost in parallel to development, starting from when the first code is ready through to final delivery. (The emerging new approach of continuous implementation to production, also called DevOps, will not be covered in this chapter.)

These five characteristics support the exchange of analytics insights between communication service providers.

### 20.1.2 Maturity Model for Analytics

Laney (2012, slide 33) describes four levels of analytic maturity, moving from information and hindsight to optimization and foresight:

1) Descriptive analytics – what happened?
2) Diagnostic analytics – why did it happen?
3) Predictive analytics – what will happen?
4) Prescriptive analytics – how can we make it happen?

Following the Gartner maturity model in Laney (2012), this chapter will focus on analytics for CSPs including descriptive analytics based on metrics, benchmarks, and data visualization, as well as diagnostic capabilities. We will then describe predictive and diagnostic analytics based on testing data warehouse and business insight solutions, with a focus on testing metrics trends. Finally, we will conclude with prescriptive analytics, based on an example of intelligent automation (IA) in CSP software testing. We will also look into analytics in CSP software testing, in phases of analytics maturity models, focusing on unique CSP needs for high quality, fast time to market, and huge, complex IT architecture.

## 20.2 Metrics and Benchmarks

Metrics and data collected from the testing process and its following production provide visibility into managing testing projects. Key performance indicators (KPIs) compare between projects that have ended and assess whether testing was successful. Progress metrics reflect the ongoing status of testing during execution, pinpointing issues, and enabling diagnosis of progress stoppers, such as blockers and lower progress areas.

### 20.2.1 KPIs for Testing Success and Industry Benchmarks

Acceptance functional testing success is typically measured based on how testing contributed to total project quality, effort, and time. Software projects for CSPs differ from other industries, as previously indicated:

1) High-quality outcomes required from testing.
2) Projects are large and complex with high testing effort.
3) Total project time is expected to be short.

The two most common KPIs used to measure quality include the overall project quality and the effectiveness of user acceptance testing on quality, which is measured by defects recovery efficiency (DRE).

- A project's overall quality is measured by the number of defects delivered to production (found when the system is live). The number of defects is divided by software size, where the most common measure is by function points. Jones (2013) found that quality in the telecom industry is relatively high. If the typical software project delivers 0.46 defects per function point, in telecom software only 0.13 defects per function point are delivered to production. According to Jones, our previous research (Shama, 2014) shows even higher quality: 0.04 defects per function point are delivered to production (based on a weighted average from 125 projects testing Amdocs systems with acceptance testing performed by Amdocs).

- DRE is the number of defects found in testing divided by the number of defects found in testing and after production (typically for the first six weeks in production). The DRE measure reflects the contribution of testing to a project's quality. Jones (2013) found that DRE in telecom operation software (97.5%) is better than the general DRE (90%).

Acceptance testing efficiency is measured by the testing effort or cost percentage out of the total project effort or cost. CSP typical software testing efforts and costs are declining. Raviart (2016) estimated a 3% annual reduction in typical testing spending versus an increase in development spending. Testing efforts are also declining in the telecom and media industry, from the past average testing effort of 30%–40% of total project effort (Custom Standish Advisory, 2013) to the current 18%–33% of total project effort (Infinity, 2015). This reduction is mainly a result of the move towards DevOps, and the increased use of automation in test execution as well as automation of the testing process. With the increasing use of automation, efficiency of testing can be measured by *automation penetration*. Infinity (2015) measured the percentage of test case executions done automatically in regression testing, and reported a median of 42%. This and other automation metrics are expected to grow significantly over time.

The last KPI relates to project time. In waterfall projects, test execution starts after development ends, and therefore testing time is the typical KPI to measure the testing contribution to project duration. *Testing time* is calculated by the number of calendar days from the start of test execution through the end divided by the total project time. With the move toward agile and DevOps, this KPI is no longer valid, as testing is executed in parallel with development. A new measure of testing impact on project duration, known as *test-only time*, is the number of days in a project where only testing is done (with no development) divided by the total project time. Agile projects, within Amdocs, measured test-only time as 15%–25% of total project time, whereas testing time on waterfall projects is 50%–60% of total project duration.

Looking at the three pillars of testing KPIs, there are changes in definition from waterfall projects to agile. These changes are summarized in Table 20.1. The waterfall KPIs were in use for many years. These KPIs are all a ratio of the same measurement once in testing, divided by the same measurement on the total project. DevOps KPIs do not require division by project size indicators, as projects are small and there is a target to encourage smaller releases.

### 20.2.2 Ongoing/Progress Metrics Versus KPIs

Ongoing measurements track the progress of testing and its relevant tasks, to ensure that KPI targets are met. The typical measurements during test execution are based on the number of test cases executed manually and automated (can be as percent of plan), number of defects found (or test case pass rate), and number (or percent) of defects fixed. Additional metrics indicating the efficiency of testing may also be collected, such as: fixing time or turnaround time, cancelled defect rate, and reopen rate. Ongoing measurements are typically measured per groups of test cases: by application,

**Table 20.1** Key performance indicators for measuring typical testing success in CSP software projects.

| Area | Waterfall KPI | Suggested DevOps KPI |
|---|---|---|
| Quality | Defects recovery efficiency (DRE) = # of defects found in testing / # of defects found in testing and production | Production quality = # of defects escaping to production |
| Efficiency | Testing efficiency = testing effort or cost / total project effort or cost | Testing automation efficiency = % work days saved by automation |
| Time | Testing time = test execution time / total project time | Test-only time = time where only testing is performed |

or by subprojects. A classification of measurement results that is unique to the communication providers' world is a classification by industry business process (Shama and Szyk, 2015).

Ongoing measurements have several objectives: (1) visibility into test management, (2) change management of testers, and (3) predictions. Ongoing automatic measurements, supported by dashboards, provide visibility and transparency for testing management and testers into project progress along with diagnostic capabilities (see the example in Figure 20.1). Change management can be achieved by measurement. For example, changing testers' behavior is creating team alignment to industry business processes, by reporting to management test success per business process. Predicting future project progress is executed using data analytics and will be discussed in detail below.

## 20.3 Testing Progress Trends and Prediction

Data collected from predefined measurements are used to construct predictive models, classified by their value to testing: (a) prediction of progress over time per selected measurement; (b) prediction at the start of testing on how it will end.

### 20.3.1 Defect Detection and Defect Resolution Trends

Predicting the progress of values over time, of a selected measurement, provides planning and control during testing. Yet, this type of prediction is subject to common limitations, such as a bias of work capacity between the days of the week. To ignore the changes by day of the week, a single value is taken per week. Agile methodologies keep injecting code to testing. If code inserted to testing is done in drops, then the timing of the drop and its size is a major input to the prediction model. In this section we describe trend predictions in two measurements over time: (1) defect detection rate, and (2) defect closure (resolution or fixing) rate.

#### 20.3.1.1 Defect Detection Rate

Defect detection rate is the number of real defects found (not canceled) over time. The defect detection rate reflects the testing result progress. The Weibull distribution, mainly in its special case of the Rayleigh distribution, is commonly used to describe the distribution of defect detection rate (some examples can be found in Simmons, 2000;

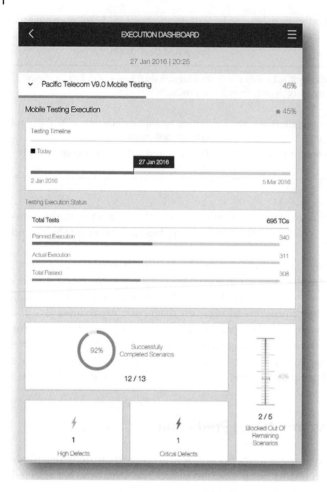

**Figure 20.1** Amdocs BEAT Mobile Reports, a solution for CSP managers to view ad hoc testing progress in real time from their mobile device.

Kececioglu, 2002; Kenett and Baker, 2010; Vladu, 2011; Hallowell 2016). In waterfall projects, where all code enters acceptance testing at the start of execution, and where there are many points of defect numbers along the test time, the parameters for the Rayleigh distribution can be matched with a maximum likelihood estimation (MLE) to describe the trend of defect detection rate (see Figure 20.2). The model enables the planning of waterfall testing: (1) the testing time required to reach a targeted quality DRE can be calculated, or, vice versa, (2) the quality per given test duration can be calculated; also, (3) testing can be enhanced to match the curve with a scale parameter of $\sigma = 0.4$ so that testing will end in time with DRE of 96%.

CSP software projects are typically agile and have many drops of developed code to acceptance testing. In projects with multiple code drops to production, the Rayleigh distribution helps as well. To diagnose a project that has ended, each drop is modeled independently with its own Rayleigh curve, starting with the drop date. The distribution model can be matched within the first three weeks of testing, matching exactly

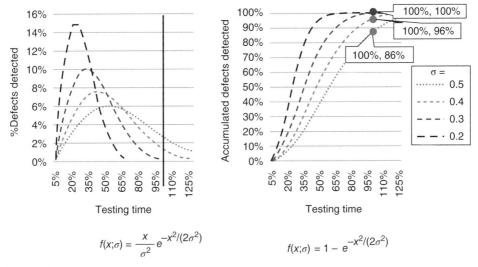

$$f(x;\sigma) = \frac{x}{\sigma^2} e^{-x^2/(2\sigma^2)}$$

$$f(x;\sigma) = 1 - e^{-x^2/(2\sigma^2)}$$

Comments:

- $x$ is the independent variable of time.
- $\sigma$ is the parameter of the distribution – the time where a maximum number of defect are found.

**Figure 20.2** Rayleigh distribution graph that can be used to model defect detection in waterfall acceptance testing.

three points of the first three weeks of defect detection. Note that this is not a statistical model, finding a distribution for a set, but rather an application of a given model using the distribution type detected on waterfall projects. Figure 20.3 shows a real example where a diagnosis was performed at the end of testing, identifying the volume of testing activities beyond the planned drops. With the matching on Figure 20.3, we were able to diagnose back the quality of each drop, the number of defects in the additional scope added, and the efficiency of additional exploratory testers added during week 9. During week 10, at the end of testing, the number of defects that will leak into production were predicted, and the decision of whether or not to go to production was made accordingly.

The same Rayleigh distribution is used to plan in advance the number of detections of defects to target in each week. The example in Figure 20.4 demonstrates how to balance between maximum defects to detect per week and quality (DRE). The closer the peak detection week is to code injection, the more defects can be found per week, and higher quality will be obtained. Deferring detection implies fewer defects to detect per week, and much lower end quality.

### 20.3.1.2 Defect Closure Rate

Defect closure rate is the number of defects that are fixed and closed (passed a retest), and as a proportion it is divided by the number of real (not canceled) defects detected. The target is to sustain a near 100% defect closure rate, to prevent fixing deferral. We examine a proposed model of the defect closure rate trend, instead of predicting the single defect fixing time (as Weiß et al., 2007). Examining 17 CSP software acceptance testing projects carried out by Amdocs Testing, it was found that a logistic distribution

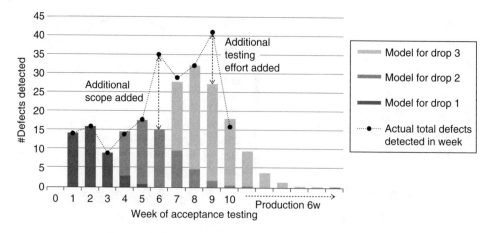

**Figure 20.3** Sum of Rayleigh functions, and the number of real defects detected per week on an Amdocs project with several drops.

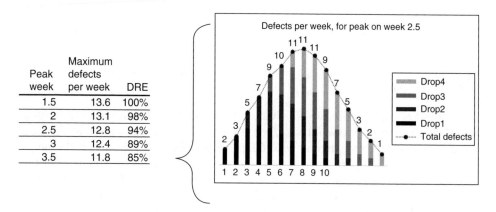

| Peak week | Maximum defects per week | DRE |
|---|---|---|
| 1.5 | 13.6 | 100% |
| 2 | 13.1 | 98% |
| 2.5 | 12.8 | 94% |
| 3 | 12.4 | 89% |
| 3.5 | 11.8 | 85% |

**Figure 20.4** Model of effort effect on quality, in an example of a total of 100 defects, and testing in ten weeks with four drops, one on each given odd week.

to model the defect closure trend gave a high fit. In 9 out of the 17 projects reviewed a logistic model fit was excellent, with Kolmogorov–Smirnov Test (KS) higher than 0.9, as in Figure 20.5; in the following five projects the fit was very good (KS higher than 0.75). The logistic modeling overrides linear modeling on large testing projects, such as CSP transformations. In small projects, there are small numbers of data points, with one measurement per week, where a linear model may match as well.

Modeling defect closure rate is useful to predict the end defect closure rate. An exponential decreasing fit means that accurate prediction can be carried out in half the testing progress to predict the closure rate by its end.

With two models, one a Rayleigh distribution for defect detection rate and the other a logistic distribution for defect closure rate, the models can be multiplied to create a new model of the number of open defects (see Figure 20.6).

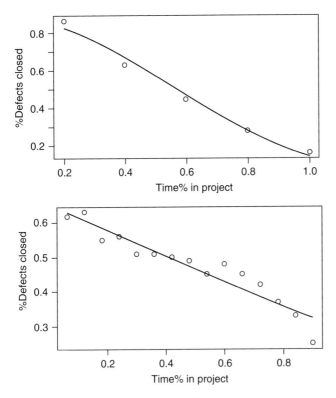

**Figure 20.5** Example of logistic trend of the percent of defects unfixed in a (one-year) CSP software testing project with agile delivery.

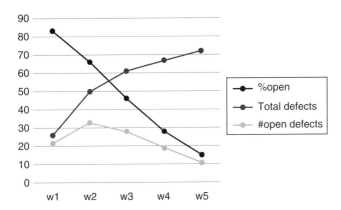

**Figure 20.6** Multiplying the two models of detection rate and closure rate creates a new model of the number of open defects.

### 20.3.2 Total Number of Defects Predicted in Legacy Systems Testing

The number of defects found in a new software project is hard to predict due to each project's uniqueness. Many papers describe the need for total defect prediction and the lack of such a model. In the CSP industry, prediction is an option. After the first few

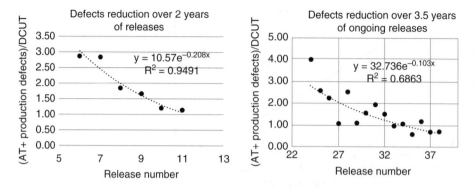

**Figure 20.7** Identified exponential decrease in defect density from release to release.

large releases a system is maintained with ongoing periodic releases, targeting ongoing requests for changes. In the ongoing change request releases, a defect pattern was identified in which the number of defects in releases of the same CSP and same line of business decreased exponentially from release to release (see the two examples in Figure 20.7). In the identified projects it is evident that prediction significantly improves when predicting the number of defects per application separately, based on the application defects trend, and then summing the prediction (Chamoli et al., 2015). This model may be further redetected in DevOps projects where, with many releases, the expectation is that the defect density per release is stabilized after several consecutive releases.

This chapter so far has presented software testing metrics and benchmarks for the telecom and media industry, and prediction models for this industry, driven by special project characteristics of some large transformation projects beside repeating ongoing small releases. We will now see how analytics predictions can be applied to intelligent automation of testing activities.

## 20.4 Intelligent Automation of the Testing Process

The highest analytics maturity level (Laney, 2012) is prescriptive analytics. Prescriptive analytics is applied to CSP software testing in two aspects: (1) recommendations on human decisions, (2) analytics as a base for automation of testing process.

### 20.4.1 Recommendations on Human Decisions

Section 20.3 demonstrated prediction models in the world of telecom and media software testing. These models and others are used for project planning.

Trend models are used to support decisions during a project: to decide if testing will end on time, to defer scope, to increase testing, or fix effort. As testing of CSP software is done by industry business processes, predictions by business processes and by applications may further support decisions, such as to move testers from one telecom business process test to another, or to increase developers' fixing efforts on one application at the expense of another.

Predicting the total number of defects in an ongoing release helps in planning the required staff. Further, prediction of the split of testing effort by the industry business processes (Shama and Szyk, 2015) helps to better plan testing team structure and assignment.

## 20.4.2 Intelligent Automation: Robotic Process Automation Based on Analytics

The testing profession has been considered one of the first adopters of automation of repetitive human tasks, in automated test execution. Automation replacing human transactions is called robotic process automation (RPA), a term is mainly used to describe computer software or a "robot" that replaces repetitive, simple human work with digital systems. Software testing can be further automated by IA – applying machine learning techniques, based on data. Intelligent automation applies a new breed of cognitive technologies with human-like capabilities, such as natural language processing and photo recognition, to automate non-routine tasks; it brings automation to testing beyond test execution, to all testing activities throughout the testing process. For example, the test design activity can be partially automated, based on risk history analytics and combinatorics, as shown below.

Intelligent automation may contribute to test design activity. Typical test design is a process of writing linear scenarios of test cases with specific values. Test design for CSP software is done by business processes (Shama and Szyk, 2015; Elgarat, 2015). In Amdocs' tool for designing tests, Amdocs BEAT Test Design Console, standard CSP business processes are uploaded from a repository. The process is displayed visually, and can be customized visually (see Figure 20.8). The process is a directed acyclic graph of activities, which can take numerous possible combinations of scenarios and values. For a defined process, a set of scenarios is selected to be tested. The typical selection of scenarios is by combinatorial methods, such as pairwise – see the vast review of combinatorial methods in testing by Nie and Leung (2011). In combinatorial scenario selection, specific paths in the process and specific values can get a higher priority (see the activity

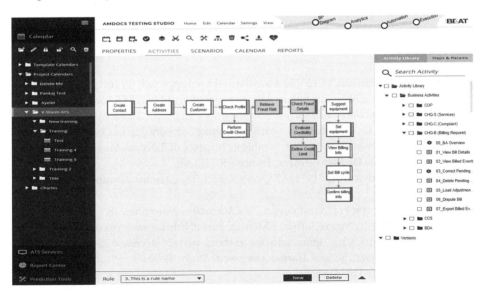

**Figure 20.8** Amdocs BEAT Design Console uses defect history to alert on risk.

coloring in Figure 20.8). Bin Noor and Hemmati (2015) define a class of metrics that estimate test case quality using their similarity to the previously failing test cases. In the CSP software industry, generic activities define similarities.

## 20.5 Summary

This chapter has presented analytics usage in the functional acceptance testing of CSP software. The analytics were based on unique CSP data: large transformation projects with vast testing data, as well as short ongoing maintenance releases with repetitive data. Other industries may benefit from adopting the models presented in this chapter, taking insights that can be viewed clearly among CSPs based on their unique attributes.

In an era of ongoing and unpredictable change within the CSP testing industry, typical functional acceptance testing is declining and is already only half of the testing outsourcing market (Raviart, 2016). New areas of specialized testing are expected to emerge, especially for communication service providers, such as digital user interface, mobile equipment, security, and internet of things, as well as DevOps methodologies which will likely dominate upcoming analytics in CSP testing. Metrics are revised to new industry needs, and analytical models are revised to DevOps development data and objectives.

Analytics is on the verge of becoming the future lead in testing within the CSP industry, in two aspects. First, the CSP industry is a Big Data industry, which is already starting to apply testing of data, as well as testing the accuracy of statistical models. Second, testing in the CSP industry needs to reach large volumes with many digital combinations. Even the testing of mobiles requires numerous combinations of equipment, operating systems, setup options, and personal user preferences. Analytics of combinations and connections of defects between the various options is a must for future testing.

The analytics presented in this chapter on software testing in the CSP world can be widely expanded to other industries, as well as to new future testing needs.

## References

Bin Noor, T. and Hemmati, H. (2015) A similarity-based approach for test case prioritization using historical failure data, in *Proceedings of the IEEE 26th International Symposium on Software Reliability Engineering (ISSRE)*.

Chamoli, S., Tenne, G., and Bhatia, S. (2015) *Analyzing Software Metrics for Accurate Dynamic Defect Prediction Models*. Symbiosis Institute of Telecom Management (SITM) and Telecom Ecole De Management, France (IRC-TEM).

Custom Standish Advisory (2013) *UAT Query Report*. The Standish Group International, Inc.

Elgarat, S. (2015) A new era in test design. *Test Magazine*, November, 18.

Hallowell, D. (2016) Six Sigma Software Metrics, Part 2. https://www.isixsigma.com/methodology/metrics/six-sigma-software-metrics-part-2/ [accessed 25 January, 2018].

Infinity (2015), *Competitive and Market Assessment Study: Telecom Software Testing*. Research sponsored by Amdocs.

Jones, C. (2013) *Approximate United States Software Industry Productivity and Quality Levels: CIRCA 2013*. Capers Jones, Namcook Analytics LLC.

Kececioglu, D. (2002) *Reliability Engineering Handbook*, vol. 1, chap. 10. Lancaster, PA: DEStech Publications.

Kenett, R. and Baker, E.R. (2010) *Process Improvement and CMMI® for Systems and Software*. Boca Raton, FL: CRC Press.

Laney, D. (2012) Information Economics, Big Data and the Art of the Possible with Analytics. https://www-01.ibm.com/events/wwe/grp/grp037.nsf/vLookupPDFs/Gartner_Doug-%20Analytics/$file/Gartner_Doug-%20Analytics.pdf [accessed 25 January, 2018].

Nie, C. and Leung, H. (2011) A survey of combinatorial testing. *ACM Computer Surveys*, 43(2), 11.

Raviart, D. (2016) Transforming Software Testing Services Through DevOps and Digital Automation. NelsonHall. https://research.nelson-hall.com/sourcing-expertise/it-services/software-testing/?avpage-views=article&id=79546&fv=1 [accessed 25 January, 2018].

Shama, G. (2014) Can you tell if your acceptance testing was successful? *Test Magazine*, October, 27.

Shama, G. and Szyk, I. (2015) Business oriented testing in the telecommunications industry. *Test Magazine*, September, 29.

Simmons, E. (2000) When will we be done testing? Software defect arrival modeling using the Weibull distribution, in *Proceedings of the Eighteenth Annual Pacific Northwest Software Quality Conference*.

Vladu, A.M. (2011) Software reliability prediction model using Rayleigh function. *UPB Scientific Bulletin Series C*, 73(4).

Weiß, C., Premraj, R., Zimmermann, T., and Zeller, A. (2007) How long will it take to fix this bug?, in *Proceedings of the 29th International Conference on Software Engineering Workshops (ICSEW'07)*.

# 21

# Measures in the Systems Integration Verification and Validation Phase and Aerospace Applications Field Experience

*Sarit Assaraf and Ron S. Kenett*

## Synopsis

Integration and testing of large-scale systems can be a long and tedious effort. This development phase is characterized by many defects and failures, not enough resources, and a significantly tight schedule. In this chapter we review the main elements of systems integration and testing methods, including systems of systems. We then present an application of a methodology called software trouble assessment matrix (STAM) that generates metrics reflecting the development process performance. The STAM metrics are driven by data from three development phases: defect injection phase, earliest possible detection phase, and defect detection phase. Using STAM metrics, the project manager and system engineers can evaluate the actual system's development maturity, plan realistic schedules, and support the achievement of quality and operational goals. The STAM measurements analysis is applied to data from a large satellite program of Israeli Aircraft Industries (IAI). The analysis of the STAM metrics indicated that the plans for the defect detection process were very effective, but that the execution was not. This implies that the organization effectively planned the tests designed to detect defects, but that the tests were performed inefficiently. This resulted in longer detection with significant cost and quality implications. The study triggered changes and reviews of the test procedures and preparation processes. A year later, results showed a significant improvement. The chapter concludes with a review of future challenges in integration testing methods and tools.

## 21.1   Introduction

Integration and testing of large-scale systems is a complex process. Typical characteristics of this phase include many defects, not enough resources, and a significantly tight schedule. In high maturity organizations one expects to find, at the integration phase, relatively few defects that are fixed and closed in minimal time. In analyzing defects, analysts can assess the stage of earliest possible detection. This reflects on the project planned inspection and testing activities. If defects are discovered in early detection stages, the rate of opening failures at integration will drop and so will costs and schedule deviations. To achieve this, one needs the right tools for managing the

*Analytic Methods in Systems and Software Testing*, First Edition.
Edited by Ron S. Kenett, Fabrizio Ruggeri, and Frederick W. Faltin.
© 2018 John Wiley & Sons Ltd. Published 2018 by John Wiley & Sons Ltd.

detection process. These tools need to provide for effective control of defect life span and for proper management of the rate of failure opening versus rate of failure closure across time (Basili et al., 1996).

The purpose of the system integration phase is to integrate the system components into a complete system. Integration encompasses a series of planning tasks and activities that bring system elements together in an orderly manner, while verifying that their relationships are in accordance with the system architecture. Integration requires nearly continuous testing. Validation and verification testing (VVT) activities, during the system integration phase, include simulation and analysis and a plan for system integration. Subsystem- and system-level tests are at the core of the integration effort, and form the basis for system qualification. The objective of the subsystem integration test is to ensure that the integrated components communicate properly through their defined interfaces and work together as intended. Integration of subsystems is an evolutionary process performed in several phases. At each phase, some additional components are integrated, on the basis of the previous integration phase. This process continues until all subsystems are integrated and the overall system environment is simulated to assess functionality and performance.

System integration involves many complex activities that are necessary in order to achieve the desired final configuration. Simulation may help to understand the interrelationships among implemented components to be integrated and drive the discovery of potential problems before going into operations. Simulations may also model the effects of different coupled physical elements. However, the user must monitor the modeling and ensure realistic simulations in order to derive valuable reliable results. For example, although commercially available finite element tools allow the simulation of complex devices, based on electro-mechanical (e.g. ultrasonic transducers) or thermo-electrical (e.g. induction furnaces) interaction, the values of the relevant material properties used as input parameters are either unknown or partially known. Parallel testing activities are typically also required. Other chapters in this book dealing with testing concepts and methods include Chapters 1–5. Chapter 6 is about teaching system and software testing in the context of a software engineering academic program.

Integration simulation involves the use of complex models, which may require expensive and extensive resources. Trade-off analysis evaluating benefits and return on investment is necessary when planning simulation activities (Cangussu and Karcich, 2005). Examples of such simulations are included in the FP5 European Research Framework Program project TITOSIM (2004), where finite element models were used to simulate a car crash and help design robust and safe car models. One of the methods used in that context was the application of emulators designed to improve the cost benefit of simulations. For more on stochastic emulators and computer experiments, see Bates et al. (2006) and Kenett and Zacks (2014). In this chapter we focus on testing activities and their outcomes. We use the terms defect, bug, and failure interchangeably. Defects exposed by testing are due to a fault in design or implementation exposed by testing. The cause(s) leading to a fault are termed "errors."

## 21.2 The "Test Cycle"

The typical test cycle consists of all the activities related to the VVT phase of the project. The first activity is planning. Test planning defines the set of test cases relevant to the system under design and the maturity of the product, and the infrastructure required for the planned testing activities. Those are tightly interconnected and often iterative in execution. A test cycle refers to planning, designing, and executing the test cases, monitoring and evaluating test results, and documenting both the process and the results.

Test cases are defined in order to test the system or its enabling products. Therefore, planning specifications regarding the stipulated test strategy and test goals must be considered. A test case consists of a set of test data for the input parameters of the test object, the expected values for the output parameters, as well as additional conditions necessary for the execution of the test case (for example triggering events). Additional data required to trigger the system's desired state need to be provided in order to carry out the test. Test case definitions explicitly state the test goal, the related system function, the requirements covered, and the related internal system structure components. For each test case, acceptance criteria must be defined in order to determine success and detect defects.

The test case design process determines the scope, type, and quality of the test. If test cases relevant to the practical application of the system are missing, the likelihood of detecting defects in the system decreases. Due to the major importance of the test case design process to successful testing results, various test methodologies have been developed during the past decades to support adequate test data selection.

The two approaches to systematic testing are white-box tests and black-box tests. White-box testing is based on internal self-test, therefore only part of all the specified requirements can be checked. Black-box tests are based on the system's functional specification and expected behavior. Usually, black-box tests are considered more effective and are more popular.

## 21.3   Testing by Simulation

Simulation in the system context refers to the combination of science and art aimed at building and implementing models of artificial and natural systems. It is important to emphasize the combination of science and art because even if common simulation is based on mathematical or statistical techniques and approaches, whose main characteristics are precision and determinism, some of its fundamental aspects, like model definition and identification, are mainly based on the user's intuition and skillfulness. The concept of simulation is strongly related to modeling. Modeling and simulation are in fact strictly joined together to include the complex activities needed to construct models representing real or ideal system behavior, and experimentation using these models to obtain required data.

If we loosely define a system as a collection of identifiable interacting parts, called components, then the state of the system at a certain time instant is known from the actual conditions of each component at that instant. Not all conditions need to be included in this description, but only the ones that are relevant for the study at hand. The time evolution of the system is then described by the time history of the states in their chronological sequence. A model of the system is then a representation of the system itself. This representation can be a physical replica or a symbolic one. In any case, the model will not represent all the aspects of the modeled system, and there will be an abstraction level in the model since some properties are omitted or approximated. Given a system and a model, simulation is the use of the model for the chronological production of a history of states of the model considered equivalent to the history of the states of the modeled system. A model, once used for simulation, is called a simulation model.

A simulation model needs to be validated and optimized. The key characteristic of a simulation model is its generalizability to the operating system and operational profiles. This investigation is critical since a concept related to a system can be typically represented by several models. The more a model can represent aspects that are relevant for the problem under study the more adequate it is to describe a system. Usually, different models reflecting different levels of detail in system description and/or different levels of knowledge about system properties are used during the system development lifecycle, depending on the needs. Thus, the adequacy of these models is determined by the different questions that simulation is supposed to answer during different phases of system development. The objective of system simulation is the assessment of system performance under different operational conditions. Simulation models allow virtual testing of system implementation under different scenarios. The use of simulation models during the system implementation phase greatly enhances the ability to discover potential defects in the system's design, early in the systems lifecycle. A valid model of the system enables the analysis of the system behavior in different scenarios and operating environments, thus validating the system's architecture and functionality. Corrective action to unexpected or unwanted responses can be implemented, and even optimization, taking into account the experience already gained with simplified/partial models used in the previous phases of the simulation.

The system complexity influences the cost and effort of the modeling and simulation; by implementing a hierarchical modeling approach, the modeling cost and effort can decrease, and the technical maturity level will gradually increase, still without compromising the simulation effectiveness. The recommended risk mitigation strategy in this case is designing an integrated simulation environment based on different tools sharing mutual information. It is important to note that system simulation should be complemented by prototype testing both for model validation and for verification under the underlying hypotheses. The next section provides an overview of a range of VVT methods.

## 21.4 Validation and Verification Testing Methods

### 21.4.1 Reviews

A formal review is one of the most common methods to access the quality of a "product" or "intermediate product" throughout the development cycle of a system. These formal reviews are often milestones in the management of a project and carry contractual obligations on both supplier and purchaser. The role of a review is to gather the most relevant people to criticize the work done, solve open issues, and decide on action items required to be fulfilled in order to pass the review.

A review is a meeting of the most qualified people: the project manager and team that develop a "product," experts in various fields, quality managers, and of course, the client/user of the "product" or "intermediate product" to be reviewed. A review consists of a series of presentations of the major activities in the system development. Project meetings and peer reviews are sometimes called "reviews," but in fact these are team activities involving a restricted set of experts of a particular discipline (software, hydraulics, acoustics, reliability, etc.), focusing on a specific set of activities and deliverables. The objective of a review is to evaluate the results of a phase and to accept or reject the current phase, and decide to proceed with the next phase of the system development. The decisions taken are critical for the project and such reviews generally represent project milestones. Issues raised based on the presentations and discussions are approved or criticized by other experts or the client; justification and solutions are suggested based on trade-off analysis (especially between cost, quality, and delay). Minutes of the meeting including the protocol, the main issues, and action items are documented and published.

One advantage of reviews is that they are relatively low-cost activities (depending on the number and the qualification skills of the participants). It is also an efficient way to facilitate information exchange, gather valuable inputs from the project team and different stakeholders, and summarize the work completed to date. Related reviews are technical reviews such as walkthroughs, inspections, and traceability analysis. These methodologies have the general purpose to prepare, guide, and formalize the analysis and the achievement of confidence about the compliance of a software system with its statutory requirements and the standards. For more on reviews, see Kenett and Baker (2010).

### 21.4.2  Evolutionary Tests

To increase the effectiveness and efficiency of the test, and thus to reduce the overall development costs for software-based systems, a test which is systematic and extensively automatable is required. Regression, or evolutionary, tests can be used for systematizing and automating the testing of functional as well as non-functional system properties and thereby ensure continuity between successive releases. They can also be applied to automatically generate test cases for conventional test methods, such as structural testing.

Evolutionary tests have a number of advantages:

- The complete automation of test case design; thus, if test oracles are available for automatic test evaluation, the entire test can be completely automated, which is not possible with classic test case design procedures.
- Due to this complete automation, the test can be performed with a large amount of error-sensitive test data, which strengthens confidence in the correct functioning of the system under test. The test can often be performed with several thousand or even millions of test data points within one day.
- The possibility to calculate an optimal time for the completion of the test by analyzing the test's convergence status; if the test has converged, the probability of ascertaining further error-sensitive test cases with the same test run is very low.
- Evolutionary tests can be used to process complex test problems which could not be covered by a human tester, the results thus exhibiting relatively fair quality in a short amount of time.
- Human errors, which testers can make during test case design, are precluded by evolutionary tests.

- Evolutionary tests are also applicable for the automation of test case design according to classical function and structure-oriented test methods.
- Evolutionary tests are suitable for a great number of test aims. Evolutionary tests can, for instance, be deployed for temporal behavior tests, safety tests, robustness tests, function tests, and structure tests.
- There are very few prerequisites for the use of evolutionary tests.
- Evolutionary tests have shown themselves to be superior to classic test procedures and static analyses in many case studies.

They do, however, have some disadvantages:

- Not all test problems can be easily translated into an optimization problem. In many cases, particularly in function tests, the definition of a suitable fitness function for the evaluation of the test data generated can be difficult.
- In order to calculate the fitness values it must be possible to monitor the test execution for the test data.

### 21.4.3 White-Box Testing

White-box testing is a method that applies exclusively to software systems. Intended to test the code for correct implementation only, it does not guarantee that the code complies with the specified requirements. This method gives information about the source code, like readability, and about how to test the software. The available techniques are:

- Cyclomatic complexity for basic path testing, where the complexity of the model is quantified with a simple metric.
- Condition testing, data flow testing, and loop testing for control structure testing, where it is evaluated if the tests perform all possible calculations.

### 21.4.4 System Integration Laboratory

A system integration laboratory (SIL) is designed with all the capabilities necessary to perform system integration, system acceptance testing procedures (ATP), and flight/road test support. The external environment of the system will be established or simulated in the designed SIL in order to provide external inputs to the system, both correct and incorrect, in order to test that the system responds appropriately. The designed SIL will be able to perform analysis of the performed tests and will be able to support system maintenance throughout the product lifecycle.

The SIL provides the engineer/operator with a real-time dynamic simulation of the physical environment. During the integration phase, related software models as required for an efficient integration process can replace the system line replacement units (LRUs).

A SIL facility usually consists of the following:

- equipment and facilities necessary to operate the SIL
- simulation of the elements necessary to operate the system in a real-time environment
- monitoring and test equipment engaged in the performance of the tests applied to the system and the operational programs
- facilities to analyze the performed tests
- system LRUs.

### 21.4.5  Failure Mode and Effects Analysis

Failure mode and effects analysis (FMEA) is an analytic approach used to identify possible failures on innovative systems and then determine the frequency and impact of the failure.

This method aims to identify the nature of failures, by dividing a system into smaller subsystems or components that consider the full range of failure modes and subsequent consequences. FMEA utilizes a qualitative analysis, with capabilities to quantify the likelihood and severity and to assess the impact of different failure modes on a system. Three indexes are assigned to each failure: probability P, gravity G, and detectability D. The product of these indexes defines the risk priority index (RPI) that initiates the corrective actions. The FMEA approach requires scheduling and managing corrective actions and responsibilities. The increased knowledge of the system helps to improve the product development. The RPI calculation permits prioritization of the top ten list of anomalies which require risk mitigation and recovery actions.

### 21.4.6  First Article Inspection

First article inspection (FAI) verifies the produced system against its specification. If performance parameters are specified and warranties and/or certificates of compliance are required, the manufacturer will perform testing to ensure that the design meets the specification. The objective is to perform testing to ensure that the produced system meets the specification.

First article testing and approval involves evaluating a contractor's initial, preproduction, or sample model or lot to ensure the contractor can furnish a product conforming to all contract requirements. The sample is normally small but demonstrates if the supplier has correctly taken into consideration all the specifications. It is particularly important when:

- the contractor has not previously delivered the product
- the contractor previously delivered the product, but:
  - there have been subsequent changes in processes or specifications
  - production has been discontinued for an extended period of time
  - the product acquired under a previous contract developed a problem during its life.
- the product is described by a performance specification.

In general, it is essential to have an approved first article that serves as a manufacturing standard of reference.

The supplier's system will provide a process, as appropriate, for the inspection, verification, and documentation of the first production article. It must be specified what component of the end item must be included in the FAI and what inspections/verifications must be conducted on those components and their associated processes. In addition, it must also be specified how the FAI should be documented.

First article testing and approval ("approval" means written notification from the contracting company to the contractor accepting the test results of the first article) ensures that the contractor can furnish a product that conforms to all contract requirements for acceptance. Before requiring first article testing and approval, the contracting company should consider:

- impact on cost or time of delivery
- risk of foregoing such a test
- availability of other, less costly, methods of ensuring the desired quality.

The integrated product team should consider the appropriate first article testing method and clearly delineate the requirements in the contract. The following illustrate the factors that should be considered in establishing the FAI requirements and state in the contract:

- who will conduct the first article test
- performance or other characteristics that must be met
- detailed technical requirements for first article testing
- first article test report data required in contractor-performed testing
- whether the approved first article will serve as the manufacturing model.

### 21.4.7 Hierarchical Testing

The goal of hierarchical testing is to optimize the test and verification plans for the complete system and for its subsystems. Using an iterative process, it tries to reduce redundant tests and to use the verification methods that cost the least.

At the project initial phase, all requirements are verified at all subsystem levels. The first versions of the test and verification plans contain many redundant and overlapping tests. Based on the criticality of the function under test and the customer tolerances, the method and the level where the requirement is to be checked are determined. The optimization may point to a validation done by a test on a bench or by analysis only if previous tests can be used as a basis. The method inputs are the test and verification plans, the system and subsystem test bench capabilities, the requirement criticality, and the available verification methods. The outputs are updated versions of the test and verification plans.

The requirement validation matrix (RVM) is used at the system and subsystem levels and correlates the VVT lifecycle with the VVT modes. Some tests can be performed at different subsystem levels, to reduce cost and time; the main objective is to avoid redundancy. The main goal of hierarchical testing is an optimization of the VVT process with two or more levels of systems and subsystems. Verification cost and time are reduced by eliminating redundant tests and deciding the best level for each test to perform. This method is used in the system design phase, and influences the system integration and system qualification stages of the lifecycle.

The advantages of this method are increased confidence levels in the VVT strategy at early stages of the project, allowing risks to be lowered. It helps identify gaps in the VVT plan early in the development process, and missing or inadequate VVT tools (test bench, model, procedure, etc.). The disadvantages of this method are the high dependency on test experts, and the expertise of the test performer.

### 21.4.8 Robust Design

The goal of robust design is the improvement of product quality. The product's ability to satisfy the needs of the customer depends on values given to the so-called quality characteristics that represent the final performance of the product. In order to establish the degree of satisfaction supplied by a quality characteristic it is necessary to know the ideal target from the users' viewpoint.

It is a common practice in industry to establish the target values of the quality characteristics in terms of nominal values and tolerance intervals. This use involves the wrong concept that the user is equally satisfied with all the values of the quality characteristic that fall within the tolerance interval, and then suddenly one is completely unsatisfied when the value of the performance fails to meet these limits. Robust design uses a different metric to measure the user's satisfaction. This metric is based on so-called quality loss: the expected value of monetary loss suffered by the user in the product's lifetime because of the differences between the system performance and the ideal performance. Robust design is a methodology to assess the stability of a process/product in terms of the variation of some key indicator (yield, quality feature, etc.) of the product/process itself.

The main goal of robust design is the evaluation and reduction of variability of the key indicator of the product/process using different alternatives, mainly based on an experimental approach, worst-case analysis approach, as well as sensitivity analysis, tolerance chain management, etc. Robust design is mostly appropriate when it is important to reduce process variability and the system model is already available. For more on robust design, see Kenett and Zacks (2014).

### 21.4.9 Design of Experiments

Design of experiments (DOE) provides a reliable basis for decision making on systems with many interacting parameters. Integration, verification, and validation normally involve a huge amount of effort and resources, it is necessary to maximize the ratio of the quantity of information achieved versus the quantity of VVT activities performed. It is also necessary to maintain the traceability of the results to focus on objectives, avoid useless reworking, assess the obtained results, and utilize to the maximum the already available information.

The advantages of the method are not only information/cost ratio maximization, the relatively high amount of information obtained relative to a small amount of test runs using many interacting factors, the traceability, and improved documentation of test activities. The disadvantages are selecting the right test during planning and scenario definition, and the difficulty in investigating and analyzing anomalies and determining the root cause. For more on DOE, see Kenett and Zacks (2014).

### 21.4.10 The Software Trouble Assessment Matrix

The software trouble assessment matrix (STAM) metrics were introduced by Kenett (1994). They provide to the project manager and system engineer the ability to evaluate the actual system's development conditions and allow for planning realistic schedules and supporting quality and operational goals. Meeting project milestones strongly depends on a system's readiness to be integrated, tested, and delivered – and, most importantly, on adequate performance at the deployment stage. Since the number of defects and non-conformities dramatically increases project costs, the "ideal" design process implies detection at early project stages and requires reaching stability towards system integration, thus lowering the projects' design costs and meeting the schedule on time. The STAM is a tool that system engineers use to evaluate the design and effectiveness of inspection and testing processes so that they can be improved; it is used to organize relationships between three dimensions: defect injection phase, earliest detection phase, and defect detection phase. Three measures are easily computed from the data collected in a STAM analysis:

- *Negligence ratio*: This ratio indicates the amount of errors that escaped through the inspection process filters. In other words, it measures inspection efficiency.
- *Evaluation ratio*: This ratio measures the delay of the inspection process in identifying errors relative to the phase in which they occurred. In other words, it measures inspection effectiveness.
- *Prevention ratio*: This ratio is an index of how early errors are detected in the development lifecycle relative to the total number of reported errors. This is a combined measure of the development and inspection processes. It assesses the system engineer's ability to generate and identify errors as early as possible in the development lifecycle.

A detailed application of STAM is provided in Section 21.7. For more on STAM, see Kenett and Baker (2010).

## 21.5   Testing Strategies, Reliability, and Failures of Systems

When manufacturers claim that their products are very reliable, they essentially mean that the products can function as required for a long period of time, when used as specified in the operating manual. When an item stops functioning satisfactorily, it is said to have failed. The time to failure or lifetime of an item is intimately linked to its reliability, and this is a characteristic that will vary from system to system even if they are identical in design and structure.

Reliability analysis enables us to answer such questions as: What is the probability that a system will fail before a given time? What percentage of systems will last longer than a certain time? What is the expected lifetime of a component within a system? What is the main failures origin? Is it internal to the system or also user generated? It is a well-known fact that failures are injected during the project lifecycle, during the requirements and the design stages, some as part of the design process and some due to human error.

Usability, on the other hand is comprised mainly of operational failures caused by human errors, and can occur during integration and testing, but also at customer site. These are a main factor of system failure, thus resources are allocated to employing procedures during system development for preventing user interface design errors and for improving system resistance to such errors. Identifying patterns of operational failures and redesigning the user interface so that these failure patterns are eliminated is one of the development activities implemented during the project lifecycle.

Thus, system reliability can be calculated and predicted, but only testing and simulation during the different development stages can validate it. Clearly, once the system is deployed in its operating environment, it is too late and too expensive to fix failures and bugs. Availability is compromised, and customer satisfaction declines accordingly.

A test strategy describes how testing will be carried out in the context of an overall development strategy. The most common test strategies implement a V-model approach (Figure 21.1). The model maps the testing activities to the various development activities, beginning from requirements definition to design implementation. The concept of

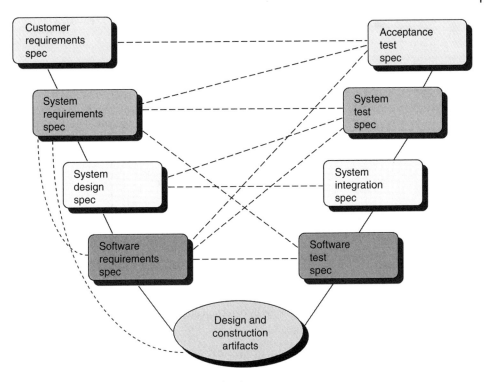

**Figure 21.1** V-model of software and system development.

the V-model applies to software and system development. In some cases, the V-model is tailored to fit the culture of the development organization. The V-model applies, in one form or another, to almost all development strategies such as waterfall (Kenett and Baker, 2010), JAD/RAD, rapid prototyping, spiral model incremental development (Boehm, 1985), and Scrum or agile development (Ambler, 2002). In all cases, testing is conducted to ensure that the design and implementation process produces a product that meets requirements. A typical aspect of the V-model is that it involves both forward and backward iteration loops. The iterations are described visually in Figure 21.1.

The levels of testing have analogs in the various levels of the design and development effort. Testing at the lowest levels can have an impact at other levels, as well. For example, when a defect is detected in a system test, fixing it requires the development engineers to revise the customer requirements, and review the requirements and related design documents. Organizations typically work on the premise that it is better to detect problems in early phases rather than later, implying that it is cost efficient to invest effort in document reviews and involve the test team in early phases. Figure 21.2 presents such data in terms of engineering hours to fix problems, by detection phase.

So how do organizations work and plan towards success? The roadmap evidently involves risk management, readiness evaluation, and aiming towards high reliability; all of these are identified using distinct goals and measurements to establish whether the criteria were met. Clearly, a satellite cannot be called back for retrofit and maintenance, thus dictating a more rigid risk management program during the development lifecycle, and a more fitted failure tolerance regime. The "high profile" in terms of availability that

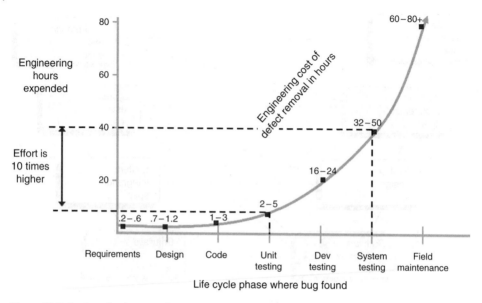

**Figure 21.2** Engineering hours to fix problems by detection phase.

satellite systems must sustain during their lifecycle (12 to 15 years in space) guides rigid test programs and development processes. In a billion dollars' worth of system, clearly high reliability and availability in space are two of the major requirements expected. We need an adequate test program and success criteria that are both relevant and measurable to verify and validate the readiness of the system to be deployed in space.

## 21.6 Readiness Success Criteria and Risk Mitigation

Readiness is a system attribute that can be determined at various points in a system's development lifecycle. Criteria can be specified at these various milestones and can be used to assess the system's progress toward meeting the users' needs, thus lowering the risk that a delivered system will be unsatisfactory. The criteria, referred to as system readiness levels (SRLs) in some circles, can be used as a project management tool for capturing evidence of system readiness. Based on this evidence, system maturity can be accessed and communicated in a consistent manner to the users and other interested stakeholders. A development organization can follow a process faithfully and still not produce a system that meets user needs. Pragmatically, there should be a number of reviews throughout the development cycle. Thus, readiness reviews are essential as part of the development process in order to determine if the deliverable system will function as intended. Readiness reviews are built into the SRL process, and are performed incrementally during the course of the development life cycle – leaving the readiness review until the time of delivery will only increase the risk of failure.

Clearly, the readiness reviews should be combined with test results and be part of a readiness roadmap, comprised of a test strategy and a readiness review plan. The results of these tests should be reviewed, together with the support and design. When there is concurrence between the development organization and the user that the system will function as intended, and that the development contractor has performed in compliance with the contract, statement of work, and customer specifications, the system is ready to be used operationally.

Project risk is an uncertain event or condition that, if it occurs, has a significant effect. A risk has a cause and, if it occurs, a consequence. Risk management consists of the following major activities: risk identification, risk analysis, risk response planning, risk monitoring, risk control, and executing risk mitigation plans.

Organizations perceive risk as it relates to threats to project success, i.e. meeting milestones, requirements, and achieving successful test results. Risks that are threats to the system's tests may be accepted if they are in balance with the reward that may be gained by taking the risk. For example, adopting a fast-track schedule that may be overrun is a risk taken to achieve an earlier completion date.

Risk control can be implemented by a new or alternate, parallel, course of action that will reduce the problem – for example, adopting less complex processes, conducting more seismic or engineering tests, or choosing a more stable vendor. It may involve changing conditions so that the probability of the risk occurring is reduced – for example, adding resources or time to the schedule. It may require prototype development to reduce the risk of scaling up from a bench-scale model. Where it is not possible to reduce probability, a control response might address the risk impact by targeting linkages that determine the severity. For example, designing redundancy into a subsystem may reduce the impact that results from a failure of the original component, and increase reliability.

Risk monitoring and control is the process of keeping track of the identified risks, monitoring residual risks and identifying new risks, ensuring the execution of risk mitigation plans, and evaluating their effectiveness in reducing risk. It involves recording the risk metrics associated with implementing contingency plans as an ongoing process during the lifecycle of the project. The risks change as the project matures, new risks develop, or anticipated risks disappear. Good risk monitoring and control processes provide information based on data and metrics, thus assisting taking effective decisions in advance of the risk's occurrence, and any mid-course correction needed to mitigate the risk. Updates to risk assessments are typically reflected by updates in FMEA evaluations and RPI scores (see Section 21.4.5).

## 21.7 The IAI Case Study

Israel Aerospace Industries (IAI) is the largest industrial company in Israel, composed of many groups and divisions that develop and manufacture large systems of systems (SOS) with over 16 000 employees. The Missiles and Space Group of IAI consists of four divisions that develop large satellite programs using SOS system engineering methodologies and processes.

Development of SOS presents several challenges to the system engineering process, such as:

- different views and meanings of system engineering activities adequate to each level of the system: SOS, system, subsystem
- large project teams geographically dispersed
- complex system integration aspects.

The SOS system engineering infrastructure has to address the various aspects of development, maintenance, and deployment of the processes. Based on this context, one needs to correlate information from different sources to the development processes. This establishes a mechanism embedded in the standard processes of the organization that provides a constant flow of and feedback on the systems' and processes' performance.

Testing methodologies and procedures are designed to gather data on system defects and non-conformities. The more valuable the data and the more valid the measurements, the more effective are considered the tests. The journey towards finding the most effective and valuable measurement methodologies at IAI encompassed several work cycles, and was not without problems. The core of the case study is based on results gathered during the VVT phase of a large satellite program. It provided the project's system engineers information on phase escape throughout the lifecycle, thus assisting risk management and the mid-course corrections needed to mitigate these risks.

The initial analysis of the STAM measures, introduced in Section 21.4.10, indicated that the process of planning defect detection was very effective, but the execution was not, meaning that IAI planned the tests effectively in order to detect defects in the required stage, but it performed them inefficiently so that the detection took longer than expected which resulted in late detection. This conclusion triggered changes and reviews of the test procedures. Moreover, it became clear that the preparation processes required a different approach of risk mitigation in order to meet the project schedule (specifically of the integration, verification, and validation phase).

As mentioned in Section 21.4.10, STAM is a method for analyzing data derived by answering three questions:

- Where were errors detected in the development cycle?
- Where were those errors actually created?
- Where could the errors have been detected?

STAM is implemented at IAI and STAM matrices are compiled for every project by determining the phases in which each error was created, detected, or could have been detected. The term "error" is used here in the sense of problem report, defect, or bug. The total number of errors is then accumulated for each successive phase. The area under the curve is determined by adding the cumulative totals. This gives a metric which represents the area under the cumulative frequency curve.

Three measures are easily computed from the data collected in a STAM analysis. Let $P$ = number of phases, $E$ = total (cumulative) number of errors, $TS_1$ be the area under the curve of accumulated errors by detection phase, $TS_2$ by earliest detection, and $TS_3$ by creation phase. Three ratios are calculated: the negligence ratio, the evaluation ratio, and the prevention ratio. The definitions and formulas for these ratios are as follows:

- *Negligence ratio* $= 100 \times (TS_2 - TS_1) / TS_1$: This ratio indicates the amount of errors that escaped through the inspection process filters. In other words, it measures inspection efficiency.
- *Evaluation ratio* $= 100 \times (TS_3 - TS_2) / TS_2$: This ratio measures the delay of the inspection process in identifying errors relative to the phase in which they occurred. In other words, it measures inspection effectiveness.
- *Prevention ratio* $= 100 \times TS_1 / (P \times E)$: This ratio is an index of how early errors are detected in the development life cycle relative to the total number of reported errors.

The prevention ratio is a combined measure of the performance of the development and inspection processes. It assesses the organization's ability to generate and identify errors as early as possible in the development life cycle. A typical error analysis begins with assessing where errors could have been detected and concludes with classifying errors into the lifecycle phases in which they were created. This procedure requires a repeat analysis of each recorded error. As previously noted, the success of an error causal analysis is highly dependent on clear entry and exit criteria for the various development phases.

Using the negligence, evaluation, and prevention ratios, developers can better understand and improve their inspection and development processes. They can also use STAM to benchmark different projects within their companies and against those of different companies.

Integration, verification, and validation (IVV) at IAI is planned and performed by the system engineers. The IAI process deployment method relies on a corporate-level process definition, followed by group/division-level tailoring and implementation. In order to improve the quality of the system the testing activities are not confined only to the IVV stage, but are interlaced in the various development lifecycle stages. Different techniques such as peer reviews, subsystem-level tests (electronic board test labs, software unit tests, Matlab and Unigraphics simulations) are performed with the purpose of detecting failures adjacent to their injection, in order to minimize escape to later phases that can cause costs and project risks to increase, not to mention the risk of not detecting these failures at all until after system delivery and deployment. At IAI the methodology was tailored and defined in the HP Quality Center tool, which was the infrastructure for managing the failures and defects during the IVV phase. This phase can take 12 to 30 months, depending on the scale and complexity of the system.

Table 21.1 describes the substages of the process, as defined in the system. Dedicated fields are defined in order to document data regarding the failures; system engineers are required to fill in the data for every failure that is recorded in the tool, in three different places: the "injection phase" – when the failure might have been created, usually during the development stage at system or software level; the "earliest detection phase," when the failure should have been detected and fixed according to the test plans; and the "detection phase," the phase when it was actually detected and fixed.

Development of SOS presents some challenges to the system engineering process, such as different views and meanings of system engineering activities adequate to each level of the system (SOS, system, and subsystem); large project teams geographically dispersed; and complex system integration. Figure 21.3 shows the typical program structure in a satellite program.

Verification and validation testing are limited, especially by the ability to simulate the environmental conditions. Once launched, repair activities are limited to software corrections and updates, therefore requiring special attention to requirements validation at early development stages, and analysis of failures found during the early stages of the IVV phase, paying special attention to improvement of the testing processes. This allows taking proactive action and improving detection during following IVV substages.

The satellite example described in this chapter refers to an IVV phase from June 2011 to August 2013 (part of the total phase), during which the STAM methodology was used to analyze and improve the IVV process. The following analysis examples show

**Table 21.1** Development and testing stages as defined in the Quality Center tool.

| Injection phase | Detection phase |
|---|---|
| 01-Scope / Requirements | 01-Review of the Design |
| 02-High-Level Design | 02-Development Tests |
| 03-Detail Design | 03-Prep for Integration & Testing |
| 04-Coding & Implementation | 04-Unit Verification |
| 05-Defect Fixing | 05-System Setup / Implementation |
| 06-Implementation | 06-System Integration (development) |
| 07-Development Tests | 07-External Interfaces Tests |
| 08-Integration | 08-System Level Tests |
| 09-System Level Tests | 09-Environmental tests |
| 10-Environmental tests | 10-Acceptance Tests |
| 11-Previous Version | 11-Validation & System Scenarios Run |
| 12-Legacy Project | 12-Prep for Launch or Developer Test |
| | 13-Developer Tests / In Orbit Tests |
| | 14-Integration at Customer Site |
| | 15-ATP at Customer Site |
| | 16-Customer Reject |

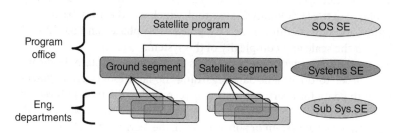

**Figure 21.3** IAI space programs – characteristic internal program structure.

data from two periods, the second one showing the improvement of the process based on the lessons learned from the first analysis period. The first data sample and analysis was performed in July 2012; Table 21.2 shows a snapshot of the data extracted from Quality Center and processed in the Minitab statistical tool. Reported errors are named "failures" here.

The analysis profile is showed in Figure 21.4. The graph shows the three profiles of "Injection," "Early Detection," and "Detection" of failures. The "Detection" profile (closed diamond with dashed line) is lagging behind the "Early Detection" profile (closed square with dashed line), implying the infectivity of the tests, meaning that failures are detected later than initially planned. Nevertheless, the "Early Detection" profile follows the "Injection" profile (closed circle with normal line), implying that the test planning process is adequate.

Calculation of the STAM metrics shows that negligence ratio = 270%, evaluation ratio = 20%, and prevention ratio = 13%. The conclusions are as follows:

1) Errors are detected on average three stages later than they should have been (had the inspection processes worked perfectly). *The inspection process is very inefficient.*
2) The theoretical testing processes considered by the analysts have detected errors 20% into the phase following their creation. *The theoretical process is very effective.*
3) Ideally, all errors are requirements errors, and they are detected in the requirements phase. In this example *only 13% of this ideal is realized, implying significant opportunities for improvement.*

**Table 21.2** Data of 135 failures recorded from June 2011 to July 2012.

| Phase | Detection phase S1 profile | Earliest detection phase S2 profile | Injection phase S3 profile | |
|---|---|---|---|---|
| 01-Review of the Design | 1 | 27 | 43 | |
| 02-Development Tests | 1 | 33 | 48 | Negligence Ratio 270.59% |
| 03-Prep for Integration & Testing | 1 | 40 | 48 | Evaluation Ratio 19.84% |
| 04-Unit Verification | 1 | 44 | 48 | Prevention Ratio 13.33% |
| 05-System Setup / Implementation | 2 | 47 | 55 | |
| 06-System Integration (development) | 62 | 61 | 60 | |
| *Cumulative* | *302* | *252* | *68* | |

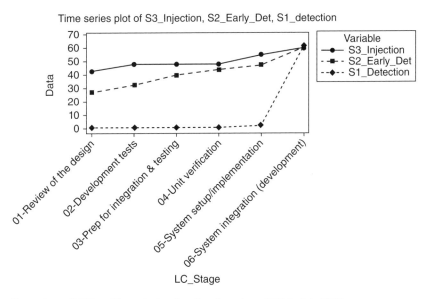

**Figure 21.4** STAM profile analysis using data from June 2011 to July 2012.

So, up until July 2012 the satellite systems engineering group's ability to identify errors as early as possible in the development lifecycle showed inefficiency and called for process improvement and changes. In other words, they planned effectively the tests in order to detect defects in the required stage, but performed inefficiently, the detection taking longer and resulting in later detection (escape).

This conclusion triggered changes and review of the test procedures and preparation processes, and required a different approach of risk mitigation to meet the project schedule (specifically of the IVV phase). By the time of the second analysis in August 2013, the analysis showed improvement as a result of the change in approach.

A second data sample and analysis was performed in August 2013. Table 21.3 shows a snapshot of the data extracted from Quality Center and processed with the Minitab software (www.minitab.com). Table 21.4 shows data for 74 failures recorded from August 2012 to July 2013.

**Table 21.3** Snapshot of the data extracted from Quality Center and processed with Minitab.

| Name | Id | Count | Missing | Type |
|------|-----|-------|---------|------|
| Extract Date | C1 | 85 | 0 | D |
| Project Name | C2 | 85 | 0 | T |
| Defect Id | C3 | 85 | 0 | N |
| Record Type | C4 | 85 | 0 | T |
| Activity | C5 | 85 | 0 | T |
| Severity | C6 | 85 | 0 | T |
| Summary | C7 | 85 | 0 | T |
| Status | C8 | 85 | 0 | T |
| Detection Phase | C9 | 85 | 0 | T |
| Earliest Detection P… | C10 | 85 | 1 | T |
| Injection Phase | C11 | 85 | 1 | T |
| Injection Reason | C12 | 85 | 2 | T |
| Cause | C13 | 85 | 0 | T |
| ___Detailed Cause | C14 | 85 | 44 | T |
| System Model | C15 | 85 | 6 | T |
| Sub System | C16 | 85 | 5 | T |
| ___Unit | C17 | 85 | 16 | T |
| _Unit Model | C18 | 85 | 13 | T |
| Detected on Date | C19 | 85 | 0 | D |
| Assigned Date | C20 | 85 | 0 | D |
| ___In Work Date | C21 | 85 | 8 | D |
| ___Fixed Date | C22 | 85 | 10 | D |
| Closing Date | C23 | 85 | 30 | D |
| DETECT MONTH | C24 | 85 | 0 | N |
| SE-08 OPENED DA… | C25 | 85 | 30 | N |
| By Detect Month | C26 | 17 | 0 | N |

**Table 21.3** (Continued)

| | C9-T<br>Detection Phase | C10-T<br>Earliest Detection Phase | C11-T<br>Injection Phase | C12-T<br>Injection Reason |
|---|---|---|---|---|
| 1 | 06-System Integration (development) | 01-Review of the Design | 03-Detail Design | Internal Communication |
| 2 | 06-System Integration (development) | 01-Review of the Design | 03-Detail Design | Internal Communication |
| 3 | 06-System Integration (development) | 04-Unit Verification | 04-Coding & Implementation | Work procedures |
| 4 | 01-Review of the Design | 01-Review of the Design | 01-Scope / Requirements | Missing Requirement |
| 5 | 06-System Integration (development) | 01-Review of the Design | 01-Scope / Requirements | Conflicting/Unclear |
| 6 | 03-Prep for Integration & Testing | 01-Review of the Design | 06-Implementation | Workmanship |
| 7 | 06-System Integration (development) | 01-Review of the Design | 03-Detail Design | Internal Communication |
| 8 | 06-System Integration (development) | 02-Development Tests | 04-Coding & Implementation | Undefined/Flawed |
| 9 | 06-System Integration (development) | 02-Development Tests | 04-Coding & Implementation | Undefined/Flawed |
| 10 | 06-System Integration (development) | 01-Review of the Design | 01-Scope / Requirements | Missing Requirement |
| 11 | 06-System Integration (development) | 01-Review of the Design | 01-Scope / Requirements | Conflicting/Unclear |
| 12 | 06-System Integration (development) | 06-System Integration (development) | 04-Coding & Implementation | Other |
| 13 | 06-System Integration (development) | 01-Review of the Design | 04-Coding & Implementation | Internal Communication |
| 14 | 06-System Integration (development) | 01-Review of the Design | 04-Coding & Implementation | Undefined/Flawed |
| 15 | 06-System Integration (development) | 03-Prep for Integration & Testing | 06-Implementation | Workmanship |
| 16 | 06-System Integration (development) | 01-Review of the Design | 03-Detail Design | Internal Communication |
| 17 | 06-System Integration (development) | 03-Prep for Integration & Testing | 03-Detail Design | Internal Communication |
| 18 | 06-System Integration (development) | 03-Prep for Integration & Testing | 04-Coding & Implementation | Work procedures |

Using this data the analysis profile show in Figure 21.5 was produced and used to recalculate the three measures. The detection profile (lower line) lag was reduced and follows the early detection profile (middle line), showing the improvement of the tests and failure detection closer to initially planned.

**Table 21.4** Data for 74 failures recorded from August 2012 to July 2013.

| Phase | Detection Phase S1 Profile | Earliest Detection Phase S2 Profile | Injection Phase S3 Profile | |
|---|---|---|---|---|
| 01-Review of the Design | 0 | 38 | 64 | |
| 02-Development Tests | 0 | 44 | 79 | |
| 03-Prep for Integration & Testing | 0 | 53 | 79 | Negligence Ratio 94.28% |
| 04-Unit Verification | 0 | 57 | 79 | Evaluation Ratio 25.65% |
| 05-System Setup / Implementation | 1 | 59 | 90 | Prevention Ratio 30.68% |
| 06-System Integration (development) | 85 | 102 | 106 | |
| 08-System Level Tests | 90 | 103 | 107 | |
| 09-Environmental Tests | 121 | 121 | 121 | |
| *Cumulative* | *297* | *577* | *725* | |

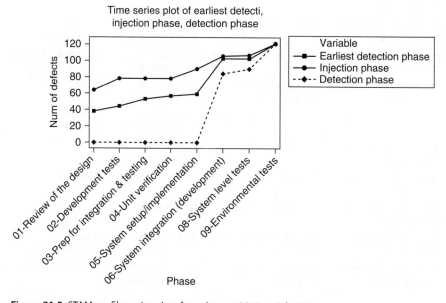

**Figure 21.5** STAM profiles using data from August 2012 to July 2013.

Calculation of the measures shows that negligence ratio = 94.28%, evaluation ratio = 25.65%, and prevention ratio = 30.68%. A significant improvement of approximately 50% occurred in the negligence ratio, implying a major improvement in the detection process:

1) Errors are detected on average one stage later than they should have been (had the inspection processes worked perfectly). *This showed an improvement in the inspection process, and an increase of approximately 65% in the efficiency relative to the first analysis.*
2) The theoretical inspection processes considered by the analysts detect errors 26% into the phase following their creation. The planning process is *30% less effective relative to the first analysis.*
3) Ideally, all failures originate from requirements errors, and they are detected in the requirements phase. *The second analysis shows a 130% improvement relative to the first analysis, meaning 30% of this ideal is realized*; opportunities for improvement still remain.

## 21.8 Summary

Integration and testing of large-scale systems can be a long and tedious phase in the system lifecycle. Typical characteristics of this phase include many defects and failures, not enough resources, and a significantly tight schedule. In high maturity organizations we expect to find relatively few defects that are fixed and closed in minimal time. In analyzing defects we can assess the stage of earliest possible detection. If defects are discovered in these early detection stages, the rate of opening failures at integration will drop and so will costs and schedule deviations. The key is to have the right tools to manage the detection process, control defect life span, and the rate of opening versus rate of closure across time. Using the STAM metrics in a large satellite program during the IVV phase from June 2011 to August 2013 to analyze and control the IVV process, we showed a significant improvement.

We performed two analyses over this period, the first in July 2012, and the second in August 2013. The second negligence ratio analysis showed that errors were detected on average one stage later than they should have been (had the inspection processes worked perfectly), relative to three stages later, as was observed during the first analysis in July 2012. This showed an improvement in the inspection process, and an increase of approximately 65% in the failure detection process efficiency, relative to the first analysis.

Another valuable conclusion was provided by the second analysis of the prevention ratio measurement, showing a 130% improvement relative to the first analysis. Ideally, all failures originate from requirements errors, and are detected in the requirements phase; this result shows that 30% of this ideal was realized; only 13% was demonstrated during the first analysis. However, the evaluation ratio analysis in August 2013 showed a 30% decrease in the effectiveness of the test planning process as considered by the analysts, detecting 26% of the errors in the phase following their creation, versus 20% in the first analysis.

In short, the case study shows a valuable impact on the IVV process in a large satellite program, and provided quantitative feedback to the system engineers, allowing them to take proactive action and influence the process and the outcome of the IVV phase. The analysis of inspection efficiency, inspection effectiveness, and development process execution over two periods showed a change that was generated according to the quantitative data, and proved that processes can be measured and controlled to improve system quality (assuming that failure detection and removal improves quality). Further implementation of the STAM methodology was performed on other Missile and Space Group projects, and a benchmarking database was established and used as part of the standard IVV processes.

## Acknowledgments

We would like to thank:

- Ron Ben-Jacob from KPA for the coaching and support he provided to our teams with the STAM metrics and Minitab analysis.
- Michael Kogan, Tsahi Shiloach, and Nitzan Solomon for the endless hours of data mining, validation, and analysis, and the support with Quality Center system administration.
- Igal Flohr, head of the Satellite Integration Department that provided valuable input and supported the correlation between the IV&V processes in our group and the STAM methodology. Furthermore, he was the driver of the process changes in his department that later on was reflected in the 2013 analysis.

## References

Ambler, S.W. (2002) *Agile Modeling: Effective Practices for Extreme Programming and the Unified Process.* New York: Wiley.

Basili, V.R., Briand, L.C., and Melo, W. (1996) A validation of object-oriented design metrics as quality indicators. *IEEE Transactions on Software Engineering*, 22, 751–776.

Bates, R., Kenett, R.S., Steinberg, D., and Wynn, H. (2006) Achieving robust design from computer simulations. *Quality Technology and Quantitative Management*, 3(2), 161–177.

Boehm, B.W. (1985) A spiral model of software development and enhancement, in *Proceedings of an International Workshop on Software Process and Software Environments, California.*

Cangussu, J.W. and Karcich, R.M. (2005) A control approach for agile processes, in *Proceedings of the 2nd International Workshop on Software Cybernetics at the 29th Annual IEEE International Computer Software and Applications Conference (COMPSAC 2005).*

Kenett, R.S. and Zacks, S., with contributions by D. Amberti (2014) *Modern Industrial Statistics: With Applications in R, MINITAB and JMP*, 2nd edn. Chichester: Wiley.

Kenett, R.S. (1994) Assessing software development and inspection errors. *Quality Progress*, October, 109–112; with correction in issue of February 1995.

Kenett, R.S. and Baker E. (2010) *Process Improvement and CMMI for Systems and Software*, Boca Raton, FL: CRC Publications.

TITOSIM (2004) Time to Market Reduction via Statistical Information Management. http://calvino.polito.it/titosim/ [accessed 26 January, 2018].

# Index

*Analytic Methods in Systems and Software Testing*, First Edition.
Edited by Ron S. Kenett, Fabrizio Ruggeri, and Frederick W. Faltin.
© 2018 John Wiley & Sons Ltd. Published 2018 by John Wiley & Sons Ltd.